PROCESSES AND CONSEQUENCES OF DEEP SUBDUCTION

PROCESSES AND CONSEQUENCES OF DEEP SUBDUCTION

Edited by

David C. Rubie

Bayerisches Geoinstitut, Universität Bayreuth, Bayreuth, Germany

Robert D. van der Hilst

Department of Earth, Atmospheric, and Planetary Sciences, Massachusetts Institute of Technology, Cambridge, MA, USA

2001

ELSEVIER

Amsterdam – London – New York – Oxford – Paris – Shannon – Tokyo

ELSEVIER SCIENCE B.V.
Sara Burgerhartstraat 25
P.O. Box 211, 1000 AE Amsterdam, The Netherlands

First edition 2001
Reprinted from Physics of the Earth and Planetary Interiors 127/1 – 4

Library of Congress Cataloging in Publication Data
A catalog record from the Library of Congress has been applied for.

British Library Cataloguing in Publication Data
A catalogue record from the British Library has been applied for.

ISBN: 0-444-50971-2

∞ The paper used in this publication meets the requirements of ANSI/NISO Z39.48-1992 (Permanence of Paper).
Printed in The Netherlands.

ELSEVIER

Physics of the Earth and Planetary Interiors 127 (2001)

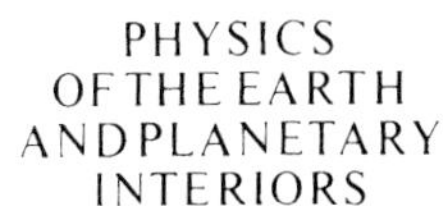

PHYSICS OF THE EARTH AND PLANETARY INTERIORS

www.elsevier.com/locate/pepi

(Abstracts/contents in: Am. Geol. Inst. Bibliogr. Index Geol.; Bull. Signalétique; Curr. Contents; Geo Abstr.; INSPEC; Pascal/CNRS; Phys. Abstr.)

Special Issue: PROCESSES AND CONSEQUENCES OF DEEP SUBDUCTION

ELSEVIER

Physics of the Earth and Planetary Interiors 127 (2001) 1–7

PHYSICS OF THE EARTH AND PLANETARY INTERIORS

www.elsevier.com/locate/pepi

Editorial

Processes and consequences of deep subduction: introduction

1. Introduction

Subduction of slabs of oceanic lithosphere into the deep mantle involves a wide range of geophysical and geochemical processes and is of major importance for the physical and chemical evolution of the Earth. For example, subduction and subduction-related volcanism are major processes through which geochemical components are recycled between the Earth's crust, lithosphere and mantle. A large proportion of the world's earthquakes and volcanoes are related to subduction. Volcanism results from a range of processes including dehydration, melting and melt migration. The deepest known earthquakes occur in subducted lithosphere at depths of 660–700 km but their cause, which has long fascinated geophysicists, is still enigmatic. In addition, the motion and velocities of lithospheric plates at the Earth's surface are controlled largely by the buoyancy forces that drive subduction.

Because of the wide variety of processes involved, subduction zones can be regarded as natural laboratories through which dynamic behavior in the Earth's mantle can be studied. Many physical and chemical processes interact during subduction, and the complexity of the system necessitates an interdisciplinary approach involving seismology, mineral physics, geochemistry, petrology, structural geology, rock mechanics, and geodynamic modeling. For example, phase transformation kinetics determine buoyancy, rates of subduction and, therefore, thermal structure, with the latter feeding back to affect the kinetics. Rheology and therefore large scale slab dynamics may also be affected by mineral transformations and their kinetics. Seismology provides constraints on the elastic properties and the geometry of the down going plate as well as on the processes involved in deep earthquakes.

In September 1999, a 5-day interdisciplinary Alfred-Wegener Workshop, to discuss the major advances that have been made in recent years was conducted. A total of 56 contributions, consisting of keynote talks and posters, were presented (see Terra Nostra 99/7; Alfred-Wegener-Stiftung Berlin, 1999, for abstracts). Topics discussed included thermal structure and buoyancy forces, rheology of mantle minerals, stabilities of hydrous minerals, partial melting and the mechanisms and rates of melt migration, kinetics of phase transformations, mechanisms of deep earthquakes, geochemical recycling, and the processes of continental collision that result from subduction. The papers contained in this volume have resulted primarily from this workshop and have been contributed mostly by keynote speakers. Many of the papers contain a strong review element and it is our intention that this volume will serve as a comprehensive reference source for researchers in the coming years.

2. Complex interaction of slabs and upper mantle transition zone

One fundamental problem concerns the depth to which subducted lithosphere penetrates into the mantle because this is related to the scale of mantle convection and the Earth's evolution over time. Although Isacks and Molnar's (1971) pioneering work on slab seismicity allowed alternative views, the cessation of deep-focus earthquakes at a depth of ~700 km has often been used in arguments against the penetration

0031-9201/01/$ – see front matter
PII: S0031-9201(01)00217-5

of subducting slabs beyond the 660 km seismic discontinuity and, thus, in favor of the layering of mantle convection at that depth. In recent years, however, seismic imaging has revealed that some subducting slabs penetrate deep into the lower mantle.

Between the mid 1970s and late 1980s, Jordan and his co-workers used anomalous travel times, projected on to the so-called "residual sphere", and wave forms of seismic body waves to argue for the presence of subducted slabs in the lower mantle beneath convergent margins in central America and the western Pacific (Jordan and Lynn, 1974; Jordan, 1975, 1977; Creager and Jordan, 1984, 1986; Fischer et al., 1991). These early claims for slab penetration across the 660 km discontinuity were confirmed by regional, high-resolution tomographic studies by van der Hilst and Spakman (1989), van der Hilst et al. (1991) and Fukao et al. (1992) (see van der Hilst et al. (1998) for a review). However, along with the detailed analysis of subduction zone seismicity (e.g. Okino et al., 1989; Ekström et al., 1990; Lundgren and Giardini, 1994) these studies also indicated that the morphology and fate of subducted slabs are more complex than expected from the end-member models of either whole mantle flow or convective layering at 660 km depth. Apparently, slabs can deflect horizontally in the upper mantle transition zone beneath some convergent margins whereas penetration to lower mantle depths can occur beneath other island arc segments. The different styles of subduction across the upper mantle transition zone, which are also borne out in recent global tomography studies (e.g. Bijwaard et al., 1998; Kárason and van der Hilst, 2000), are illustrated in Fig. 1 and have been interpreted in terms of the combined effects of relative plate motion and downward flow in a stratified mantle (Kincaid and Olson, 1987; Gurnis and Hager, 1988; van der Hilst and Seno, 1993; Griffiths et al., 1995; Guillou-Frottier et al., 1995; Davies, 1995; Zhong and Gurnis, 1995; Christensen, 1996, 2001), the weakening of the slab due to grain size reduction during phase transformations (Riedel and Karato, 1997; Karato et al., 2001), or the interaction of the subducting slab with localized upwellings (Gurnis et al., 2000). van der Hilst et al. (1997) and Grand et al. (1997) pointed out that many slabs have sunk into the lower mantle, in particular beneath America and southern Asia and Indochina, but Fig. 1 suggests that in the current snap-shot of convection not all of these deep slabs reach the core mantle boundary (van der Hilst and Kárason, 1999; Kárason and van der Hilst, 2000).

3. Slab dynamics and rheology

The first four papers in this volume are concerned with the dynamics of subduction, the interaction of the slab with upper mantle discontinuities, and the influence of slab mineralogy. King (2001) focuses on several classical subduction issues, including the variation in dip angle of subducted slabs in the upper mantle, the unbending of the downgoing plate, and the topography of deep sea trenches. Christensen reviews numerical and analog (laboratory) models of deep subduction and discusses the mechanisms that can produce the complex morphologies of lithospheric slabs as inferred from tomographic images. An increase of intrinsic density or viscosity with depth as well as phase transitions with a negative Clapeyron slope can all inhibit or delay deep subduction, and the tectonic conditions and the relative motion of the tectonic plates at Earth's surface can also play an important role. Collier et al. (2001) present observational evidence from converted seismic waves for topography on the upper mantle discontinuities that mark the phase transitions of olivine to wadsleyite (at a global average depth of 410 km) and ringwoodite to perovskite and magnesiowustite (near 660 km depth). For the convergent margins studied, their data suggest that the olivine → wadsleyite transition occurs under (thermodynamic) equilibrium conditions. While the location of the discontinuity cannot be constrained everywhere in the slab, this result argues against the existence of a wedge of metastable olivine. Bina et al. (2001) discuss effects of slab mineralogy and phase chemistry on subduction dynamics (e.g. buoyancy, stress field), kinematics (e.g. rate of subduction and plate motion), elasticity (e.g. deformation and seismic wave speed), thermometry (effects of, e.g. latent heat, isobaric superheating), and seismicity (e.g. as due to adiabatic shear instabilities).

The next two papers deal with the rheology of mantle minerals, which is clearly of considerable importance in controlling the subduction process. Because of experimental limitations, rheology is exceedingly difficult to study experimentally at pressures of the transition zone and lower mantle (e.g. Karato and Rubie,

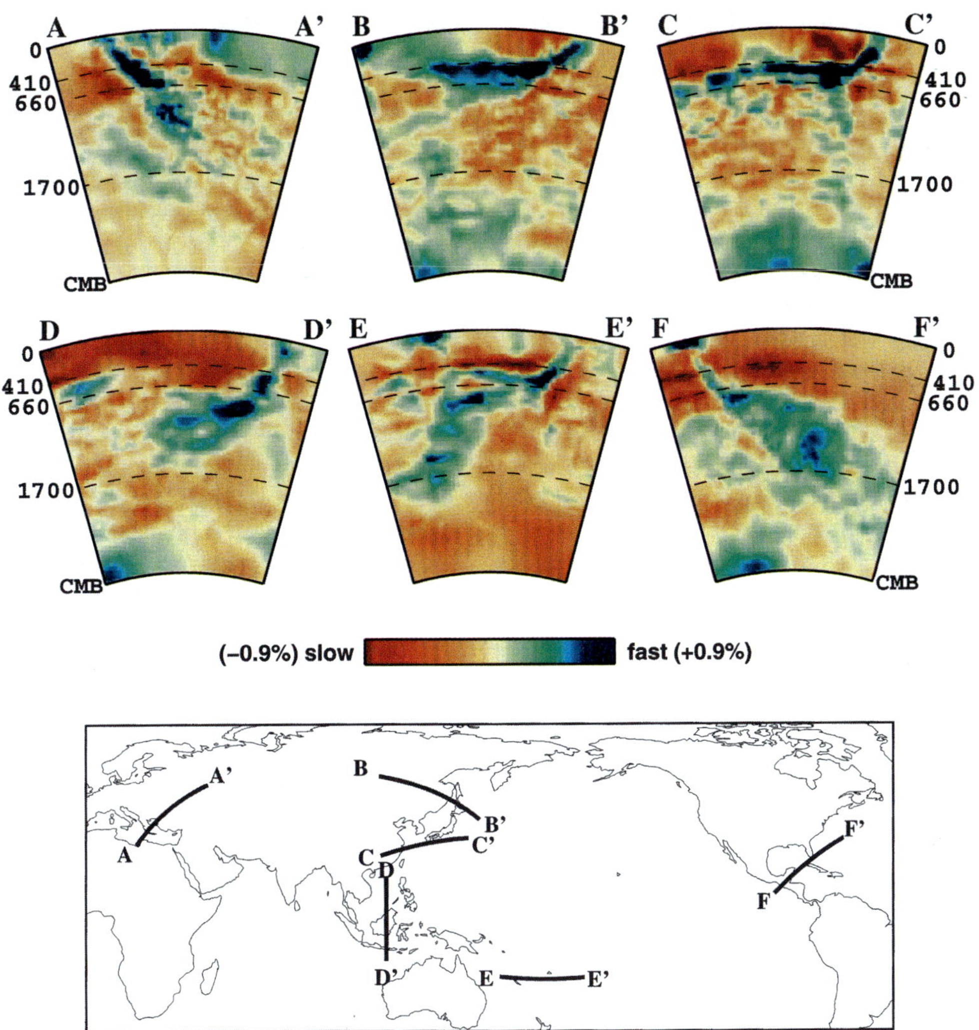

Fig. 1. Slab structure illustrated by vertical mantle sections across: (A) the Hellenic (or Aegean) arc; (B) the southern Kurile arc; (C) Izu Bonin; (D) the Sunda arc (Java); (E) the northern Tonga arc, and (F) central America (see Kárason and van der Hilst (2000) for more details).

1997). Weidner et al. (2001) have utilized an indirect method based on in situ stress measurements determined from the broadening of X-ray diffraction peaks. Sample strains can also be determined directly by in situ X-ray imaging techniques. Although the strains involved in these experiments are small, so that derivations of "steady-state" flow laws might be uncertain, the experimental approach has considerable potential. They review data for major mantle minerals and conclude that increases in strength are likely to occur as olivine transforms first to wadsleyite and ringwoodite and subsequently to perovskite + magnesiowüstite. Based on correlations between strength and temperature, they suggest that deep seismicity is likely to result from plastic instabilities, with the distribution of earthquakes being related to strength distribution and therefore slab mineralogy. Karato et al. (2001) propose a complex rheological structure for subducting slabs due to the effects of grain size reduction during phase transformations. Based on their model, rapidly-subducting and therefore relatively cold slabs should be weaker than slowly-subducting warm slabs.

They apply this model to explain the observed interactions of slabs with the 660 km discontinuity and, for example, how some slabs flatten out at the 660 km discontinuity — a feature previously explained by trench migration (e.g. van der Hilst and Seno, 1993).

4. Deep earthquakes

In their characterization of subduction zone seismicity, Isacks and Molnar (1971) noticed that not all seismic zones in subducted slabs are continuous and that some deep events seem to occur isolated from those at shallower depths. The continuity of the slab across gaps in the Wadati–Benioff seismic zone between deep earthquake clusters and shallow seismicity has been debated. Following an approach similar to the study of high-frequency wave propagation in slabs by, for instance, Barazangi et al. (1972), Snoke et al. (1974), Gubbins and Snieder (1991), van der Hilst and Snieder (1996) and Okal (2001) uses the observation and triggering mechanism of high-frequency oceanic "T" waves to argue for the mechanical continuity of (most) slabs across the deep seismicity gap. He argues that only the deep earthquakes beneath the north Fiji basin (the so-called Vityaz cluster) and the deep events beneath New Zealand may occur as detached events with no mechanical connection to the surface.

A problem that has concerned geophysicists for many years is how to explain the occurrence of deep earthquakes (e.g. Kirby, 1987). About 10 years ago, "transformation faulting" (or "anticrack" mechanism) was proposed as the mechanism for deep earthquakes largely on the basis of experimental observations (Kirby, 1987; Green and Burnley, 1989; Kirby et al., 1991; Burnley and Green, 1991). This mechanism involves a shear instability that develops during the incipient transformation of metastable olivine to wadsleyite or ringwoodite. The high-pressure phase is postulated to form in the shear zone with an ultra-small grain size that allows high-strain rate deformation to occur by grain size sensitive creep. Although this mechanism appeared to provide an appealing explanation for deep earthquakes (e.g. Kirby et al., 1996), most evidence for its operation is based on experiments performed on analogue materials at low pressures. There has been only one brief report, over 10 years ago, that transformational faulting might occur at transition zone pressures (Green et al., 1990) but this result has not been confirmed subsequently. Several papers in this volume address this issue from the perspectives of seismology and mineral physics.

Wiens (2001) reviews empirical parameters (e.g. fault dimension, focal mechanism, stress drop, temporal and spatial aspects of after shock sequences, magnitude–frequency relationships (b-values)) of deep earthquakes ($z > 300$ km) in comparison with shallow ($z < 70$ km) and intermediate depth events (70 km $< z <$ 300 km). He documents the lack of a seismically-detectable "wedge" of metastable olivine (see also Collier et al., 2001) and argues against both transformational faulting and dehydration-induced faulting based on the large fault dimensions of some deep earthquakes. Instead, he presents evidence for the sensitivity of deep earthquakes to the temperature of the slab and, thus, favors a temperature-activated phenomenon. Based on extrapolations of experimental kinetic data combined with two-dimensional thermal models of subduction zones, Mosenfelder et al. (2001) conclude that the maximum depths of olivine metastability are considerably less (perhaps by 200 km or more) than the depths of the deepest seismicity; the latter cannot therefore be explained by transformational faulting. This conclusion contrasts with results of previous studies in which thermokinetic models were based on olivine → spinel kinetic data from analog compositions such as Mg_2GeO_4 and Ni_2SiO_4 (e.g. Rubie and Ross, 1994; Kirby et al., 1996). Several papers in the volume thus reach a consensus, based on different approaches, that deep earthquakes result from thermal shear instabilities involving thermal run away and possibly localized melting, rather than from transformational faulting.

5. The effects of water and hydrous phases

Recently, the effect of water and hydrous phases on rheology, melting, phase chemistry, and earthquake triggering has received substantial attention. The stability of hydrous phases in subduction zones is reviewed by Angel et al. (2001). This topic is critical for evaluating a number of issues, including the causes of subduction-related volcanism through hydrous melting, dehydration embrittlement as a possible cause of intermediate and deep earthquakes, the reduced seismic wave speed in the mantle wedge above the slab beneath and beyond the volcanic arc (see also Zhao, 2001), and the possibility that water is transported into and stored in the lower mantle. Angel et al. (2001) conclude that slabs are likely to dehydrate as they enter the lower mantle. Based on elasticity data, they further conclude that the seismic detection of even significant amounts of hydrous phases in slabs may be difficult or impossible. The subsequent three papers consider some of these issues in the context of shallow processes in the slab and overlying mantle wedge.

Zhao (2001) reviews recent findings on the structure, magnetism, and dynamics of subduction zones. In particular, he presents evidence for the continuation of the low wave speed, high attenuation (low Q), and seismically anisotropic wedge to near 400 km depth beneath some back arcs. He argues that this may be explained by the persistence of hydrous phases to (and perhaps beyond) that depth and that magnetic systems are not limited to near-surface regions of the mantle. He also discusses links between dehydration of the slab, rupture nucleation and crustal earthquake triggering. The role of water in promoting partial melting of peridotite and how its presence affects the composition of partial melts are reviewed in detail by Ulmer (2001). In addition to significantly reducing the melting temperature, small amounts (0.1–0.5 wt.%) of H_2O in the source region can explain both the major element (magnesian-poor and silica-rich) and trace element characteristics of arc magmas and results in 1–7 wt.% H_2O in basaltic to picritic primary liquids. The necessity of high temperatures (1250–1300°C) in the melting regions is emphasized. Isotopic fractionation and transport during dehydration and fluid–rock interaction are discussed by Nakano and Nakamura (2001) using boron isotopes as an example. Boron contents and boron isotopic compositions of metasediments from the Sambagawa metamorphic belt, Japan, indicate a lack of bulk fluid–rock boron isotope fractionation during devolatilization. Boron that is contained in muscovite and chlorite at lower grades is found in tourmaline in the high-grade rocks.

6. Subduction and orogeny

The final two papers, by Ernst (2001) and O'Brien (2001), respectively, review the effects of subduction of crustal rocks during continental collision. Following subduction to depths greater than 90 km, slices of continental crust have been exhumed back to the surface very rapidly, thus ensuring the survival of the high-pressure minerals. The exhumation mechanism is not fully understood but is currently of considerable interest for geodynamicists and metamorphic petrologists. Ernst (2001) reviews several collision belts in which crustal fragments were exhumed from conditions of around 2.8 GPa and 600–900°C. He emphasizes the role of fluids in catalyzing eclogite-forming reactions and in-forming andesitic melts. He argues that when hydrous phases fail to dehydrate to produce such fluids, the consequences can be the metastable preservation of low-pressure mineral assemblages and a lack of volcanism. He notes that crustal slices in which ultrahigh-pressure assemblages are preserved are normally very thin (1–5 km thick) and were exhumed back to the surface at rates as high as 10 mm per year. Finally, O'Brien (2001) compares the Alpine and Himalayan chains. As well as discussing the mesozoic and tertiary evolution of these collision belts, he reviews evidence from seismic tomography for the present-day deep structure. In the Himalayan region, the subduction of several thousand kilometers of oceanic lithosphere may explain the presence of magmatic arcs, in contrast to the Alps where the volume of subducted lithosphere was relatively small. It is suggested that the rapid exhumation of crustal rocks from great depths (>90 km) is a consequence of slab break-off, a process that can be identified below both mountain belts.

Acknowledgements

We would like to thank all the authors for contributing to this volume, the reviewers for dedicating their valuable time and assistance and the Alfred Wegener Foundation for supporting the Workshop on "Processes and Consequences of Deep Subduction" that was held in Verbania, Italy, in September 1999.

References

Angel, R.J., Frost, D.J., Ross, N.L., Hemley, R., 2001. Stabilities and equations of state of dense hydrous magnesium silicates, this issue.

Barazangi, M., Isacks, B., Oliver, J., 1972. Propagation of seismic waves through and beneath the lithosphere that descends under the Tonga island arc. J. Geophys. Res. 77, 952–958.

Bijwaard, H., Spakman, W., Engdahl, E.R., 1998. Closing the gap between regional and global travel time tomography. J. Geophys. Res. 103, 30055–30078.

Bina, C.R., Stein, S., Marton, F.C., Van Ark, E.M., 2001. Implications of slab mineralogy for subduction dynamics, this issue.

Burnley, P.C., Green II, H.W., 1991. Faulting associated with the olivine to spinel transformation in Mg_2GeO_4 and its implications for deep-focus earthquakes. J. Geophys. Res. 96, 425–443.

Christensen, U.R., 1996. The influence of trench migration on slab penetration into the lower mantle. Earth Planet. Sci. Lett. 140, 27–39.

Christensen, U.R., 2001. Geodynamic models of deep subduction, this issue.

Collier, J.D., Helffrich, G.R., Wood, B.J., 2001. Seismic discontinuities and subduction zones, this issue.

Creager, K.C., Jordan, T.H., 1984. Slab penetration into the lower mantle. J. Geophys. Res. 89, 3031–3049.

Creager, K.C., Jordan, T.H., 1986. Slab penetration into the lower mantle below the Mariana and other island arcs of the northwest Pacific. J. Geophys. Res. 91, 3573–3589.

Davies, G.F., 1995. Penetration of plates and plumes through the mantle transition zone. Earth Planet. Sci. Lett. 133, 507–516.

Ekström, G., Dziewonski, A.M., Ibanez, J., 1990. Deep earthquakes outside slabs. EOS Trans. AGU, 71, p. 1462 (Abstract).

Ernst, W.G., 2001. Subduction, ultrahigh-pressure metamorphism, and regurgitation of buoyant crustal slices — implications for arcs and continental growth, this issue.

Fischer, K.M., Creager, K.C., Jordan, T.H., 1991. Mapping the Tonga slab. J. Geophys. Res. 96, 14403–14427.

Fukao, Y., Obayashi, M., Inoue, H., Nenbai, M., 1992. Subducting slabs stagnant in the mantle transition zone. J. Geophys. Res. 97, 4809–4822.

Grand, S.P., van der Hilst, R.D., Widiyantoro, S., 1997. High resolution global tomography: a snapshot of convection in the Earth. Geol. Soc. Am. Today 7, 1–7.

Green II, H.W., Burnley, P.C., 1989. A new self-organizing mechanism for deep-focus earthquakes. Nature 341, 733–737.

Green II, H.W., Young, T.E., Walker, D., Scholz, C.H., 1990. Anticrack-associated faulting at very high pressure in natural olivine. Nature 348, 720–722.

Griffiths, R.W., Hackney, R., van der Hilst, R.D., 1995. A laboratory investigation of effects of trench migration on the descent of subducted slabs. Earth Planet. Sci. Lett. 133, 1–17.

Gubbins, D., Snieder, R.K., 1991. Dispersion of P waves in subducted lithosphere: evidence for an eclogite layer. J. Geophys. Res. 96, 6321–6333.

Guillou-Frottier, L., Buttles, J., Olson, P., 1995. Laboratory experiments on the structure of subducted lithosphere. Earth Planet. Sci. Lett. 133, 19–34.

Gurnis, M., Hager, B.H., 1988. Controls of the structure of subducted slabs. Nature 335, 317–321.

Gurnis, M., Ritsema, J., van Heijst, H.J., Zhong, S.J., 2000. Tonga slab deformation: the influence of a lower mantle upwelling on a slab in a young subduction zone. Geophys. Res. Lett. 27, 2373–2376.

Isacks, B., Molnar, P., 1971. Distribution of stresses in the descending lithosphere from a global survey of focal mechanism solutions of mantle earthquakes. Rev. Geophys. Space Phys. 9, 103–173.

Jordan, T.H., 1975. Lateral heterogeneity and mantle dynamics. Nature 257, 745–750.

Jordan, T.H., 1977. Lithospheric slab penetration into the lower mantle beneath the sea of Okhotsk. J. Geophys. Res. 43, 473–496.

Jordan, T.H., Lynn, W.S., 1974. A velocity anomaly in the lower mantle. J. Geophys. Res. 79, 2679–2685.

Kárason, H., van der Hilst, R.D., 2000. Constraints on mantle convection from seismic tomography. In: Richards, M.R., Gordon, R., van der Hilst, R.D. (Eds.), The History and Dynamics of Global Plate Motion. American Geophysical Union, Washington, DC. Geophys. Mon. 121, 277–288.

Karato, S., Rubie, D.C., 1997. Towards an experimental study of deep mantle rheology: a new multi-anvil sample assembly for deformation studies under high pressures and temperatures. J. Geophys. Res. 102, 20111–20122.

Karato, S., Riedel, M.R., Yuen, D.A., 2001. Rheological structure and deformation of subducted slabs in the mantle transition zone: implications for mantle circulation and deep earthquakes, this issue.

Kincaid, C., Olson, P., 1987. An experimental study of subduction and slab migration. J. Geophys. Res. 92, 13832–13840.

King, S., 2001. Subduction zones: observations and geodynamic models, this issue.

Kirby, S.H., 1987. Localized polymorphic phase transformations in high-pressure faults and applications to the physical mechanism of deep earthquakes. J. Geophys. Res. 92, 13789–13800.

Kirby, S.H., Durham, W.B., Stern, L.A., 1991. Mantle phase changes and deep-earthquake faulting in subducted lithosphere. Science 252, 216–225.

Kirby, S.H., Stein, S., Okal, E.A., Rubie, D.C., 1996. Deep earthquakes and metastable phase transformations in subducting oceanic lithosphere. Rev. Geophys. 34, 261–306.

Lundgren, P., Giardini, D., 1994. Isolated deep earthquakes and the fate of subduction in the, mantle. J. Geophys Res. 99, 15833–15842.

Mosenfelder, J.L., Marton, F.C., Ross II, C.R., Kerschhofer, L, Rubie, D.C., 2001. Experimental constraints on the depth of olivine metastability in subducting lithosphere, this issue.

Nakano, T., Nakamura, E., 2001. Boron isotope geochemistry of metasedimentary rocks and tourmalines in a subduction zone metamorphic suite, this issue.

O'Brien, P.J., 2001. Subduction followed by collision: alpine and Himalayan examples, this issue.

Okal, E.A., 2001. "Detached" deep earthquakes: are they really? This issue.

Okino, K., Ando, M., Kaneshima, S., Hirahara, K., 1989. A horizontally lying slab. Geophys. Res. Lett. 16, 1059–1063.

Riedel, M.R., Karato, S., 1997. Grain-size evolution in subducted oceanic lithosphere associated with the olivine–spinel transformation and its effects on rheology. Earth Planet. Sci. Lett. 148, 27–43.
Rubie, D.C., Ross II, C.R., 1994. Kinetics of the olivine–spinel transformation in subducting lithosphere: experimental constraints and implications for deep slab processes. Phys. Earth Planet. Int. 86, 223–241.
Snoke, J.A., Sacks, S., Okada, H., 1974. Empirical models for anomalous high-frequency arrivals from deep-focus earthquakes in South America. Geophys. J.R. Astron. Soc. 37, 133–139.
Ulmer, P., 2001. Partial melting in the mantle wedge — the role of H_2O in the genesis of mantle-derived "arc-related" magmas, this issue.
van der Hilst, R.D., Kárason, H., 1999. Compositional heterogeneity in the bottom 1000 km of Earth's mantle: towards a hybrid convection model. Science 283, 1885–1888.
van der Hilst, R.D., Seno, T., 1993. Effects of relative plate motion on the deep structure and penetration depth of slabs below the Izu-Bonin and Mariana island arcs. Earth Planet. Sci. Lett. 120, 375–407.
van der Hilst, R.D., Snieder, R.K., 1996. High-frequency precursors to P-wave arrivals in New Zealand: implications for slab structure. J. Geophys. Res. 101, 8473–8488.
van der Hilst, R.D., Spakman, W., 1989. Importance of the reference model in linearized tomography and images of subduction below the Caribbean plate. Geophys. Res. Lett. 16, 1093–1096.
van der Hilst, R.D., Engdahl, E.R., Spakman, W., Nolet, G., 1991. Tomographic imaging of subducted lithosphere below northwest Pacific island arcs. Nature 353, 37–42.
van der Hilst, R.D., Widyantoro, S., Engdahl, E.R., 1997. Evidence for deep mantle circulation from global tomography. Nature 386, 578–584.
van der Hilst, R.D., Widiyantoro, S., Creager, K.C., McSweeney, T., 1998. Deep subduction and aspherical variations in P-wave speed at the base of Earth's mantle. In: Gurnis, M., Wysession, M.E., Knittle, E., Buffett, B.A. (Eds.), Observational and Theoretical Constraints on The Core Mantle Boundary Region. American Geophysical Union. Geodyn. Ser. 28, 5–20.
Weidner, D.J., Chen, J., Xu, Y., Wu, Y., Vaughan, M.T., Li, L., 2001. Subduction zone rheology, this issue.
Wiens, D.A., 2001. Seismological constraints on the mechanism of deep earthquakes: temperature dependence of deep earthquake source properties, this issue.
Zhao, D., 2001. Seismological structure of subduction zones and its implications for arc magmatism and dynamics, this issue.
Zhong, S., Gurnis, M., 1995. Mantle convection with plates and mobile, faulted plate margins. Science 267, 838–843.

David C. Rubie*
Bayerisches Geoinstitut, Universität Bayreuth
D-95440 Bayreuth, Germany
E-mail address: dave.rubie@uni-bayreuth.de
(D.C. Rubie)

Rob D. van der Hilst
Department of Earth, Atmospheric, and Planetary Sciences, Massachusetts Institute of Technology
Cambridge, MA 02139, USA
E-mail address: hilst@mit.edu (R.D. van der Hilst)

*Corresponding author. Tel.: +49-921-55-3711
fax: +49-921-55-3769

ELSEVIER

Physics of the Earth and Planetary Interiors 127 (2001) 9–24

PHYSICS
OF THE EARTH
AND PLANETARY
INTERIORS

www.elsevier.com/locate/pepi

Subduction zones: observations and geodynamic models

Scott D. King*

Department of Earth and Atmospheric Science, Purdue University, West Lafayette, IN 47907-1397, USA

Received 15 November 1999; accepted 28 July 2000

Abstract

This review of subduction and geodynamic models is organized around three central questions: (1) Why is subduction asymmetric? (2) Are subducted slabs strong or weak? (3) How do subducted slabs interact with phase transformations, changes in mantle rheology, and possibly chemical boundaries in the mantle? Based on laboratory measurements of the temperature dependence of olivine, one would conclude that the core of a subducting slab is at least 10,000 times more viscous than ambient mantle; however, there are a number of complementary but independent observations that suggest that slabs are much weaker than this. Slabs undergo significant deformation in the upper mantle and may thicken to twice their original width by the time they reach the base of the transition zone. The lack of a clear correlation between the observed dip angle of deep slabs and plate velocity, rate of trench migration, and slab age in modern subduction zones is consistent with hypothesis that subduction is a time-dependent phenomenon. Both tank and numerical convection experiments with plates conclude that subduction is not a steady phenomenon, but that slabs bend, thicken, stretch, and change dip through time. This is at odds with the assumptions used in steady-state slab thermal models, where slab deformation is not considered.

Keywords: Subduction zones; Slabs; Mantle convection; Mantle rheology; Plate tectonics

1. Introduction

The formation, evolution, and eventual subduction of oceanic lithosphere dominates the large-scale dynamics of the mantle–lithosphere system and the heat budget of the mantle (e.g. Davies and Richards, 1992). Plates organize flow within the mantle, with descending flow at subduction zones and ascending flow in the intervening regions. Subducted slabs are the major source of buoyancy (in this case negative buoyancy) that drives mantle flow (Hager, 1984). Thus, it becomes clear that understanding the process of subduction is important for furthering our understanding of the thermal and chemical evolution of the Earth. Plate motions also provide the most direct observations of the scale of mantle motion as evidenced by the agreement between a kinematic flow model using plate reconstructions to advect subducting slabs for the last 200 Ma and long-wavelength global seismic tomography (Lithgow-Bertelloni and Richards, 1995).

It is important to properly incorporate subduction in geodynamic models because: (1) plate motions provide the most direct observation of the scale of mantle convection (e.g. Hager and O'Connell, 1978, 1981; Richards and Davies, 1992); (2) the negative buoyancy of subducting slabs provides the dominant driving force for plate motions (e.g. Forsyth and Uyeda, 1975; Hager, 1984); (3) seismic tomographic images of subduction zones provide the most direct observation of mass flux between the upper and lower mantle

* Corresponding author. Tel.: +1-765-494-3696; fax: +1-765-497-1620.
E-mail address: sking@purdue.edu (S.D. King).

0031-9201/01/$ – see front matter © 2001 Elsevier Science B.V. All rights reserved.
PII: S0031-9201(01)00218-7

(e.g. Grand et al., 1997; van der Hilst et al., 1997); (4) plates and slabs impose a long-wavelength platform to the pattern of convection in the mantle that would not otherwise develop in uniform-viscosity or stagnant-lid mode of convection at the Rayleigh numbers appropriate for the mantle (e.g. Gurnis, 1989; Lowman and Jarvis, 1996); (5) the long-wavelength platform of mantle convection has a significant impact on the heat flow and thermal evolution of the mantle (Gurnis, 1989; Lowman and Jarvis, 1996). The key question becomes what is the correct way to incorporate subduction in convection models? The answer is that this depends on the nature of the problem being addressed. For some global investigations, it may be sufficient to impose plate velocities as surface boundary conditions because the most important affect of plates and slabs at the longest wavelength is organization, or platform that plates and slabs impose on mantle flow. In order to understand problems at the scale of crustal recycling and/or volatile evolution, it is probably necessary to model the deformation associated with the bending and unbending of the slab in the shallow subduction zone.

2. Why is subduction asymmetric?

This is one of the most basic features of plate tectonics and has presented a challenge for geodynamic models because the unique, one-sided motion of the two plates that come together at a subduction zone is not a feature of classical Rayleigh–Benard convection. Several solutions to the asymmetric subduction problem have been proposed and these will be examined in detail below. It is important first to clarify the relationship between features in convection models and observable features on Earth.

The surface of the Earth and the core-mantle boundary are both isothermal, zero-stress boundaries for the mantle. The Rayleigh number is a dimensionless parameter that can be thought of as a ratio of the magnitude of the buoyancy force, resulting from temperature variations within the fluid, to viscous stresses. For classical Rayleigh–Benard convection (convection in an incompressible fluid with isothermal top and bottom boundaries, uniform viscosity, and constant thermodynamic properties where the density assumed constant in every term of the equation except for the buoyancy force term) the Rayleigh number and the geometry (i.e. a Cartesian box of specified length or spherical shell with a specified ratio of inner and outer radii, for example) uniquely describe the problem. While Rayleigh–Benard convection is a useful starting point for understanding the physics of fluid flow, there are a number of assumptions in Rayleigh–Benard convection that can be removed in order to better approximate the Earth (e.g. the mantle is slightly compressible, the mantle is not isoviscous and the thermodynamic properties of the mantle vary with temperature and pressure). The Rayleigh number remains a useful measure of the vigor of convection even though it is no longer sufficient to uniquely describe the conditions of more complex fluid problems (see Table 1). For example, in a vigorously convecting fluid the temperature variations in the interior of the fluid are small and most of the temperature difference between the surface and bottom of the fluid is confined to thin layers near the top and bottom of the fluid (and thin vertical columns that separate the nearly isothermal interiors). These layers are called thermal boundary

Table 1
Values for mantle Rayleigh number

Symbol	Quantity	Value	Reference
ρ_0	Density	3.7×10^3 kg/m^3	Dziewonski and Anderson (1981)
g_0	Magnitude of gravity	9.8 m/s^2	
h	Depth of the core–mantle boundary	2.891×10^6 m	
ΔT	Superadiabatic temperature increase	2×10^3 K	
α_0	Thermal expansivity	(5–20) $\times 10^{-6}$ K^{-1}	Chopelas and Boehler (1989)
κ_0	Thermal diffusivity	(5.4–24.0) $\times 10^{-7}$ m^2/s	Brown (1986)[a]
η_0	Viscosity	(3–10) $\times 10^{21}$ Pa s	Mitrovica (1996)
$Ra = \rho_0 g_0 \alpha_0 \Delta T h^3 / \kappa_0 \eta_0$	Rayleigh number	(0.365–21.6) $\times 10^6$	

[a] For thermal conductivity, $c_p = 1.25$ kJ/kg K, density.

layers and their thickness scales with the $-1/4$ to $-1/3$ power of the Rayleigh number (McKenzie et al., 1974). The relationship between thermal boundary layer thickness and Ralyleigh number has the same form (although different coefficients) for Rayleigh–Benard and compressible convection (Jarvis and Peltier, 1989).

The lithosphere of the Earth is the cold, top thermal boundary layer of the mantle. The conceptual picture of a plate as a passive raft that sits on top of the mantle neglects the fact that oceanic plates are created from and return to the mantle. It is generally accepted that the cold, negatively-buoyant slabs provide the dominant force that drives plate motions, historically referred to as 'slab-pull' after Forsyth and Uyeda (1975). The terms plate, lithosphere, and thermal boundary layer are often used synonymously when describing convection calculations with dynamic plates and/or subduction. The lithosphere is comprised of chemically and rheologically distinct crustal and residual mantle components, or layers. The average thickness of the oceanic crust, 6 km, is four to five times smaller than the distance between adjacent grid points in most numerical calculations. Compositional and rheological components of the lithosphere have been included in some convection calculations (Gaherty and Hager, 1994; Lenardic and Kaula, 1996; Van Keken et al., 1996; Christensen, 1997). The general consensus is that buoyant crust is not able to separate from the denser lithosphere (e.g. Gaherty and Hager, 1994) unless there is a low viscosity region separating the crust and lithosphere (Van Keken et al., 1996). Thus, most mantle convection calculations ignore the compositional and rheological variations that would be introduced by including a chemically distinct crust.

Mantle convection studies began with the examination of the Rayleigh–Benard problem described above. Symmetric downwellings (e.g. the thermal boundary layers on both sides of the downwelling participate in the formation and development of the downwelling limb) are always observed Rayleigh–Benard problems (e.g. Gaherty and Hager, 1994). Downwellings associated with subduction do not behave in this manner; instead one plate descends into the mantle beneath the other. The importance of this difference cannot be over-emphasized. When both downwelling limbs participate in the formation of the downwelling, usually the coldest, near-surface part of the boundary layer remains at the surface and the lower part of the boundary layers participate in the formation of the downwelling. In our current view of subduction, most, if not all, of the subducting plate becomes the subducting slab. In this case, although there is only one boundary layer forming the downwelling, the temperatures within that downwelling may be colder than the case of symmetric downwellings which have more material participating in the downwelling in total but less of the coldest parts of the downwelling. There has been very little systematic study of the details of slab thermal structure, so it is not possible to state with certainty which case will produce the greatest density anomaly. Different formulations of asymmetric subduction may produce different results.

Corner flow models of subduction have shown that the angle of subduction could be controlled by imposing asymmetric velocity boundary conditions (Stevenson and Turner, 1977; Tovish et al., 1978). These studies show that for a given slab buoyancy, there is a critical angle at which a dipping slab will be in steady equilibrium. Angles shallower than the critical angle are unstable, while angles steeper than the critical angle will evolve toward the critical angle. Tank experiments with migrating trenches (Griffiths et al., 1995; Guillou-Frottier et al., 1995) and numerical experiments (Scott and King, in preparation) confirm that downwellings with angles shallower than a critical value undergo significant deformation and often break up.

Davies (1977) suggested that plate motions are an important component in the development of slab geometries. Using observed plate velocities to drive a kinematic flow in a 3D spherical model Hager and O'Connell (1978) predicted the geometry of subduction zones surprisingly well. The flow model in that study did not include buoyancy forces which would have increased the vertical component of slab motion relative to the kinematic flow because the cold, dense slabs will sink in a viscous fluid. This would have reduced the fit of the predicted slab dips to the observed slab geometries. This leaves a rather odd paradox because, it is clear that slabs are more dense that their surrounding (e.g. Hager, 1984) and provide a major driving force for plate motions (Forsyth and Uyeda, 1975) yet some observations (like slab geometry) appear to be better fit by models that ignore slab

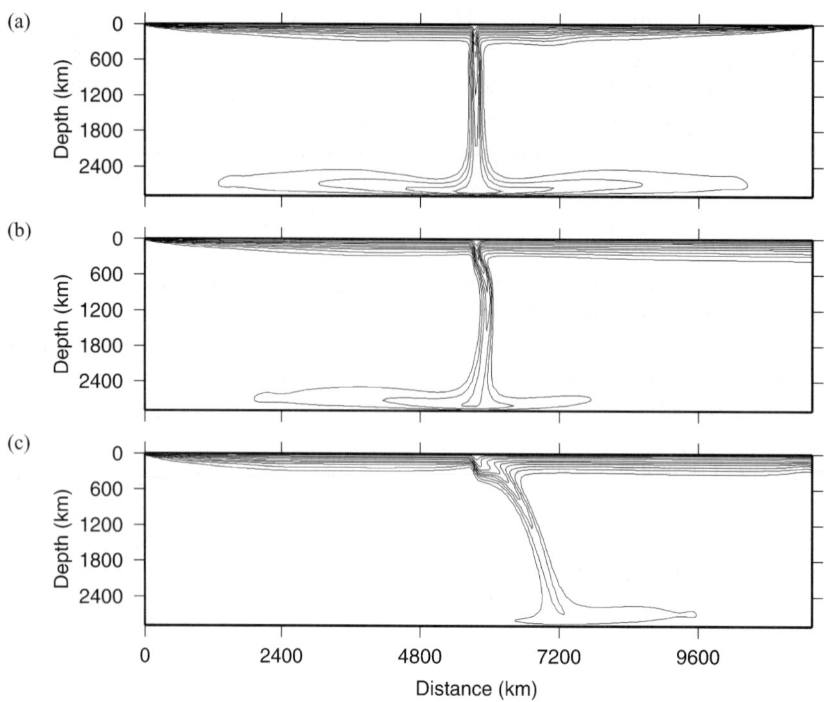

Fig. 1. Thermal fields from two convection calculations with plates. In both calculations the Rayleigh number is 10^6, the top and bottom are isothermal free-slip boundaries, the fluid is internally heated, and the viscosity is temperature dependent. The initial temperature field has a square root of age plate thermal field on the left side of the box and a uniform boundary layer on the right side of the box. Weak material zones allow the plate to deform at the trench and the ridge. The 'ridge' is in the left corner of the box and the 'trench' is in the center of the box. Further details on this type of calculation can be found in Chen and King (1998). (a) The side velocity boundary conditions are free-slip (material is free to move vertically along the edge). (b) Identical to (a) except that the velocity of a single node on the right-hand side is set to zero, breaking the symmetry of the top boundary layer velocities. (c) Identical to (b) except that the side velocity boundary conditions are an imposed horizontal velocity of 20 mm per year. This is identical to moving the trench and plate relative to the center of mass of the mantle.

density. As we will see below, this remains a problem for convection models.

Without imposing some form of asymmetry in the velocity boundary conditions, it is difficult, if not impossible, to generate an asymmetric downwelling. In Fig. 1a, the temperature field from a convection calculation that attempts to reproduce subduction is shown. The top surface of the model has a free-slip boundary condition, and there is a weak zone at the center of the box that keeps the downwelling form migrating. Even though the right-hand side of the box has a uniform thermal boundary layer and the left-hand side of the box has a square-root of age temperature profile characteristic of uniform plates, both sides of the upper boundary layer participate in the downwelling. The development of a square-root of age temperature profile on the right-hand side of the box between the non-dimensional distances of 3.5 and 4.0 illustrates that this side of the thermal boundary layer is moving like a uniform plate and is moving toward the central downwelling. When asymmetric velocity boundary conditions are applied, the resulting downwelling starts out with a dip that is approximately the critical-angle dip predicted by half-space theory (Fig. 1b); however, the downwelling quickly steepens to a near-90° dip within a distance of about 1–2 times the thickness of the boundary layer. In this case, the right-hand side of the top boundary layer is stationary, as shown by the parallel isotherms because, a zero velocity boundary condition is imposed on one node, breaking the symmetry between the left- and right-hand sides. A limitation of 2D geometries is illustrated in Fig. 1b. Because the flow is confined to the 2D plane, the upwelling flow driven by the separation of the plates at the ridge draws fluid from the right side of the box. This flow pulls the slab back under the moving plate. In a 3D geometry, the upwelling flow will draw from the 3D region and the stress on the slab induced by the upwelling flow will not be as strong as it is in the 2D geometry. Models with the most realistic slab geometries generally have both an asymmetric surface condition and trench migration, as shown in the model in Fig. 1c.

Attempts to circumvent the problem of return flow in a 2D geometry have included the use of periodic boundary conditions (e.g. Gurnis and Hager, 1988; Lowman and Jarvis, 1996; Chen and King, 1998; Han and Gurnis, 1999) or a cylindrical geometry (Puster et al., 1995; Zhong and Gurnis, 1995; Ita and King, 1998). Yet, even 3D geometries are not sufficient to explain the deep structure of subducted slabs (e.g. Bercovici et al., 1989; Tackley et al., 1993, 1994; Ratcliff et al., 1995, 1996; Bunge et al., 1996, 1997). In 3D constant viscosity calculations (e.g. Tackley et al., 1993, 1994) the global platform of the flow could be described as linear, sheet-like downwellings and cylindrical, plume-like upwellings. Although these features do not resemble plumes or slabs in any detail, the results are encouraging. When temperature-dependent rheology is included in 3D spherical calculations, the platform changes, i.e. cylindrical downwellings at the pole and sheet-like upwellings (Ratcliff et al., 1995, 1996). Even though calculations have moved from 2D Cartesian geometries to more a appropriate geometry for the mantle (3D spherical models), the results suggest that there is still the need for an additional symmetry-breaking condition to produce asymmetric downwellings.

A series of investigations of convection with temperature-dependent and/or non-Newtonian rheologies documented a transition from the free-slip platform found in constant viscosity calculations through a sluggish-lid platform to a stagnant-lid platform of convection as the stiffness of the boundary layer increases (Van den Berg et al., 1991; Solomatov, 1993; Christensen, 1984; Solomatov, 1995; Moresi and Solomatov, 1995; Solomatov and Moresi, 1996; Tackley, 1996, 1998). The sluggish and stagnant-lid convective planforms do not include asymmetric downwellings or piecewise-uniform surface velocities, both of which are characteristics of subduction zones that we would like to be able to model (Weinstein and Christensen, 1991). While the structure of the top thermal boundary layer might seem somewhat removed from the issue of subduction, it is really quite important. In stagnant- and sluggish-lid convection, most of the top boundary layer remains at the surface and only the weakest part of the boundary layer participates in the active flow. Thus, most of the cold boundary layer is never recycled into the interior of the fluid. This is quite different from our picture of subduction, where most, if not all, of the subducting plate descends into the interior of the fluid. This has important implications for the plate driving force due to slabs and for gravity anomalies over subduction zones because the temperature structure is related to the density structure through the coefficient of thermal expansion.

In the stagnant lid platform, a significant amount of the negative buoyancy in the top thermal boundary layer does not participate in the downwelling (Moresi and Solomatov, 1995; Conrad and Hager, 1999). Furthermore, as a result of the one-sided nature of subduction, the surface heat flow from stagnant-lid convection calculations is significantly smaller than the heat flow from models with temperature-dependent rheology when a plate formulation is included (Gurnis, 1989).

The addition of a strain-weakening (non-Newtonian) component to the rheology weakens parts of the cold thermal boundary layer, especially at regions of high stress (such as in the corners of the computational domain). When coupled with temperature-dependent rheology, this can produce nearly uniform surface velocities (Weinstein and Olson, 1992; King et al., 1992; Larsen et al., 1993); however, power-law rheology calculations are typically unstable and the plate-like behavior quickly breaks down as these calculations evolve away from the carefully chosen initial conditions. Some researchers have argued that subduction forms at pre-existing zones of weakness in the lithosphere (e.g. Toth and Gurnis, 1998). Rheologies based solely on the properties of mantle minerals have not successfully produced piecewise-uniform surface velocities and slab-like downwellings. As a result, in order to generate slab-like asymmetric downwellings, convection calculations have used imposed velocity boundary conditions (e.g. Stevenson and Turner, 1977; Davies, 1986, 1988, 1989; Christensen, 1996), a priori specified mechanically weak zones (e.g. Gurnis and Hager, 1988; King and Hager, 1994), and/or dipping viscous faults (e.g. Zhong and Gurnis, 1992, 1994a, 1995, 1996, 1997). It is beyond the scope of this review to address the strengths and limitations of each of these approaches in detail. The reader is referred to King et al. (1992), Zhong et al. (1998), Han and Gurnis (1999), Bercovici et al. (1999) and references therein for further discussion of plate generation methods.

A criticism of the calculations discussed above is that they are limited, for the most part, to two-dimensional geometries, or three-dimensional geometries with reflecting side-wall boundary conditions. Symmetric, near-90°-dipping downwellings are also the observed platform in three-dimensional spherical convection models (e.g. Tackley et al., 1994; Bunge et al., 1997). Even when temperature-dependent rheology is included in spherical calculations, dipping slab-like features are not observed (Ratcliff et al., 1995, 1996). The use of periodic boundary conditions in two-dimensions eliminates the effect of the side walls and two-dimensional calculations can be formulated in such a way that they are formally equivalent to calculations that explicitly allow the trench to move relative to the grid (cf. Han and Gurnis, 1999).

Trench migration has been one of the most studied mechanisms for producing dipping slabs (Davies, 1986; Gurnis and Hager, 1988; Zhong and Gurnis, 1995; Griffiths et al., 1995; Christensen, 1996; Ita and King, 1998). It is interesting to note that while numerical models models and tank experiments have demonstrated that trench migration is an effective mechanism for generating shallow dipping slabs, a compilation of data from various subduction zones (Jarrard, 1986) shows almost no correlation between the dip of a slab in the 100–400 km depth range and the rate of trench migration (Fig. 2). There are several cases where three-dimensional models have been employed to study specific regions with considerable success matching seismic structure (van der Hilst and Seno, 1993; Moresi and Gurnis, 1996). In these cases, the focus of attention was in the western Pacific, where slabs are old and steeply dipping. In addition, these cases have focused on subduction zones where a significant amount of trench migration has been documented. We can reconcile the apparent lack of correlation of the global trench migration

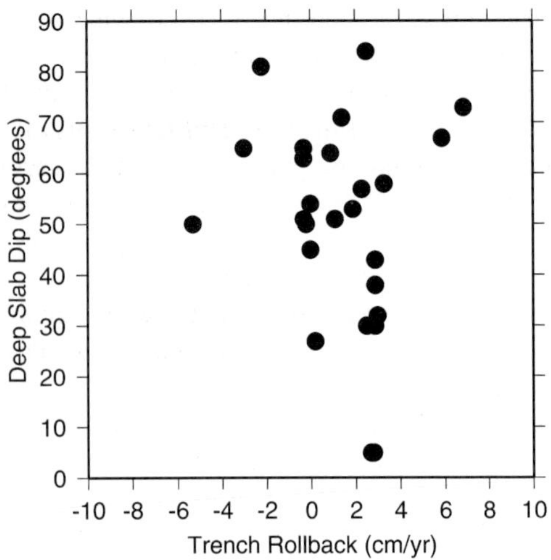

Fig. 2. Deep slab dip, as measured by the shape of the Wadati-Benioff zone in the 100-400 km depth range, vs. trench rollback using the Minster and Jordan plate motion model. Data taken from (Jarrard, 1986).

observations with slab geometry and the success of the regional subduction studies by recognizing that subduction is a time-dependent phenomenon and the shape of subducting slabs evolves with time. Both numerical and tank experiments show that buoyancy driven subduction is a time-dependent phenomenon (Gurnis and Hager, 1988; Griffiths et al., 1995; Becker et al., 1999). The geometries of present day subduction zones provide a single snap shot of a time-dependent phenomenon and each subduction zone is at a different stage in this process.

There are several observations that have not been fully exploited in convection modeling. The first of these is the difference in dip between eastward dipping and westward dipping slabs (e.g. Le Pichon, 1968; Ricard et al., 1991; Doglioni, 1993; Marotta and Mongelli, 1998; Doglioni et al., 1999) which is illustrated in Fig. 3. The difference between Chilean style subduction zones and Mariana style subduction zones has been recognized for some time (e.g. Uyeda and Kanamori, 1979). However, it is not easy to separate the effects of dip direction from other factors because most of the eastward dipping slabs are being over-ridden by the North and South American plates (i.e. continents) while many of the slabs in the Pacific are being over-ridden by island arcs or oceanic plates. Yet, a series of flexure calculations that include the effect of motion of the slab-trench system relative to the underlying mantle fit slab geometries remarkably well (Marotta and Mongelli, 1998). One of the reasons that this observation has not been explored is that many of the techniques used to generate subduction features in two dimensions are not practical to implement in a complete sphere in three dimensions calculations. Furthermore, the grid resolution that is required to study subduction problems exceeds what can be practically achieved in a three-dimensional spherical shell at present.

There is also a difference in the average dip angle between the populations of slabs where the overriding plate is continental or oceanic plate (Furlong et al., 1982; Jarrard, 1986). While the average of the two populations is distinctly different (40° versus 66° from Jarrard's data), there are notable exceptions including the shallow west-dipping Japan slab and the steep eastward-dipping Solomon and New Hebrides slabs (see Fig. 3). Studies of supercontinent breakup observe shallow dipping slabs under continents (e.g. Lowman and Jarvis, 1996). Because we are currently in a phase where continents are generally moving away from the site of the former supercontinent and over-riding oceanic plates, it is not clear whether this correlation once again reflects the effect of trench migration, or mechanical differences in the overriding plate.

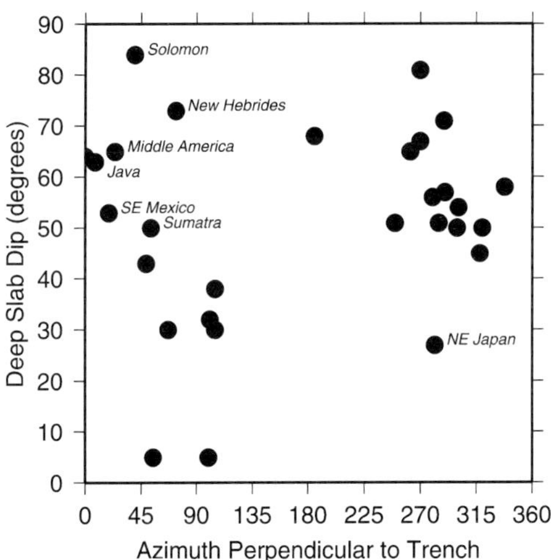

Fig. 3. Deep slab dip, as measured by the shape of the Wadati–Benioff zone in the 100–400 km depth range, vs. azimuth perpendicular to the trench. Data taken from Jarrard (1986).

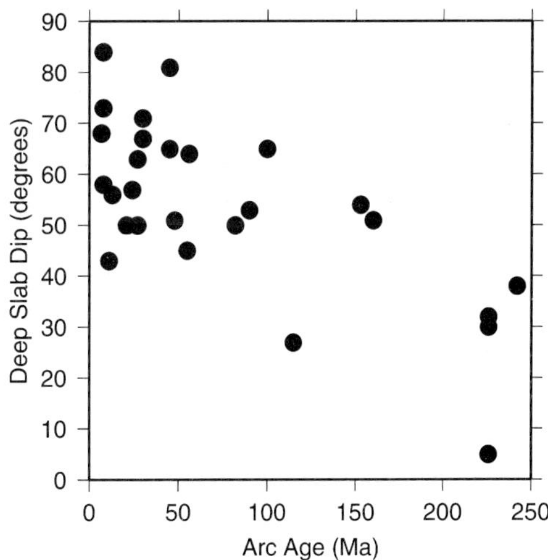

Fig. 4. Deep slab dip, as measured by the shape of the Wadati-Benioff zone in the 100–400 km depth range, vs. duration of subduction. Data taken from Jarrard (1986).

It is interesting to note that the strongest correlation in the data set compiled by Jarrard (1986) is the correlation of the duration of subduction with dip of the deep part of the slab, defined as 100–400 km (Fig. 4). This was the only significant correlation that Jarrard found in the global compilation of subduction zone data. While it is not possible to isolate this from the other factors that seem to influence slab geometry, this correlation is consistent with the assumption that subducted slabs are evolving, time-dependent features that are not at steady-state equilibrium. Furthermore, it is encouraging that the observed subduction zone geometries shown in Fig. 4 agree with the general findings of the models, that slab dip angles become more shallow with time. In both numerical and laboratory experiments, young buoyancy driven slabs steepen with age from the time of the initiation of subduction until the slab reaches the transition zone (Gurnis and Hager, 1988; Griffiths et al., 1995; Becker et al., 1999). Once the slabs reach the top of the lower mantle slab dip angles become progressively shallower if there is ocean-ward trench migration.

The area where the asymmetry of subduction is most apparent is the subduction of young lithosphere. The most striking example is the subduction of the Pacific plate underneath Alaska. Perhaps, not coincidentally, this is an area where convection calculations have had little success. It is highly probable that the subduction of young lithosphere is not driven by the negative buoyancy of the subducting slab, which is the underlying assumption in most convection subduction studies. Convection models have focused on the case of older slabs where the negative buoyancy of the slab itself provides the driving motion for the plate system (e.g. slab-pull). There are a few examples of studies of the subduction of young lithosphere (e.g. England and Wortel, 1980; Vlaar, 1983). Young slabs have less negative buoyancy than older slabs, and thus, younger slabs should have shallower dip angles than older slabs, all other things being equal.

It is important to recognize that while buoyancy within a slab is dominated by the effect of temperature, phase transformations can have significant effects (e.g. Daessler and Yuen, 1996; Bina, 1996, 1997; Christensen, 1997; Ita and King, 1998; Schmeling et al., 1999; Marton et al., 1999). The impact of phase transformations on buoyancy is most pronounced if the olivine to spinel phase transformation is kinetically hindered in cold slabs (see Kirby et al., 1996 and references therein). The presence of metastable olivine can reduce slab decent velocities by as much as 30% (Kirby et al., 1996; Schmeling et al., 1999; Marton et al., 1999) and will have the greatest effect on the oldest and coldest slabs. The stresses induced by the differential buoyancy are consistent with the pattern of deep earthquakes (Bina, 1997).

Many researchers have taken an approach which bypasses some of the problems of the development of the asymmetric downwelling by solving for the thermal structure of a subducting slab using a kinematic model where the velocity field is specified (e.g. Minear and Toksoz, 1970, Peacock, 1991, 1996, Peacock et al. 1994; Molnar et al., 1979; Helffrich et al., 1989; Kirby et al., 1996). The attraction of this approach is that the required input to the kinematic model includes plate age, plate velocity, and a specified slab dip. These quantities are easily observed (or measured), making it practical to set up a thermal structure calculation for specific subduction zones. In a convection calculation, the plate velocity, slab velocity and slab dip are not input controls, making it more difficult to set up a calculation with a geometry that resembles specific subduction zones. There are limitations to the kinematic approach. First, the velocity of the slab is not driven by the buoyancy structure of the slab. A result of the independence of slab buoyancy and plate/slab velocity that is built into the kinematic model is that old slabs and young slabs can descend with the same velocity, in apparent violation of the generally held theory that slab density is a major driving force for plate motion (e.g. Forsyth and Uyeda, 1975). In most, if not all kinematic models, the velocity of the slab is the same as the velocity of the plate (or some fraction thereof). Second, the deformation that occurs as a plate bends, breaks, and generally deforms as it passes through a subduction zone (Conrad and Hager, 1999) is not accounted for in kinematic models. It is reasonable to ask how our limited understanding of the deformation at subduction zones affects buoyancy driven slab models approximation the deformation at a trench. A third, and perhaps the most important, limitation of kinematic slab thermal models is that the slab behaves rigidly, with no internal deformation. There are a number of observations which show that slabs deform as they sink, especially in the region of the transition zone (see Lay, 1994 and references therein). Because there has been no attempt to directly compare the thermal structures from buoyancy driven slabs and kinematic slabs, it is not possible to quantify the degree to which the different assumptions impact the thermal models.

3. Are slabs strong or weak?

The strength of subducted slabs is an important mechanical property of the mantle–lithosphere system, yet flow models can reproduce some observations at subduction zone with strong or weak slabs. One example is the global distribution of seismicity with depth. Seismicity rates decay exponentially with depth in the upper mantle, reaching a minimum between the depths of 300 and 500 km. Seismicity rate then increases until 700 km at which point all seismicity abruptly ends. The pattern of seismicity with depth has been modeled using both strong (Vassiliou et al., 1984) and weak (Tao and O'Connell, 1993) slabs meeting an increase in viscosity at about 670 km depth.

Laboratory investigations of the deformation mechanism of mantle minerals demonstrate that the creep strength of mantle minerals is a strong function of temperature, pressure, strain-rate, and grain-size (e.g. Karato and Li, 1992; Karato and Wu, 1993; Karato and Rubie, 1997; Riedel and Karato, 1997). Any reasonable estimate of the temperature difference between the core of a subducting slab and average mantle leads to the conclusion that the majority of a slab must be much stronger than the mantle. However, if great earthquakes periodically cut through the entire thickness of the elastic oceanic lithosphere (Kanamori, 1971), then the mechanical properties of individual minerals may not be representative of the properties of the subducted slab as a whole. The role of water adds an additional complicating factor as laboratory work has shown that the viscosity of olivine can be reduced by as much as a factor of 100 in the presence of water (Hirth and Kohlstedt, 1996). The consideration of water raises additional problems because there are a series of dehydration reactions that take place in the slab (e.g. Peacock, 1996) and the partitioning of water between phases is not well constrained.

There are numerous seismic studies which show that slabs can bend, kink and thicken. These include the location of Wadati–Benioff zones (WBZ) as mapped through earthquake hypocenters (Isacks and Barazangi, 1977; Giardini and Woodhouse, 1984, 1986; Fischer et al., 1988, 1991), the study of earthquake focal mechanisms (Isacks and Molnar, 1968, 1971; Vassiliou, 1984; Giardini and Woodhouse, 1986; Holt, 1995), the study of moment tensors (Bevis, 1988; Fischer and Jordan, 1991; Holt, 1995) and tomography (van der Hilst et al., 1991, 1993, 1995; Fukao et al., 1992). The seismic observations have been extensively reviewed by (Lay, 1994) and here I will only highlight a few examples.

The rate of accumulation of seismic moment in WBZ can be used to estimate the average down dip strain rate in subducting slabs (Bevis, 1986; Fischer and Jordan, 1991; Holt, 1995). This assumes that the amount of aseismic deformation is small and that the viscous deformation is parallel with the brittle layer with the same deformation-rate tensor. Thus, the seismic deformation rate probably represents a minimum estimate of slab deformation. Because slabs are generally in down-dip compression or down-dip extension, the average change in the length of a slab due to the cumulative seismicity can be related to the sum of the seismic moments of the individual seismic events (Bevis, 1988). Even after accounting for the scatter in focal mechanisms, in the depth range of 75–175 km the average strain rate in subducting slabs is estimated to be $10^{-15}\,s^{-1}$ (Bevis, 1988). This compares with a characteristic asthenospheric strain rate of $3 \times 10^{-14}\,s^{-1}$ (Turcotte and Schubert, 1982; Hager and O'Connell, 1981). This leads to the surprising suggestion that the difference in strain rates between the asthenosphere and subducting slabs is no more than a factor of 3. Taking the average decent rate of slabs this strain rate corresponds to a total accumulated strain in the depth range of 75–175 km of 5%. Fischer and Jordan (1991) find that the seismic data require a thickening factor of 1.5 or more in the seismogenic core of the slab in central Tonga and complex deformation in northern and southern Tonga. Holt (1995) argues that the average seismic strain rate in Tonga represents as much as 60% of the total relative vertical motion between the surface and 670 km.

The geometry of WBZ also indicates that slabs are significantly deformed. Subducting slabs bend to accommodate over-ridding plates and the descend into the mantle; they also unbend, otherwise WBZs would curve back on themselves (Lliboutry, 1969). The dip of a subducting slab changes over a narrow depth interval, from 10–20° in the inter-plate thrust zone to 30–65° at depths near 75 km (Isacks and Barazangi, 1977). This change in shallow dip is a feature that is not well modeled by any slab model. In addition to deformation as the slab descends into the mantle, there is also deformation in the horizontal plane at subduction zones. Because oceanic plates themselves are spherical caps, the degree of misfit between subducting slab and the best fit spherical cap provides an estimate of the amount of slab deformation (Bevis, 1986). Based on this analysis, the Alaska-Aleution, Sumatra-Java-Flores, Caribbean, Scotia, Ryukyu, and Mariana arcs undergo a minimum of 10% strain (Bevis, 1986; Giardini and Woodhouse, 1986).

While convection models can reproduce many subduction zone observations with both strong and weak slabs, there are several indications from convection studies that slabs are weak. Houseman and Gubbins (1997) use a dynamic model of the lithosphere that assumes that the properties of the slab are uniform through out the entirety of the slab; the slab is both more viscous and more dense than the surrounding mantle. They find that the shape of the deformed slab is a strong function of the viscosity of the slab. They also observe a buckling mode that produces slab geometries similar to the Tonga slab. The effective viscosity of the slab needed to produce this slab geometry is 2.5×10^{22} Pa s, no more than 200 times the upper mantle viscosity and maybe only a factor of two greater than the lower mantle viscosity (cf. King and Ita, 1995).

The regional gravitational potential, or geoid, high over subduction zones has provided at important constraint on mantle rheology (Hager, 1984). In the most simple terms, the geoid at subduction zones is a balance between two large terms that are opposite in sign and almost cancel. Those terms come from the positive gravitational potential due to the dense, slab and the negative gravitational potential due to the deformation of the surface (i.e. dynamic topography). If the mantle were static, and the slab did not sink, then this term would be zero. Hager (1984) showed that in an isoviscous mantle, the dynamic topography generated by a reasonable slab density anomaly creates a dynamic topography contribution to the geoid that exceeds the geoid anomaly due to the slab, resulting in geoid lows over the subduction zone, opposite of the observation. The dynamic topography contribution can be reduced, resulting a in geoid high, with an increase in viscosity at approximately 670 km. There have been a number of efforts to improve the uniform viscosity flow model used by Hager by including: both depth dependent and temperature dependent viscosity; the effect of phase transformations; and plate formulations with temperature-dependent rheology (Zhong and Gurnis, 1992; King and Hager, 1994; Moresi and Gurnis, 1996; Zhong and Davies, 1999). The surprising result from this work is that weak slabs provide a much better fit to the geoid than strong (high viscosity) slabs.

It would be easy to dismiss any one of these observations taken by itself; however, the number and diversity of the observations that slabs are not significantly stronger than ambient mantle suggests that this should receive serious consideration. A weak slab is not necessarily inconsistent with the laboratory observations. While the viscosity of olivine is a strong function of temperature, it is also a strong function of grain size (Riedel and Karato, 1997). Grain size reduction resulting from recrystalization of minerals at phase boundaries may counter balance the effects of temperature on viscosity. In addition, the brittle component of the slab may be mechanically weakened by the faulting and thus the single crystal measurements of viscosity may not correctly describe the deformation of the slab.

It is important to recognize that assumptions regarding slab viscosity are built into some slab thermal models. For example, in the Minear and Toksoz (1970) and Peacock et al. (1994) models, the kinematic velocity field used to advect the temperatures is uniform except in the corner where the plate becomes the slab. Because the slab has a uniform velocity, there is zero strain, and hence zero deformation. The evidence reviewed above suggests that slabs thicken with depth. Because the interior of a slab warms primarily by diffusion, this suggests that the interior of thickening slabs will be colder than the kinematic models predict, all other things being equal. Because there have been no direct comparison between kinematic model, which do not include slab deformation and dynamic models, which can include slab deformation, it is impossible to make a quantitative statement.

4. How do slabs interact with phase, rheology, and (possibly) chemical boundaries in the mantle?

The eventual fate of deep slabs has been a topic of debate for several decades (see Lay, 1994 and Christensen, 1995 for reviews). One of the original arguments for a barrier to convection at 670 km depth was the cessation of earthquakes at this depth. While this is approximately the depth that corresponds to the pressure of the olivine-spinel to perovskite plus ferro-periclase phase transformation, both a chemical boundary and rheological boundary have been proposed to explain the cessation of earthquakes.

Early attempts to model subduction zones used a non-Newtonian fluid with an endothermic phase transformation. This approach produced stiff downwelling limbs that descended at a 90° dip angle as a direct consequence of the free-slip (reflecting) boundary conditions on the sides of the domain (Christensen and Yuen, 1984). The symmetry of the downwelling (i.e. material from both sides of the thermal boundary layer at the downwelling) is the major shortcoming of this approach. Christensen and Yuen identified three planforms of deep slabs at a phase boundary: slab penetration, slab stagnation, and partial slab penetration. Fluid experiments with corn syrup which used a setup designed to produce an asymmetric downwelling confirm these basic planforms (Kincaid and Olson, 1987). The tank experiments induced an asymmetric

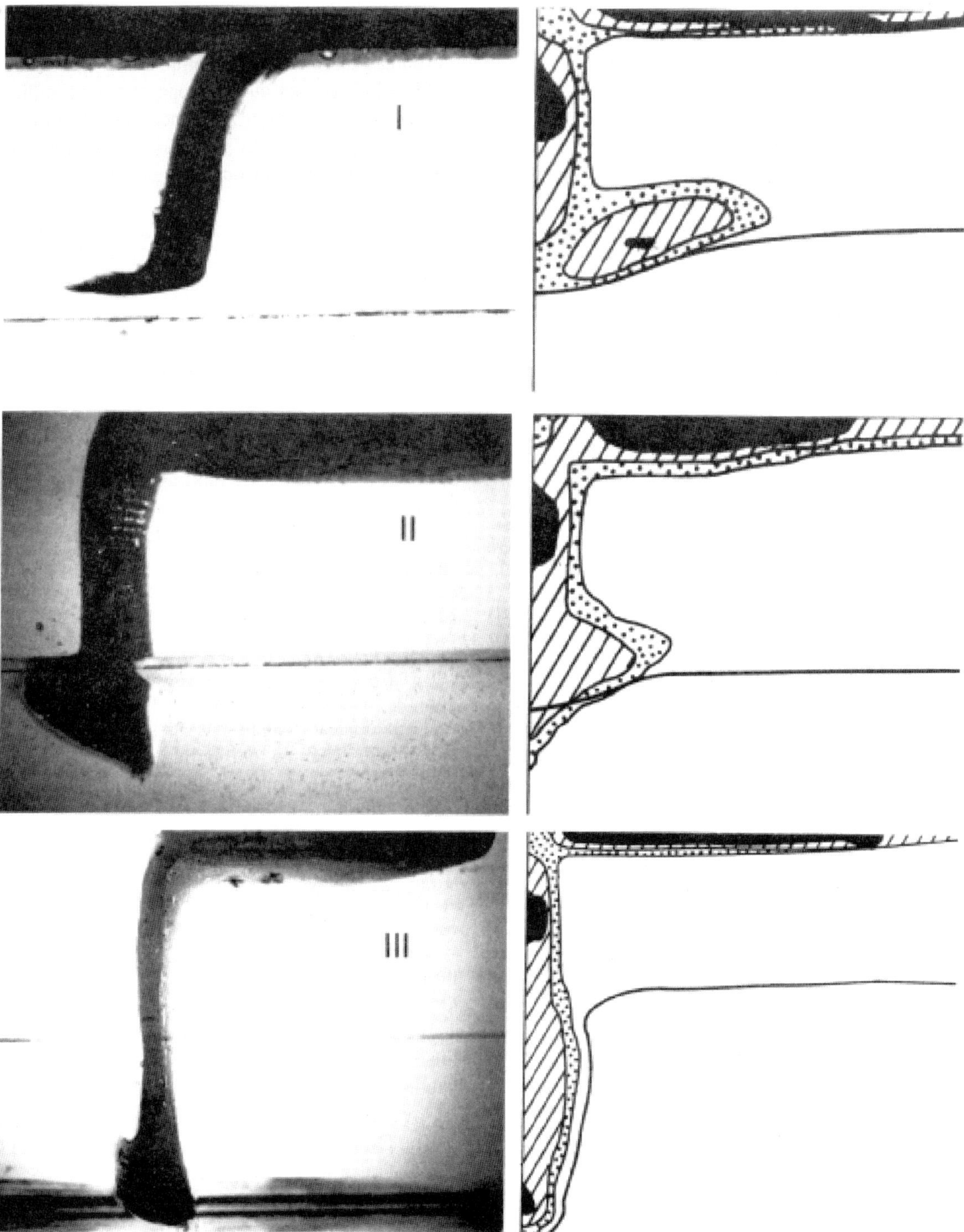

Fig. 5. Experimentally observed styles of slab penetration through a density discontinuity (left) compared with two-dimensional calculation by Christensen and Yuen (1984) (right). Part I: Slab deflection with $R \approx -0.2$. Part II: Partial slab penetration with $R \approx 0.0$. Part III: Complete slab penetration with $R \approx 0.5$. From Kincaid and Olson (1987) with permission.

downwelling by placing two cold sheets of concentrated sucrose solution on the top surface of the fluid in the tank. The larger of the two sheets was introduced with a shallow dipping bend at its leading edge; this provided the instability that allowed this plate to subduct under the other plate. The agreement between the tank and numerical experiments (Fig. 5) suggests that the three planforms discussed above are robust features of cold downwellings interacting with phase transformations and/or chemical boundaries. Subsequent numerical (Zhong and Gurnis, 1994b; King and Ita, 1995; Ita and King, 1998) and tank (Griffiths et al., 1995; Guillou-Frottier et al., 1995) experiments have confirmed the basic results and added additional insight into the interaction between subducting slabs and phase transformations.

Many investigators have studied subduction with one or more phase changes. There is a general consensus that has emerged from both 2D and 3D calculations in both Cartesian and spherical geometries. The evolution of the flow can be described by three distinct stages: (1) the initial impingment of the leading edge of the subducting slab on the spinel to perovskite plus ferro-periclase phase boundary; (2) a period of increased trench roll-back and draping of the slab on the phase boundary and; (3) virtual cessation of trench roll-back as the slab penetrates into the lower mantle. The interaction of subduction, trench-migration, and phase changes is now reasonably well understood. Several numerical studies have clearly demonstrated that the effect of a phase transformation on the pattern of flow in a convecting fluid is sensitive to the Rayleigh number, initial conditions, boundary conditions, and the equation of state approximation (Ita and King, 1994; King et al., 1997).

5. Conclusion

It is well established from both laboratory and numerical experiments, that trench migration can produce shallow dipping downwelling features. Detailed plate reconstructions have been successfully used to explain slab geometries in the western Pacific, lending support to the hypothesis that trench migration is a controlling factor in slab geometry. There may be other factors, including the properties of the overriding plate and the direction of subduction (i.e. eastward versus westward dipping slabs). The fact that slab dip angles and trench migration are not strongly correlated in the global compilation of subduction zones observations at first appears to contradict the findings of subduction zone modeling; however it is important to remember that both the numerical and laboratory experiments and the global compilation of subduction zone parameters indicate that subduction is a time-dependent phenomenon.

A more surprising result, consistent with a variety of observations, is that slabs appear to be much weaker than the predictions of slab viscosity based on temperature dependent viscosity alone. At the base of the transition zone, the viscosity contrast between the slab and lower mantle may be as small as a factor of three. This can be reconciled with laboratory creep data from single crystals if the effective viscosity of a subducting slab is weakened by faulting or as a result of the reduction of mineral grain sizes due to dynamic recrystalization as a result of phase transformations or faulting.

Both of these results require reconsideration of various slab thermal models because, the most popular and widely used slab thermal models do not account for slab deformation or time-dependent behavior. If a slab thickens by as much as a factor of two, then thermal models that assume that slabs cool by conduction are underestimating the cooling time for subducting slabs to reach thermal equilibrium with the mantle.

The subduction of young lithosphere has received little attention (e.g. England and Wortel, 1980; Vlaar, 1983). Convection models assume that subduction is driven by the negative buoyancy of the slab and it is not clear whether young, warm slabs subduct due to their own negative buoyancy or are being forced under other plates by a global force/mass balance. The Cocos plate subducting under Central America and the Juan de Fuca plate subducting under North America are both examples where young lithosphere is being subducted. In these cases, subducting may be driven by the over-riding North American continental landmass.

The Earth appears to be the only terrestrial body on which the process of subduction is currently active. This is an important but often neglected observation. A complete understanding of the process of subduction should not only explain the observations from subduction zones on Earth, it must also explain why subduction is presently not occurring on Venus or Mars. The most popular explanation, that plate tectonics requires water to weaken the crust, has only been tested at the most basic level (Lenardic and Kaula, 1994).

Acknowledgements

This work was supported by NSF grant EAR-9903002. Reviews by S. Zhong and R. van der Hilst greatly improved the manuscript. Discussions with participants at the Processes and Consequences of Deep Subduction Workshop were valuable in forming many of the ideas in this review.

References

Becker, T.W., Faccenna, C., O'Connell, R.J., Giardini, D., 1999. The development of slabs in the upper mantle: insights from numerical and laboratory experiments. J. Geophys. Res. 104, 15207–15226.

Bercovici, D., Ricard, Y., Richards, M.A., 1999. The relation between mantle dynamics and plate tectonics: a primer. In: Richards, M.A., Gordon, R., van der Hilst, R. (Eds.), AGU Monograph Series: History and Dynamics of Global Plate Motions, in press.

Bercovici, D., Schubert, G., Glatzmaier, G.A., 1989. Three-dimensional spherical models of convection in the Earth's mantle. Science 244, 950–955.

Bevis, M., 1986. The curvature of Wadati-Benioff Zones and the torsional rigidity of subducting plates. Nature 323, 52–53.

Bevis, M., 1988. Seismic slip and down-dip strain rates in Wadati-Benioff zones. Science 240, 1317–1319.

Bina, C.R., 1996. Phase transition buoyancy contributions to stresses in subducting lithosphere. Geophys. Res. Lett. 23, 3563–3566.

Bina, C.R., 1997. Patterns of deep seismicity reflect buoyancy stresses due to phase transitions. Geophys. Res. Lett. 24, 3301–3304.

Brown, J.M., 1986. Interpretation of the D$''$ zone at the base of the mantle: dependence on assumed values of thermal conductivity. Geophys. Res. Lett. 13, 1509–1512.

Bunge, H.P., Richards, M.A., Baumgardner, J.R., 1996. The effect of depth-dependent viscosity on the platform of mantle convection. Nature 379, 436–438.

Bunge, H.P., Richards, M.A., Baumgardner, J.R., 1997. A sensitivity study of 3-dimensional spherical mantle convection at 10^8 Rayleigh number — effects of depth-dependent viscosity, heating mode, and an endothermic phase-change. J. Geophys. Res. 102, 11991–12007.

Chen, J., King, S.D., 1998. The influence of temperature and depth dependent viscosity on geoid and topography profiles from models of mantle convection. Phys. Earth Planet. Int. 106, 75–91.

Chopelas, A., Boehler, R., 1989. Thermal expansion measurements at very high pressure, systematics, and a case for a chemically homogeneous mantle. Geophys. Res. Lett. 16, 1347–1350.

Christensen, U.R., 1984. Heat transfer by variable viscosity convection and implications for the Earth's thermal evolution. Phys. Earth Planet Int. 35, 264–282.

Christensen, U.R., 1995. Effects of phase transitions on mantle convection. Annu. Rev. Earth Planet Sci. 23, 65–87.

Christensen, U.R., 1996. The influence of trench migration on slab penetration into the lower mantle. Earth Planet. Sci. Lett. 140, 27–39.

Christensen, U.R., 1997. Influence of chemical buoyancy on the dynamics of slabs in the transition zone. J. Geophys. Res. 102, 22435–22443.

Christensen, U.R., Yuen, D.A., 1984. The interaction of a subducting lithospheric slab with a chemical or phase boundary. J. Geophys. Res. 89, 4389–4402.

Conrad, C., Hager, B.H., 1999. The effect of plate bending and fault strength at subduction zones on plate dynamics. J. Geophys. Res. 104, 17551–17571.

Daessler, R., Yuen, D.A., 1996. The metastable wedge in fast subducting slabs: constraints from thermokinetic coupling. Earth Planet. Sci. Lett. 137, 109–118.

Davies, G.F., 1977. Whole mantle convection and plate tectonics. Geophys. J.R. Astron. Soc. 49, 459–486.

Davies, G.F., 1986. Mantle convection under simulated plates: effects of heating modes, ridge and trench migration, and implications for the core-mantle boundary, bathymetry, the geoid and Benioff zones. Geophys. J.R. Astron. Soc. 84, 153–183.

Davies, G.F., 1988. Ocean bathymetry and mantle convection. 1. Large-scale flow and hotspots. J. Geophys. Res. 93, 10467–10480.

Davies, G.F., 1989. Mantle convection model with a dynamic plate: topography, heat-flow and gravity anomalies. Geophys. J. 98, 461–464.

Davies, G.F., Richards, M.A., 1992. Mantle convection. J. Geol. 100, 151–206.

Doglioni, C., 1993. Geological evidence for a global tectonic polarity. J. Geol. Soc. Lond. 150, 991–1002.

Doglioni, C., Harabaglia, P., Merlini, S., Mongelli, F., Peccerillo, A., Piromallo, C., 1999. Orogens and slabs vs. their direction of subduction. Earth Sci. Rev. 45, 167–208.

Dziewonski, A.M., Anderson, D.L., 1981. Preliminary reference earth model (PREM). Phys. Earth Planet. Int. 25, 297–356.

England, P., Wortel, R., 1980. Some consequences of the subduction of young slabs. Earth Planet. Sci. Lett. 47, 403–415.

Fischer, K.M, Jordon, T.H., Creager, K.C., 1988. Seismic constraints on the morphology of deep slabs. J. Geophys. Res. 93, 4773–4783.

Fischer, K.M., Jordan, T.H., 1991. Seismic strain rate and deep slab deformation in Tonga. J. Geophys. Res. 96, 14429–14444.

Fischer, K.M., Creager, K.C., Jordan, T.H., 1991. Mapping the Tonga slab. J. Geophys. Res. 96, 14403–14427.

Forsyth, D.W., Uyeda, S., 1975. On the relative importance of the driving forces of plate motion. Geophys. J. Royal Astron. Soc. 43, 163–200.

Fukao, Y., Obayashi, M., Inoue, H., Nenbai, M., 1992. Subducting slabs stagnant in the mantle transition zone. J. Geophys. Res. 97, 4809–4822.

Furlong, K.P., Chapman, D.S., Alfeld, P.W., 1982. Thermal modeling of the geometry of subduction with implications for the tectonics of the overriding plate. J. Geophys. Res. 87, 1786–1802.

Gaherty, J.B., Hager, B.H., 1994. Compositional vs. thermal buoyancy and the evolution of subducted lithosphere. Geophys. Res. Lett. 21, 141–144.

Giardini, D., Woodhouse, J.H., 1984. Deep seismicity and modes of deformation in the Tonga subduction zone. Nature 307, 505–509.

Giardini, D., Woodhouse, J.H., 1986. Horizontal shear flow in the mantle beneath the Tonga arc. Nature 319, 551–555.

Grand, S.P., Vander Hilst, R.D., Widiyantoro, S., 1997. Global seismic tomography: a snapshot of convection in the Earth. GSA Today 7 (4), 1–7.

Griffiths, R.W., Hackney, R.I., Vander Hilst, R.D., 1995. A laboratory investigation of effects of trench migration on the descent of subducted slabs. Earth Planet. Sci. Lett. 133, 1–17.

Guillou-Frottier, L., Buttles, J., Olson, P., 1995. Laboratory experiments on the structure of subducted lithosphere. Earth Planet. Sci. Lett. 133, 19–34.

Gurnis, M., 1989. A reassessment of the heat transport by variable viscosity convection with plates and lids. Geophys. Res. Let. 16, 179–182.

Gurnis, M., Hager, B.H., 1988. Controls on the structure of subducted slabs and the viscosity of the lower mantle. Nature 335, 317–321.

Hager, B.H., 1984. Subducted slabs and the geoid: constraints on mantle rheology and flow. J. Geophys. Res. 89, 6003–6015.

Hager, B.H., O'Connell, R.J., 1978. Subduction zone dip angles and flow driven by plate motion. Tectonophys. 50, 111–133.

Hager, B.H., O'Connell, R.J., 1981. A simple global model of plate dynamics and mantle convection. J. Geophys. Res. 86, 4843–4867.

Han, L., Gurnis, M., 1999. How valid are dynamic models of subduction and convection when plate motions are prescribed? Phys. Earth Planet. Int. 110, 235–246.

Helffrich, G., Stein, S., Wood, B.J., 1989. Subduction zone thermal structure and mineralogy and their relationship to seismic wave reflections and conversions at the slab/mantle interface, J. Geophys. Res. 94, 753–763.

Hirth, G., Kohlstedt, D.L., 1996. Water in the oceanic upper mantle: implications for rheology, melt extraction and the evolution of the lithosphere. Earth Planet. Sci. Lett. 144, 93–108.

Holt, W.E., 1995. Flow fields within the Tonga slab determined from the moment tensors of deep earthquakes. Geophys. Res. Lett. 22, 989–992.

Houseman, G.A., Gubbins, D., 1997. Deformation of subducted oceanic lithosphere. Geophys. J. Int. 131, 535–551.

Isacks, B., Molnar, P., 1968. Mantle earthquake mechanisms and the sinking of the lithosphere. Nature 233, 1121–1124.

Isacks, B., Molnar, P., 1971. Distribution of stresses in the descending lithosphere from a global survey of focal-mechanism solutions of mantle earthquakes. Rev. Geohys. Space Phys. 8, 103–174.

Isacks, B.L., Barazangi, M., 1977. Talwani, M., Pitman III, W.C. (Eds.), Geometry of Benioff Zones: Lateral Segmentation and Downwards Bending of the Subducted Lithosphere, in Island Arcs, Deep Sea Trenches and Back Arc Basins. American Geophysical Union, Washington, DC, 1977, pp. 99–114.

Ita, J.J., King, S.D., 1994. Sensitivity of convection with an endothermic phase change to the form of governing equations, initial conditions, boundary conditions and equation of state. J. Geophys. Res. 99, 15919–15938.

Ita, J.J., King, S.D., 1998. The influence of thermodynamic formulation on simulations of subduction zone geometry and history. Geophys. Res. Lett. 25, 1463–1466.

Jarrard, R.D., 1986. Relations among subduction parameters. Rev. Geophys. 24, 217–284.

Jarvis, G., Peltier, W.R., 1989. Convection models and geophysical observations. In: Peltier, W.R. (Ed.), Mantle Convection: Plate Tectonics and Global Dynamics; Fluid Mech. Astrophys. Geophys. 4, 479–593.

Kanamori, H., 1971. Great earthquakes at island arcs and the lithosphere. Tectonophys. 12, 187–198.

Karato, S., Li, P., 1992. Diffusive creep in perovskites: implications for the rheology of the lower mantle. Science 255, 1238–1240.

Karato, S., Rubie, D., 1997. Toward an experimental study of deep mantle rheology: a new multianvil sample assembly for deformation studies under high pressures and temperatures. J. Geophys. Res. 102, 20111–20120.

Karato, S., Wu, P., 1993. Rheology of the upper mantle: a synthesis. Science 260, 771–778.

Kincaid, C., Olson, P.L., 1987. An experimental study of subduction and slab migration. J. Geophys. Res. 92, 13832–13840.

King, S.D., Balachandar, S., Ita, J.J., 1997. Using eigen-functions of the two-point correlation function to study convection with multiple phase transformations. Geophys. Res. Lett. 24, 703–706.

King, S.D., Gable, C., Weinstein, S., 1992. Models of convection driven tectonic plates: a comparison of methods and results. Geophys. J. Int. 109, 481–487.

King, S.D., Hager, B.H., 1994. Subducted slabs and the geoid. 1. Numerical experiments with temperature-dependent viscosity. J. Geophys. Res. 99, 19843–19852.

King, S.D., Ita, J.J., 1995. The effect of slab rheology on mass transport across a phase transition boundary. J. Geophys. Res. 100, 20211–20222.

Kirby, S., Engdahl, E.R., Denlinger, R., 1996. Intermediate-depth intraslab earthquakes and arc volcanism as physical expressions of crustal and uppermost mantle metamorphism in subducting slabs, In: Bebout, G.E., School, D.W., Kirby, S.H., Platt, J.P. (Eds.), Subduction: Top to Bottom, Geophysical Monograph 96. American Geophysical Union, Washington, DC. pp. 195–222.

Larsen, T.B., Malevsky, A.V., Yuen, D.A., Smedsmo, J.L., 1993. Temperature-dependent Newtonian and non-Newtonian convection: implications for lithospheric processes. Geophys. Res. Lett. 20, 2595–2598.

Lay, T., 1994. Seismological constraints on the velocity structure and fate of subducting slabs: 25 years of progress. Adv. Geophys. 35, 1–180.

Lenardic, A., Kaula, W.M., 1994. Self-lubricated convection: two-dimensional models. Geophys. Res. Lett. 21, 1707–1710.

Lenardic, A., Kaula, W.M., 1996. Near-surface thermal/chemical boundary layer convection at infinite Prandtl number; two-dimensional numerical experiments. Geophys. J. Int. 126, 689–711.

Le Pichon, X., 1968. Sea floor spreading and continental drift. J. Geophys. Res. 73, 3661–3697.

Lliboutry, L., 1969. Seafloor spreading, continental drift and lithosphere sinking with an asthenosphere at melting point. J. Geophys. Res. 74, 6525–6540.

Lithgow-Bertelloni, C., Richards, M.A., 1995. Cenozoic plate driving forces. Geophys. Res. Lett. 22, 1317–1320.

Lowman, J.P., Jarvis, G.T., 1996. Continental collisions in wide aspect ratio and high Rayleigh number two-dimensional mantle convection models. J. Geophys. Res. 101 (11), 25485–25497.

Marton, F., Bina, C.R., Stein, S., Rubie, D.C., 1999. Effects of slab mineralogy on subduction rates. Geophys. Res. Lett. 26, 119–122.

Marotta, A.M., Mongelli, F., 1998. Flexure of subducted slabs. Geophys. J. Int. 132, 701–711.

McKenzie, D.P., Roberts, J.M., Weiss, N.O., 1974. Convection in the earth's mantle: towards a numerical simulation. J. Fluid Mech. 62, 465–538.

Minear, J.W., Toksöz, N.M., 1970. Thermal regime of a downgoing slab and new global tectonics. J. Geophys. Res. 75, 1397–1419.

Mitrovica, J.X., 1996. Haskell (1935) revisited. J. Geophys. Res. 101, 555–569.

Molnar, P., Freedman, D., Shih, J.S.F., 1979. Lengths of intermediate and deep seismic zones and temperatures in downgoing slabs of lithosphere. Geophys. J. Royal Astron. Soc. 56, 41–54.

Moresi, L., Gurnis, M., 1996. Constraints on the lateral strength of slabs from 3-dimensional dynamic flow models. Earth Planet. Lett. 138, 15–28.

Moresi, L.N., Solomatov, V.S., 1995. Numerical investigation of 2D convection with extremely large viscosity variations. Phys. Fluids 7, 2154–2162.

Peacock, S.M., 1991. Numerical-simulation of subduction zone pressure temperature time paths: constraints on fluid production and arc magmatism. Philos. Trans. R. Soc. Lond. A 335, 341–353.

Peacock, S.M., 1996. Thermal and petrologic structure of subduction zones. In: Bebout, G.E., Scholl, D.W., Kirby, S.H., Platt, J.P. (Eds.), Subduction: Top to Bottom, Geophysical Monograph 96. American Geophysical Union, Washington, DC, pp. 119–133.

Peacock, S.M., Rushmer, T., Thompson, A.B., 1994. Partial melting of subducting oceanic crust. Earth Planet. Sci. Lett. 121, 227–244.

Puster, P., Hager, B.H., Jordan, T.H., 1995. Mantle convection experiments with evolving plates. Geophys. Res. Lett. 22, 2223–2226.

Ratcliff, J.T., Schubert, G., Zebib, A., 1995. Three-dimensional variable viscosity convection of an infinite Prandtl number Bousinessq fluid in a spherical shell. Geophys. Res. Lett. 22, 2227–2230.

Ratcliff, J.T., Schubert, G., Zebib, A., 1996. Effects of temperature-dependent viscosity on thermal convection in a spherical shell. Physica D. 97, 242–252.

Ricard, Y., Doglioni, C., Sabadini, R., 1991. Differential rotation between lithosphere and mantle: a consequence of lateral mantle viscosity variations. J. Geophys. Res. 96, 8407–8416.

Riedel, M.R., Karato, S., 1997. Grain-size evolution in subducted oceanic lithosphere associated with the olivine-spinel transformation and its effect on rheology. Earth Planet. Sci. Lett. 148, 27–43.

Schmeling, H., Monz, R., Rubie, D.C., 1999. The influence of olivine metastability on the dynamics of subduction. Earth Planet. Sci. Lett. 165, 55–66.

Solomatov, V.S., 1993. Parameterization of temperature- and stress-dependent viscosity convection and the thermal evolution of Venus. In: Stone, D.B. Runcorn, S.K. (Eds.), Flow and Creep in the Solar System: Observations, Modeling and Theory. Kluwer Academic Publishers, London, pp. 133–145.

Solomatov, V.S., 1995. Scaling of temperature- and stress-dependent viscosity convection. Phys. Fluids 7, 266–274.

Solomatov, V.S., Moresi, L.N., 1996. Stagnant lid convection on Venus. J. Geophys. Res. 101, 4737–4753.

Stevenson, D.J., Turner, J.S., 1977. Angle of subduction. Nature 270, 334–336.

Tackley, P.J., 1996. On the ability of phase transitions and viscosity layering to induce long wavelength heterogeneity in the mantle. Geophys. Res. Lett. 23, 1985–1988.

Tackley, P.J., 1998. Self-consistent generation of tectonic plates in three-dimensional mantle convection. Earth Planet. Sci. Lett. 157, 9–22.

Tackley, P.J., Stevenson, D.J., Glatzmaier, G.A., Schubert, G., 1993. Effects of an endothermic phase transition at 670 km depth in a spherical model of convection in the Earth's mantle. Nature 361, 699–704.

Tackley, P.J., Stevenson, D.J., Glatzmaier, G.A., Schubert, G., 1994. Effects of multiple phase transitions in a 3-D spherical model of convection in the Earth's mantle. J. Geophys. Res. 99, 15877–15901.

Tao, W.C., O'Connell, R.J., 1993. Deformation of a weak slab and variation of seismicity with depth. Nature 361, 626–628.

Toth, J., Gurnis, M., 1998. Dynamics of subduction initiation at preexisting fault zones. J. Geophys.Res. 103, 18053–18067.

Tovish, A., Schubert, G., Luyendyk, B.P., 1978. Mantle flow pressure and the angle of subduction; non-Newtonian corner flows. J. Geophys. Res. 83, 5892–5898.

Turcotte, D.L., Schubert, G., 1982. Geodynamics: Applications of Continuum Physics to Geological Problems. Wiley, New York, 450 pp.

Uyeda, S., Kanamori, H., 1979. Back arc opening and the mode of subduction. J. Geophys. Res. 84, 1049–1061.

Van den Berg, A.P., Yuen, D.A., Van Keken, P.E., 1991. Effects of depth-variations in creep laws on the formation of plates in mantle dynamics. Geophys. Res. Lett. 18, 2197–2200.

van der Hilst, R.D., Engdahl, E.R., Spakman, W., Nolet, G., 1991. Tomographic imaging of subducted lithosphere below Northwest Pacific island arcs. Nature 353, 37–43.

van der Hilst, R.D., Engdahl, E.R., Spakman, W., 1993. Tomographic inversion of P and pP data for aspherical mantle structure below the Northwest Pacific region. Geophys. J. Int. 115, 264–302.

van der Hilst, R.D., Widiyantoro, S., Engdahl, E.R., 1997. Evidence for deep mantle circulation from global tomography. Nature 386, 578–584.

van der Hilst, R.D., Seno, T., 1993. Effects of relative plate motion on the deep-structure and penetration depth of slabs below the Izu-Bonin and Mariana Island arcs. Earth Planet. Sci. Lett. 120, 395–407.

Van Keken, P.E., Karato, S., Yuen, D.A., 1996. Rheological control of oceanic crust separation in the transition zone. Geophys. Res. Lett. 23, 1821–1824.

Vassiliou, M.S., 1984. The state of stress in subducting slabs as revealed by earthquakes analyzed by moment tensor inversion. Earth Planet. Sci. Lett. 69, 195–202.

Vassiliou, M.S., Hager, B.H., Raefsky, A., 1984. The distribution of earthquakes with depth and stress in subduction slabs. J. Geodyn. 1, 11–28.

Vlaar, N.J., 1983. Thermal anomalies and magmatism due to lithospheric doubling and shifting. Earth Planet. Sci. Lett. 65, 322–330.

Weinstein, S., Christensen, U., 1991. Convection plan-forms in a fluid with a temperature-dependent viscosity beneath a stress-free upper boundary. Geophys. Res. Lett. 18, 2035–2038.

Weinstein, S., Olson, P.L., 1992. Thermal convection with non-Newtonian plates. Geophys. J. Int. 111, 515–530.

Zhong, S., Davies, G.F., 1999. Effects of plate and slab viscosities on the geoid. Earth Planet. Sci. Lett. 170, 487–496.

Zhong, S., Gurnis, M., 1992. Viscous flow model of a subduction zone with a faulted lithosphere: long and short wavelength topography, gravity, and geoid. Geophys. Res. Lett. 19, 1891–1894.

Zhong, S.J., Gurnis, M., 1994a. Controls on trench topography from dynamic-models of subducted slabs. J. Geophys. Res. 99, 15683–15695.

Zhong, S.J., Gurnis, M., 1994b. Role of plates and temperature-dependent viscosity in phase change dynamics. J. Geophys. Res. 99, 15903–15917.

Zhong, S.J., Gurnis, M., 1995. Mantle convection with plates and mobile, fractured plate margins. Science 267, 838–843.

Zhong, S., Gurnis, M., 1996. Interaction of weak faults and non-Newtonian rheology produces plate-tectonics in a 3D model of mantle flow. Nature 383, 245–247.

Zhong, S., Gurnis, M., 1997. Dynamic interaction between tectonic plates, subducting slabs, and the mantle, Earth Interactions, 1.

Zhong, S., Gurnis, M., Moresi, L., 1998. Role of faults, nonlinear rheology, and viscosity structure in generating plates from instantaneous mantle flow models. J. Geophys. Res. 103, 15255–15268.

ELSEVIER

Physics of the Earth and Planetary Interiors 127 (2001) 25–34

PHYSICS
OF THE EARTH
AND PLANETARY
INTERIORS

www.elsevier.com/locate/pepi

Geodynamic models of deep subduction

Ulrich Christensen*

Institut für Geophysik, Universität Göttingen Herzberger Landstrasse 180, 37075 Göttingen, Germany

Received 22 April 2000; accepted 28 July 2000

Abstract

Numerical and laboratory models that highlight the mechanisms leading to a complex morphology of subducted lithospheric slabs in the mantle transition zone are reviewed. An increase of intrinsic density with depth, an increase of viscosity, or phase transitions with negative Clapeyron slope have an inhibiting influence on deep subduction. The impingement of slabs on a viscosity and density interface has been studied in laboratory tanks using corn syrup. Slab interaction with equilibrium and non-equilibrium phase transitions has been modelled numerically in two dimensions. Both the laboratory and the numerical experiments can reproduce the variety of slab behaviour that is found in tomographic images of subduction zones, including cases of straight penetration into the lower mantle, flattening at the 660-km discontinuity, folding and thickening of slabs, and sinking of slabs into the lower mantle at the endpoint of a flat-lying segment. Aside from the material and phase transition properties, the tectonic conditions play an important role. In particular, the retrograde motion of the point of subduction (trench-rollback) has an influence on slab penetration into the lower mantle. A question that still needs to be clarified is the mutual interaction between plate kinematics and the subduction process through the transition zone.

Keywords: Numerical models; Laboratory models; Mantle phase boundaries; Transition zone

1. Introduction

The question of how deep slabs of subducting oceanic lithosphere penetrate into the mantle has been debated controversially ever since it was accepted that subduction does occur. The fate of slabs is intimately linked to the style of mantle convection — whole mantle flow or convection in separate layers — because slabs are believed to provide most of the buoyancy forces that drive the mantle circulation (Davies and Richards, 1992; Lithgow-Bertelloni and Richards, 1995). In addition, the deep seismicity and the anomalous seismic velocity inside slabs provide a means to track the descending part of mantle convection. The early observations of the cessation of seismicity at 700 km depth and of down-dip compressive stresses inferred from focal mechanisms of deep earthquakes (Isacks and Molnar, 1971) suggested that slabs experience a resistance in the transition zone that may prevent them from sinking into the lower mantle. Some scientists, such as Davies (1977) and Jordan (1977), found the arguments for convective layering inconclusive and championed the idea of whole mantle convection and deep slab penetration. For about two decades, seismologists, geochemists, mineral physicists and geodynamists have tried to find compelling evidence for one or the other model. The debate was only recently brought to a conclusion, when high-resolution seismic tomography established beyond reasonable doubt that several slabs do indeed penetrate to at least 1700 km

* Fax: +49-551-397459.
E-mail address: urc@uni-geophys.gwdg.de (U. Christensen).

0031-9201/01/$ – see front matter
PII: S0031-9201(01)00219-9

depth and possibly deeper (Grand, 1994; van der Hilst et al., 1997). Nonetheless, the tomographic images also show that the deep slab structure can be quite variable and surprisingly complex near the bottom of the transition zone. In some cases, the slab penetrates straight through the seismic 660-km discontinuity, in other cases, it flattens and lies subhorizontally above this boundary, though it may eventually sink into the lower mantle at larger distance from the trench (van der Hilst, 1995; Widiyantoro and van der Hilst, 1996; van der Hilst et al., 1997; Bijwaard et al., 1998). Slabs may become kinked or thicken substantially when entering the lower mantle. This confirms that processes in the transition zone or at the 660-km seismic discontinuity tend to inhibit subduction, even if there is no impenetrable barrier.

Geodynamic modelling, though hampered by incomplete knowledge of rheological and mineralogical properties or by the inability to account for them properly in simple models, can provide the link between material properties and the structures observed by seismology. This paper reviews both numerical and laboratory work which determines the key parameters controlling the style of subduction and the circumstances that can give rise to the observed complexity of slab structure.

2. Dynamical influences on slab descent

Slabs sink into the mantle mainly because they are colder, hence denser, than the surrounding mantle. Thermal diffusion in the mantle is inefficient, so that slabs will thermally equilibrate with the mantle only on the time-scale of 100 million years or more. The compositional differentiation of oceanic lithosphere into a basaltic (at depth eclogitic) crust and a depleted harzburgite layer could give rise to additional buoyancy forces. In a wide depth range, the composite slab is nearly neutrally buoyant at mantle temperature. However, garnet-majorite, which is the dominant mineral in the crustal component at transition zone pressures, transforms to perovskite at a higher pressure than the gamma-phase of olivine, which transforms at 660 km. Therefore, the slab is compositionally buoyant within some depth interval below the 660-km discontinuity. Ringwood and Irifune (1988) suggested that this mechanism inhibits slab penetration into the lower mantle.

Another possible mechanism for layered mantle convection that has been invoked by the proponents of this model is a difference of bulk composition between upper and lower mantle, such as a higher FeO or SiO_2-content in the lower mantle, leading to intrinsically higher density than in the upper mantle (Jeanloz and Thompson, 1983). Although there is still some ambiguity concerning a possible difference between upper mantle and lower mantle compositions from the comparison of seismic data and thermoelastic laboratory data obtained for the relevant minerals, a uniform mantle composition is well in the range of acceptable solutions (Jackson, 1998). The evidence for widespread slab penetration into the lower mantle and the implied mixing make it quite unlikely that the 660-km discontinuity is a major compositional boundary.

Among the various potential mechanisms for layered convection, the inhibiting influence of equilibrium transformations between low-pressure and high-pressure phases has received most attention. The endothermic transition from the gamma-phase of olivine to silicate perovskite and magnesiowüstite, which is held responsible for the 660-km seismic discontinuity and which has a negative Clapeyron slope $dp/dT \approx -2.8$ MPa/K (Ito and Takahashi, 1989), is of particular importance. Because of the negative dp/dT, the equilibrium phase boundary is shifted to higher pressure (greater depth) in the cold slab compared to the warm ambient mantle. A strong buoyancy force is associated with the depression of the lower-density gamma-phase into the surrounding high-density perovskite-dominated assemblage, which opposes further sinking of the slab. However, while the olivine component accounts for roughly two-third in a mantle of uniform composition, other components must not be neglected. More recently, it has been determined that the transformation of majorite-rich garnet, which accounts for one-third of the material in the deeper parts of the transition zone, to perovskite has a distinctly positive Clapeyron slope (Akaogi and Ito, 1999). This would compensate for part of the inhibiting effect.

While the phase transformations are assumed to occur close to thermodynamic equilibrium in the warm mantle, slow reaction kinetics could retard them substantially at temperatures below 700°C. This would lead to the formation of metastable wedges of olivine or other low-pressure phases in the cold core of slabs (Rubie and Ross, 1994; Kirby et al., 1996). While such metastable wedges have been discussed mainly in connection with possible mechanisms for deep earthquakes (Green and Burnley, 1988), their existence would obviously imply a buoyancy force opposing the sinking of the slab. Metastable phase transitions may also influence the mechanical properties of the slab. Riedel and Karato (1997) use a nucleation and growth model to infer a strong grain-size reduction during the kinematically retarded transition of olivine to the β- or γ-phase. They suggest that the resulting small grain-size in connection with diffusion creep mechanisms could soften the slab substantially.

The viscosity structure of the mantle is another factor that can influence the dynamics of deep subduction. There is growing consensus for a substantial increase of viscosity from the upper mantle to the lower mantle by a factor of typically 30, both from the analysis of postglacial rebound data and from the study of long-wavelength geoid anomalies (Hager, 1984; Forte and Mitrovica, 1996). While a viscosity increase alone cannot prevent a slab from sinking into the lower mantle, it will slow down its descent rate and could shape the slab morphology, in particular, when acting in concert with other effects.

Finally, plate kinematics can play a role for the subduction behaviour of individual slabs. While the sum of all forces that act on an entire plate including its subducted parts must be in equilibrium, this is not necessarily true for a single slab. There could be a disequilibrium between local driving forces and resistive forces, which include thermal, compositional and phase-change related buoyancy, viscous resistance against the surrounding mantle and resistance against slab deformation. Subduction can be forced against dominating resistive forces, for example, if there is a strong trench pull from other slabs connected to the same plate. Another important factor is trench-rollback, the progressive motion of the point of subduction away from the overriding plate. van der Hilst and Seno (1993) suggested that differences in the deep structure of the Mariana slab, which penetrates into the lower mantle, and the Izu-Bonin slab, which flattens in the transition zone, are caused by the tectonically inferred faster rate of trench-rollback for the latter suduction zone.

3. Laboratory experiments

How some of the effects discussed in the previous section can lead to complex slab morphology has been nicely illustrated and partly quantified in laboratory experiments with corn syrup. Corn syrup has a high and strongly temperature-dependent viscosity, and its density and viscosity can be easily modified by adding small amounts of water or, to influence viscosity and density in different ways, of salt solutions. In the first such experiment by Kincaid and Olson (1987), a strongly cooled sheet of syrup was initially placed on top of a tank filled with two layers of syrup at room temperature. The lower layer was more viscous and had a slightly higher intrinsic density than the upper layer. The cold sheet could sink freely under its own weight, and the interaction with the interface depends on the density and viscosity ratios. The slab was found to flatten at the interface when the resistance for penetration was high, either from the viscosity or the density contrast (which have not been varied independently). Partial or complete penetration was observed for a weaker resistance. A retrograde motion of the inflexion point of the sheet, i.e. trench-rollback, occurred in most experiments and was particularly pronounced when the slab flattened at the interface. Griffiths and Turner (1988) extruded a slab at a fixed point into a viscous two-layer fluid, and demonstrated how it folded and buckled when it impinges onto the density and/or viscosity interface.

The emergence of well-resolved tomographic images of slabs prompted Guillou-Frottier et al. (1995) and Griffiths et al. (1995) to refine on these experiments and to address in particular the role of trench-rollback. In order to allow for controlled rates of subduction and of trench migration, the model slabs were injected at a fixed rate from an extruder that moved relative to the tank. Fig. 1 shows results from such an experiment. The strongly cooled and dyed slab is 10^5 times more viscous and 5% denser than the upper layer, but only 1% denser than the lower layer. The viscosity between both layers differs by a factor of 40. Initially, fast trench-rollback was imposed, leading to a flat-lying slab (Fig. 1A and B).

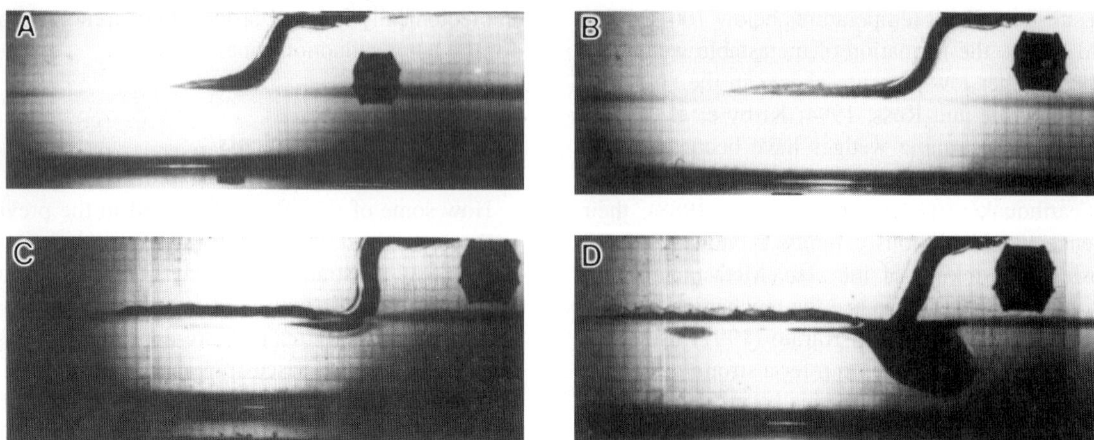

Fig. 1. Snapshots from a laboratory experiment by Guillou-Frottier et al. (1995). A sheet of chilled and dyed corn syrup is extruded into a tank with two layers of syrup. Initial rapid trench migration leads to a flat slab. After (B), trench-rollback is stopped and slab material accumulates in a pile that sinks into the lower layer. The whole sequence corresponds to approximately 400 million years of subduction.

Then the retrograde motion was stopped, with the result that slab material accumulated and sank slowly as a big pile into the lower layer (Fig. 1D). Similar effects were observed by Griffiths et al. (1995). In most of their experiments, the slab was only slightly more viscous than the ambient fluid, with the result that flattened slabs lying above the interface eventually sank into the lower layer through a diapiric (drop-like) instability. Both sets of experiments demonstrated that, aside from the contrast of material properties at the interface, the rate of trench migration is an important control parameter. When the trench velocity, scaled by some free (Stokes) sinking velocity of the slab, exceeds a critical value, the slab flattens at the interface.

The major shortcoming of these experiments, when applying them to the Earth, is the inability to incorporate the dynamic effects of phase boundaries. The intrinsic density difference between the layers in the experiments reduces the driving buoyancy force acting on the slab in the entire lower layer. In contrast, the buoyancy effects of phase boundary deflections are much stronger, but locally restricted.

4. Numerical models

4.1. Models with a viscosity or density interface

The first dynamical model of a subducting slab interacting with an intrinsic density interface (Christensen and Yuen, 1984) employed temperature-dependent non-Newtonian viscosity. It was found that the intrinsic density contrast has to exceed the thermal density contrast between slab and ambient mantle (roughly it needs to be larger than 3%), to preclude deep penetration of the slab into the lower mantle, although a significant depression of the boundary would still develop at a slightly larger density difference. According to current knowledge, any intrinsic density difference between the upper and lower mantle, if it exists, is likely to be small. A somewhat unrealistic aspect of this model was that subduction was forced to occur vertically at the sidewall of the box, so that the slab was not free in its lateral motion. Gurnis and Hager (1988) used a weak 'fault zone' in the strong surface layer of their model to initiate subduction under an oblique angle in the middle of their model box. They studied the interaction of the model slab with a 30-fold viscosity increase at mid-depth, finding that it develops a kink at the boundary and thickens upon entering the lower mantle. Gaherty and Hager (1994) showed in models with a slab composed of a dense eclogitic layer and a buoyant harburgitic layer, which descends vertically onto a viscosity interface, that the thickening is not uniform, but occurs through folding and buckling. They also showed that the compositional contribution to buoyancy has little effect on subduction as long as the whole assemblage is neutrally buoyant. A separation of the two components is inhibited by the high viscosity of the slab.

The rheological properties of subducting slabs are probably complex and only partly understood. In many numerical models involving slabs, a simple (and often too weak) temperature-dependence of viscosity has been assumed. Houseman and Gubbins (1997) addressed the influence of slab rheology on its deformation behaviour in models where a slab with uniform properties impinges onto a density interface. They define a 'buoyancy number' F as the major control parameter for the deformation style, which for Newtonian rheology is given by

$$F = \frac{L^2 \Delta\rho g}{U\eta} \quad (1)$$

where L is the slab thickness, η its viscosity, $\Delta\rho$ the density contrast to the ambient mantle, g gravity, and U is the subduction velocity. F measures the ratio of stresses generated by the density anomaly of the slab and the stress associated with the viscous deformation of the slab at a scaling strain rate given by U/L. For low F, less than 0.05 or 0.2 depending on whether the rheology is Newtonian or non-Newtonian, the slab bends into a smooth arc when it meets a resistance, whereas at high values of F the slab buckles and folds. The authors conclude that slabs in the Earth's mantle would be close to the transitional value of F. They show that the effects of trench migration on the slab structure are more pronounced in the regime of high F (softer slabs). Trench migration was also studied by Olbertz et al. (1997) in a convection model with a viscosity interface at 660 km, who concluded that even small rates of trench-rollback can have a significant effect.

4.2. Models with phase changes

The thermally induced deflection of an endothermic phase boundary in a cold slab (or hot plume) can be thought of as if the thermal expansion coefficient were strongly negative in a thin depth interval some tens of kilometres wide (Christensen and Yuen, 1985). Therefore, cold material is buoyant and would rise unless strong viscous coupling with the parts of the slabs above or below the phase transition region forces it to sink along with the rest.

Numerical modelling of isoviscous convection with phase boundaries started with Richter (1973), who demonstrated the impeding influence of a phase change with negative $\mathrm{d}p/\mathrm{d}T$. Christensen and Yuen (1985) showed that the phase boundary effect depends to first-order on the "phase buoyancy parameter"

$$P = \frac{\Delta\rho \, \mathrm{d}p/\mathrm{d}T}{\rho^2 \alpha g h} \quad (2)$$

where $\Delta\rho$ is the density contrast between the two phases, ρ the mean density, α the thermal expansion coefficient and h is the height of the convecting layer. P measures the ratio of the localised buoyancy force by phase boundary deflection relative to the driving thermal buoyancy that is distributed over the entire depth range. The effects of latent heat release associated with the phase transformation are of second-order. It was demonstrated that layered convection does occur when P becomes sufficiently negative, depending on the Rayleigh number. At higher Rayleigh number, more moderate values of P are sufficient to make convection layered. Based on the values for the thermodynamic parameters that were available at that time, Christensen and Yuen (1985) concluded that layering caused by the phase change at the boundary between upper and lower mantle was a remote possibility.

The interest in such models was revived in the early 1990s, when 2D and 3D numerical models at high Rayleigh number showed that the phase change can lead to an intermittent style of convection (Machetel and Weber, 1991; Tackley et al., 1993; Solheim and Peltier, 1994). Here cold sinking fluid accumulates above the phase change boundary during a period of layered convection. The ponded material is than destabilised in a short catastrophic 'avalance' event and sinks rapidly into the deep mantle. This finding caused some excitement, because the intermediate style of mantle convection offered seemingly the potential to satisfy both the requirements for some degree of isolation between various parts of the mantle demanded by geochemical data (Hofmann, 1997) and the seismological evidence for the penetration of some slabs into the lower mantle. The avalanche model offered also a possible explanation for the episodicity of large magmatic or tectonic events in Earth's history, such as episodes of continental crust formation (Stein and Hofmann, 1994). A review of the influence of phase transition on mantle convection with emphasis on the avalanche models is given by Christensen (1995).

However, most of the models showing avalanches used somewhat extreme values for the thermodynamic parameters, such as a Clapeyron slope of −4 MPa/K, which must be considered as lower bound, and they ascribed the total density jump at the 660-km discontinuity to the endothermic transition from the γ-phase of olivine to perovskite and magesiowüstite. Furthermore, the isoviscous calculations did not model slabs (or plumes) properly, because they ignore the differences in their rheological properties compared to the ambient mantle.

The first numerical models of the interaction of rheologically distinct slabs with a phase change boundary (Christensen and Yuen, 1984; Zhong and Gurnis, 1994) assumed a stationary trench with subduction occurring at the side wall of the model box. Nonetheless, the latter work suggested that the higher viscosity of the slab favours penetration through a phase boundary with negative Clapeyron slope and makes it more steady in comparison to the results from isoviscous models. Davies (1995) determined in models where subducting occurred at a fixed point in the middle of the convection box that high-viscosity slabs penetrate the phase boundary more easily than hot low-viscosity plumes, while King and Ita (1995) found only moderate differences between isoviscous convection and the case of temperature-dependent viscosity in a similar setup. All the models with temperature-dependent viscosity agree that the temporal evolution is less spasmodic than what the isoviscous avalanche models suggested.

In a more elaborate convection model, Zhong and Gurnis (1995) used a non-Newtonian, temperature- and depth-dependent rheology with a moveable fault zone in the strong surface layer to allow for plate motion and trench migration in a largely unconstrained way. When a phase boundary with a negative Clapeyron slope was included, they observed initial flattening of the slab in the transition zone that was accompanied by rapid trench-rollback. When subsequently the slab penetrates into the lower mantle, the retrograde motion of the trench almost ceases.

In a survey of the influence of various control parameters on the style of subduction through the transition zone, I have used a somewhat simpler model with imposed rates of convergence and trench migration (Christensen, 1996). The viscosity is strongly temperature-dependent and increases with depth in a continuous way. The average viscosities of the upper and lower mantle differ by a factor of about 30. Some models have an additional jump of viscosity at the phase boundary at 660 km depth. A large variety of deep slab structures has been found (Fig. 2). Straight penetration of the slab (Fig. 2a) is typically found when the rate of trench migration is low (0–2 cm/a), even when the Clapeyron slope at the 660-km discontinuity is as negative as −4.0 MPa/K. Flat-lying slabs above the 660-km boundary (Fig. 2b) are typically found for rates of trench-rollback of 3 cm/a or more. Often the slab develops a flat-lying segment after it penetrated initially into the lower mantle. This segment is still connected to the part of the slab in the lower mantle and it continues to descend below the phase boundary at the 'knee' in the structure (Fig. 2c). In some cases, the development of this double kink may be an artifact of the two-dimensional (2D) model: with trench-rollback, there must be a net flow of mantle from below the subducting plate to underneath the overriding plate. While in the real Earth, this flow would occur mainly around the lateral edges of the slab, this is not possible in the 2D model and the resulting excess pressure below the subducting plate tends to lift the slab and reduce its dip angle. A shallow dip angle favours the formation of flattened slab segments. However, in other cases, for example when the rate of trench-rollback increases from an initially low value, the formation of structures such as in Fig. 2c could be real.

The flat-lying slabs are not stable in the long term. A diapiric instability (Fig. 2d), which was also observed in the laboratory experiments by Griffiths et al. (1995) is only found when the slab has a viscosity not much higher than that of the surrounding mantle. In some cases, the slab is briefly arrested and flattens above the 660-km boundary just after the start of subduction, but after some million years, it breaks through the discontinuity and drags the flat segment into the lower mantle (Fig. 2e). Occasionally, this breakthrough occurs close to the current position of the trench after a considerable delay (Fig. 2g) and leaves the flat slab segment in place for a rather long time. Slab folding and buckling can be observed when the resistance against slab penetration is particularly strong, either because of a jump of viscosity (Fig. 2f) or by an extreme value of the

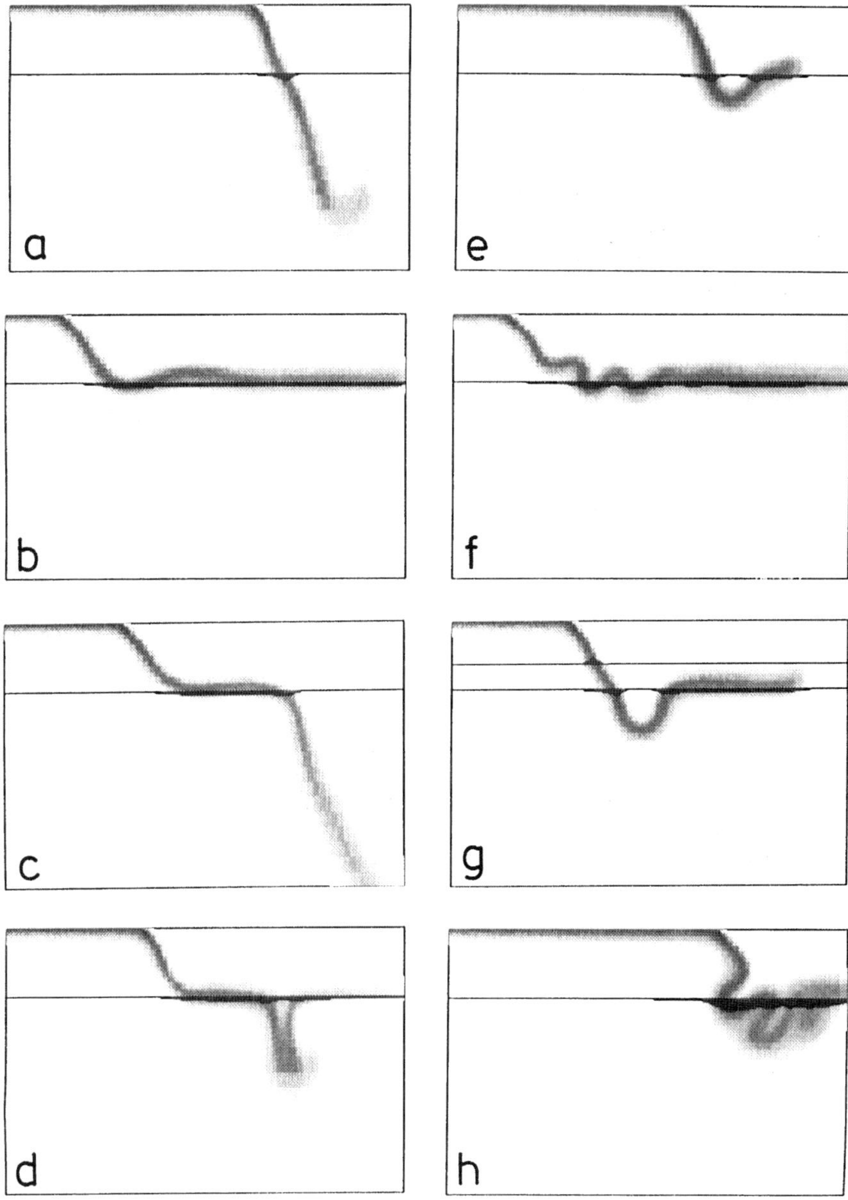

Fig. 2. Variety of slab morphology from numerical modelling with temperature- and depth-dependent viscosity and a phase boundary with negative Clapeyron slope (shown as horizontal line). Depressions of the phase boundary below its normal depth are shown as black regions. In part (g), a phase boundary at 410 km with positive $\mathrm{d}p/\mathrm{d}T$ is also modelled. The plate velocity (typically 5 cm/a) and the trench migration rate are imposed, but at depth the slab sinks subject to the various buoyancy forces acting on it. (a) Straight penetration at a trench velocity of 1 cm/a; (b) flattening of the slab at 3 cm/a; (c) flat slab segment descending into the lower mantle at its end for a case where the trench migration rate has been increased with time; (d) diapir-like instability of flattened slab for small viscosity contrast between slab and mantle; (e) retarded penetration of slab into lower mantle for 2 cm/a of trench-rollback; (f) buckling of flat slab in a model with an additional viscosity jump at the discontinuity; (g) breakthrough of initially flattened slab into the lower mantle for 3 cm/a of trench-rollback and the driving buoyancy of the elevated 410-km boundary added; (h) folding and piling of slab for an (unrealistic) Clapeyron slope of −5.6 MPa/K and slow trench velocity.

Clapeyron slope (Fig. 2h). Buckling can be observed both for penetrating and for non-penetrating slabs.

When the results of these numerical experiments are quantified, they predict that the immediate penetration of slabs through the 660-km discontinuity is prevented when the velocity of trench-rollback is larger than about one-fifth of the free sinking velocity of the slab. This can only be a rough guide, and other factors play a role. For example, forced subduction, for which the rate of convergence that is faster than the slab's free sinking velocity, favours deep penetration. Furthermore, the positive Clapeyron slope of the garnet–perovskite transition was not included in these models, so that the critical trench migration rate for slab flattening is probably underestimated by the figure given above.

Some of the structures seen in the geodynamical models can be directly compared with seismic images of the mantle below subduction zones. Qualitatively, the structure of the slab beneath central America resembles Fig. 2a, while the slab geometry under Japan and northern China is similar to that seen in Fig. 2b or f, and the Tonga slab has a structure like the model in Fig. 2c (van der Hilst et al., 1997; Bijwaard et al., 1998). A consequence of slab flattening is the broad-scale depression of the 660-km discontinuity by some 20–30 km, shown in Fig. 2 as dark stripe below the thin line which indicates the equilibrium depth of the phase boundary. For a slab that penetrates straight and steeply through the phase boundary, the peak depression is up to 50 km. But it is also very localised and has little influence on the regional average of the transition depth. By analysing SS-precursors caused by underside reflections off the 660-km boundary, Flanagan and Shearer (1998) calculated a global map of the topography on this boundary and found broad-scale depressions of about 15 km amplitude in many major subduction zones. This may indicate that complex slab structures involving large subhorizontal segments, which depress the 660-km phase boundary, are more the rule rather than the exception.

Additional complexities of the phase transitions modulate the slab morphology and the rates of descent, but do not lead to qualitatively new phenomena. Christensen (1997) modelled the influence of different phase relations in the basaltic and harzburgitic components of subducted lithosphere; in particular, the greater depth for the garnet–perovskite transition in the crustal component. Even when it is assumed that this transition is only completed several hundred kilometers below the 660-km boundary, which is no longer favoured by recent experimental data (Hirose et al., 1999), its influence was found to be relatively minor. Only for subducting lithosphere of young age, in which thermal buoyancy is weaker and composition-related buoyancy plays a larger role, a significant retarding influence could be found. In a similar model setup, but with a plate whose rate of subduction and trench-rollback were dynamically determined rather than imposed, Schmeling et al. (1999) and Tetzlaff and Schmeling (2000) studied the effect of a metastable olivine core in the slab on the subduction velocity and on slab penetration through the 660-km phase boundary. In their simplified model, olivine remains metastable below the equilibrium phase boundaries at temperatures lower than 600°C and transforms at the equilibrium pressure for temperatures above 700°C. Significant regions of metastable olivine occur only for old lithosphere (age at trench >100 Ma). Metastable olivine slows down the subduction rate by up to 20% compared to the case of transformation at thermodynamic equilibrium. Because the reduced rates of descent allow for warming of the slab at shallower depth and consequently the disappearance of metastable olivine, there is a negative feedback that could stabilize the subduction rate ('parachute effect', see also Kirby et al., 1996 and Marton et al., 1999). Slabs that are old (thick) enough to develop a significant metastable olivine region flatten out and are temporarily arrested above the 660-km boundary in the model. After a time lapse of order 8 million years, diffusive warming lets the metastable olivine disappear in the flattened slab, which then sinks into the lower mantle.

5. Summary and outlook

In the past 10 years, geodynamic modelling of deep subduction has developed in step with the increasingly more detailed tomographic images of slabs in the mantle transition zone. The models can explain the observed variety in deep subduction styles, but various different mechanisms are capable to produce complex slab morphology. Among them are the inhibiting influence of the endothermic phase change at the 660-km discontinuity, the possible survival of metastable olivine down to this depth, or the increase of viscosity from the upper to the lower mantle, which could either alone or in combination prevent straight slab penetration into the lower mantle. The modelling also suggests that these inhibiting factors may be insufficient by themselves and that in addition some tectonic control is required for arresting slabs temporarily in the transition zone. The age of the subducting plate can play a role, in particular, in conjunction with metastable olivine. A very important factor is the rate of trench-rollback, as was clearly demonstrated in several laboratory and numerical experiments. The evidence from observations for the influence of trench-rollback is less obvious, but often it is difficult to quantify its rate. Global plate models predict rather low rates for most trenches that do not exceed a few centimetres per year. However, backarc-spreading could lead in some instances to much larger rates, for example in the case of the Tonga subduction zone (Bevis et al., 1995), where tomography finds a flat-lying slab segment.

A chicken-and-egg problem is associated with trench migration — does fast trench-rollback lead to flattened slabs or does trench-rollback result from a strong resistance against deep slab penetration? Several 2D numerical models in which the slab is free to migrate seem to suggest that the latter may be the case. However, on a three-dimensional (3D) Earth with complex interlocked plate boundaries, the rates of subduction and of rollback at individual trenches are actually less free than in the 2D models. Also the conditions for the occurrence of backarc spreading are not well understood. The exact roles of tectonic control versus mechanisms acting on the slabs in the transition zone still need to be clarified.

Within the last few years, slab penetration into the lower mantle, which was an unresolved issue for decades, has become generally accepted and geodynamics has provided a qualitative understanding for some of the complications that slabs experience on their way down. But as often in science, when one question is settled, the next one follows immediately. Do slabs sink all the way to the core-mantle boundary, or are they stopped at a shallower level, as suggested by the loss of slab-like anomalies in the tomographic images around 1500–2000 km depth (van der Hilst and Kárason, 1999)? The suggested mechanisms for preventing slabs from reaching the core-mantle boundary are similar to those that have been discussed previously for layering at the upper-lower mantle boundary: a weak chemical stratification (Kellogg et al., 1999) or compositional buoyancy of the various constituents of oceanic lithosphere (Kesson et al., 1998). Eventually, the combination of seismology, mineral physics and geodynamic modelling will also decide this question, but this is a story of the future.

Acknowledgements

Supported by the Deutsche Forschungsgemeinschaft (Grant Ch77/8).

References

Akaogi, M., Ito, E., 1999. Calorimetric study on majorite-perovskite transition in the system $Mg_4Si_4O_{12}$–$Mg_3Al_2Si_3O_{12}$: transition boundaries with positive pressure–temperature slopes. Phys. Earth Planet. Inter. 114, 129–140.

Bevis, M., Taylor, F.W., Schult, B.E., Recy, J., Isacks, B.L., Helu, S., Singh, R., Kendrick, E., Stowell, J., Taylor, B., Calmant, S., 1995. Geodetic observations of very rapid convergence and back-arc extension at the Tonga arc. Nature 374, 249–251.

Bijwaard, H., Spakman, W., Engdahl, R., 1998. Closing the gap between regional and global travel time tomography. J. Geophys. Res. 103, 30055–30078.

Christensen, U.R., 1995. Effects of phase transitions on mantle convection. Annu. Rev. Earth Planet. Sci. 23, 65–87.

Christensen, U.R., 1996. The influence of trench migration on slab penetration into the lower mantle. Earth Planet Sci. Lett. 140, 27–39.

Christensen, U.R., 1997. Influence of chemical buoyancy on the dynamics of slabs in the transition zone. J. Geophys. Res. 102, 22435–22443.

Christensen, U.R., Yuen, D.A. 1984. The interaction of a subducting lithospheric slab with a chemical or phase boundary. J. Geophys. Res. 89, 4389-4402.

Christensen, U.R., Yuen, D.A., 1985. Layered convection induced by phase transitions. J. Geophys. Res. 89, 4389–4402.

Davies, G.F., 1977. Whole-mantle convection and plate tectonics. Geophys. J. R. Astron. Soc. 49, 459–486.

Davies, G.F., 1995. Penetration of plates and plumes through the mantle transition zone. Earth Planet. Sci. Lett. 133, 507–516.

Davies, G.F., Richards, M.A., 1992. Mantle convection. J. Geol. 100, 151–206 .

Flanagan, M.P., Shearer, P.M., 1998. Global mapping of topography on transition zone velocity discontinuities by stacking SS precursors. J. Geophys. Res. 103, 2673–2692.

Forte, A.M., Mitrovica, J.X., 1996. New inferences on mantle viscosity from joint inversion of long-wavelength mantle

convection and postglacial rebound data. Geophys. Res. Lett. 23, 1147–1150.
Gaherty, J.B., Hager, B.H., 1994. Compositional versus thermal buoyancy and the evolution of subducted lithosphere. Geophys. Res. Lett. 21, 141–144.
Grand, S.P., 1994. Mantle shear structure beneath the Americas and surrounding oceans. J. Geophys. Res. 99, 11591–11621.
Green, H.W., Burnley, P.C., 1988. A self-organizing mechanism for deep-focus earthquakes. Nature 341, 733–737.
Griffiths, R.W., Turner, J.S., 1988. Folding of viscous plumes impinging on a density or viscosity interface. Geophys. J. 95, 397–419.
Griffiths, R.W., Hackney, R.I., van der Hilst, R.D., 1995. A laboratory investigation of effects of trench migration on the descent of subducted slabs. Earth Planet. Sci. Lett. 133, 1–17.
Guillou-Frottier, L., Buttles, J., Olson, P., 1995. Laboratory experiments on the structure of subducted oceanic lithosphere. Earth Planet. Sci. Lett. 133, 19–34.
Gurnis, M., Hager, B.H., 1988. Controls on the structure of subducted slabs. Nature 335, 317–321.
Hager, B.H., 1984. Subducted slabs and the geoid: constraints on mantle rheology and flow. J. Geophys. Res. 89, 6003–6016.
Hirose, K., Fei, Y., Ma, Y., Mao, H.-K., 1999. The fate of subducted basaltic crust in the Earth's lower mantle. Nature 397, 53–56.
Hofmann, A.W., 1997. Mantle geochemistry: the message from oceanic volcanism. Nature 385, 219–229.
Houseman, G.A., Gubbins, D., 1997. Deformation of subducted oceanic lithosphere. Geophys. J. Int. 131, 535–551.
Isacks, B., Molnar, P., 1971. Distribution of stresses in the descending lithosphere from a global survey of focal-mechanism solutions of mantle earthquakes. Rev. Geophys. Space Phys. 9, 103–174.
Ito, E., Takahashi, E., 1989. Postspinel transformations in the system Mg_2SiO_4–Fe_2SiO_4 and some geophysical implications. J. Geophys. Res. 94, 10637–10646.
Jackson, I., 1998. Elasticity, composition and temperature of the Earth's lower mantle: a reappraisal. Geophys. J. Int. 134, 291–311.
Jeanloz, R., Thompson, A.B., 1983. Phase transitions and mantle discontinuities. Rev. Geophys. Space Phys. 21, 51–74.
Jordan, T.H., 1977. Lithospheric slab penetration into the lower mantle beneath the sea of Okhotsk. J. Geophys. 43, 473–496.
Kellogg, L.H., Hager, B.H., van der Hilst, R.D., 1999. Compositional stratification of the deep mantle. Science 283, 1881–1884.
Kesson, S.E., Fitz Gerald, D.D., Shelley, J.M., 1998. Mineralogy and dynamics of a pyrolite lower mantle. Nature 393, 252–255.
Kincaid, C., Olson, P., 1987. An experimental study of subduction and slab migration. J. Geophys. Res. 92, 13832–13840.
King, S.D., Ita, J., 1995. Effect of slab rheology on mass transport across a phase boundary. J. Geophys. Res. 100, 20211–20222.
Kirby, S.H., Stein, S., Okal, E.A., Rubie, D.C., 1996. Metastable mantle phase transformations and deep earthquakes in subducting oceanic lithosphere. Rev. Geophys. 34, 261–301.
Lithgow-Bertelloni, C., Richards, M.A., 1995. Cenozoic plate-driving forces. Geophys. Res. Lett. 22, 1317–1320.
Machetel, P., Weber, P., 1991. Intermittent layered convection in a model mantle with an endothermic phase change at 670 km. Nature 350, 55–57.
Marton, F., Bina, C.R., Stein, S., 1999. Effects of slab mineralogy on subduction rates. Geophys. Res. Lett. 26, 119–122.
Olbertz, D., Wortel, M.J.R., Hansen, U., 1997. Trench migration and subduction zone geometry. Geophys. Res. Lett. 24, 221–224.
Richter, F.M., 1973. Finite amplitude mantle convection through a phase boundary. Geophys. J. R. Astron. Soc. 35, 265–276.
Riedel, M., Karato, S.-I., 1997. Grain-size evolution in subducted oceanic lithosphere associated with the olivine-spinel transformation and its effect on rheology. Earth Planet. Sci. Lett. 148, 27–43.
Ringwood, A.E., Irifune, T., 1988. Nature of the 650-km seismic discontinuity: implications for mantle dynamics and differentiation. Nature 331, 131–136.
Rubie, D.C., Ross, C.R., 1994. Kinetics of olivine-spinel transformation in subducting lithosphere: experimental constraints and implications for deep slab processes. Phys. Earth Planet. Inter. 86, 223–241.
Schmeling, H., Monz, R., Rubie, D.C., 1999. The influence of olivine metastability on the dynamics of subduction. Earth Planet. Sci. Lett. 165, 55–66.
Solheim, L.P., Peltier, W.R., 1994. Avalanche effects in phase transition modulated thermal convection: a model of the Earth's mantle. J. Geophys. Res. 99, 6997–7018.
Stein, M., Hofmann, A.W., 1994. Mantle plumes and episodic crustal growth. Nature 372, 63–68.
Tackley, P.J., Stevenson, D.J., Glatzmeier, G.A., Schubert, G., 1993. Effects of an endothermic phase transition at 670 km depth in a spherical model of convection in the Earth's mantle. Nature 361, 699–704.
Tetzlaff, M., Schmeling, H., 2000. The influence of olivine metastability on deep subduction of oceanic lithosphere. Phys. Earth Planet. Inter., in press.
van der Hilst, R.D., 1995. Complex morphology of subducted lithosphere in the mantle beneath the Tonga trench. Nature 374, 154–157.
van der Hilst, R.D., Seno, T., 1993. Effects of relative plate motion on the deep structure and penetration depth of slabs below the Izu-Bonin and Mariana island arcs. Earth Planet. Sci. Lett. 120, 395–407.
van der Hilst, R.D., Widiyantoro, S., Engdahl, E.R., 1997. Evidence for deep mantle cicrulation from global tomography. Nature 386, 578–584.
van der Hilst, R.D., Kárason, H., 1999. Compositional heterogeneity in the bottom 1000 km of Earth's mantle: toward a hybrid convection model. Science 283, 1885–1888.
Widiyantoro, S., van der Hilst, R.D., 1996. Structure and evolution of lithospheric slab beneath the Sunda arc, Indonesia. Science 271, 1566–1570.
Zhong, S., Gurnis, M., 1994. Role of plates and temperature-dependent viscosity in phase change dynamics. J. Geophys. Res. 99, 15903–15917.
Zhong, S., Gurnis, M., 1995. Mantle convection with plates and mobile, faulted plate margins. Science 267, 838–843.

ELSEVIER

Physics of the Earth and Planetary Interiors 127 (2001) 35–49

PHYSICS
OF THE EARTH
AND PLANETARY
INTERIORS

www.elsevier.com/locate/pepi

Seismic discontinuities and subduction zones

Jonathan D. Collier*, George R. Helffrich, Bernard J. Wood

Department of Earth Sciences, University of Bristol, Wills Memorial Building, Queen's Road, Bristol BS8 1RJ, UK

Received 27 October 1999; accepted 27 July 2000

Abstract

The seismic discontinuities at 410 and 660 km depth in the mantle are generally believed to be due to phase transformations in its major component, olivine. An increasing observational base of short-period regional seismic network data supports this view. Here, we analyse cross-sections of discontinuity topography in the northwest Pacific subduction zones to infer the thermal and chemical state of the mantle in a subduction environment. Most of the data is from the southern part of the Izu-Bonin subduction zone. Penetration of the lower mantle by the steeply dipping subducting slab is clearly indicated by depression of the 660 km discontinuity. Elevation of the 410 km discontinuity in this region by up to 60 km implies a thermal anomaly of approximately 900–1000°C relative to the ambient mantle. Observation of the elevated 410 in the seismically active part of the slab indicates that the olivine → wadsleyite phase transformation occurs under essentially equilibrium conditions there, which argues against the transformational faulting hypothesis as the mechanism for deep-earthquakes. Upgoing P-wave reflections from the 410 km discontinuity indicate that it is sharp in the mantle far from subducting slabs. Reflection coefficient variability in and near slabs indicates either a chemical effect or the consequences of topographic focusing. © 2001 Elsevier Science B.V. All rights reserved.

Keywords: upper-mantle; Discontinuities; Phase transitions; Seismology; Transition zone

1. Introduction

The seismic discontinuities demarking the top and bottom of the upper mantle transition zone are at nominal global depths of around 410 and 660 km (hereafter, referred to as the 410 and 660) in spherical velocity models (Fig. 1a). These discontinuities are believed to be caused by two of the phase transformations in the olivine component (Fig. 1b) of the mantle (Bernal, 1936; Ringwood, 1969):

$$\underset{\text{olivine}(\alpha)}{(\mathrm{Mg,Fe})_2\mathrm{SiO}_4} \rightarrow \underset{\text{wadsleyite}(\beta)}{(\mathrm{Mg,Fe})_2\mathrm{SiO}_4} \quad \text{at 410 km} \qquad (1)$$

$$\underset{\text{wadsleyite}(\beta)}{(\mathrm{Mg,Fe})_2\mathrm{SiO}_4} \rightarrow \underset{\text{ringwoodite}(\gamma)}{(\mathrm{Mg,Fe})_2\mathrm{SiO}_4} \quad \text{at 520 km} \qquad (2)$$

$$\underset{\text{ringwoodite}(\gamma)}{(\mathrm{Mg,Fe})_2\mathrm{SiO}_4} \rightarrow \underset{\text{magnesiowustite}}{(\mathrm{Mg,Fe})\mathrm{O}} + \underset{\text{perovskite}}{(\mathrm{Mg,Fe})\mathrm{SiO}_3} \quad \text{at 660 km} \qquad (3)$$

To confirm this interpretation, experimental and seismological observations and predictions must be carefully analysed for consistency with one another. The wadsleyite to ringwoodite transformation (Eq. (2)) has been suggested as the origin for a discontinuity near 520 km (Shearer, 1996) but the transformation occurs over a large pressure range and is, therefore, difficult to observe seismically.

One way of distinguishing the phase transformation explanation for the 410 and 660 from the hypothesis

* Corresponding author. Present address: Qinetiq, St. Andrews Road, Malvern, Worcester-Shire, WR14 3PS, UK.
E-mail address: jcollier@qinetiq.com (J.D. Collier).

PII: S0031-9201(01)00220-5

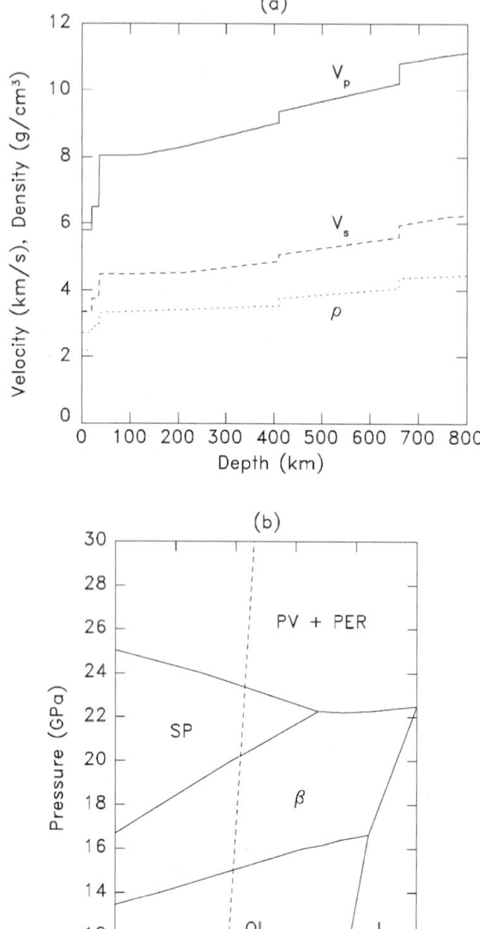

Fig. 1. (a) upper-mantle velocity structure from IASP91 (Kennett and Engdahl, 1991); (b) calculated phase-equilibrium boundaries in pure Mg_2SiO_4 after (Fei et al., 1990). The geotherm is calculated with a gradient of 0.3°K/km. OL, olivine-phase, SP, spinel; L, liquid; PER, periclase; and PV, perovskite.

that the discontinuities mark compositional changes in the mantle is to examine how the discontinuities respond to lateral changes in temperature such as those occurring in subduction zones. A compositional boundary should be largely indifferent to a lateral temperature (T) change whereas the thermodynamic properties of a phase transformation prescribe elevation or depression of the discontinuity to different pressures (P) depending on the sign and magnitude of the Clapeyron slope, dP/dT (e.g. Fig. 1b). On the other hand, a compositional boundary will respond dynamically to mantle convective flows. Thus, regions in the mantle where thermal anomalies and flow kinematics are known are ideal areas to test the competing hypotheses for the seismic discontinuities. Subduction zones are good locations for conducting seismological investigations since they are regions where both large negative thermal anomalies (Fig. 2) as well as chemical heterogeneities exist. If the phase transformation interpretation can be confirmed then experimental knowledge of the properties of these phase transitions enables us to use the discontinuities as probes of the earth's chemical and thermal structure.

The $\alpha \rightarrow \beta$ phase transformation (Eq. (1)) has a positive Clapeyron slope of greater magnitude (+2.9 MPa/K) than the negative slope (−1.9 MPa/K) for the breakdown of γ-spinel (Eq. (3)) (Bina and Helffrich, 1994). Therefore, in subduction zones, where temperatures are low, the 410 is expected

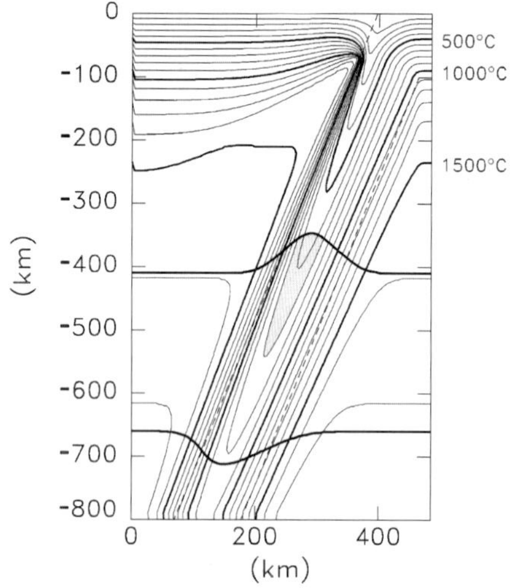

Fig. 2. Thermal model of the Izu-Bonin subducting slab at 29°N using a finite difference method (Minear and Toksož, 1970; Toksož et al., 1973). Subduction parameters used are a convergence rate of 4.2 mm per year, a slab age of 140 million years and dip of 68° with a basal lithosphere temperature of 1450°C (Stein and Stein, 1992). Equilibrium boundaries for the 410 and 660 km discontinuities are shown using Clapeyron slopes of +2.9 and −1.9 MPa/°C, respectively (Bina and Helffrich, 1994). The shaded region illustrates a possible metastable region bounded by the 700°C isotherm.

to be elevated by a greater amount than the 660 is depressed relative to their positions in the ambient mantle. Thermodynamic calculations show that low-temperatures may also be expected to broaden the $\alpha \rightarrow \beta$ phase transition (Fig. 3) and change the phase transition series at a particular composition from $\alpha \rightarrow \alpha + \beta \rightarrow \beta \rightarrow \beta + \gamma \rightarrow \gamma$ (Fig. 3c) to $\alpha \rightarrow \alpha + \gamma \rightarrow \beta + \gamma \rightarrow \gamma$ (Fig. 3a) in the coldest part of the slab (Bina, 1997). The calculated phase diagram at 800°C (Fig. 3a) is, however, based on extrapolated experimental data obtained between 1200 and 1600°C (e.g. Katsura and Ito 1989) and is, therefore, very uncertain.

Chemical differences in subduction zones such as the presence of water (Wood, 1995; Wood et al., 1995; Helffrich and Wood, 1996) may affect both the depth and sharpness of the discontinuities. Water is very soluble in wadsleyite, the partitioning of water between olivine and wadsleyite being 1:10 (Young et al., 1993; Kohlstedt et al., 1996), and so the 410 is predicted to be broadened substantially in the presence of a few hundred ppm water in the mantle (Wood, 1995) and move to a shallower depth. It has been shown (Ito and Takahashi, 1989; Wood, 1990) that the breakdown of ringwoodite (Eq. (3)) occurs over a depth range sufficiently sharp to explain the 660 km discontinuity. The sharpness of the discontinuity may, however, be dependent on the aluminium content of the mantle (Wood and Rubie, 1996; Weidner and Wang, 1998). Seismic discontinuity observations may, therefore, constrain upper-mantle compositions. Thus, seismic studies of the transition zone discontinuities, when combined with experimental observations, may yield valuable information about the composition of the mantle and the structure of subduction zones.

Observations of the 410 may also have important implications for theories of the mechanism of deep-earthquakes. The mechanism of shallow earthquakes is one of brittle deformation, but deeper earthquakes (>100 km) must have a different mechanism since at these depths the temperatures and confining pressures are so large that the shear stress necessary to drive frictional sliding is expected to exceed the brittle strength of the slab materials. The mechanisms that cause deep earthquakes (Kirby, 1987; Green and Burnley, 1989; Furukawa, 1994; Lundgren and Giardini, 1994; Rundle and Klein, 1995) are, however, not yet fully understood. deep-earthquakes are only

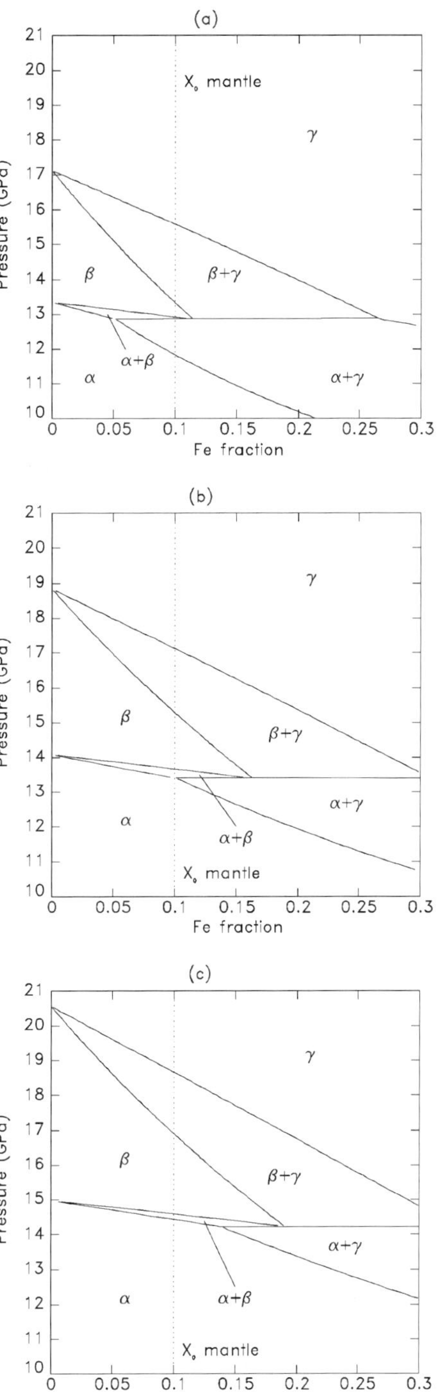

Fig. 3. Phase diagrams calculated from the data of (Fei et al., 1991) for the binary system Mg_2SiO_4–Fe_2SiO_4 (a) at 800°C; (b) at 1200°C and (c) at 1600°C.

observed in slabs older than 80–90 million years, so this must mean that lower temperatures are the most important factor. Strain rate, asymmetric heating of the slab, internal phase transitions and resistance to downward motion may, in any particular region, combine to produce the observed seismicity. One actively investigated effect relating to deep-earthquakes is a possible link between the transition of metastable olivine and earthquake occurrence. Metastable olivine may exist in the coldest subducting slabs since the kinetics of the $\alpha \rightarrow \beta$ reaction at low-temperatures could mean that the reaction is delayed (Sung and Burns, 1976; Rubie, 1984; Kirby, 1987; Green and Burnley, 1989; Kirby et al., 1991, 1996) and that, therefore, the reaction needs to be overstepped to higher pressures before transformation occurs. Observation of the 410 km discontinuity inside a subducting slab should, therefore, place constraints on the presence or absence of a metastable wedge of unreacted olivine at depths below the equilibrium 410 depth.

2. Imaging upper-mantle discontinuities

The depths and sharpness of the transition zone discontinuities can be imaged by observing seismic waves which interact with them. Using deep earthquakes, upgoing P to S (or S to P) conversions (Fig. 4) at depths of around 410 km to local receivers can sometimes be seen (Sacks and Snoke, 1977; Collier and Helffrich, 2001). These observations, however, are sparse and arrivals are often difficult to pick in individual records where the low amplitude arrivals are frequently close to background noise levels. In addition, the upgoing converted phase and the reference phase (usually P) follow different paths through the upper-mantle and crust, thus, introducing travel time interpretation errors due to local heterogeneities.

Receiver function analysis has been used for detailed investigations of the 410 and 660 depths beneath receivers using teleseismic data (Gurrola et al., 1994; Niu and Kawakatsu, 1996; Bostock and Cassidy, 1997; Dueker and Sheehan, 1997; Shen et al., 1998; Neal and Pavlis, 1999). These studies use upgoing P to S conversions at the discontinuities beneath the receivers to image their depths. Such studies are consequently mainly confined to regions with a reasonable density of receivers. The most widely used techniques to image upper mantle discontinuities in fact involve using stacks of data from large numbers of receivers to enhance the weak discontinuity interactions and suppress the background noise. Stacking of data is often performed in receiver function analyses as well as near-source and global studies.

Since both the 410 and 660 are expected to be worldwide features global studies map them by searching for reflections at the discontinuities which appear as long-period precursors to stronger phases such as SS or PP (Fig. 4) (Shearer, 1991, 1993; Shearer and Masters, 1992; Gossler and Kind, 1996; Estanrook and Kind, 1996; Gu et al., 1998; Flanagan and Shearer, 1998a). These studies find peak-to-peak topography on the 410 and 660 km discontinuities of about 20–30 km. This peak-to-peak topography is

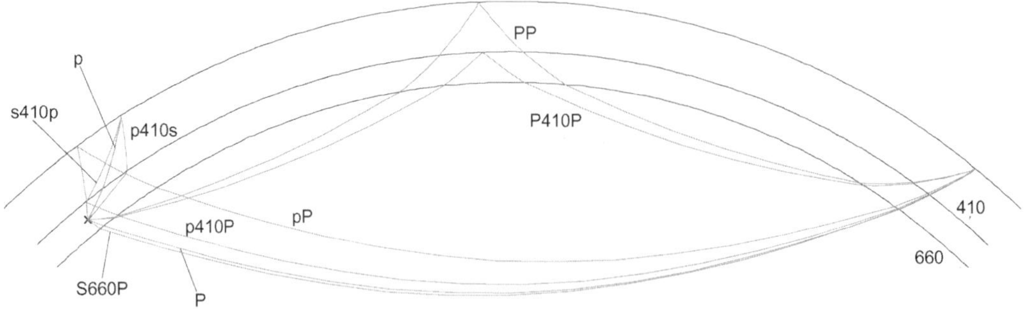

Fig. 4. Schematic diagram showing some of the seismic phases used to study the transition zone discontinuities. An upgoing P-wave to a local receiver is denoted p whilst P to S and S to P conversions at the 410 are denoted p410s and s410p. A near source P reflection from the 410 to a teleseismic receiver is called p410P (similarly s410P for an S to P conversion). A precursor to PP from the 410 km discontinuity is P410P (similarly P660P for the 660 and S410S and S660S for SS precursors).

much smaller than that expected from Clapeyron slope estimates (between 100 and 150 km peak-to-peak) and the low-temperature anomalies expected in subduction zones (800–1000°C). There is, however, an important lack of resolution resulting from a large Fresnel zone size of typically 10–20 degrees. This problem has been addressed by Neele et al. (1997), Neele and de Regt (1997), Chaljub and Tarantola (1997), and Shearer et al. (1999) who have shown that small-scale, sub-Fresnel zone size topography, associated with cold subducting slabs, cannot be imaged reliably with long-period SS and PP precursors. With improvements to the data analysis technique, however, they find that it is possible to resolve lateral structures smaller than the Fresnel zone size but larger than the dominant wavelength used. Typically frequencies are around 0.04 Hz (Shearer, 1991) leading to wavelengths for transition zone discontinuity precursors of around 250 km for PP and 130 km for SS. The important point is that long wavelength precursors to SS and PP are unable to image any structure of mantle discontinuities which is less than 100–200 km wide. Thus, features arising from cold subducting slabs are only resolved qualitatively. An additional problem associated with using precursors to phases such as PP and SS comes from the need to have a good knowledge of the upper-mantle velocity structure (Flanagan and Shearer, 1998b) when interpreting travel time differences. Despite these problems of resolution global mapping of the 410 and 660 using these methods are extremely useful in showing that they are truly worldwide features.

Although not observable in long-period studies small-scale topography (<100 km wide) on the transition zone discontinuities can be imaged by observing near source pP and sP precursors and downgoing S to P conversions in short-period teleseismic regional network data (Figs. 4 and 5). These phases are very weak; their amplitudes are often only a few percent of the main P phase and very close to the background noise level. In order to observe these weak phases we can slant stack the records from a large number of seismometers in a single array. A point in the slant stack (also called a vespagram) represents the sum along a line through a seismic section (e.g. Fig. 5) and is identified by its time at a reference distance and its slowness (the reciprocal of the apparent velocity) across the array. Slant stacking enhances plane wave arrivals whilst minimising incoherent signals, thus, increasing the signal to noise ratio sufficiently

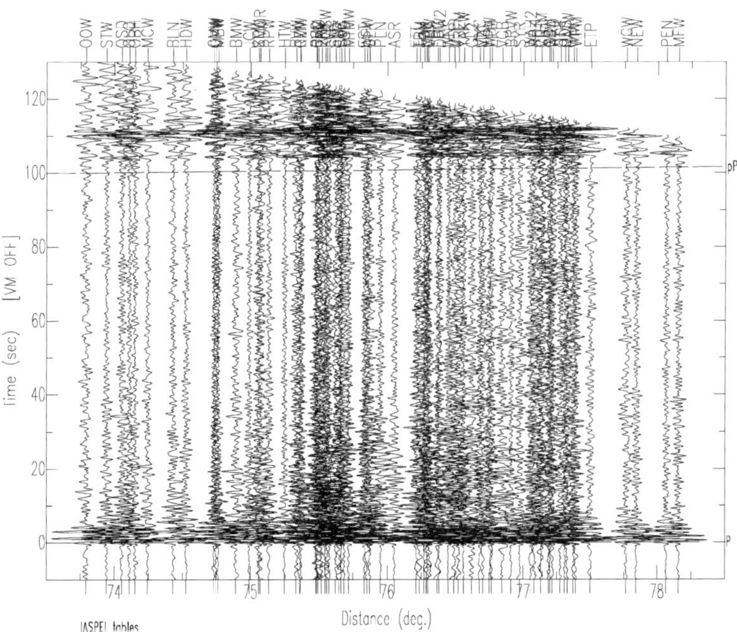

Fig. 5. Record section for event c. Records are aligned on the P arrival, the theoretical pP time from IASP91 is marked.

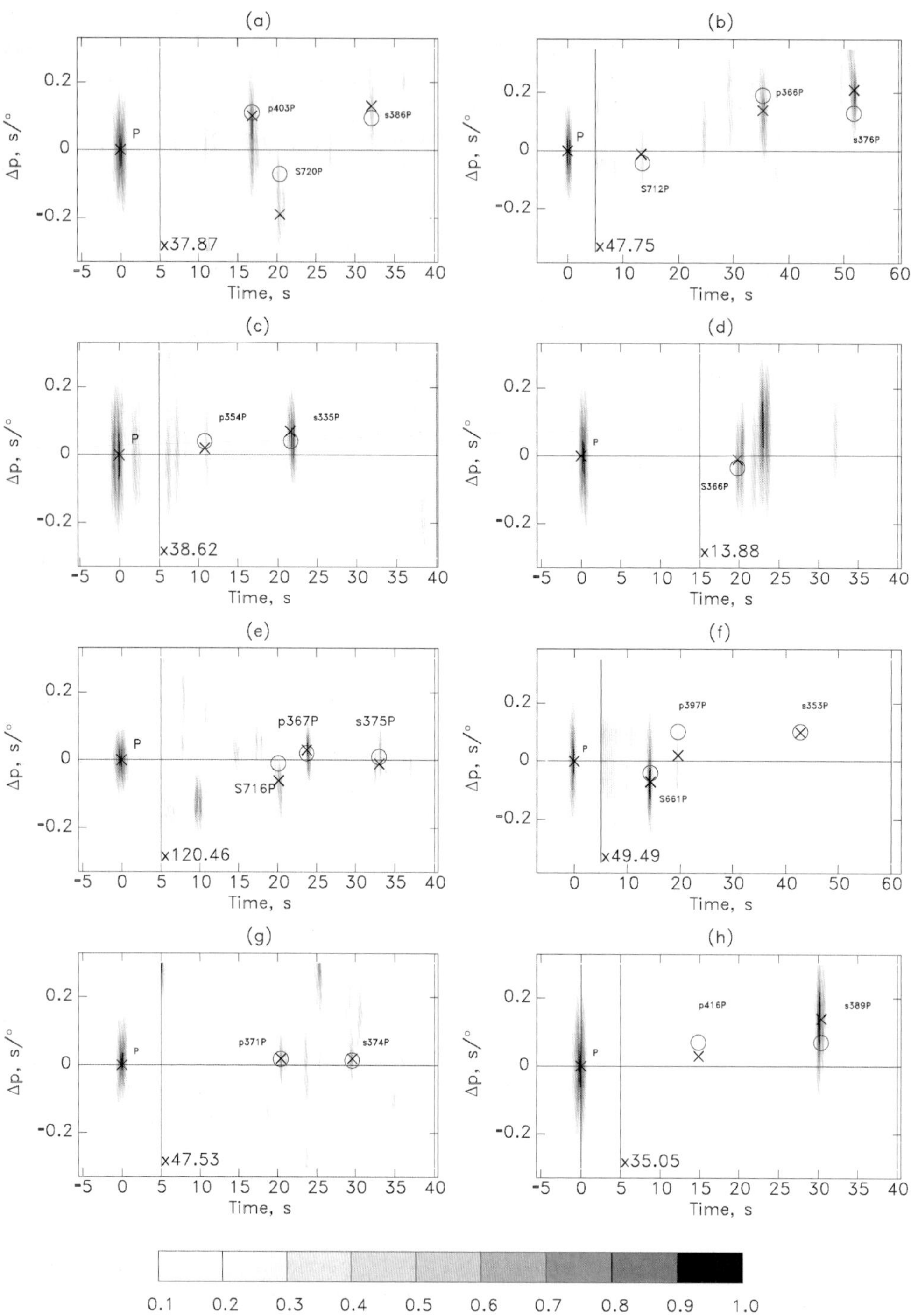

Fig. 6. Examples of Nth-root (with $N = 3$) vespagrams for the event/network combinations given in Table 1. Horizontal and vertical axes show the time (s) and slowness (s/°) relative to the P-wave arrival. The region of the stack to the right of the vertical line has been magnified by the factor shown. Arrivals interpreted to be from transition zone discontinuities are marked. Theoretical times using the IASP91 model are shown by circles, picks are shown by crosses.

Table 1
Earthquakes and networks used in the vespagrams in Fig. 6[a]

Event	Date	Depth	Latitude	Longitude	mb	Stations	Network
a	6 February 1981	506	48.18	146.37	5.3	27	UW
b	8 October 1983	566	44.14	130.95	5.6	50	UW
c	10 April 1985	409	29.99	139.02	5.7	89	CA
d	12 December 1987	178	29.78	140.37	6.0	64	UW
e	5 August 1990	490	29.35	137.77	5.9	54	UK
f	21 July 1994	509	41.27	132.43	5.8	71	UW
g	26 June 1996	479	27.82	139.85	5.5	45	UK
h	31 January 1997	498	26.20	140.96	5.6	70	UW

[a] UK: United Kingdom network; UW: University of Washington network; CA: northern Californian network.

to observe and identify weak arrivals based on their arrival time and slowness (Fig. 6). Both the slowness resolution and the discrimination of arrivals from different backazimuths are important factors in identifying weak phases. The slowness resolution of the regional networks depends greatly on the station configuration used but, is generally ± 0.1 s/° (Collier, 1999). In our studies, the beam power for the combination of stations in each slant stack is calculated and, thus, the slowness resolution obtained. We only positively identify arrivals if the difference between the observed and theoretical slowness is less than the resolution of the array. Azimuthal deviations are monitored using the method of (Kaneshima and Helffrich, 1998) and for our interpreted arrivals are less than the resolution of the technique (Table 1).

With regional arrays of seismometers slant stacking is the preferred method for the analysis of teleseismic events because of the ability to identify weak arrivals. Vidale and Benz (1992) used this method to enhance near source P reflections and S to P conversions in studies of the upper-mantle discontinuities, and later to place constraints on both the depth and thickness of the mantle discontinuities that they observed in different geographical regions (Benz and Vidale, 1993). This technique has also been used with California network data by Ritsema et al. (1995) and Niu and Kawakatsu (1995), and using the Japan array by Yamazaki and Hirahara (1994), Kawakatsu and Niu (1994) and Niu and Kawakatsu (1995, 1997). More recently, west Pacific events have been studied extensively using the University of Washington array (Castle and Creager, 1997; Collier and Helffrich, 1997; Kaneshima and Helffrich, 1998; Collier, 1999) and also using the United Kingdom network (Collier and Helffrich, 1997; Collier, 1999).

3. Discontinuity topography in subduction zones

Most studies of the 410 and 660 discontinuities in subduction zones concentrate on the regions surrounding the Pacific Ocean because these are the best illuminated due to frequent earthquakes and favourable regional network locations. Amongst the earliest of such studies, Vidale and Benz (1992) found 410 elevation near subduction zones surrounding the Pacific while others found downward deflection of the 660 km discontinuity in the Tonga region (Richards and Wicks, 1990; Niu and Kawakatsu, 1995). Collier (1999) and Collier and Helffrich (2001) observed topography on both the 410 and 660 beneath South America. The northwest Pacific and particularly the Izu-Bonin region (Fig. 7) remains, however, the most intensively studied region.

Beneath southern Izu-Bonin (Fig. 8a and b) depression of the 660 is well resolved with stacked short-period network data (Barley et al., 1982; Wicks and Richards, 1993; Castle and Creager, 1997, 1998a; Collier and Helffrich, 1997). Examples of the data used can be seen in Fig. 6. In this region the depression of the 660 is imaged on either side of the deepest seismicity with the maximum depression appearing to occur beneath the deepest earthquakes (Castle and Creager, 1998b). Further north it is only possible to image the discontinuity on the oceanward side of the seismicity (Fig. 8c–h) because of the limited source/receiver geometry available. Well away from

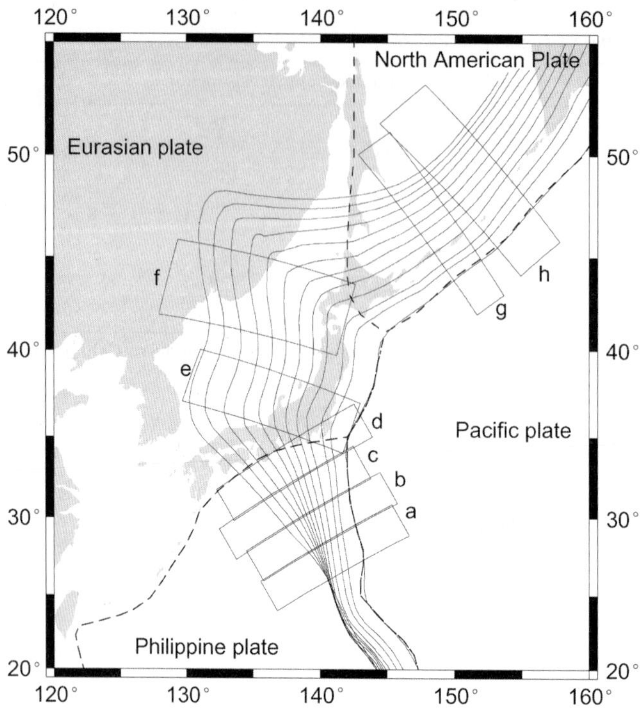

Fig. 7. Map of the northwest Pacific showing the locations of the cross-sections in Fig. 8. The dashed line shows the NUVEL-1 plate boundary model (De Mets et al., 1990) and the solid lines show slab surface contours from (Gudmundsson and Sambridge, 1998) from 0 to 550 km every 50 km.

the slab in northern Izu-Bonin (Fig. 8c and d) and beneath the Japan Sea (Fig. 8e and f) the discontinuity can clearly be seen to be at its global average depth (660 km), moving deeper as the subducting slab is approached. Further to the west receiver functions image a complicated discontinuity structure and stagnation of the slab above the 660 (Niu and Kawakatsu, 1996). In the cross-sections of the Kurile Islands (Fig. 8g and h) there is evidence for topography on the 660 near to the subducting slab although the data are sparse. Both the locations and magnitudes of topography on the 660 in these areas are consistent with a phase change origin for the discontinuity (Castle and Creager, 1997; Collier and Helffrich, 1997). They are, on the other hand, inconsistent with a chemical change in the mantle due to the fact that the observed 660 depression of 50 km is smaller than the 100–300 km predicted for a chemical change by dynamical models (Christensen and Yuen, 1984; Kincaid and Olson, 1987).

Previous workers report a slight elevation of the 410 by up to 25 km (Revenaugh and Jordan, 1991; Vidale and Benz, 1992) in the vicinity of subduction zones. This is consistent with a phase change origin hypothesis with a positive Clapeyron slope for the $\alpha \rightarrow \beta$ transition although smaller than expected. The sparseness of the observations in these early studies may, however, indicate that the larger topography expected from experimentally determined Clapeyron slopes is not being imaged. Recently, Thirot et al. (1998) report up to 60 km elevation of the 410 in the Japan subduction zone which would be consistent with a temperature depression of ~900°C relative to the ambient mantle. By mapping the discontinuity depths across a subduction zone where the thermal structure is well constrained, Collier and Helffrich (1997) and Collier (1999) attempted to find the maximum topography and to find evidence for the existence or absence of a metastable wedge of olivine in the coldest part of the slab. Evidence of discontinuity topography,

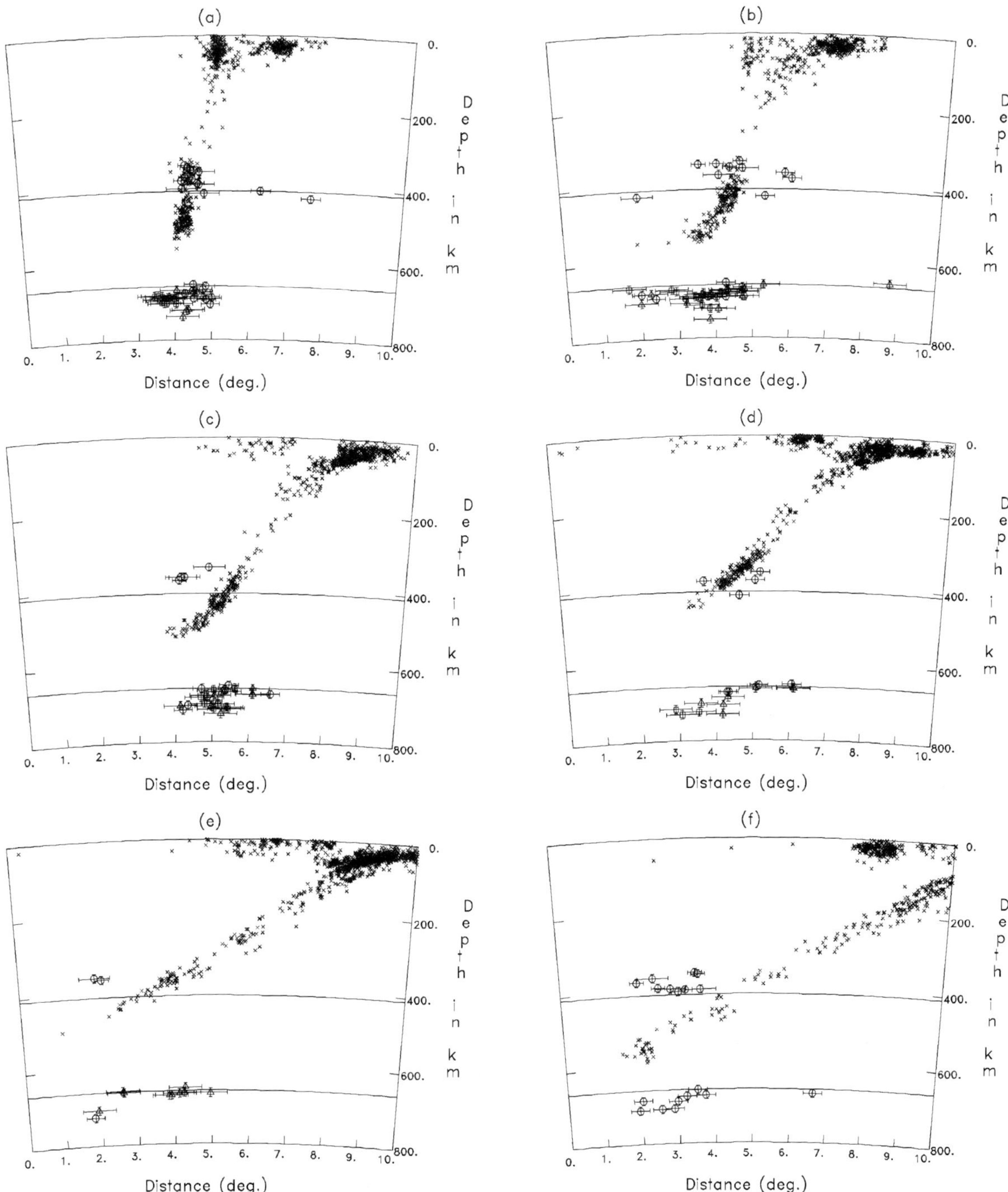

Fig. 8. Cross-sections of the northwest Pacific subduction zones. The 410 km discontinuity data (circles) is from Collier and Helffrich (1997) and Collier (1999). The 660 km discontinuity data is from Wicks and Richards (1993) (squares), Castle and Creager (1997, 1998a,b) (triangles) and Collier and Helffrich (1997), Collier (1999) (circles). Horizontal error bars derive from typical Fresnel zone sizes, the largest cause of uncertainty, with a source-discontinuity vertical separation of 200 km. Vertical error bars come from three sources: earthquake depth errors, relative arrival time picking errors and velocity anomaly errors (Collier and Helffrich, 1997). Earthquake locations (crosses) are from Engdahl et al. (1998).

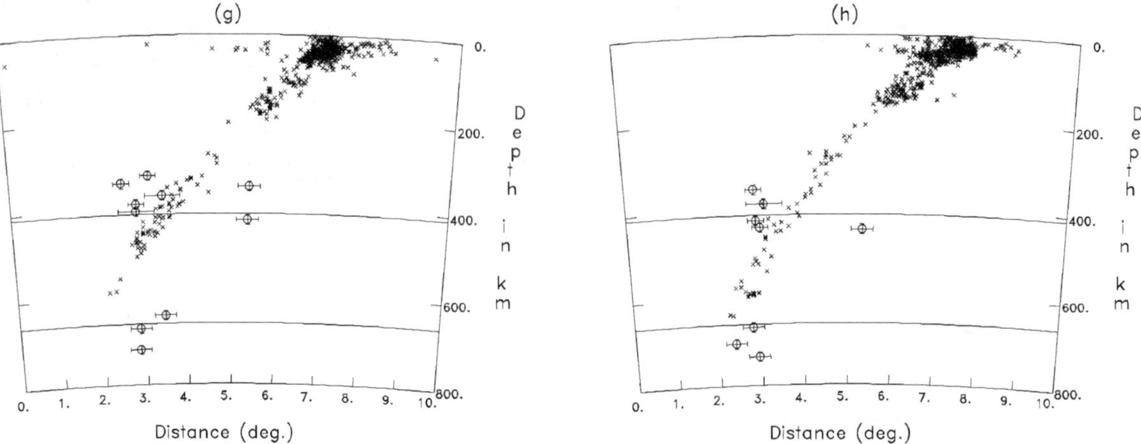

Fig. 8. (*Continued*).

indicating a thermal anomaly, in aseismic regions would also be evidence of slab continuity where there are no earthquakes.

The southern cross-sections (Fig. 8a and b) clearly show interaction points at the very heart of the subducting slab with elevations of up to 60–70 km (Collier and Helffrich, 1997; Collier, 1999). That these observations are made both with upgoing discontinuity interactions to the western United States networks and downgoing interactions to the United Kingdom with earthquakes both below and above the discontinuity, respectively, rules out systematic errors or mis-interpretation of slab/mantle interface interactions. The larger deflection seen in these data compared to previous work is due to the sampling of the discontinuity close to and inside the subducting slab where the elevation should be greatest (Bina and Helffrich, 1994).

Thermal modelling (Fig. 2) indicates an absolute temperature of 550°C in the interior of the Izu-Bonin slab at 350 km depth. This is consistent with a value of 600 ± 75°C for the coldest part of the slab calculated using a seismic Clapeyron slope of 2.04 MPa/K (Helffrich and Bina, 1994) and our observations of the 410 at 350 km depth. These temperature estimates are in the same range as that assumed by Furukawa (1994) to be the kinetic cut-off temperature for the olivine–spinel transformation in his discussion of the increases of the numbers of earthquakes with depth in some subduction zones. The observation of 410 km discontinuity interactions in the same location as slab seismicity implies, however, that the $\alpha \rightarrow \beta$ phase change is occurring at or near equilibrium in the slab interior. The maximum elevation is seen in the region which coincides with the slab seismicity at a depth of 350 km (Fig. 8a and b). This marks the cold core of the slab and is, therefore, significant evidence against the existence of a metastable olivine wedge.

Thermal modelling (Fig. 2), calculated for a slab thickness of 100 km, gives a thermal halo, where the mantle is cooled significantly by the slab, of width 170 km determined by the breadth of the seismically resolvable uplift of approximately 11 km. The data (Fig. 8b) suggests, however, in the central Izu-Bonin region, a much larger anomaly, up to 300 km wide, but narrower farther south. Wider thermal anomalies than expected could indicate that the slab is thicker than the 100 km modelled; or that it is cooling the mantle more effectively than previously thought; or that there is some influence from trench migration on the thermal structure (van der Hilst and Seno, 1993). An evaluation of the results of numerical modelling indicates that the temperature in a subducting slab is mainly controlled by thermal diffusion in the direction perpendicular to the slab (Furukawa, 1994). Values for the thermal diffusivity are not well known and have both temperature and pressure dependence (Kanamori et al., 1968; Fujisawa et al., 1968; Keiffer, 1982; Chai et al., 1996). Thus, further experimental work on the diffusivity of mantle minerals at relevant temperatures and pressures can improve the predictions of thermal models which might then yield broader thermal anomalies. To this end, Hofmeister (1999) has recently developed a theoretical framework describing variations of the conductivity of mantle minerals with temperature and pressure. Using Hofmeister's results Hauck et al. (1999) examined the effects of variable diffusivity on a model of the thermal structure of a subducting slab, and showed that it does lead to a broader thermal anomaly than a similar model with constant diffusivity.

Beneath the Japan Sea all of the 410 km interaction points are elevated and are near or above the surface of the subducting slab (Fig. 8e and f). The elevation of the discontinuity indicates the presence of an extended region of the mantle cooled by the subducting slab. Some individual data points yield 410 elevation of 50–60 km, unexpected since they are not in the cold core of the slab, defined by both seismicity and tomography (Zhao et al., 1994; van der Hilst et al., 1997; Bijwaard et al., 1998), but in the surrounding mantle. The observations are, however, in agreement with the elevation seen in this region by Thirot et al. (1998). The elevation of the 410 far from the slab beneath Japan may imply that it is regionally shallow or indicate that the breadth of the elevation is much larger than expected from thermal models and similar to some observations in Izu-Bonin.

4. Discontinuity sharpness

Previous short-period (0.5–5 s) studies of transition zone discontinuities (Lees et al., 1983; Vidale and Benz, 1992; Yamazaki and Hirahara, 1994; Neele, 1996) have shown that their reflectivities can vary between very small values of 5% to as large as 13%, implying discontinuity thicknesses as small as 2–5 km. The wide variability and magnitude of these reflections have caused problems when interpreting them in terms of mantle composition, mineralogy and existing global seismic velocity models which mainly suggest larger transition thicknesses or smaller velocity contrasts. One possible explanation for such discrepancies is that the strongest reflections occur only under special conditions, possibly where small-scale topography causes local focusing of seismic waves. Detailed reflectivity estimates for the 410 km discontinuity are sparser than for the 660, qualitatively suggesting that it is a weaker reflector.

The abundance of short-period discontinuity observations such as those in Fig. 8 provides an opportunity for accurate determinations of both their sharpness and variability in subduction zones. To extract these properties, however, the raw data must be corrected

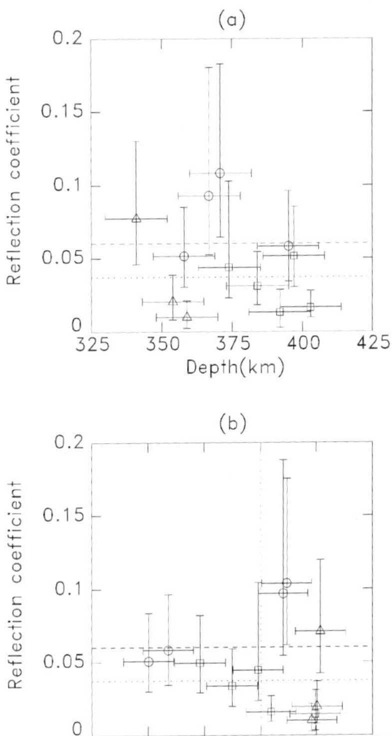

Fig. 9. The p410P reflection coefficients for interactions where neither P or p410P source radiation is near nodal plotted against (a) discontinuity interaction depth and (b) lateral offset from seismicity (positive offset is in the oceanward direction). Reflection coefficients are calculated using envelope amplitude measurements and are corrected for source mechanism, attenuation and geometric spreading. Vertical error bars are calculated from uncertainties in focal mechanisms (Helffrich, 1997). Horizontal error bars derive from typical Fresnel zone sizes. Circles — United Kingdom data, squares — University of Washington data, triangles — northern Californian data. The horizontal dotted line represent the theoretical first-order discontinuity reflection using IASP91 (Kennett and Engdahl, 1991), the dashed line AK135 (Kennett et al., 1995) (Montagner and Kennet, 1996). The vertical dotted line at an offset of — 100 km separates the normal reflection coefficients away from the slab from the highly scattered values near to the slab.

for attenuation within the earth, source radiation patterns and the effects of discontinuity topography and 3D velocity structure. Fig. 9 shows calculated p410P reflection coefficients for the discontinuity observations of Collier and Helffrich (1997) and Collier (1999) in the northwest Pacific. The data show large scatter and no observable trend with depth (Fig. 9a), a linear fit to the data is not significantly different from a constant weighted mean value at the 87% level. Some useful observations become apparent, however, when the data are plotted as a function of offset from the slab seismicity (Fig. 9b). Far from the centre of the slab (>100 km) the observed reflection coefficients are entirely consistent with values expected from a first order discontinuity. However, near the surface of and inside the slab the amplitudes are far more variable; this is also the region where the largest reflection coefficients are seen. Although the variability in these amplitudes may be the result of chemical and thermal heterogeneities in the slab the wide range of values at small offsets indicate that topography near the slab cores may play an important role in affecting amplitudes through focusing/de-focusing effects. Kirchhoff waveform modelling of underside 410 reflections (Collier, 1999) suggests that topographic effects may be corrected for if the discontinuity topography can be modelled, permitting some of the scatter to be sensibly interpreted.

Estimates of 660 km discontinuity sharpness from short-period network data present a similar problem to those of the 410. In the past, high noise levels, and the effects of discontinuity topography have prevented analysis of the shear velocity contrast at the 660 from S to P and P to S conversions (Paulssen, 1985; Richards and Wicks, 1990; van der Lee, 1994). Recently, however, Castle and Creager (2000) analysed the S660P phase in Izu-Bonin data and found the discontinuity to be sharp (<10 km) with an S-wave contrast of 0.60 ± 0.11 km/s, 70% higher than in the global model IASP91 (Kennett and Engdahl, 1991). Castle and Creager reconcile this higher value with experimental data if the mantle a few hundred kilometres from the slab is at near normal temperatures and contains between 3 and 5% aluminium. Thus, discontinuity sharpness estimates are starting to constrain both the temperature and composition of the mantle. In the future it should be increasingly possible to map small-scale chemical heterogeneities, such as those associated with subducting slabs, using these methods.

5. Conclusions

Observations of the topography and sharpness of the 410 and 660 km discontinuities in subduction zones are consistent with those expected for phase transformation origins in the olivine component of the mantle. Elevation of the 410 of 60–70 km in the coldest part of the Izu-Bonin slab indicates that the $\alpha \rightarrow \beta$ phase transformation occurs close to equilibrium there. This is evidence against the existence of a significant metastable wedge and argues against the transformational faulting hypothesis as the mechanism for deep-earthquakes. Observations of 410 reflectivity variations in the vicinity of northwest Pacific subducting slabs suggests a topographic focusing effect from the topography of the discontinuity in addition to any chemical and thermal effects of the slab.

Acknowledgements

We thank the British Geological Survey and the IRIS DMC for access to UK and US network data and Doug Neuhauser for help obtaining northern Californian data. We also thank John Castle for Izu-Bonin data and valuable discussions. Luisa Branã and Ed Hill provided useful comments on an early version of the manuscript. John Castle and an anonymous reviewer provided constructive criticisms of the manuscript and Rob van der Hilst provided useful comments. J.C. was funded by a NERC studentship and PDRA during this work.

References

Barley, B.J., Hudson, J.A., Douglas, A., 1982. S to P scattering at the 650 km discontinuity. Geophys. J. R. Astro. Soc. 69, 159–172.

Benz, H.M., Vidale, J.E., 1993. Probing earth's interior using seismic arrays. Geotimes 38 (7), 20–22.

Bernal, J.D., 1936. Commentary. Observatory 59, 268.

Bijwaard, H., Spakman, W., Engdahl, E.R., 1998. Closing the gap between regional and global travel time tomography, J. Geophys. Res. 30055–30078.

Bina, C.R., 1997. Patterns of deep seismicity reflect buoyancy stresses due to phase transitions. Geophys. Res. Lett. 24, 3301–3304.

Bina, C.R., Helffrich, G.R., 1994. Phase transition Clapeyron slopes and transition zone seismic discontinuity topography. J. Geophys. Res. 99, 15853–15960.

Bostock, M.G., Cassidy, J.F., 1997. upper-mantle stratigraphy beneath the southern Slave craton. Can. J. Earth Sci. 34, 577–587.

Castle, J.C., Creager, K.C., 1997. Seismic evidence against a mantle chemical discontinuity near 660 km depth beneath Izu-Bonin. Geophys. Res. Lett. 24 (3), 241–244.

Castle, J.C., Creager, K.C., 1998a. Topography of the 660 km seismic discontinuity beneath Izu-Bonin: Implications for tectonic history and slab deformation. J. Geophys. Res. 103, 12511-12527.

Castle, J.C., Creager, K.C., 1998b. NW Pacific slab rheology, the seismicity cut-off, and the olivine to spinel phase change. Earth Planets and Space 50, 977–985.

Castle, J.C., Creager, K.C., 2000. Sharpness and shear wave velocity jump across the 660 km discontinuity. J. Geophys. Res. 105, 6191–6200.

Chai, M., Brown, J.M., Slutsky, L.J., 1996. Thermal diffusivity of mantle minerals. Phys. Chem. Minerals 23, 470–475.

Chaljub, E., Tarantola, A., 1997. Sensitivity of SS precursors to topography on the upper-mantle 660 km discontinuity. Geophys. Res. Lett. 24, 2613–2616.

Christensen, U.R., Yuen, D.A., 1984. The interaction of a subducting lithospheric slab with a chemical or phase boundary. J. Geophys. Res. 89, 4389–4402.

Collier, J.D., Helffrich, G.R., 1997. Topography of the 410 and 660 km seismic discontinuities in the Izu-Bonin subduction zone. Geophys. Res. Lett. 24, 1535–1538.

Collier, J.D., 1999. An investigation of the depths and properties of the mantle's seismic discontinuities in subduction zones. Ph.D. Thesis, University of Bristol, United Kingdom.

Collier, J.D., Helffrich, G.R., 2001. The thermal influence of the subducting slab beneath South America from 410 to 660 km discontinuity observations. Geophys. J. Int., in press.

De Mets, C., Gordon, R.G., Argus, D.F., Stein, S., 1990. Current plate motions. Geophys. J. Int. 101, 425–478.

Dueker, K.G., Sheehan, A.F., 1997. Mantle discontinuity structure from midpoint stacks of converted P to S waves across the Yellowstone hotspot track. J. Geophys. Res. 102, 8313–8327.

Engdahl, E.R., van der Hilst, R., Buland, R., 1998. Global teleseismic earthquake relocation with improved travel times and procedures for depth determinations. Bull. Seismol. Soc. Am. 88, 722–743.

Estabrook, C.H., Kind, R., 1996. The nature of the 660 km upper-mantle seismic discontinuity from precursors to the PP phase. Science 274, 1179–1182.

Fei, Y., Saxena, S.K., Navrotsky, A., 1990. Internally consistent thermodynamic data and equilibrium phase relations for compounds in the system $MgO–SiO_2$ at high-pressure and high-temperature. J. Geophys. Res. 95, 6915 6928.

Fei, Y., Mao, H.-K., Mysen, B.O., 1991. Experimental determination of element partitioning and calculation of phase relations in the $MgO–FeO–SiO_2$ system at high-pressure and high-temperature. J. Geophys. Res. 96, 2157–2169.

Flanagan, M.P., Shearer, P.M., 1998a. Global mapping of topography on transition zone velocity discontinuities by stacking SS precursors. J. Geophys. Res. 103, 2673–2692.

Flanagan, M.P., Shearer, P.M., 1998b. Topography on the 410 km seismic velocity discontinuity near subduction zones from stacking sS, sP, and pP precursors. J. Geophys. Res. 103, 21165–21182.

Fujisawa, H., Fuji, N., Mitzutani, H., Kanamori, H., Akimoto, S.-I., 1968. Thermal diffusivity of Mg_2SiO_4, Fe_2SiO_4 and NaCl at high-pressures and temperatures. J. Geophys. Res. 73, 4727–4733.

Furukawa, Y., 1994. Two types of deep seismicity in subducting slabs. Geophys. Res. Lett. 21, 1181–1184.

Gossler, J., Kind, R., 1996. Seismic evidence for very deep roots of continents. Earth Planet. Sci. Lett. 138, 1–13.

Green, H.W., Burnley, P.C., 1989. A new self-organising mechanism for deep-focus earthquakes. Nature 341, 733–737.

Gu, Y., Dziewonski, A.M., Agee, C.B., 1998. Global de-correlation of the topography of transition zone discontinuities. Earth Planet. Sci. Lett. 157, 57–67.

Gudmundsson, Sambridge, 1998. A regionalised upper-mantle (RUM) seismic model, J. Geophys. Res. 103, 7121–7136.

Gurrola, H., Minster, J.B., Owens, T., 1994. The use of velocity spectrum for stacking receiver functions and imaging upper-mantle discontinuities. Geophys. J. Int. 117, 427–440.

Hauck, S.A., Phillips, R.J., Hofmeister, A.M., 1999. Variable conductivity: effects on the thermal structure of subducting slabs. Geophys. Res. Lett. 26, 3257–3260.

Helffrich, G.H., 1997. How good are routinely determined focal mechanisms? Empirical statistics based on a comparison of Harvard, USGS and ERI moment tensors. Geophys. J. Int. 131, 741–750.

Helffrich, G., Bina, C.R., 1994. Frequency dependence of the visibility and depths of mantle seismic discontinuities. Geophys. Res. Lett. 21, 2613–2616.

Helffrich, G.R., Wood, B.J., 1996. The 410 km discontinuity sharpness and the form of the phase diagram: resolution of apparent seismic contradictions. Geophys. J. Int. 126, 7–12.

Hofmeister, A.M., 1999. Mantle values of thermal conductivity and the geotherm from phonon lifetimes. Science 283, 1699–1707.

Ito, E., Takahashi, E., 1989. Postspinel transformations in the system $Mg_2SiO_4–Fe_2SiO_4$ and some geophysical implications. Geophys. J. Res. 94, 10637–10646.

Kanamori, H., Fuji, N., Mizutani, H., 1968. Thermal diffusivity measurement of rock forming minerals from 300 to 1100 K. J. Geophys. Res. 73, 595–605.

Kaneshima, S., Helffrich, G.R., 1998. Detection of lower mantle scatterers northeast of Mariana subduction zone using short-period array data. J. Geophys. Res. 103, 4825–4838.

Katsura, T., Ito, E., 1989. The system $Mg_2SiO_4–Fe_2SiO_4$ at high-pressures and temperatures: precise determination of stabilities of olivine, modified spinel, and spinel. J. Geophys. Res. 94, 15663–15670.

Kawakatsu, H., Niu, F., 1994. Seismic evidence for a 920 km discontinuity in the mantle. Nature 371 (6495), 301–305.

Kennett, B.L.N., Engdahl, E.R., 1991. Travel times for global earthquake location and phase identification. Geophys. J. Int. 105, 429–465.

Kennett, B.L.N., Engdahl, E.R., Buland, R., 1995. Constraints on seismic velocities in the earth from traveltimes. Geophys. J. Int. 122, 108–124.

Kieffer, S.W., 1982. Thermodynamics and lattice-vibrations of minerals. Part 5. Applications to phase-equilibria, isotopic fractionation, and high-pressure thermodynamic properties. Rev. Geophy. 20, 827–849.

Kincaid, C., Olson, P., 1987. An experimental study of subduction and slab migration. J. Geophys. Res. 92, 13832–13840.

Kirby, S., 1987. Localised polymorphic phase transformations in high-pressure faults and applications to the physical mechanism of deep-earthquakes. J. Geophys. Res. 92, 13789–13800.

Kirby, S.H., Durham, W.B., Stern, L.A., 1991. Mantle phase changes and deep-earthquake faulting in subducted lithosphere. Science 252, 216–225 .

Kirby, S.H., Stein, S., Okal, E.A., Rubie, D.C., 1996. Metastable mantle phase transformations and deep-earthquakes in subducting oceanic lithosphere. Rev. Geophy. 34, 261–306.

Kohlstedt, D.L., Keppler, H., Rubie, D.C., 1996. Solubility of water in the α, β, γ phases of $(Mg, Fe)_2SiO_4$. Contrib. Mineral Petrol. 123, 345–357.

Lees, A.C., Bukowinski, M.S.T., Jeanloz, R., 1983. Reflection properties of phase transition and compositional change models of the 670 km discontinuity. J. Geophys. Res. 88 (10), 8145–8159.

Lundgren, P., Giardini, D., 1994. Isolated deep-earthquakes and the fate of subduction in the mantle. J. Geophys. Res. 99, 15833–15842.

Minear, J.W., Toksož, M.N., 1970. Thermal regime of a downgoing slab and new global tectonics. J. Geophys. Res. 75, 1397–1419.

Montagner, J.-P., Kennett, B.L.N., 1996. How to reconcile body-wave and normal-mode reference earth models. Geophys. J. Int. 125, 229–248.

Neal, S.L., Pavlis, G.L., 1999. Imaging P to S conversions with multichannel receiver functions. Geophys. Res. Lett. 26, 2581–2584.

Neele, F., 1996. Sharp 400 km discontinuity from short-period P reflections. Geophys. Res. Lett. 23, 419–422.

Neele, F., de Regt, H., 1997. Imaging upper-mantle discontinuity topography using underside-reflection data. Geophys. J. Int. 137, 91–106.

Neele, F., de Regt, H., VanDecar, J., 1997. Gross errors in upper-mantle discontinuity topography from underside-reflection data. Geophys. J. Int. 129, 194–207.

Niu, F., Kawakatsu, H., 1995. Direct evidence for the undulation of the 660 km discontinuity beneath Tonga: comparison of Japan and California array data. Geophys. Res. Lett. 22, 531–534.

Niu, F., Kawakatsu, H., 1996. Complex structure of mantle discontinuities at the tip of the subducting slab beneath northeast China —a preliminary investigation of broadband receiver functions. J. Phys. Earth 44, 701–711.

Niu, F., Kawakatsu, H., 1997. Depth variation of mid-mantle seismic discontinuity. Geophys. Res. Lett. 24, 429–432.

Paulssen, H., 1985. upper-mantle converted waves beneath NARS array. Geophys. Res. Lett. 12, 709–712.

Revenaugh, J., Jordan, T.H., 1991. Mantle layering from ScS reverberations. Part 2. The transition zone. J. Geophys. Res. 96, 19763–19780.

Richards, M.A., Wicks Jr., C.W., 1990. S-P conversion from the transition zone beneath Tonga and the nature of the 670 km discontinuity. Geophys. J. Int. 101, 1–35.

Ringwood, A.E., 1969. Phase transformations in the mantle. Earth Planet. Sci. Lett. 5, 401–412.

Ritsema, J., Hagerty, M., Lay, T., 1995. Comparison of broad-band and short-period seismic wave-form stacks —implications for upper-mantle discontinuity structure. Geophys. Res. Lett. 22/23, 3151–3154.

Rubie, D.C., 1984. The olivinespinel transformation and the rheology of subducting lithosphere. Nature 308, 505–508.

Rundle, J.B., Klein, W., 1995. New ideas about the physics of earthquakes, Rev. Geophys. (Suppl.) 283–286.

Sacks, I.S., Snoke, J.A., 1977. The use of converted phases to infer the depth of the lithosphere-asthenosphere boundary beneath South America. J. Geophys. Res. 82, 2011–2017.

Shearer, P., 1991. Constraints on upper-mantle discontinuities from observations of long-period reflected and converted phases. J. Geophys. Res. 96, 18147–18182.

Shearer, P., 1993. Global mapping of upper-mantle reflectors from long-period SS precursors. Geophys. J. Int. 115, 878–904.

Shearer, P.M., 1996. Transition zone velocity gradients and the 520 km discontinuity. J. Geophys. Res. 101, 3053–3066.

Shearer, P.M., Masters, T.G., and Masters, 1992. Global mapping of topography on the 660 km discontinuity. Nature 355, 791–796.

Shearer, P.M., Flanagan, M.P., Hedlin, M.A.H., 1999. Experiments in migration processing of SS precursor data to image upper-mantle discontinuity structure. J. Geophys. Res. 104, 7229–7242.

Shen, Y., Sheehan, A.F., Dueker, K.G., de Groot-Hedlin, C., Gilbert, H., 1998. Mantle discontinuity structure beneath the southern East Pacific rise from P to S converted phases. Science 280, 1232–1235.

Stein, C.A., Stein, S., 1992. A model for the global variation in oceanic depth and heat flow with lithospheric age. Nature 359, 123–129.

Sung, C.M., Burns, R.G., 1976. Kinetics of the olivine–spinel transition: implications to deep-focus earthquake genesis. Earth Planet. Sci. Lett. 32, 165–170.

Thirot, J.-L., Montagner, J.-P., Vinnik, L., 1998. upper-mantle seismic discontinuities in a subduction zone (Japan) investigated from P to S converted waves. Phys. Earth Planet. Interiors 108, 61–80.

Toksož, M.N., Sleep, N.H., Smith, A.T., 1973. Evolution of the downgoing lithosphere and the mechanisms of deep focus earthquakes. Geophys. J. R. Astronom. Soc. 35, 285–310.

van der Hilst, R., Seno, T., 1993. Effects of relative plate motion on the deep structure and penetration depth of slabs beneath the Izu-Bonin and Mariana island arcs. Earth Planet. Sci. Lett. 120, 395–407.

van der Hilst, R.D., Widiyantoro, S., Engdahl, E.R., 1997. Evidence for deep mantle circulation from global tomography. Nature 386, 578–584.

van der Lee, S., Paulssen, H., Nolet, G., 1994. Variability of P660s phases as a consequence of topography of the 660 km discontinuity. Phys. Earth Planet. Interiors 86, 147–164.

Vidale, J.E., Benz, H.M., 1992. Upper-mantle seismic discontinuities and the thermal structure of subduction zones. Nature 356, 678–683.

Weidner, D.J., Wang, Y., 1998. Chemical and Clapeyron induced buoyancy at the 660 km discontinuity. J. Geophys. Res. 103, 7431–7441.

Wicks, C.W., Richards, M.A., 1993. A detailed map of the 660 km discontinuity beneath the Izu-Bonin subduction zone. Science 261, 1424–1427.

Wood, B.J., 1990. Postspinel transformations and the width of the 670 km discontinuity: a comment on postspinel transformations in the system Mg_2SiO_4–Fe_2SiO_4 and some geophysical implications. In: Ito, E., Takahashi, E. (Eds.), J. Geophys. Res. 95, 12681–12685.

Wood, B.J., 1995. The effect of H_2O on the 410 km seismic discontinuity. Science 268, 74–76.

Wood, B.J., Rubie, D.C., 1996. The effect of alumina on phase-transformations at the 660 km discontinuity from Fe–Mg partitioning experiments. Science 273, 1522–1524.

Wood, B.J., Pawley, A.P., Frost, D.R., 1995. Water and carbon in the earth's mantle. Philos. Trans. R. Soc., London 354, 1495-1511.

Yamazaki, A., Hirahara, K., 1994. The thickness of upper-mantle discontinuities, as inferred from short-period J-array data. Geophys. Res. Lett. 21, 1811–1814.

Young, T.E., Green, H.W., Hofmeister, A.M., Walker, D., 1993. Infra-red spectroscopic investigation of hydroxyl in β-$(Mg, Fe)_2SiO_4$ and coexisting olivine: implications for mantle evolution and dynamics. Phys. Chem. Minerals 19, 409–422.

Zhao, D., Hasegawa, A., Kanamori, H., 1994. Deep structure of Japan subduction zone as derived from local, regional and teleseismic events. J. Geophys. Res. 99, 22313–22329.

ELSEVIER

Physics of the Earth and Planetary Interiors 127 (2001) 51–66

PHYSICS OF THE EARTH AND PLANETARY INTERIORS

www.elsevier.com/locate/pepi

Implications of slab mineralogy for subduction dynamics

Craig R. Bina*, Seth Stein, Frederic C. Marton[1], Emily M. Van Ark

Department of Geological Sciences, Northwestern University, Evanston, IL, USA

Received 30 March 2000; accepted 14 December 2000

Abstract

Phase relations among mantle minerals are perturbed by the thermal environment of subducting slabs, both under equilibrium and disequilibrium (metastable) conditions. Such perturbations yield anomalies not only in seismic velocities but also in density. The buoyancy forces arising from these density anomalies may exert several important effects. They contribute to the stress field within the slab, in a fashion consistent with observed patterns of seismicity. They may affect subduction rates, both by inducing time-dependent velocity changes under equilibrium conditions and by imposing velocity limits through a thermal feedback loop under disequilibrium conditions. They may affect slab morphology, possibly inhibiting penetration of slabs into the lower mantle and allowing temporary stagnation of deflected or detached slabs. Latent heat release from phase transitions under disequilibrium conditions in slabs can yield isobaric superheating, which may generate adiabatic shear instabilities capable of triggering deep seismicity.

Keywords: Slab mineralogy; Subduction; Morphology

1. Introduction

Subducting slabs sink into the mantle because they are negatively buoyant. Much of this negative buoyancy is thermal, arising from the temperature difference between cold slab and warm mantle material. Thus colder slabs, such as those of greater age and/or faster subduction rates, should possess greater negative thermal buoyancies (McKenzie, 1969; Elsasser, 1969; Minear and Toksöz, 1970; Forsyth and Uyeda, 1975; Becker et al., 1999). There are, however, additional important sources of slab buoyancy. Temperature contrasts between subducting slab and ambient mantle imply that the equilibrium depths of chemical reactions, such as subsolidus phase transformations between low- and high-pressure mineral phases, will differ between slab and mantle, and the consequent lateral juxtaposition of phase assemblages of differing densities must generate additional "petrological" buoyancy forces. Furthermore, the low temperatures within slabs also may give rise to disequilibrium effects due to kinetic hindrance of reactions, such as metastable persistence of low-pressure minerals into the stability fields of high-pressure phases, and the resulting spatial variations in mineralogy will also contribute to petrological buoyancy forces.

These buoyancy forces, both thermal and petrological, have important effects on the physical behavior of subducting slabs. The spatial distribution of buoyancy forces contributes to the stress field within slabs, and these stresses may be reflected in patterns of seismicity. Additionally, spatial and temporal variations in the net buoyancies of slabs may be reflected in varying subduction rates, through changes in the important "slab pull" driving forces for surface plate

* Corresponding author. Fax: +1-847-491-8060.
E-mail address: craig@earth.northwestern.edu (C.R. Bina).
[1] Present address: Geophysical Laboratory, Carnegie Institution of Washington, Washington, DC, USA.

0031-9201/01/$ – see front matter
PII: S0031-9201(01)00221-7

motions. Furthermore, these buoyancy forces may affect slab morphology, such as the lateral deflection of some slabs at the top of the lower mantle. Finally, the thermal perturbation of phase relations which gives rise to the petrological buoyancy forces may also influence the thermal evolution of slab material, through latent heats of reaction.

2. Phase relations

To study these effects, we begin by constructing a simple thermal model of a subduction zone. For the example shown in Fig. 1, a two-dimensional numerical model of the temperature distribution within a subduction zone was calculated on a 120×90 grid for lithosphere of 140-Ma age subducting at 8 cm per year at a dip angle of 60°, using a simple finite difference algorithm (Minear and Toksöz, 1970; Toksoz et al., 1973) that neglects both shear and radiogenic heating, with initial lithospheric temperatures derived from the plate model GDH1 (Stein and Stein, 1992). To construct a mineralogical model that corresponds to this thermal structure, we first supplement this thermal structure with simple representations of pressure and composition. Here, the variation of pressure with depth is obtained by radial integration of the reference density model PEMC (Dziewonski et al., 1975). For simplicity, the bulk chemical composition is assumed to be uniformly that of $(Mg_{0.9}Fe_{0.1})_2SiO_4$ mantle olivine (Ringwood, 1982), although more sophisticated models might take into account compositional layering within slabs (Helffrich et al., 1989). From such a map of temperature, pressure, and bulk composition at each point of interest, we can determine the stable mineral assemblages at each corresponding point. Mineral phase assemblages are determined by free energy minimization, using a simulated annealing algorithm (Bina, 1998a) and thermodynamic parameters (Fei et al., 1991) for olivine (α), its polymorphs wadsleyite (β) and ringwoodite (γ), and its disproportionation products magnesiowüstite (mw) and ferromagnesian silicate perovskite (pv).

Two resulting models are shown in Fig. 2: the equilibrium case, and a disequilibrium case in which α olivine persists metastably at temperatures below 1000 K (Sung and Burns, 1976a,b; Green and Burnley, 1989; Kirby et al., 1991; Rubie and Ross, 1994; Green and Houston, 1995; Kirby et al., 1996; Kerschhofer et al., 1996; Wang et al., 1997). While this latter representation of the limits of metastability as a simple isothermal cutoff is an oversimplification of the complex kinetics of the olivine transformations (Liu et al., 1998), it is sufficient to illustrate the first-order buoyancy effects. As expected, perturbation of equilibrium phase relations in the mantle (Turcotte and Schubert, 1971, 1972; Schubert et al., 1970, 1975; Bina, 1996) by the cold slab results in upward deflection (due to their positive Clapeyron slopes) of the $\alpha \rightarrow \alpha+\beta \rightarrow \beta$ and the $\beta \rightarrow \beta+\gamma \rightarrow \gamma$ transitions from their nominal depths of ~410 and ~530 km, respectively, as well as downward deflection (due to its negative Clapeyron slope) of the $\gamma \rightarrow \gamma + \mathrm{pv} + \mathrm{mw} \rightarrow \mathrm{pv} + \mathrm{mw}$ transition from its nominal depth of ~660 km (Navrotsky, 1980; Akaogi et al., 1989; Katsura and Ito, 1989; Ito and Takahashi,

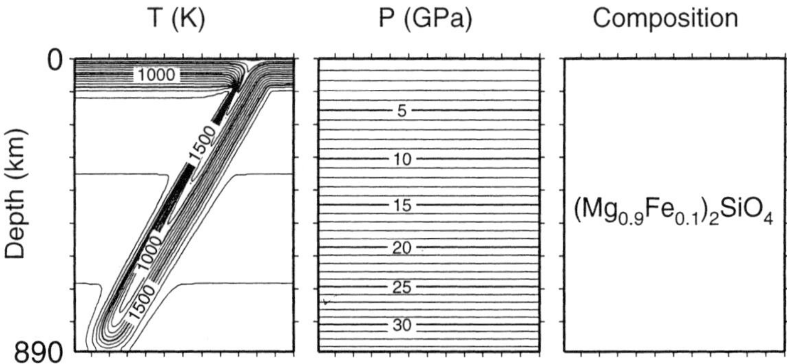

Fig. 1. Sample pressure, temperature, and compositional model for a simple model of a subducting slab (140-Ma lithosphere subducting at 8 cm per year at 60° dip) (modified from Bina (1998a)).

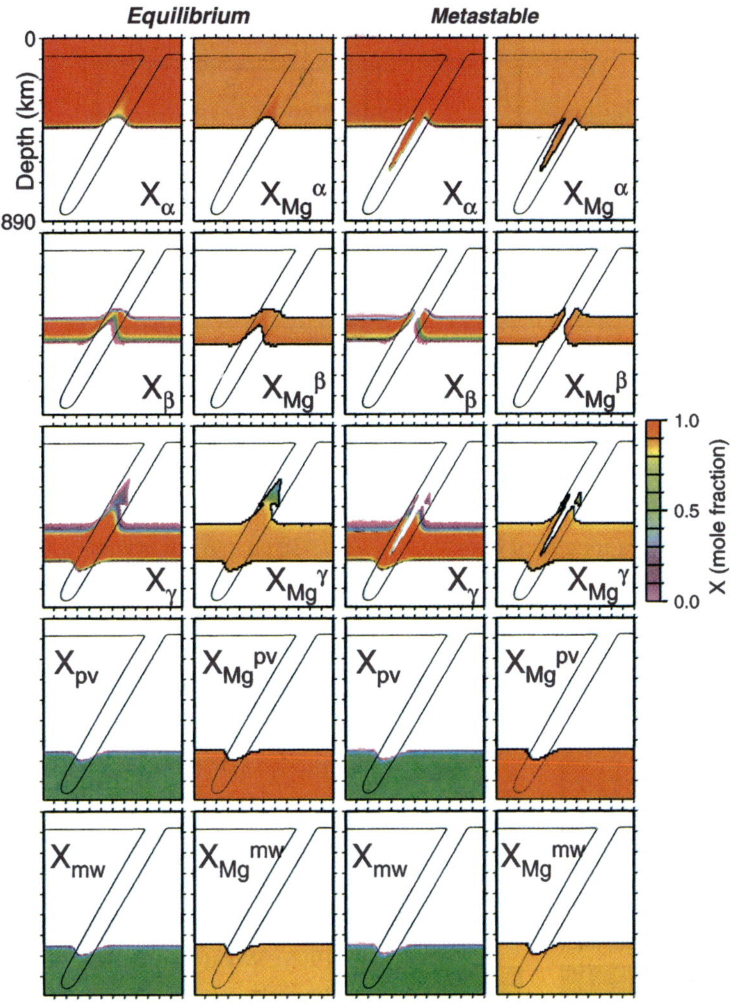

Fig. 2. Molar phase distribution (X_ϕ) and composition (X^ϕ_{Mg}) for equilibrium and disequilibrium (metastable) cases. Also shown is 1500 K isotherm (modified from Bina (1996, 1998a)).

1989; Ito et al., 1990; Akaogi and Ito, 1993; Bina and Helffrich, 1994; Green and Houston, 1995; Bina, 1996, 1997; Helffrich, 2000). However, solid-solution effects in the $(Mg_{0.9}Fe_{0.1})_2SiO_4$ bulk composition dictate that these are not simple deflections of univariant phase boundaries. In these divariant reactions, the relative proportions of the phases change continuously over narrow depth intervals, rather than discontinuously as would occur in a pure Mg_2SiO_4 composition. Furthermore, within the coldest core of the slab, the usual $\alpha \rightarrow \alpha+\beta \rightarrow \beta \rightarrow \beta+\gamma \rightarrow \gamma$ sequence of upper mantle transitions is replaced by the $\alpha \rightarrow \alpha+\gamma \rightarrow \beta+\gamma \rightarrow \gamma$ series (Green and Houston, 1995), in which Fe-rich γ becomes stable along with Fe-poor α at relatively shallow depths (Bina, 1996), prior to subsequent transformation of this $\alpha+\gamma$ to a $\beta+\gamma$ assemblage. These effects also can be seen in Fig. 3, which shows the non-linear variation (Helffrich and Bina, 1994; Stixrude, 1997) of phase proportions and compositions along down-dip profiles through the temperature minimum of the slab as well as a vertical profile through the surrounding mantle. The results for the disequilibrium case are broadly similar to those for the equilibrium

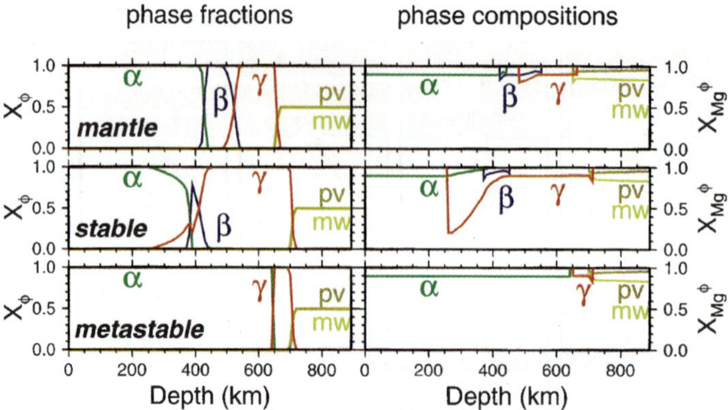

Fig. 3. Profiles of molar phase distribution (X_ϕ) and composition (X^{ϕ}_{Mg}) vs. depth, for ambient mantle and along slab temperature minimum for equilibrium (stable) and disequilibrium (metastable) cases (modified from Bina (1996)).

case, except that in the former a tongue of α extends metastably to great depths within the coldest core of the slab (Fig. 3). The portions of Figs. 2 and 3 which depict the phase proportions for these two cases are coarsely summarized in the upper panels of Fig. 4.

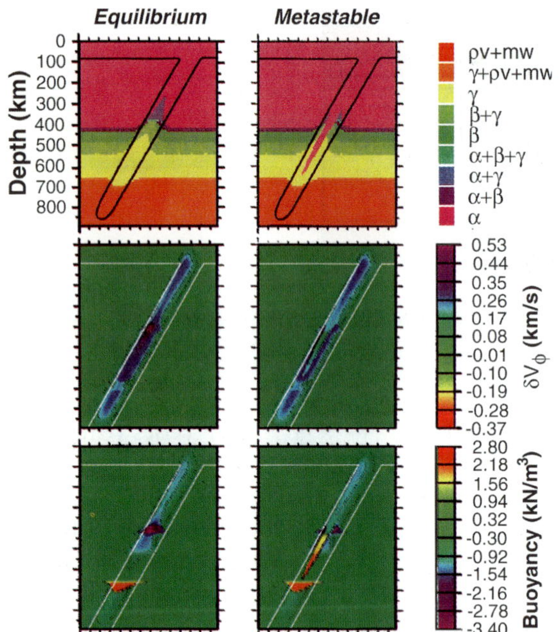

Fig. 4. Phase assemblages (top), bulk sound velocity anomalies (middle), and buoyancy anomalies (bottom) for equilibrium and disequilibrium (metastable) cases. Also shown is 1500 K isotherm (top) and slab boundary (middle and bottom). $V_\phi^2 = V_P^2 - (4/3)V_S^2$ (modified from Bina (1997, 1998a)).

Given our choice of a bulk composition of pure olivine stoichiometry, this simple analysis treats only reactions within $(Mg, Fe)_2SiO_4$ olivine, which forms perhaps 55–65% by volume of a peridotite mantle. The remaining $(Ca, Mg, Fe)SiO_3 \cdot Al_2O_3$ pyroxene (px) and garnet (gt) phases also undergo phase transformations, to silicate ilmenite (il), majoritic garnet (mjgt), and both calcic and ferromagnesian silicate perovskites (pv). However, such transitions as px + gt → px + mjgt → mjgt and mjgt → mjgt + pv → pv should be broader in depth-extent by about an order of magnitude than the transitions in olivine (Bina and Wood, 1984; Akaogi et al., 1987; Bina and Liu, 1995; Wood and Rubie, 1996; Vacher et al., 1998), and their contributions to local density anomalies and hence buoyancy forces should be considerably more diffuse. Indeed, the breadth of transitions involving garnet-bearing assemblages reflects their highly multivariant nature, so that in a natural multicomponent system they may not possess meaningful Clapeyron slopes (which are strictly defined only for univariant reactions).

3. Buoyancy anomalies

The lateral juxtaposition of high- and low-pressure phase assemblages resulting from this perturbation of mantle phase relations by the slab's thermal field implies anomalies in both seismic velocity and density relative to the surrounding mantle (Richter, 1973;

Schubert et al., 1975; Christensen and Yuen, 1984, 1985; Ito and Sato, 1992; Bina, 1996, 1997, 1998a; Yoshioka et al., 1997; Koper and Wiens, 2000), and the action of gravity upon these density anomalies gives rise to buoyancy anomalies (Fig. 4) which are superimposed on the negative thermal buoyancy of the slab (also reflected in overall fast seismic velocities). The upward equilibrium deflection of the $\alpha \rightarrow \beta \rightarrow \gamma$ phase transitions yields negative buoyancies (also reflected in locally faster velocity anomalies) at shallow depths, by stabilizing dense high-pressure phases in the cold slab that are surrounded by less dense lower-pressure phases in the ambient mantle. Furthermore, the downward equilibrium deflection of the $\gamma \rightarrow \mathrm{pv} + \mathrm{mw}$ transition yields positive buoyancies (also reflected in locally slower velocity anomalies) at greater depths, by allowing less dense phases in the cold slab to persist stably to depths where they are surrounded by denser higher-pressure phases in the mantle. On the other hand, in the disequilibrium case where α persists metastably beyond its stability field, the resulting tongue of metastable α implies positive buoyancy anomalies at depth, which should also be reflected in locally slow velocity anomalies.

It is worth noting that despite the attractive features of the disequilibrium metastability model, seismological studies have failed to clearly detect direct evidence of a metastable wedge (Iidaka and Obara, 1997; Helffrich, 1998; Koper et al., 1998; Collier and Helffrich, 1997, 2000). Although the low seismic-velocity wedge (Fig. 5) within the complex geometry of the high-velocity slab is likely to be an elusive target (years of study having been required to observe trapped seismic waves for the analogous case of low-velocity fault interiors), whose detection will be further complicated by additional non-olivine components (Hogrefe et al., 1994; Vacher et al., 1999), current seismological results are most easily explained by the absence of such a coherent wedge. On the other hand, recent failures to detect fast seismic velocity anomalies in a slab (using earthquakes outboard of a subduction zone) do suggest chemical or petrological counterweights to low slab temperatures (Brudzinski et al., 1997; Brudzinski and Chen, 2000), such as partial melt or volumes (either coherent or distributed) of metastable olivine.

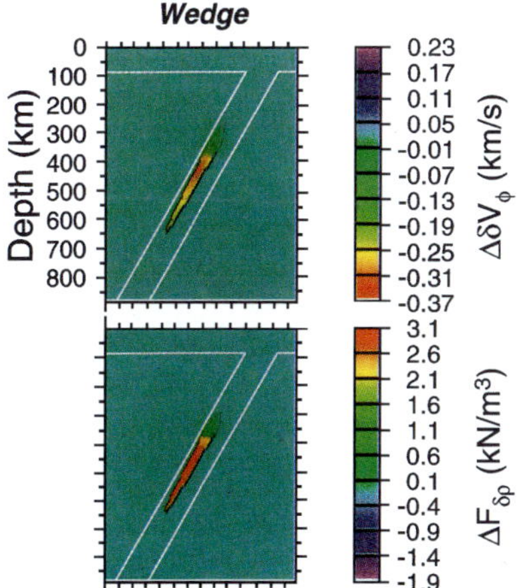

Fig. 5. Differential bulk sound velocity anomalies (top) and buoyancy anomalies (bottom) between disequilibrium and equilibrium cases, showing predicted effect of metastable olivine wedge. Also shown is slab boundary (modified from Bina (1998a)).

4. Stress fields

The spatial distribution of the buoyancy anomalies arising from the thermal perturbation of equilibrium phase relations (Fig. 4) suggests that material between the regions of deflected phase transitions may be compressed by the opposed buoyancy forces (Griggs, 1972; Goto et al., 1987; Ito and Sato, 1992; Bina, 1996, 1997; Yoshioka et al., 1997). We can calculate the principal stresses arising from these buoyancy forces, using a finite-element method (Gobat and Atkinson, 1995) with 4-node isoparametric elements for an elastic medium with spatially varying moduli. For this simplest case, we model both the slab and mantle as purely elastic media, albeit with elastic moduli which vary spatially due to variations in pressure, temperature, and phase assemblage. Solution for the displacements and stresses which minimize the strain energy at static equilibrium (Segerlind, 1984), yields significant down-dip compressional stresses in this region (Fig. 6). It is the opposition of these buoyancy forces through the depth interval of the transition zone, rather than any single buoyancy anomaly alone, which gives rise to this stress pattern in the slab (Bina, 1997).

The down-dip slab stresses change from tensional to compressional near ~400 km depth, with down-dip compressional stresses persisting to ~690 km, below which compressional stresses cease. This pattern is

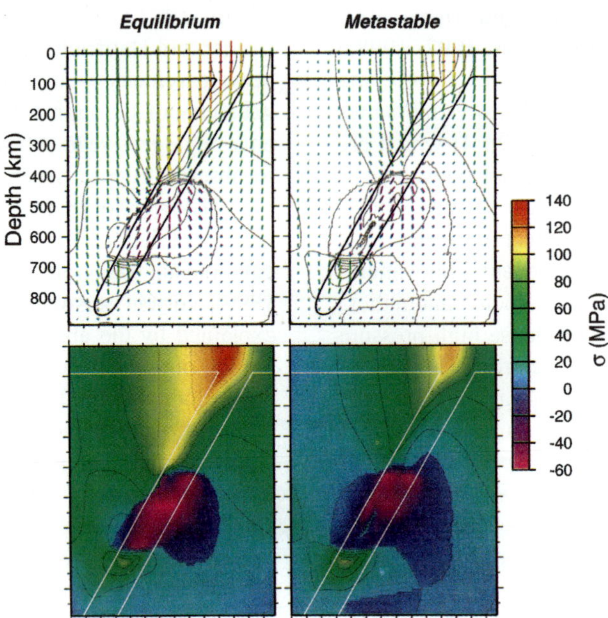

Fig. 6. Principal stresses, shown as vectors (top) and as maximum absolute magnitudes (bottom), calculated for an elastic rheology, for equilibrium and disequilibrium (metastable) cases. Negative stresses are compressional; positive stresses are extensional. Also shown is 1500 K isotherm (top) and slab boundary (bottom) (modified from Bina (1997)).

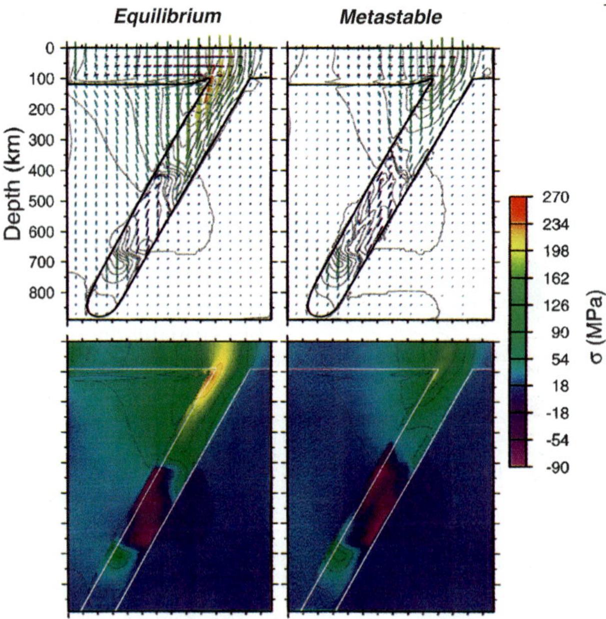

Fig. 8. Principal stresses, shown as vectors (top) and as maximum absolute magnitudes (bottom), calculated for an elastic slab within a Maxwell viscoelastic mantle ($t = 6$ ky), for equilibrium and disequilibrium (metastable) cases. Also shown is 1700 K isotherm (top), which delimits elastic–viscoelastic transition, and slab boundary (bottom).

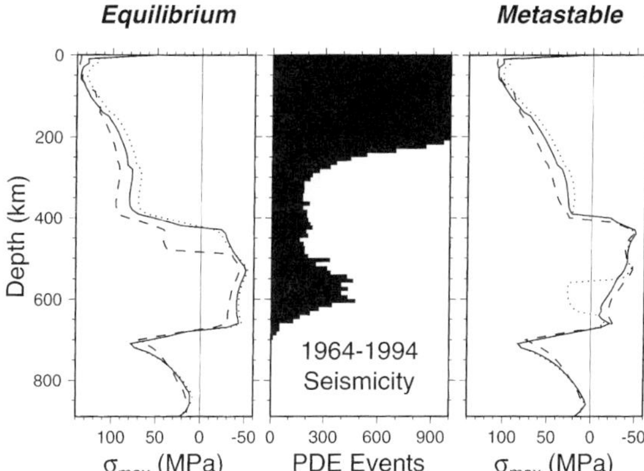

Fig. 7. Magnitude of the absolute maximum principal stress profiled down-dip along slab temperature minimum (solid), for equilibrium and disequilibrium (metastable) cases. Also shown are down-dip profiles along lines 25 km above (dashed) and below (dotted) the temperature minimum, measured normal to the slab. Center panel shows depth distribution of global seismicity for 1964–1994 from PDE catalogue (modified from Bina (1997)).

in good agreement with observed patterns of deep seismicity (Isacks and Molnar, 1971; Vassiliou et al., 1984; Rees and Okal, 1987; Vassiliou and Hager, 1988; Frohlich, 1989; Okal and Bina, 1998), for which down-dip compressional focal mechanisms are generally observed, as can be seen by comparing the depth distribution (NEIC, 1994) of global seismicity with down-dip profiles of maximum principal stress along the temperature minimum of the slab's interior (Fig. 7). Furthermore, the larger compressive stresses are of the same order as the larger stress drops estimated for deep earthquakes (Houston and Williams, 1991). Moreover, the transition from down-dip extensional to compressional stress near 400 depth corresponds roughly to the observed minimum in deep seismicity around 300–400 km depth, and the termination of down-dip compression near 700 km depth is in accord with the observed cessation of seismicity below that depth. The large extensional stresses in the wedge-slab corner and the large compressional stresses in the mantle outside the deep slab are boundary artifacts of our simple model of static stresses in a purely elastic medium, whose relative magnitudes will begin to decline when we introduce a more realistic rheology below.

Calculated stress patterns in this simple elastic model for the disequilibrium case of metastable persistence of α are similar, except that the presence of a positively buoyant tongue of α within the negatively buoyant interior of the slab results in a narrow interior region in which the absolute maximum principal stress is tensional and not oriented down-dip (Fig. 6). This effect, together with slightly rotated stresses of decreased compressional magnitude in the outer portions of the slab, shifts the compressive stress maximum toward somewhat shallower depths (Fig. 7).

In these simple stress models for purely elastic media, the region of large compressive stresses extends beyond the boundaries of the slab and into the mantle. However, the rheology of the cold slab may differ significantly from that of the warmer mantle, and a viscous mantle is unlikely to respond purely elastically (Karato, 1997). This issue may be addressed by incorporating a more complex rheology, such as viscoelasticity, into the finite-element model (Northwest, 1998). For example, we may take the same purely elastic slab but embed it in a viscoelastic mantle, retaining the same spatially varying elastic moduli. Here, the mantle exhibits Maxwell viscoelasticity in which the shear moduli relax with a viscosity of 10^{21} Pa s, and the cutoff between elastic and viscoelastic behavior is modeled as the 1700 K isotherm. In this case (Fig. 8), the down-dip compressive stresses become concentrated within the slab itself (Bina and Kirby, 1999). Furthermore, in the disequilibrium case the zone of maximum compressive stress divides into two regions lying above and below the metastable wedge. Some anomalous shallow wedge stresses and bleeding of compressional stresses into the mantle below the slab remain, as boundary artifacts of our simple static model, but they continue to decline with greater relaxation times (Ken Kumayama, personal communication). A more thorough analysis might incorporate a fully viscoelastic model in which both mantle and slab viscosities vary with both temperature and pressure, but ultimately a fully dynamical model involving mantle return flow would be necessary to resolve fine structure. Moreover, use of a variety of regionally appropriate thermal models would allow comparison of consequent stress patterns with seismicity profiles for various individual subduction zones (Helffrich and Brodholt, 1991), rather than simply with a global ensemble.

Other studies (Goto et al., 1987; Devaux et al., 1999) have suggested that transformational strain associated with heterogeneous volume changes along the boundaries of a metastable wedge would generate stresses much larger than those associated with thermal strain and buoyancy forces. However, these studies have not incorporated the effects of stress relaxation across phase changes (Karato, 1997; Riedel and Karato, 1997), nor have they included the $\gamma \rightarrow$ pv + mw transition whose buoyancy force plays an important role in generating large down-dip compressive stresses in the slab (Bina, 1997).

5. Subduction rates

The spatial distribution of buoyancy anomalies associated with the thermal perturbation of phase relations (Fig. 4) also suggests the potential for temporal variations in subduction rates (Bassett, 1979; Rubie, 1993; Kirby et al., 1996; Marton et al., 1999a; Schmeling et al., 1999). Shortly after initiation of subduction, a slab will be negatively buoyant due simply to negative thermal buoyancy. Some time thereafter, the sinking slab will reach the equilibrium depth of the first of the uplifted phase transitions, gaining additional negative buoyancy due to the petrological contribution, which should increase the slab's rate of descent. Eventually the slab will reach the equilibrium depth of the downwardly deflected transition, where it will acquire a positive petrological buoyancy contribution. Although the slab may remain negatively buoyant in the aggregate, the attendant decrease in the downward buoyancy force should decrease its rate of descent (Marton et al., 1999a; Schmeling et al., 1999).

This effect can be seen in Fig. 9, where the terminal velocity of our model slab (from Fig. 1) as a function of penetration depth is estimated by equating the down-dip component of the slab's aggregate (mean summed) buoyancy force with the total viscous drag force. The latter is determined for the two sides of the slab, given by the product of length and shear stress, and for a semicircular slab tip, given by the half the Stokes's law value (Batchelor, 1967), for a slab of fixed geometry sinking through a mantle of viscosity 3.7×10^{20} Pa s (Mitrovica and Peltier, 1993). The resulting variations in subduction rate are on the order of a few cm per year as the slab penetrates the transition zone (Marton et al., 1999a). While

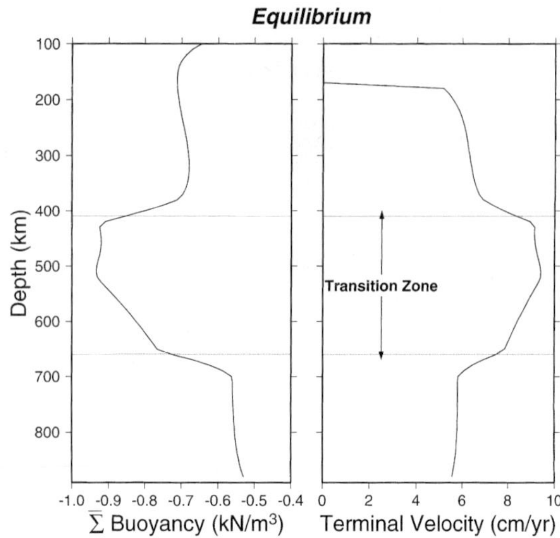

Fig. 9. Variation of slab's mean summed buoyancy force (left) and calculated terminal velocity (right) with slab penetration depth, for equilibrium case (modified from Marton et al. (1999a)).

most proposed mechanisms for changes in surface plate motions involve relatively gradual changes in velocity (Lithgow-Bertelloni and Richards, 1998), such time-dependent changes in bulk slab buoyancy, and hence subduction rate, with depth of penetration may provide a mechanism for more rapid changes. Note, however, that this mechanism generates changes in descent velocities only when a slab first encounters the transition zone, so that subduction occurs at a steady rate after the slab penetrates into the lower mantle.

In the disequilibrium case of metastable persistence of α within slabs, the positive buoyancy of a wedge of metastable α should act to further decrease subduction rates. Moreover, colder slabs, which should sink faster due to their greater negative thermal buoyancy, should have their descent rates slowed more by their larger metastable wedges (Bassett, 1979; Rubie, 1993; Kirby et al., 1996; Schmeling et al., 1999; Marton et al., 1999a). This "parachute effect", in which olivine metastability reduces subduction velocities for colder slabs, can be seen in Fig. 10. Here, we abandon our simple isothermal-cutoff model of metastability in favor of simple rate laws of transformation kinetics which are experimentally constrained (Rubie and Ross, 1994).

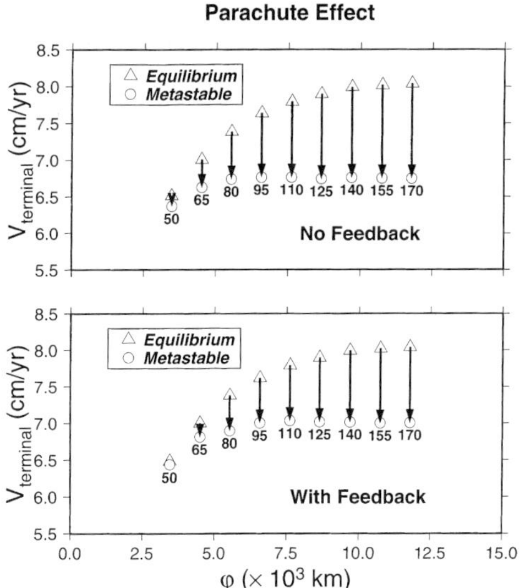

Fig. 10. Calculated terminal velocities, for equilibrium case (triangles) and disequilibrium (metastable) case (circles), showing the parachute effect (arrows) by which larger metastable wedges slow colder slabs to a greater extent. The thermal parameter (ϕ) is controlled by varying the age of lithosphere at the trench (numbers, in Ma). Magnitude of effect predicted by simple calculation (top) is damped by inclusion of feedback phenomenon (bottom) (modified from Marton et al. (1999a,b)).

The thermal state of the slab is represented simply by its "thermal parameter" (ϕ) — given by the product of the trench-normal convergence rate, the sine of the slab dip, and the age of the lithosphere entering the trench — so that larger thermal parameters correspond to colder (older and/or more rapidly subducting) slabs (Molnar et al., 1979; Gorbatov et al., 1996; Marton et al., 1999a). Slab thermal models are calculated using the same simple finite difference method (Minear and Toksöz, 1970). Colder slabs, which would sink faster in the equilibrium case, are slowed more than warmer slabs by their metastable olivine "parachutes" (Marton et al., 1999a; Schmeling et al., 1999). Moreover, such slowing, of the order of one cm per year, enters into a negative feedback loop. While colder slabs have larger metastable wedges and are slowed more due to the parachute effect, this deceleration exposes them to greater conductive warming which thermally erodes their wedges, so the resulting loss of positive buoyancy will accelerate their sinking somewhat. This feedback damps the amplitude of the parachute effect (Fig. 10), but the net effect is still to slow the rates of slab descent (Marton et al., 1999b; Tetzlaff and Schmeling, 2000). Such feedback damping may act to limit (Gordon, 1991) the allowable range of subduction rates (Marton et al., 1999a,b; Tetzlaff and Schmeling, 2000). However, it should be noted that the nature of such speed limits remains a matter of some uncertainty. The model shown here predicts that the velocities of slabs containing metastable olivine should approach a limiting value with increasing age of the slab (Marton et al., 1999a). Other, fully dynamical, models have suggested that these velocities should decrease (Schmeling et al., 1999) or continue to increase (Tetzlaff and Schmeling, 2000) with increasing age of the slab. Resolution of this issue awaits more sophisticated modeling and better understanding of rheological constraints (Tetzlaff and Schmeling, 2000).

6. Slab morphology

For extremely cold slabs, a metastable wedge may be so large as to render the slab positively buoyant in the aggregate for penetration below some critical depth. This can be seen in Fig. 11, in which a thermal model for the very cold Tonga slab (Bevis et al., 1995) yields a wedge of metastable α whose kinetically limiting 1000 K isotherm extends so deeply that the slab is positively buoyant if it penetrates directly into the lower mantle (Bina and Kirby, 1999). Thus, the metastable wedge imposes a "petrological buoyancy cost" negating the negative thermal buoyancy that drives subduction. These buoyancy forces should produce very large slab compressive stresses in the transition zone as well as large bending moments (Lundgren and Giardini, 1992, 1994; Lay, 1994; van der Hilst, 1995; Bina, 1996, 1997; Bina and Kirby, 1999) associated with upward deflection of the slab below 660 km. Equilibrium depression of the $\gamma \rightarrow \text{pv} + \text{mw}$ transition alone contributes a significant positive buoyancy increment to the slab at the base of the upper mantle (Fig. 11), but it is the large magnitude and extent of the positive buoyancy associated with a metastable wedge (the same buoyancy that limits subduction velocities through the parachute effect discussed above) which allows the

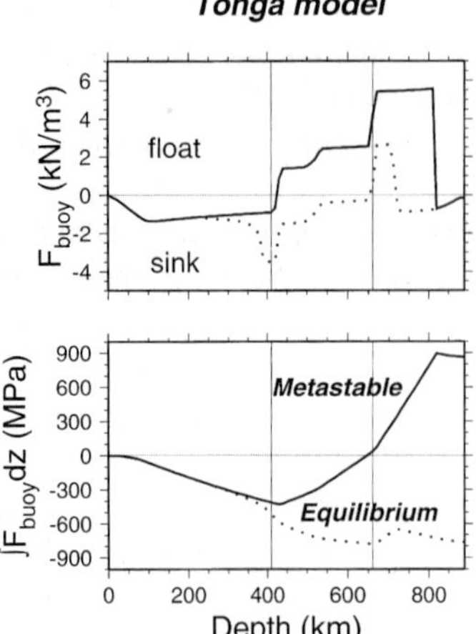

Fig. 11. Buoyancy force (top) and depth-integrated buoyancy force (bottom), profiled down-dip along slab temperature minimum, for a cold Tonga slab model (120 Ma lithosphere at 24 cm per year and 40°), for equilibrium and disequilibrium (metastable) cases. Vertical lines delimit the transition zone. Slab with large metastable wedge becomes positively buoyant upon entering the lower mantle (modified from Bina and Kirby (1999)).

net buoyancy of the slab to become neutral or even positive. The net effect of these buoyancy forces is to deflect the slab toward the horizontal at the base of the transition zone (Tetzlaff and Schmeling, 2000). Stagnation of slab material near the depth of neutral buoyancy should persist until thermal erosion of the positively buoyant metastable wedge by conductive warming allows the slab to resume sinking under the influence of its residual negative thermal buoyancy (Bina and Kirby, 1999; Tetzlaff and Schmeling, 2000).

Such increased resistance to slab descent into the lower mantle beneath Tonga is consistent with tomographic images and the distribution of deep seismicity, which together suggest the presence of recumbent slab material up to 750 km west of the base of the inclined Wadati–Benioff zone (van der Hilst, 1995; Zhao et al., 1997; Okal and Kirby, 1998; Deal and Nolet, 1999; Fukao et al., 2001), as well as with apparent depression of the $\gamma \rightarrow \text{pv} + \text{mw}$ transition to the west of the base of the inclined Wadati–Benioff zone (Roth and Wiens, 1999). It is also consistent with the observed (Po-Fei Chen, personal communication) unusual upward extension of the down-dip compressional stress regime of deep earthquakes to intermediate depths in Tonga. Interestingly, the strongest evidence for subhorizontal deflection of the Tonga slab derives from the northern end of the Tonga arc (Deal and Nolet, 1999). This appears to correspond to the region of highest subduction rates (Bevis et al., 1995) and thus presumably to the region of lowest slab temperatures and greatest potential for mineral metastability. Just as in the case of the parachute effect, a feedback loop may be involved, if slab deflection also acts to slow descent rates thereby raising slab temperatures and limiting the extent of metastability, but dynamical models suggest that the amount of such slowing should be small even for large subhorizontal deflections of the slab (Tetzlaff and Schmeling, 2000).

In addition to the case of deflected slabs, there is that of detached slabs. The deep seismicity beneath the North Fiji Basin, for example, is thought to occur in a slab of former Pacific plate that was subducted at the Vitiaz trench and then detached during a late Miocene arc reversal (Okal and Kirby, 1998). Thermokinetic modeling suggests that a metastable wedge of significant extent is unlikely to survive the thermal evolution of most such pieces of detached slab (Van Ark et al., 1998). In such models, combining slab thermal evolution (Marton et al., 1999a) with olivine transformation kinetics (Rubie and Ross, 1994), a cold wedge of metastable α olivine rapidly achieves a steady-state size. This initial size is characteristic of the thermal parameter of subduction, in that colder slabs develop larger wedges. Upon detachment of the lower reaches of the slab from its overlying portion, modeled as occurring in the uppermost mantle near the depth of the base of the overriding lithosphere, the metastable wedge remains of roughly constant size until the detachment surface enters the transition zone. Thereafter, enhanced thermal erosion effectively inverts the association of large metastable wedges with initially cold slabs. Because colder slabs sink faster, they spend more time surrounded by hotter material (at greater depths), and their metastable wedges thermally erode more rapidly. Thus, colder slabs form larger wedges, but upon slab detachment these large wedges decay faster. For a range of simple thermal models of Vitiaz plate geometry and evolution (Gordon and Jurdy, 1986; Hamburger and Isacks, 1987; Greene et al., 1994), calculated using the same finite difference method

(Minear and Toksöz, 1970) with Tonga-like parameters (100 Ma lithosphere at 14 cm per year and 60°), the large metastable wedge is reduced in size to a few tens of percent of its initial extent by ~10 My after detachment. Thus, any faulting nucleated within any metastable material persisting in the Vitiaz detached slab probably must propagate beyond the boundaries of the metastable region in order to account for the observed seismic activity (Van Ark et al., 1998).

7. Thermal effects

To the extent that disequilibrium conditions allow metastable persistence of lower pressure phases, the thermal evolution of slab material may also be affected. For example, latent heat release by exothermic phase transitions in subducting slab material is significantly enhanced (Fig. 12) under disequilibrium relative to equilibrium conditions (Daessler and Yuen, 1993, 1996; Rubie and Ross, 1994; Green and Houston, 1995; Green and Zhou, 1996; Kirby et al., 1996; Daessler et al., 1996; Devaux et al., 1997; Bina, 1998b). Hence, delayed transformation of metastable material may lead to local superheating, potentially accompanied by shear instability and seismicity (Griggs and Handin, 1960; Ogawa, 1987; Hobbs and Ord, 1988; Karato, 1997; Regenauer-Lieb and Yuen, 1998; Bina, 1998b; Regenauer-Lieb et al., 1999; Raterron et al., 1999). Furthermore, while recent attempts to

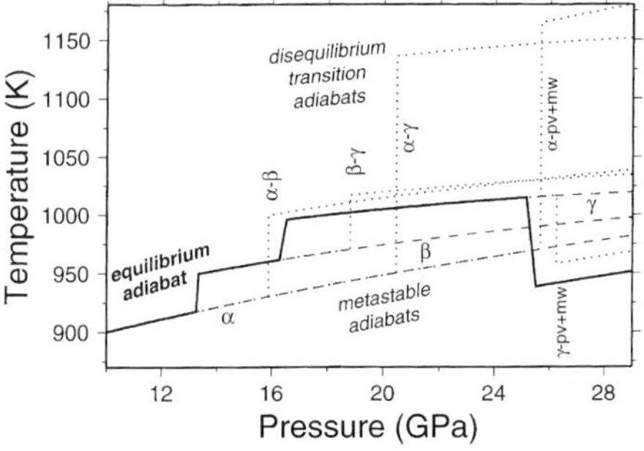

Fig. 12. For exothermic phase transitions in subducting slab material, latent heat release under disequilibrium conditions (dotted) from metastably persisting phases (dashed) is significantly enhanced relative to equilibrium conditions (solid). Adiabats calculated for transformations in forsterite (modified from Bina, 1998b).

incorporate spatially varying (temperature-dependent) thermal conductivity into slab models predict somewhat warmer slabs (Hauck et al., 1999), models employing temperature-dependent thermal diffusivity do predict enhanced development of such shear instability within the stability fields of the high-pressure olivine polymorphs in cold subducting slabs (Branlund et al., 2000).

In addition to perturbing subsolidus phase transformations, the low temperatures in subducting slabs may also affect dehydration reactions. Progressive dehydration of subducting slab material generally liberates water as a fluid phase which plays important roles in island-arc magmatism and intermediate-depth seismicity. In the coldest subduction zones, however, the stable phase of water may be solid ice VII rather than a fluid. For example, finite difference (Minear and Toksöz, 1970) thermal models for slabs possessing a variety of thermal parameters (Kirby et al., 1996) suggest that the Tonga slab may remain sufficiently cold for dehydration to occur within the ice VII stability field. The resulting shifts in reaction boundaries and relative buoyancies would significantly affect H_2O transport in subduction zones, altering patterns of seismicity, magmatism, and geochemistry (Bina and Navrotsky, 2000). The lower mobility of water in solid form could allow transport of greater amounts of H_2O below the nominal depths of dehydration reactions. Even in the absence of ice, however, H_2O can be transported by slabs deep within the mantle through the vehicle of dense hydrous magnesium silicate phases, stabilized by the low temperatures in slabs (Wood et al., 1996; Bose and Navrotsky, 1998).

8. Conclusions

Perturbation of mantle phase relations by the cold thermal environment of a subducting slab yields anomalies in the spatial distribution not only of seismic velocities but also of density, and these density anomalies give rise to significant buoyancy forces. These buoyancy forces manifest themselves in a number of ways. They contribute to the stress field within the slab in a manner reflected in observed patterns of the distribution of both moment release and focal mechanisms of deep seismicity. They may affect subduction rates, both by introducing time-dependent velocity changes as slabs encounter equilibrium phase boundaries and by limiting the subduction rates of cold slabs in which buoyant metastable mineral phases persist under disequilibrium conditions. These buoyancy forces also may induce changes in slab morphology, allowing the temporary stagnation of deflected or detached slabs. While equilibrium buoyancy forces should remain important even in recumbent or detached slabs, any metastably persisting material may be subject to rapid thermal erosion in such environments.

While latent heats of phase transitions in subducting slabs under equilibrium conditions are simply manifested by refraction of adiabats along phase boundaries, latent heat release from exothermic phase transformations under disequilibrium conditions results in isobaric superheating above the equilibrium adiabat. Indeed, even nominally endothermic transitions can yield heating under sufficient metastable overpressure (Bina, 1998b). Thus, eventual transformation of any pods of metastable material should lead to local superheating which, if coupled to a strongly temperature-dependent viscosity, may lead to thermal runaway by strain localization and viscous shear heating. Such adiabatic instability may allow seismogenesis to nucleate in metastable material yet propagate well beyond its confines. The occurrence of such adiabatic shear instability in lithosphere near the surface is supported by field observations of pseudotachylite (Obata and Karato, 1995), and it is also manifest in numerical models of lithospheric necking (Regenauer-Lieb and Yuen, 1998, 2000). The role of shear instabilities in subducting lithosphere, on the other hand, remains speculative (Ogawa, 1987; Hobbs and Ord, 1988; Bina, 1998b), but recent numerical models of such processes do suggest its plausibility at the low temperatures of slab interiors (Regenauer-Lieb and Yuen, 1998; Branlund et al., 2000), and recent experimental results do suggest a role for plastic deformation in stress instabilities observed in olivine at high-pressures (Raterron et al., 1999).

Acknowledgements

We thank Th. Becker, O. Čadek, P.-F. Chen, W.-P. Chen, H. Green, A. Hook, E. Okal, S. Karato, S. Kirby, K. Kumayama, L. Leffler, C. Parks, D. Rubie, R. Russo, H. Schmeling, N. Sleep, V. Steinbach, H. Weertman, M. Woods, D. Yuen, and S. Yoshioka for stimulating discussions of various portions this work. For helpful discussions of software packages, we are grateful to J. Gobat (FElt), R. Foerch (Zebulon), and P. Wessel (GMT). All figures were produced using the GMT graphics package (Wessel and Smith, 1995). R. van der Hilst provided thoughtful comments on the manuscript. CRB acknowledges the support of the US National Science Foundation (EAR-9158594, EAR-9316396, INT-9603234, EAR-9706152), Princeton University (CHiPR), and the University of Tokyo (ERI). We thank the Alfred–Wegener–Stiftung for support of the 1999 Conference on Processes and Consequences of Deep Subduction in Verbania-Pallanza, Italy.

References

Akaogi, M., Ito, E., 1993. Refinement of enthalpy measurement of $MgSiO_3$ perovskite and negative pressure–temperature slopes for perovskite-forming reactions. Geophys. Res. Lett. 20, 1839–1842.

Akaogi, M., Ito, E., Navrotsky, A., 1989. Olivine modified spinel–spinel transitions in the system Mg_2SiO_4–Fe_2SiO_4: calorimetric measurements, thermochemical calculation, and geophysical application. J. Geophys. Res. 94, 15671–15685.

Akaogi, M., Navrotsky, A., Yagi, T., Akimoto, S., 1987. Pyroxene-garnet transformation: thermochemistry and elasticity of garnet solid solutions, and application to a pyrolite mantle. In: Manghnani, M.H., Syono, Y. (Eds.), High-Pressure Research in Mineral Physics. American Geophysical Union, Washington, DC, pp. 251–260.

Bassett, W.A., 1979. The diamond cell and the nature of the Earth's mantle. Ann. Rev. Earth Planet. Sci. 7, 357–384.

Batchelor, G.K., 1967. An Introduction to Fluid Dynamics. Cambridge University, New York.

Becker, Th.W., Faccenna, C., O'Connell, R.J., Giardini, D., 1999. The development of slabs in the upper mantle: insight from numerical and laboratory experiments. J. Geophys. Res. 104, 15207–15226.

Bevis, M., Taylor, F.W., Schutz, B.E., Recy, J., Isacks, B.L., Helu, S., Singh, R., Kendrick, E., Stowell, J., Taylor, B., Calmant, S., 1995. Geodetic observations of very rapid convergence and back-arc extension at the Tonga arc. Nature 374, 249–251.

Bina, C.R., 1996. Phase transition buoyancy contributions to stresses in subducting lithosphere. Geophys. Res. Lett. 23, 3563–3566.

Bina, C.R., 1997. Patterns of deep seismicity reflect buoyancy stresses due to phase transitions. Geophys. Res. Lett. 24, 3301–3304.

Bina, C.R., 1998a. Free energy minimization by simulated annealing with applications to lithospheric slabs and mantle plumes. Pure Appl. Geophys. 151, 605–618.

Bina, C.R., 1998b. A note on latent heat release from disequilibrium phase transformations and deep seismogenesis. Earth Planets Space 50, 1029–1034.

Bina, C.R., Helffrich, G., 1994. Phase transition Clapeyron slopes and transition zone seismic discontinuity topography. J. Geophys. Res. 99, 15853–15860.

Bina, C.R., Kirby, S.H., 1999. Model estimates of the petrological buoyancy cost of descent into the transition zone and lower mantle of slabs containing metastable olivine: results for the Tonga subduction zone. Eos Trans. Am. Geophys. Union 80, Spring Suppl., S31B–09.

Bina, C.R., Liu, M., 1995. A note on the sensitivity of mantle convection models to composition-dependent phase relations. Geophys. Res. Lett. 22, 2565–2568.

Bina, C.R., Navrotsky, A., 2000. Possible presence of high-pressure ice in cold subducting slabs. Nature 408, 844–847.

Bina, C.R., Wood, B.J., 1984. The eclogite to garnetite transition: experimental and thermodynamic constraints. Geophys. Res. Lett. 11, 955–958.

Bose, K., Navrotsky, A. 1998. Thermochemistry and phase equilibria of hydrous phases in the system $MgO–SiO_2–H_2O$: implications for volatile transport to the mantle. J. Geophys. Res. 103, 9713-9719.

Branlund, J.M., Kameyama, M.C., Yuen, D.A., Kaneda, Y., 2000. Effects of temperature-dependent thermal diffusivity on shear instability in a viscoelastic zone: implications for faster ductile faulting and earthquakes in the spinel stability field. Earth Planet. Sci. Lett. 182, 171–185.

Brudzinski, M.R., Chen, W.-P., 2000. Variations of P-wave speeds and outboard earthquakes: evidence for a petrologic anomaly in the mantle transition zone. J. Geophys. Res. 105, 21661–21682.

Brudzinski, M.R., Chen, W.-P., Nowack, R.L., Huang, B.-S., 1997. Variations of P-wave speeds in the mantle transition zone beneath the Northern Philippine Sea. J. Geophys. Res. 102, 11815–11827.

Christensen, U.R., Yuen, D.A., 1984. The interaction of a subducting lithospheric slab with a chemical or phase boundary. J. Geophys. Res. 89, 4389–4402.

Christensen, U.R., Yuen, D.A., 1985. Layered convection induced by phase transitions. J. Geophys. Res. 90, 10291–10300.

Collier, J.D., Helffrich, G.R., 1997. Topography of the 410 and 660 km seismic discontinuities in the Izu–Bonin subduction zone. Geophys. Res. Lett. 24, 1535–1538.

Collier, J., Helffrich, G., 2000. Seismic discontinuities and subduction zones. In: Processes and Consequences of Deep Subduction, Rubie, D., van der Hilst, R.D. (Eds.), Phys. Earth Planet. Inter., this volume.

Daessler, R., Yuen, D.A., 1993. The effects of phase transition kinetics on subducting slabs. Geophys. Res. Lett. 20, 2603–2606.

Daessler, R., Yuen, D.A., 1996. The metastable olivine wedge in fast subducting slabs: constraints from thermo-kinetic coupling. Earth Planet. Sci. Lett. 137, 109–118.

Daessler, R., Yuen, D.A., Karato, S., Riedel, M.R., 1996. Two-dimensional thermo-kinetic model for the olivine-spinel phase transition in subducting slabs. Phys. Earth Planet. Int. 94, 217–239.

Deal, M.M, Nolet, G., 1999. Slab temperature and thickness from seismic tomography. Part 1. Method and application to Tonga. J. Geophys. Res. 104, 28789–28802.

Devaux, J.P., Schubert, G., Anderson, C., 1997. Formation of a metastable olivine wedge in a descending slab. J. Geophys. Res. 102, 24627–24637.

Devaux, J.P., Fleitout, L., Schubert, G., Anderson, C., 1999. Stresses in a subducting slab in the presence of a metastable olivine wedge. J. Geophys. Res. 105, 13365–13374.

Dziewonski, A.M., Hales, A.L., Lapwood, E.R., 1975. Parametrically simple Earth models consistent with geophysical data. Phys. Earth Planet. Int. 10, 12–48.

Elsasser, W.M., 1969. Convection and stress propagation in the upper mantle. In: Runcorn, S.K. (Ed.), The Application of Modern Physics to the Earth and Planetary Interiors. Wiley, New York, pp. 223–246.

Fei, Y., Mao, H.-K., Mysen, B.O., 1991. Experimental determination of element partitioning and calculation of phase relations in the $MgO–FeO–SiO_2$ system at high-pressure and high temperature. J. Geophys. Res. 96, 2157–2170.

Forsyth, D.W., Uyeda, S., 1975. On the relative importance of the driving forces of plate motion. Geophys. J.R. Astron. Soc. 43, 163–200.

Frohlich, C., 1989. The nature of deep-focus earthquakes. Ann. Rev. Earth Planet. Sci. 17, 227–254.

Fukao, Y., Widiyantoro, S., Obayashi, M., 2001. Stagnant slabs in the upper and lower mantle transition region. Rev. Geophys. 39, 291–323.

Gobat, J.I., Atkinson, D.C., 1995. The FElt System: User's Guide and Reference Manual, Computer Science Technical Report. University of California, San Diego, pp. CS94–376.

Gorbatov, A., Kostoglodov, V., Burov, E., 1996. Maximum seismic depth versus thermal parameters of subducted slab: application to deep earthquakes in Chile and Bolivia. Geofis. Int. 35, 41–50.

Gordon, R.G., 1991. Plate tectonic speed limits. Nature 349, 16–17.

Gordon, R.G., Jurdy, D.M., 1986. Cenozoic global plate motions. J. Geophys. Res. 91, 12389–12406.

Goto, K., Suzuki, Z., Hamaguchi, H., 1987. Stress distribution due to olivine-spinel phase transition in descending plate and deep focus earthquakes. J. Geophys. Res. 92, 13811–13820.

Green, H.W. II, Burnley, P., 1989. A new self-organizing mechanism for deep-focus earthquakes. Nature 338, 753.

Green, H.W. II, Houston, H., 1995. The mechanics of deep earthquakes. Ann. Rev. Earth Planet. Sci. 23, 169–213.

Green, H.W. II, Zhou, Y., 1996. Transformation-induced faulting requires an exothermic reaction and explains the cessation of earthquakes at the base of the mantle transition zone. Tectonophys. 256, 39–56.

Greene, H.G., Collot, J.Y., Fisher, M.A., Crawford, A.J., 1994. Neogene tectonic evolution of the New Hebrides island-arc: a review incorporating ODP drilling results. Proc. ODP Sci. Results 134, 19–46.

Griggs, D., 1972. The sinking lithosphere and the focal mechanism of deep earthquakes. In: Robertson, E.C. (Ed.), The Nature of the Solid Earth. McGraw-Hill, New York, pp. 361–384.

Griggs, D., Handin, J., 1960. Observations on fracture and a hypothesis of earthquakes. In: Griggs, D., Handin, J. (Eds.), Rock Deformation, Memoir 79. Geological Society America, New York, pp. 347–364.

Hamburger, M.W., Isacks, B.L., 1987. Deep earthquakes in the southwest Pacific: a tectonic interpretation. J. Geophys. Res. 92, 13841–13854.

Hauck, S.A., II, , Phillips, R.J., Hofmeister, A.M., 1999. Variable conductivity: effects on the thermal structure of subducting slabs. Geophys. Res. Lett. 26, 3257–3260.

Helffrich, G., 1998. Comment on 'seismological evidence for the existence of anisotropic zone in the metastable wedge inside the subducting Izu-Bonin slab' by T. Iidaka and K. Obara. Geophys. Res. Lett. 25, 3243–3244.

Helffrich, G., 2000. Topography of the transition zone seismic discontinuities. Rev. Geophys. 38, 141–158.

Helffrich, G., Bina, C.R., 1994. Frequency dependence of the visibility and depths of mantle seismic discontinuities. Geophys. Res. Lett. 21, 2613–2616.

Helffrich, G., Brodholt, J., 1991. Relationship of deep seismicity to the thermal structure of subducted lithosphere. Nature 353, 252–255.

Helffrich, G., Stein, S., Wood, B.J., 1989. Subduction zone thermal structure and mineralogy and their relationship to seismic wave reflections and conversions at the slab/mantle interface. J. Geophys. Res. 94, 753–763.

Hobbs, B.E., Ord, A., 1988. Plastic instabilities: implications for the origin of intermediate and deep focus earthquakes. J. Geophys. Res. 93, 10521–10540.

Hogrefe, A., Rubie, D.C., Sharp, T.G., Seifert, F., 1994. Metastability of enstatite in deep subducting lithosphere. Nature 372, 351–353.

Houston, H., Williams, Q., 1991. Fast rise times and the physical mechanism of deep earthquakes. Nature 352, 520–522.

Iidaka, T., Obara, K., 1997. Seismological evidence for the existence of anisotropic zone in the metastable wedge inside the subducting Izu–Bonin slab. Geophys. Res. Lett. 24, 3305–3308.

Isacks, B., Molnar, P., 1971. Distribution of stresses in the descending lithosphere from a global survey of focal-mechanism solutions of mantle earthquakes. Rev. Geophys. Space Phys. 9, 103–174.

Ito, E., Akaogi, M., Topor, L., Navrotsky, A., 1990. Negative pressure–temperature slopes for reactions forming $MgSiO_3$ perovskite from calorimetry. Science 249, 1275–1278.

Ito, E., Sato, H., 1992. Effect of phase transformations on the dynamics of the descending slab. In: Syono, Y., Manghnani, M.H. (Eds.), High-Pressure Research: Application to Earth and Planetary Sciences. American Geophysical Union, Washington, DC, pp. 257–262.

Ito, E., Takahashi, E., 1989. Post-spinel transformations in the system Mg_2SiO_4–Fe_2SiO_4 and some geophysical implications. J. Geophys. Res. 94, 10637–10646.

Karato, S.-I., 1997. Phase transformations and rheological properties of mantle minerals. In: Crossley, D.J. (Ed.), Earth's Deep Interior: The Doornbos Memorial Volume. Gordon and Breach, Amsterdam, pp. 223–272.

Katsura, T., Ito, E., 1989. The system Mg_2SiO_4–Fe_2SiO_4 at high-pressures and temperatures: precise determination of stabilities of olivine, modified spinel, and spinel. J. Geophys. Res. 94, 15663–15670.

Kerschhofer, L., Sharp, T.G., Rubie, D.C., 1996. Intracrystalline transformation of olivine to wadsleyite and ringwoodite under subduction zone conditions. Science 274, 79–81.

Kirby, S.H., Durham, W.B., Stern, L., 1991. Mantle phase changes and deep-earthquake faulting in subducting slabs. Science 252, 216–225.

Kirby, S.H., Stein, S., Okal, E.A., Rubie, D.C., 1996. Metastable mantle phase transformations and deep earthquakes in subducting oceanic lithosphere. Rev. Geophys. 34, 261–306.

Koper, K.D., Wiens, D.A., 2000. The waveguide effect of metastable olivine in slabs. Geophys. Res. Lett. 27, 581–584.

Koper, K.D., Wiens, D.A., Dorman, L.M., Hildebrand, J.A., Webb, S.C., 1998. Modeling the Tonga slab: can travel time data resolve a metastable olivine wedge? J. Geophys. Res. 103, 30079–30100.

Lay, T., 1994. The fate of subducting slabs. Ann. Rev. Earth Planet. Sci. 22, 33–61.

Lithgow-Bertelloni, C., Richards, M.A., 1998. The dynamics of Cenozoic and Mesozoic plate interiors. Rev. Geophys. 36, 27–78.

Liu, M., Kerschhofer, L., Mosenfelder, J.L., Rubie, D.C, 1998. The effect of strain energy on growth rates during the olivine-spinel transformation and implications for olivine metastability in subducting slabs. J. Geophys. Res. 103, 23897–23909.

Lundgren, P., Giardini, D., 1992. Seismicity, shear failure and modes of deformation in deep subduction zones. Phys. Earth Planet. Int. 74, 63–74.

Lundgren, P.R., Giardini, D., 1994. Isolated deep earthquakes and the fate of subduction in the mantle. J. Geophys. Res. 99, 15833–15842.

Marton, F.C., Bina, C.R., Stein, S., Rubie, D.C., 1999a. Effects of slab mineralogy on subduction rates. Geophys. Res. Lett. 26, 119–122.

Marton, F.C., Bina, C.R., Stein, S., Rubie, D.C., 1999b. Mineralogy and the regulation of subduction rates. Eos Trans. Am. Geophys. Union 80, Fall Suppl., T11G–08.

McKenzie, D.P., 1969. Speculations on the consequences and causes of plate motions. Geophys. J.R. Astrol. Soc. 18, 1–32.

Minear, J., Toksöz, M.N., 1970. Thermal regime of a downgoing slab and new global tectonics. J. Geophys. Res. 75, 1379–1419.

Mitrovica, J.X., Peltier, W.R., 1993. The inference of mantle viscosity from an inversion of the Fennoscandian relaxation spectrum. Geophys. J. Int. 114, 45–62.

Molnar, P., Freedman, D., Shih, J.S.F., 1979. Lengths of intermediate and deep seismic zones and temperatures in downgoing slabs of lithosphere. Geophys. J. Int. 6, 41–54.

NEIC, 1994. Catalogue of Preliminary Determinations of Epicenters (PDE), 1964–1994. National Earthquake Information Center, US Geological Survey, Denver.

Navrotsky, A., 1980. Lower mantle phase transitions may generally have negative pressure–temperature slopes. Geophys. Res. Lett. 7, 709–711.

Northwest Numerics, 1998. Zebulon Handbook, Version 7.2/8.0. Northwest Numerics and Modeling, Seattle.

Obata, M., Karato, S., 1995. Ultramafic pseudotachylite from the Balmuccia Peridotite, Ivrea-Verbano Zone, northern Italy. Tectonophysics 242, 313–328.

Ogawa, M., 1987. Shear instability in a viscoelastic material as the cause of deep focus earthquakes. J. Geophys. Res. 92, 13801–13810.

Okal, E.A., Bina, C.R., 1998. On the cessation of seismicity at the base of the transition zone. J. Seism. 2, 65–86. Erratum, J. Seism., 2, 377–381.

Okal, E.A., Kirby, S.H., 1998. Deep earthquakes beneath the Fiji Basin, SW Pacific: Earth's most intense deep seismicity in stagnant slabs. Phys. Earth Planet. Int. 109, 25–63.

Raterron, P., Wu, Y., Weidner, D.J., 1999. Olivine plastic instability as an alternative process for deep focus earthquakes? TEM investigation of high-pressure olivine samples exhibiting stress instabilities above 400°C. Eos Trans. Am. Geophys. Union 80, Fall Suppl., T22E–06.

Rees, B.A., Okal, E.A., 1987. The depth of the deepest historical earthquakes. Pure Appl. Geophys. 125, 699–715.

Regenauer-Lieb, K., Petit, J.P., Yuen, D.A., 1999. Adiabatic shear bands in the lithosphere: numerical and experimental approaches. Electron. Geosci. 4, 2.

Regenauer-Lieb, K., Yuen, D.A., 1998. Rapid conversion of elastic energy into plastic shear heating during incipient necking of the lithosphere. Geophys. Res. Lett. 25, 2737–2740.

Regenauer-Lieb, K., Yuen, D.A., 2000. Quasi-adiabatic instabilities associated with necking processes of an elasto–viscoplastic lithosphere. Phys. Earth Planet. Int. 118, 89–102.

Riedel, M.R., Karato, S., 1997. Grain-size evolution in subducted oceanic lithosphere associated with the olivine-spinel transformation and its effect on rheology. Earth Planet. Sci. Lett. 148, 27–43.

Richter, F.M., 1973. Finite amplitude convection through a phase boundary. Geophys. J.R. Astron. Soc. 35, 265–287.

Ringwood, A.E., 1982. Phase transformations and differentiation in subducted lithosphere: implications for mantle dynamics, basalt petrogenesis, and crustal evolution. J. Geol. 90, 611–643.

Roth, E.G., Wiens, D.A., 1999. Depression of the 660 km discontinuity beneath the Tonga slab determined from near-vertical ScS reverberations. Geophys. Res. Lett. 26, 1223–1226.

Rubie, D.C., 1993. Mechanisms and kinetics of solid state reconstructive phase transformations in the Earth's mantle. In: Luth, R.W. (Ed.), Short Course Handbook on Experiments at High-Pressure and Applications to the Earth's Mantle, Vol. 21. Mineral. Association Canada, Edmonton, pp. 247–303.

Rubie, D.C., Ross, C.R. II, 1994. Kinetics of the olivine-spinel transformation in subducting lithosphere: experimental constraints and implications for deep slab processes. Phys. Earth Planet. Int. 86, 223–241.

Schmeling, H., Monz, R., Rubie, D.C., 1999. The influence of olivine metastability on the dynamics of subduction. Earth Planet. Sci. Lett. 165, 55–66.

Schubert, G., Turcotte, D.L., Oxburgh, E.R., 1970. Phase change instability in the mantle. Science 169, 1075–1077.

Schubert, G., Yuen, D.A., Turcotte, D.L., 1975. Role of phase transitions in a dynamic mantle. Geophys. J.R. Astrol. Soc. 42, 705–735.

Segerlind, L.J., 1984. Applied Finite Element Analysis. Wiley, New York.

Stein, C.A., Stein, S., 1992. A model for the global variation in oceanic depth and heat flow with lithospheric age. Nature 359, 123–129.

Stixrude, L., 1997. Structure and sharpness of phase transitions and mantle discontinuities. J. Geophys. Res. 102, 14835–14852.

Sung, C.-M., Burns, R.G., 1976a. Kinetics of the high-pressure phase transformations: implications to the evolution of the olivine-spinel phase transition in the downgoing lithosphere and its consequences on the dynamics of the mantle. Tectonophysics 31, 1–32.

Sung, C.-M., Burns, R.G., 1976b. Kinetics of the olivine-spinel transition: implications to deep-focus earthquake genesis. Earth Planet. Sci. Lett. 32, 165–170.

Tetzlaff, M., Schmeling, H., 2000. The influence of olivine metastability on deep subduction of oceanic lithosphere. Phys. Earth Planet. Int. 120, 29–38.

Toksoz, M.N., Sleep, N.H., Smith, A.T., 1973. Evolution of the downgoing lithosphere and the mechanisms of deep focus earthquakes. Geophys. J.R. Astrol. Soc. 35, 285–310.

Turcotte, D.L., Schubert, G., 1971. Structure of the olivine-spinel phase boundary in the descending lithosphere. J. Geophys. Res. 76, 7980–7987.

Turcotte, D.L., Schubert, G., 1972. Correction. J. Geophys. Res. 77, 2146.

Vacher, P., Moquet, A., Sotin, C., 1998. Computation of seismic profiles from mineral physics: the importance of the non-olivine components for explaining the 660 km discontinuity. Phys. Earth Planet. Int. 106, 275–298.

Vacher, P., Spakman, W., Wortel, M.J.R., 1999. Numerical test on the seismic visibility of metastable minerals in subduction zones. Earth Planet. Sci. Lett. 170, 335–349.

Van Ark, E.M., Marton, F.C., Bina, C.R., Stein, S.A., Rubie, D., 1998. Survival of metastable olivine in detached slabs: difficult but not impossible. Eos Trans. Am. Geophys. Union 79, Spring Suppl., S164.

Vassiliou, M.S., Hager, B.H., 1988. Subduction zone earthquakes and stress in slabs. Pure Appl. Geophys. 128, 547–624.

Vassiliou, M.S., Hager, B.H., Raefsky, A., 1984. The distribution of earthquakes with depth and stress in in subducting slabs. J. Geodynam. 1, 11–28.

Wang, Y., Martinez, I., Guyot, F., Liebermann, R.C., 1997. The breakdown of olivine to perovskite and magnesiowüstite. Science 275, 510–513.

Wessel, P., Smith, W.H.F., 1995. New version of the generic mapping tools released. Eos Trans. Am. Geophys. Union 76, 329.

Wood, B.J., Pawley, A., Frost, D.R., 1996. Water and carbon in the Earth's mantle. Phil. Trans. R. Soc. Lond. A 354, 1495–1511.

Wood, B.J., Rubie, D.C., 1996. The effect of alumina on phase transformations at the 660-km discontinuity from Fe–Mg partitioning experiments. Science 273, 1522–1524.

Yoshioka, S., Daessler, R., Yuen, D.A., 1997. Stress fields associated with metastable phase transitions in descending slabs and deep-focus earthquakes. Phys. Earth Planet. Int. 104, 345–361.

van der Hilst, R.D., 1995. Complex morphology of subducted lithosphere in the mantle beneath the Tonga trench. Nature 374, 154–157.

Zhao, D., Xu, Y., Wiens, D.A., Dorman, L., Hildebrand, J., Webb, S., 1997. Depth extent of the Lau back-arc spreading center and its relationship to subduction processes. Science 278, 254–257.

ELSEVIER

Physics of the Earth and Planetary Interiors 127 (2001) 67–81

PHYSICS OF THE EARTH AND PLANETARY INTERIORS

www.elsevier.com/locate/pepi

Subduction zone rheology☆

Donald J. Weidner*, Jiuhua Chen, Yaqin Xu, Yujun Wu, Michael T. Vaughan, Li Li

CHiPR and Department of Geosciences, State University of New York, Stony Brook, NY 11794-2100, USA

Received 17 March 2000; accepted 5 November 2000

Abstract

Rheological flow laws can be obtained from studies using multi-anvil high-pressure systems with synchrotron-based piezometers and strain metrics. The high flux X-ray source provides minute-scale time resolution with accurate measurement of diffraction patterns and direct sample images. Measurements of length changes with an accuracy of one part in 10^4 are being developed and will provide a new generation of rheological tools. Flow laws derived from peak broadening agree well with literature data for corundum, spinel, and olivine.

Properties of several mantle phases are compared for the temperature and pressure regime appropriate to a subducting slab. Temperature dependence of these properties exhibits a strong, temperature insensitive low temperature region, a thermally softened region and a weak high temperature region. The middle of these could be related to the seismogenic zone of a subduction zone. The progression of the temperature for softening with mineral phase suggests that earthquakes deeper than 400 km correspond to higher temperatures than for olivine in the upper 400 km. Plastic instabilities are suggested by these data as the origin of deep earthquakes.

Keywords: Rheology; Mantle minerals; Subduction zone; Deep earthquakes; Strength; Synchrotron; Stress measurements; Olivine; Wadsleyite; Ringwoodite; Perovskite; Majorite

1. Introduction

Subduction zones are unique regions of the earth's mantle as evidenced by the large lateral variations in seismic velocity and high level of seismicity. Indeed the low temperatures expected in such regions provide the basis for the distinction from 'normal mantle'. The rheology of the materials in subduction zones dictates many of the dynamic processes that occur in this region. Deformation of the subducting slab must occur by means of plastic deformation. Earthquakes sample the rheology of the region. The material must be strong enough to hold significant stress over time, yet capable of catastrophic stress release in these earthquakes. While the instability may be associated with dehydration, phase transformation, or plastic shear instability, the rheology of the material significantly influences the process. Plastic processes in a low-temperature, high-stress, high-pressure environment for the mantle suite of minerals have not yet been studied owing to experimental limitations. New advances in synchrotron X-ray studies at high pressure offer in situ piezometers and strain metrics that allow the determination of flow laws in the pressure and temperature regime of a subducting slab for both low- and high-pressure phases. Here, we examine the current state of development of these tools. We demonstrate that the flow laws determined by these methods agree with those in the literature for a few standard materials and

☆ This is a Mineral Physics Institute publication # MPI-276.

* Corresponding author. Fax: +1-631-632-8140.
E-mail address: donald.weidner@sunysb.edu (D.J. Weidner).

0031-9201/01/$ – see front matter
PII: S0031-9201(01)00222-9

provide preliminary results for the major minerals of the mantle.

2. Experimental

In this study, we use a multi-anvil high-pressure apparatus to create pressures up to 20 GPa and deviatoric stress in the sample along with synchrotron generated X-rays to characterize the stress–strain field in the sample. Temperature is increased up to 1500°C during the experiment with special attention to the temperature range appropriate to the subducting slab. Yield is determined if the stress that is supported by the sample saturates during loading (i.e. no longer elastic) and decreases with heating as the strength is reduced.

It is convenient to distinguish two scales of deviatoric stress relative to the X-ray sampling volume. We use the term macro-scale stress to indicate the deviatoric stress field that is uniform over the sampling volume and micro-scale stress to indicate the stress field that is heterogeneous in this volume. Both stress fields can be exploited to yield rheology information, but will require different experimental protocols as they will have different manifestations in the diffracted X-ray signal and they have different expressions of the plastic strain field. Micro-scale stress broadens the diffraction peak with the resulting plastic strain occurring at the expense of elastic strain of the sample. In this regard, micro-scale deformation experiments are relaxation experiments in that the volume of the sample remains quite constant as the stress relaxes by converting the elastic strain to plastic strain. Samples recovered from different stages of yield have been examined with transmission electron microscopy to determine the deformation mechanism. Time resolution of the relaxation process has been quantified and the results interpreted in terms of flow laws. To date, the observations are in good agreement with existing data. In the case of macro-scale stress, a uniform deviatoric stress is generated over the entire sample by appropriate design of the cell assembly. The magnitude of stress can be determined by quantifying the strains recorded by the individual diffraction lines, while strain is monitored using direct images of the sample with X-ray shadowgraphs. Time resolution allows definition of the stress–strain rate relation.

2.1. High-pressure apparatus

We use two multi-anvil high-pressure toolings with a 250 ton press (SAM-85) on the superconducting beam line (X-17) of the National Synchrotron Light Source (NSLS) and at the GSECARS beamline at the Advanced Light Source. The DIA cubic anvil system is described by Weidner et al. (1992a,b), but we will describe a few of the salient features here. With this system, the force of a uniaxial ram is transmitted to six anvils that simultaneously advance into the solid pressure medium that is cubic in shape. Gaps between the anvils of about 0.5 mm afford space for X-rays to enter the sample chamber and diffracted X-rays to exit. The cross-section of the pressure medium is illustrated in Fig. 1. A cylindrical hole in the sample chamber contains a graphite furnace and the sample. The dimension of the sample volume is about 1 mm in diameter and 1.5 mm long. A thermocouple that passes through one of the anvil gaps monitors the temperature. A diffraction standard with a well-known equation of state is used to define the pressure. With 4 mm truncated, tapered anvils this system operates routinely to 10 GPa and on occasion to 13 GPa.

The second tooling, the T-cup, is a 6–8 style system using 10 mm cubes for the second stage (Vaughan et al., 1998). Pressures to 23 GPa have been achieved with tungsten carbide anvils. The general operation of both pressure cells is similar with comparable X-ray optics. Together they afford the possibility of exploring the rheological properties to pressures in excess of 20 GPa and temperatures up to 1500°C.

White X-radiation yields an energy dispersive analysis with energies between 15 and 100 keV. We typically use a 2θ angle of 5–10°. Data gathering times as short as 30 s from an X-ray beam of dimensions of $100\,\mu m \times 100\,\mu m$ provide robust X-ray patterns that can be analyzed for position and width. The accuracy of the lattice spacing determinations is often 0.00008 nm. This rapid mode of data acquisition enables analysis of the evolution of the stress–strain rate relationship during fast relaxation of the stress.

Monochromatic radiation at selectable energies up to 80 keV can be used with an imaging plate recording system or with a charge coupled device (CCD) recording system. Monochromatic radiation offers an increase peak-width resolution of about a factor of 4

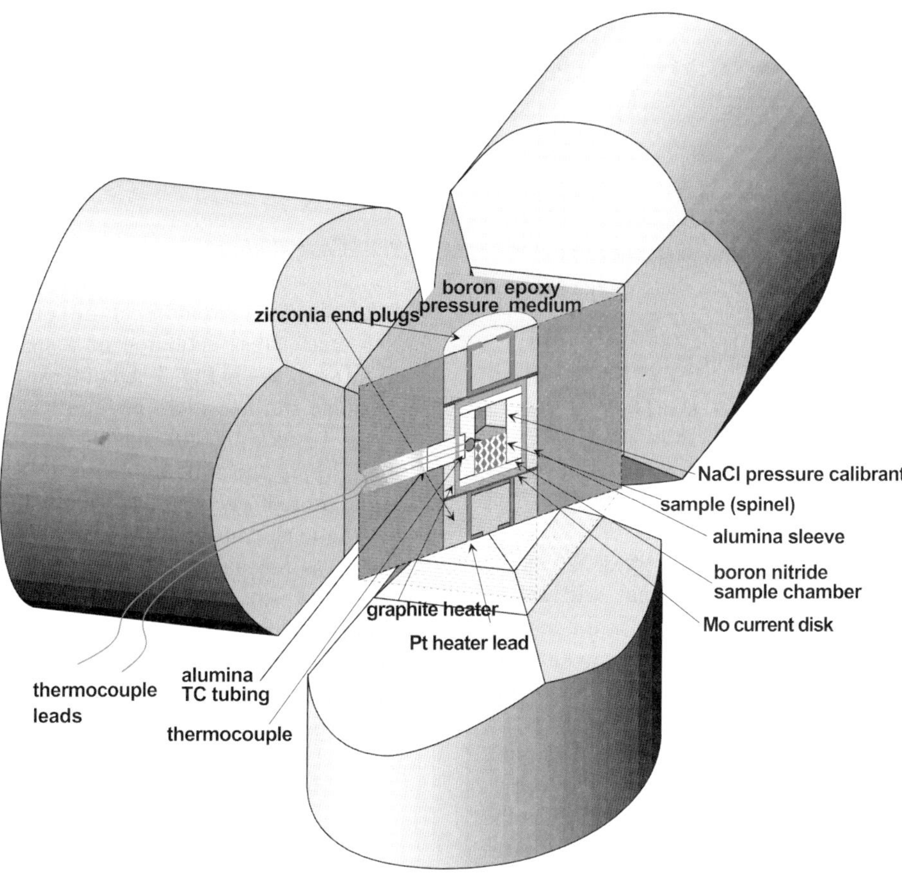

Fig. 1. Cross-section of pressure cell for cubic DIA apparatus.

over the white radiation, but at a cost in time resolution of about a factor of 6.

By analyzing the lattice parameters of samples at several vertical positions within the sample chamber, we have determined that the temperature variation over the entire sample length is of the order of 20 at 1200°C. Using multiple thermocouples, we find that the radial temperature gradient is less than 5°C at these conditions. We find no evidence of a vertical pressure gradient at room temperature.

2.2. *Micro-scale stress*

Deviatoric stress generally accompanies hydrostatic pressure in all high-pressure devices unless each individual grain is surrounded by a fluid medium that has no strength. This stress field can be uniform throughout the sample, with little or no variation from grain to grain, or it can be heterogeneous, varying in magnitude and orientation from point to point as long as mechanical equilibrium is satisfied. Homogeneous stresses arise from anisotropy of the elastic and plastic properties of the pressurizing medium or from the symmetry of the loading system. A heterogeneous stress field results from heterogeneities within the sample such as a two-phase mixture or randomly oriented elastically anisotropic grains. A sample comprised of loosely packed grains will experience deviatoric stresses as it is compressed owing to the void space between grains and the point contacts of the grains. These micro-scale stresses will occur even if the sample is externally loaded hydrostatically.

Our experimental technique using micro-scale stress for rheological measurements is detailed in

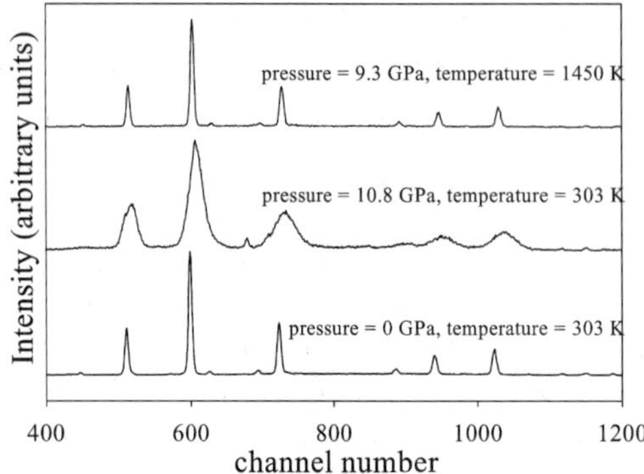

Fig. 2. Diffraction signal from spinel at ambient conditions, at room temperature and 10 GPa, and at 10 GPa and 1450 K. Note the evolution of peak width.

Weidner (1998) and Weidner et al. (1998). We model the manner that the diffraction signal will reflect this stress field. Each diffraction peak is the sum of diffraction from only a subset of grains within the specimen. These are the grains that have the particular orientation that aligns the specific set of lattice planes with the diffraction vector. The micro-scale deviatoric stress field will broaden the X-ray diffraction lines with the amount of line broadening determined by the distribution of longitudinal elastic strain parallel to the diffraction vector. Fig. 2 displays the diffraction signal from $MgAl_2O_4$, spinel at ambient conditions, after the sample was compressed to 10.8 GPa at room temperature, and after it was heated to 1450 K at this pressure. The amount of broadening of the peaks on compression reflects the magnitude of the elastically supported deviatoric stress. On heating, the sample weakens, the amount of deviatoric stress that can be supported reduces, and the peaks narrow. During this process, the total sample-strain remains constant, with elastic strain being relieved by plastic strain. Thus, the amount of peak broadening yields the magnitude of the deviatoric stress and the amount of narrowing with time or temperature reflects the plastic strain. Differentiating the plastic strain with time yields strain rate.

Weidner (1998) and Weidner et al. (1998) demonstrate the effects of sample grain size and elastic anisotropy on peak shapes and how these can be removed from the strain signal. Strain broadening results from stresses generated by forces at the surface of the grain as well as dislocations internal to the grains. Both sources of stress induce dislocation flow as evidenced by the fact that internal strains can be eliminated by annealing at high temperature in the absence of applied stress by dislocation mediated processes.

2.2.1. Case studies

2.2.1.1. Spinel. Weidner et al. (1998) report the micro-scale stress evolution of a spinel ($MgAl_2O_4$) sample as shown in Fig. 3. All of these data were taken in a single run in which pressure was first increased to over 10 GPa and the sample was then heated to 400°C where the time evolution of the peak width was monitored. Then the sample was heated to 600°C with continued monitoring of peak width. This procedure was repeated at 800 and 1100°C. On further heating, the peak width was entirely dominated by the instrumental peak width with strain broadening lost in the noise. The stress values in this figure represent the differential stress and are obtained by deconvolving the instrument response from the diffraction peak shape and multiplying by Young's modulus (Weidner et al., 1998). These relaxation experiments contain information about strain rate as well as stress because

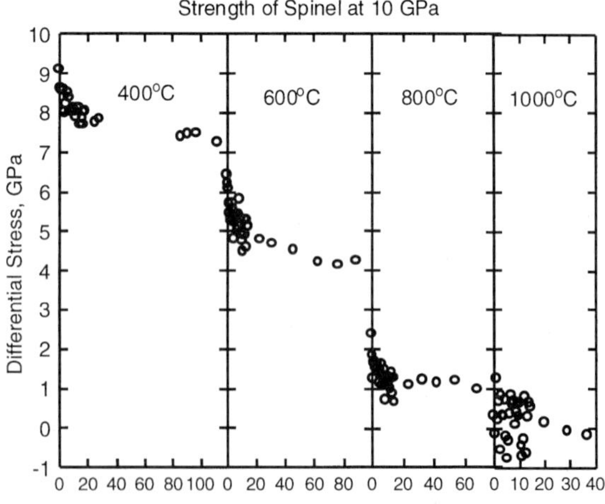

Fig. 3. Differential stress in spinel ($MgAl_2O_4$) as a function of time for a sample compressed to 10 GPa. The sample is heated in steps from room temperature to 1100°C. The heating time is about 1 s; data collection times are about 30 s.

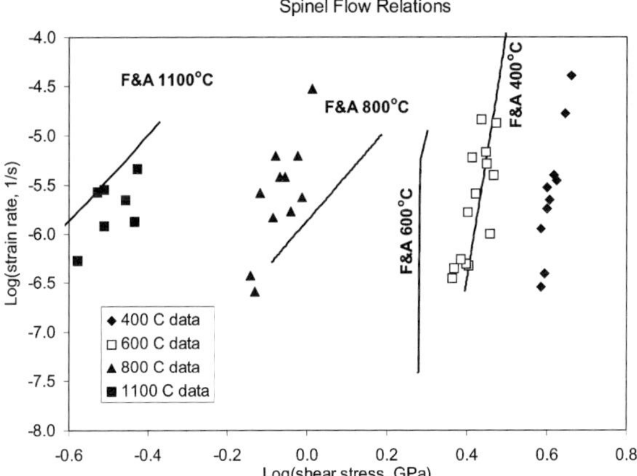

Fig. 4. Stress–strain rate relations for spinel. Observed data are shown by symbols and the experimental temperature is noted. Lines are from the model of Frost and Ashby (1982).

the total sample strain remains constant or

$$\left(\frac{\partial \varepsilon}{\partial t}\right)_{\text{tot}} = \left(\frac{\partial \varepsilon}{\partial t}\right)_{\text{plastic}} + \left(\frac{\partial \varepsilon}{\partial t}\right)_{\text{elastic}} = 0 \qquad (1)$$

Thus, the plastic sample strain rate is equal to the negative of the elastic sample strain rate. Relaxation data, by virtue of this equation, contain information on both strain rate and stress. Only elastic strain is directly observed in the experiment. Stress is deduced from the magnitude of elastic strain and strain rate is defined by the time variation of the elastic strain. For these spinel data, we fit a parametric relation to the elastic strain versus time ($\ln(\varepsilon) = a + b\ln(t)$), and differentiate to obtain strain rate. The resulting shear strain rate versus shear stress is illustrated in Fig. 4. The slopes of the lines yield values, n, which are the effective power for a power law stress strain rate model. This value is generally close to 3 in the region where dislocation recovery is operative and to 1 in the region where diffusion processes dominate. High values suggest that dislocation glide is the deformation mechanism.

The lines shown in Fig. 4 were obtained from the flow law model of Frost and Ashby (1982) for this temperature–stress–strain rate region. Their model has three mechanisms active in this temperature regime. At the lowest temperature and highest stress, their model includes a dislocation glide region where flow is limited by lattice resistance and governed by the rate equation

$$\dot{\varepsilon} = A\left(\frac{\sigma}{\mu}\right)^2 \exp -\left\{\frac{\Delta F}{kT}\left[1-\left(\frac{\sigma}{\tau}\right)^{3/4}\right]^{4/3}\right\} \qquad (2)$$

where σ is the shear stress, μ the shear modulus, ΔF the activation free energy, and τ is the flow stress at 0 K. According to their model, this relation dominates the flow at temperatures up to 600°C. At 600°C as stress reduces, flow becomes limited by obstacle resistance given by the rate equation

$$\dot{\varepsilon} = A\exp -\left\{\frac{\Delta F}{kT}\left[1-\left(\frac{\sigma}{\tau}\right)\right]\right\} \qquad (3)$$

At higher temperatures, the obstacles are overcome by recovery processes and the flow law transforms to a power law relation controlled by oxygen diffusion. They derived the parameters for the low temperature region from the hardness data of Westbrook (1966) after some corrections for cracking and for elastic distortion. The power law region is modeled on oxygen diffusion of Ando and Oishi (1974) and constrained by compression tests of Choi (1965). The agreement between our observations and the Frost and Ashby model in Fig. 4 is quite good. Indeed the agreement suggests that our observations, taken in a relaxation mode, reflect the steady state flow law. As with the model, 600–800°C appears to represent a transition from glide to power law creep as reflected by a change in slope of the curves and the increased temperature sensitivity of the flow law. Indeed, the new data can be used to refine the flow parameters used in these models. A change of (τ/μ) from 0.02 to 0.15 for the obstacle parameter and from 0.085 to 0.12 for the lattice resistance brings the model flow law closer to the data as shown in Fig. 5. Such changes are well within the latitude of the data used to constrain the Frost and Ashby (1982) model.

2.2.1.2. Corundum. Fig. 6 illustrates the stress–strain rate relations observed for corundum at 200, 600, 700, and 1200°C from several relaxation experiments in which the temperature was held constant and the time dependence of stress and strain were monitored. These data require a small temperature dependence of strength for temperatures less than

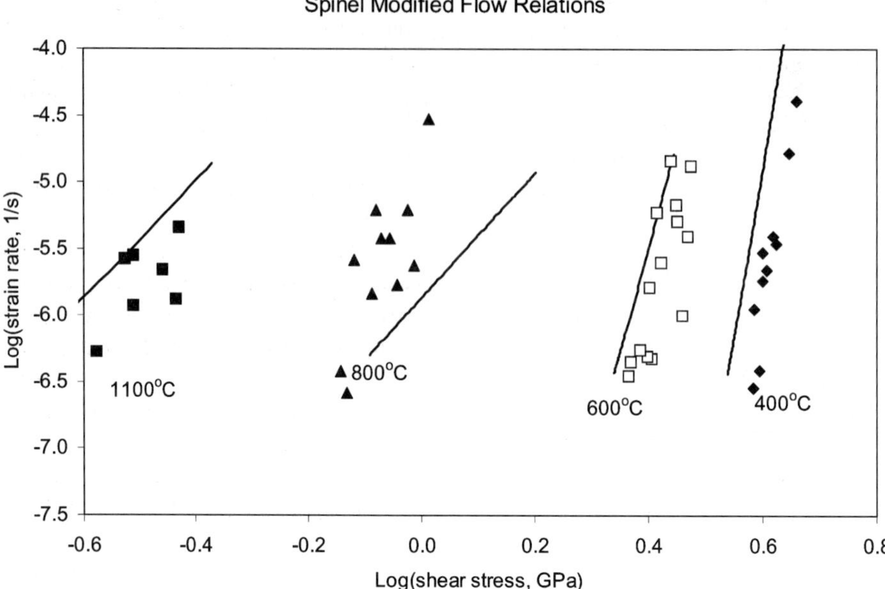

Fig. 5. Stress–strain rate relations for spinel. Observed data are shown by symbols and the experimental temperature is noted. Lines are from the model modified from that of Frost and Ashby (1982) by changing (τ/μ) from 0.02 to 0.15 for the obstacle stress parameter and from 0.085 to 0.12 for the lattice resistance stress parameter.

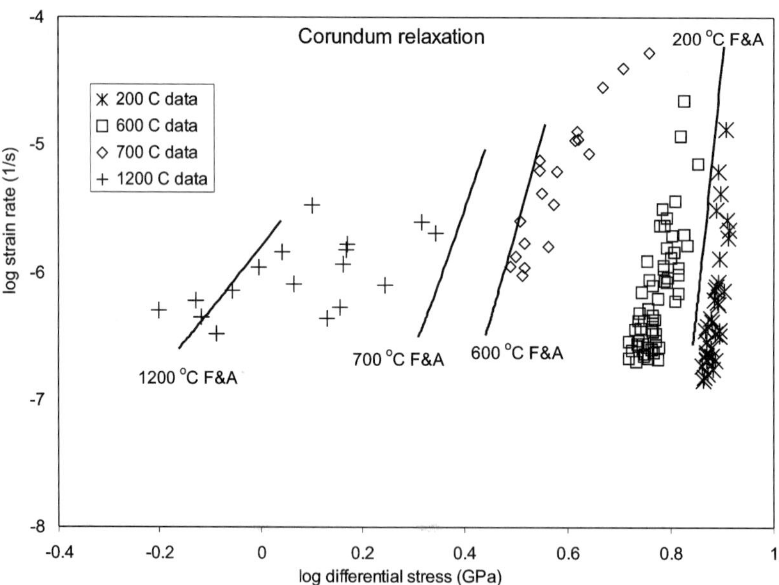

Fig. 6. Stress–strain rate relations for corundum. Observed data are shown by symbols and the experimental temperature is noted. Lines are from the model of Frost and Ashby (1982).

600°C, but a rapid evolution of strength at higher temperatures. The lines in this figure represent the model of Frost and Ashby (1982) for corundum. Their low temperature curves (700°C and less) are based on a lattice resistance controlled glide relation and their high temperature curve is for a power law creep relation controlled by oxygen and aluminum diffusion. In the region where their model has the strongest data control, namely from low temperature indentation experiments and high temperature creep experiments, the model is in excellent agreement with our results. Furthermore, the slopes of the curves are in good agreement with the data, thus, yielding similar stress dependence of the strain rate. The differences between model and data in the dislocation glide region may suggest slight variations in the model parameters. Again the strong agreement between the model and our data suggest that the data are providing a excellent sampling of the flow laws.

2.2.1.3. Olivine. Despite the abundance of studies of the rheology of olivine, there are relatively few in the low temperature regime. Phakey et al. (1972) obtained stress–strain curves for single crystal forsterite at 600 and 800°C under 1 GPa confining pressure. Their test at 800°C with different crystallographic orientations gave yield stresses that ranged from 0.57 to 1.3 GPa. Goetze (1978) and Evans and Goetze (1979) measured indentation hardness to study the strength of olivine up to 800°C. Meade and Jeanloz (1990) studied the room temperature strength of olivine up to 30 GPa with the diamond anvil cell. We have studied olivine in a manner similar to that described above for corundum and spinel. Our room temperature data are in excellent agreement with these previous workers. Fig. 7 presents the differential stress calculated from line broadening as a function of loading pressure for San Carlos olivine. The stress increases rapidly with pressure at low pressure, presumably indicating elastic loading, until it saturates at about 5.5 GPa. With further loading, it increases only moderately, indicating a saturation of stress and hence plastic flow. Room temperature results from Evans and Goetze (1979) and Meade and Jeanloz (1990) are also included in the figure. Their data suggest a slight increase in strength with increasing pressure and are completely consistent with our measurements. Strength as a function of temperature from indentation experiments of Evans and

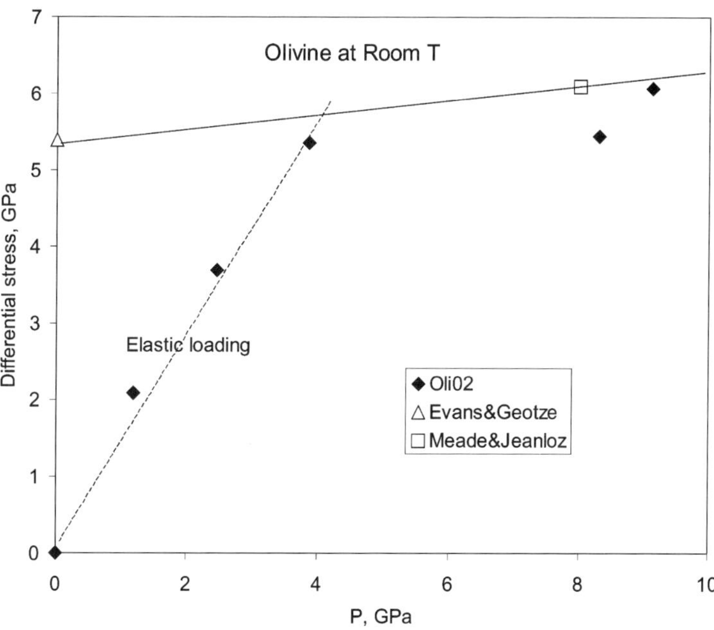

Fig. 7. Differential stress in San Carlos olivine as a function of pressure. Data from this study are compared with that of Evans and Goetze (1979) at room pressure and Meade and Jeanloz (1990) at elevated pressure.

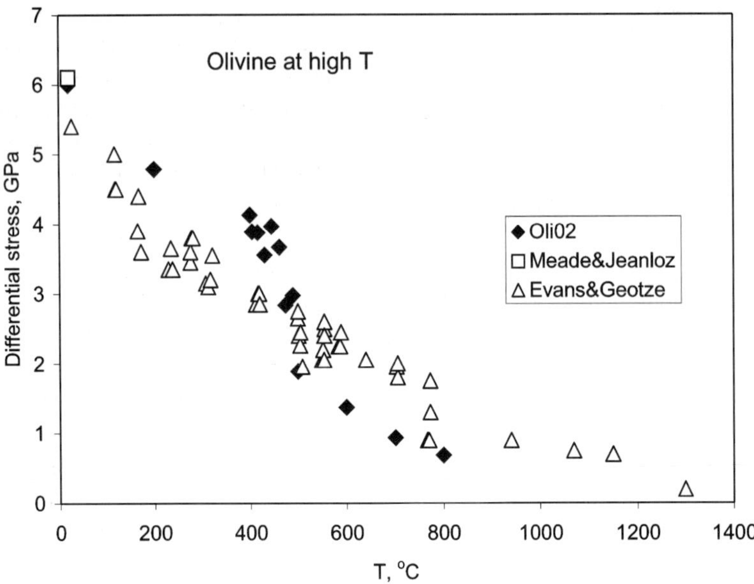

Fig. 8. Differential stress as a function of temperature. Data correspond to the stress after 1000 s and are compared with data of Evans and Goetze (1979) that was collected at room pressure using indentation techniques.

Goetze (1979) is compared with our data in Fig. 8. Our data correspond to strength at temperature measured about 1000 s after relaxation at that temperature has begun. The sample will usually have a strain rate of about $10^{-7}\,s^{-1}$ at this time. Our stresses are slightly larger than that of Evans and Goetze (1979) in the low temperature regime ($T < 500°C$) as expected from the pressure effect indicated in Fig. 7. At higher temperatures, our strength is lower than their values, suggesting a change in the flow mechanism. Nonetheless, it is clear that the values of stress as well as flow laws that are measured with this technique are consistent with measured values that come from more traditional techniques. The ability to resolve time in the low temperature range of these measurements adds a significant dimension to results provided by standard indentation experiments.

2.3. Macro-scale stress

It is possible to generate non-hydrostatic stress that is uniform over the dimension of the X-ray scattering volume, by introducing hard end plugs above and below the sample chamber. The anisotropy of the cell is thus, responsible for causing deviatoric stress in the sample. For the X-ray diffraction vector parallel to the cylinder axis (as in the DIA apparatus), a stress field with a cylindrical symmetry does not broaden the diffraction peaks, but displaces the peaks by differing amounts. The displacement of each peak depends on the magnitude of the stress field and the elastic moduli of the sample (see discussions by Kinsland and Bassett, 1977; Singh and Kennedy, 1974; Singh, 1993; Singh and Balasingh, 1994; Weidner et al., 1992b, 1994, 1998; Weidner, 1998).

In order to measure rheological properties, strain metrics are needed to monitor plastic strain of the sample as a function of time. Monitoring of sample length or strain markers such as that used by Karato and Rubie (1997) hold a possible application for in situ, time-dependent stress–strain determinations. We have conducted a study on an MgO sample using X-ray shadowgraphs that demonstrate the feasibility of doing this during a high pressure–high temperature experiment. The cell was constructed to have a very hard axial direction, with the sample sandwiched between two corundum pistons. A thin gold foil was placed at each end of the sintered MgO sample. X-rays that bathe the sample were converted to visible light by a fluorescence screen down-stream of the sample. This

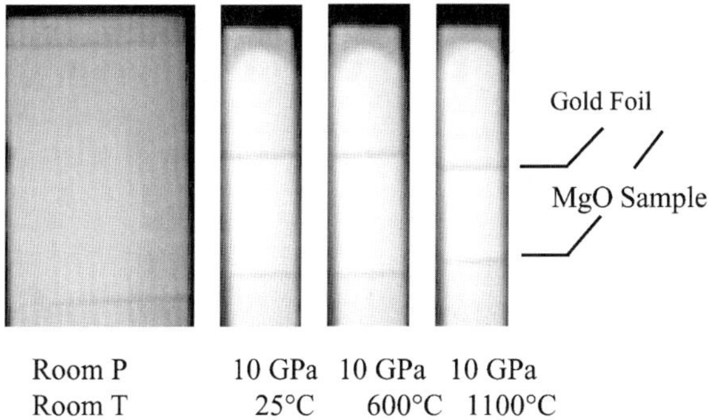

Fig. 9. X-ray shadowgraph of MgO sample at different P–T–time conditions. Horizontal lines are due to gold foils placed at top and bottom of sample. The width of the shadowgraph is defined by the opening between anvils.

image was magnified and recorded on videotape. A typical image is displayed in Fig. 9. The two dark lines are from the gold foil at the top and bottom of the sample while the width of the illuminated region is determined by the size of the gap between the anvils. Captured digital images were analyzed to determine length, strain, and strain rate. Length as a function of time is illustrated in Fig. 10 in units of pixels, where one pixel is about 4 μm. The regions with no data were periods when the X-rays were used for diffraction. The sample was compressed during the first 150 min of the experiment. During this time, the length was reduced to half of its initial value. At the final pressure, the sample was heated to 600°C at the 210 min point and held until the 300 min point, when it was further heated to 1100°C. Both heating cycles evidenced a further shortening even though no further loading of the press occurred. Thus, differential stress was maintained in the sample and assembly throughout the experiment. Fig. 11 shows the strain rate after the temperature increases from 600 to 1100°C. In this experiment, we estimate an uncertainty in strain of 0.001. Recent improvements in the camera system have improved the resolution to 0.0001 for samples of 1 mm length.

The X-ray diffraction pattern has the possibility to determine the time dependence of many aspects of the experiment. For example, peak heights are sensitive to any preferred orientations that develop within the sample. For the MgO sample reported here, a strong lattice preferred orientation developed during cold compression as shown by the X-ray intensities of Fig. 12. The top and bottom of the sample demonstrate unusually small intensities in the (2 0 0) and (4 0 0) peaks compared to the (1 1 1), (2 2 0), and (2 2 2) peaks, while in the center, the reverse is true. This suggests that dislocation flow was responsible for a significant amount of the shortening as the equant-shaped grains should not develop a fabric if the flow occurred by rotation. Furthermore, the orientations of the finite strain ellipses and hence the flow lines vary with position in the sample. Once in place, the fabric remained throughout the experiment.

The magnitude of the deviatoric stress may be determined from the elastic strain recorded by individual

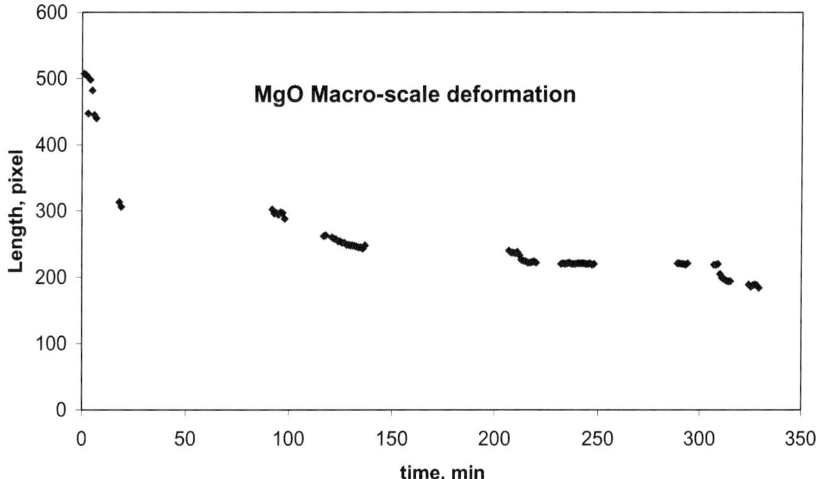

Fig. 10. Length of MgO sample as a function of time measured in pixels (~4 μm) of digital image.

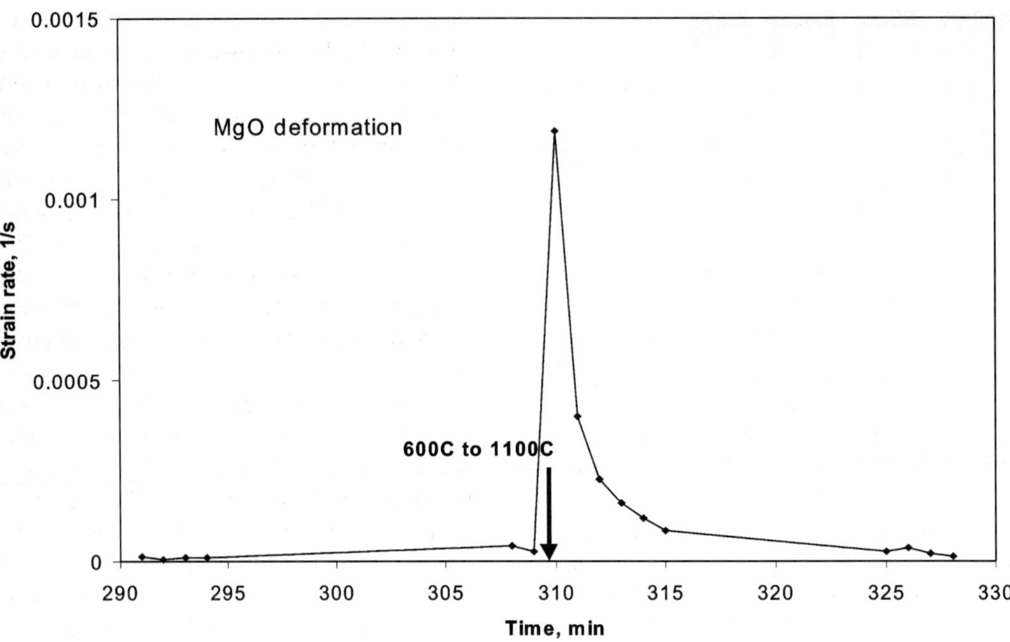

Fig. 11. Strain rate as a function of time of MgO sample determined from digital images as the temperature is increased from 600 to 1100°C.

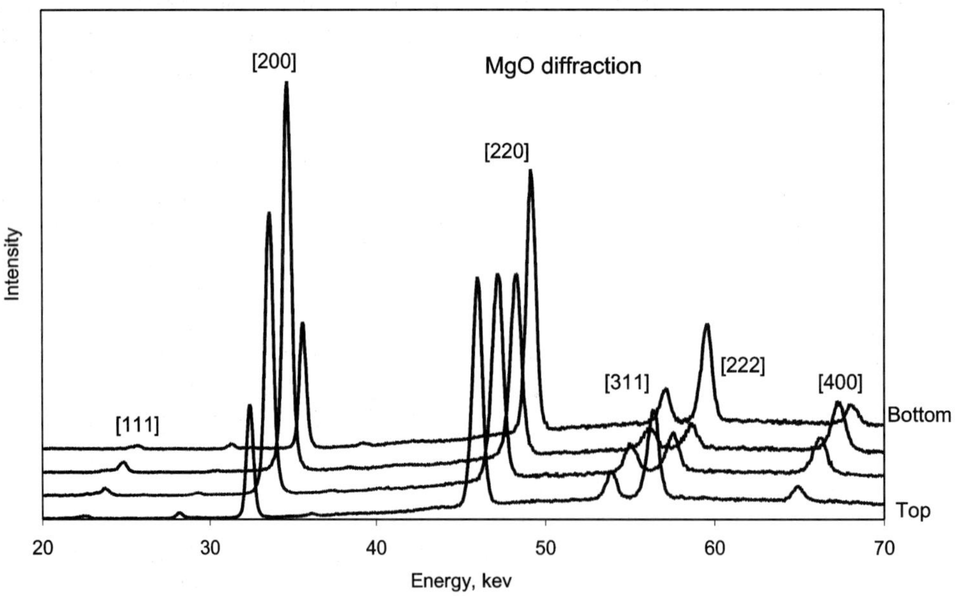

Fig. 12. X-ray diffraction patterns measured along the length of the MgO sample during cold compression. The patterns have been displaced along both axes to aid the viewing of the signals.

diffraction peaks $(h\,k\,l)$, as is given by

$$\varepsilon_{(h\,k\,l)} = \frac{\Delta d_{(h\,k\,l)}}{d_{(h\,k\,l)}} = S_{11}\sigma_1 + (S_{12} + S_{13})\sigma_3 \qquad (4)$$

where the compliance tensor, S, is rotated to the coordinate system of the stress system. The important point here is that each peak is generated from a subset of the crystals in the sample. This subset are all oriented with the particular crystallographic axis, $(h\,k\,l)$, parallel to the diffraction vector, which in our case is also the maximum stress axis. Since each peak corresponds to a different orientation, the values of the Ss will change with $(h\,k\,l)$. However, the magnitude of the variation of the Ss with $(h\,k\,l)$ depends on the elastic anisotropy of the sample. An isotropic material will have the same value of S_{ij} for all crystallographic orientations. Thus, an isotropic material will manifest strains for each diffraction peak that are shifted from the hydrostatic value, but the magnitude of the shift will be the same for each diffraction peak. A material that is elastically anisotropic, will exhibit strains that vary with orientation (or $(h\,k\,l)$), and the magnitude of the variation will be proportional to the magnitude of the differential stress.

Eq. (4) has been used to define the stress field in the case where the elastic anisotropy is known (Weidner et al., 1994) and to define the elastic properties of the material (i.e. the S_{ij}s) in the case where the stress state is defined (Mao et al., 1998). In both cases, it is assumed that the stress is uniform and does not vary with orientation of the grains. Indeed, the grains that contribute to a (2 2 2) X-ray peak do not contribute to the (2 0 0) peak, as mutually exclusive orientations are required for each peak. The local stress field is influenced by the orientation of the particular grain since some directions in the grain are stiffer than others. A Reuss type assumption would assert that each grain samples the local stress field, while a Voigt type assumption would assert that each grain samples the local strain field. The Voigt state would result in exactly the same displacement (strain) of each of the diffraction peaks and there would be no signal originating from the deviatoric stress. Eq. (4) is actually appropriate for the Reuss state. An intermediate state, such as the Hill state, would imply that the stress calculated in this equation is about a factor of 2, too small.

A more profound effect can result from an anisotropy in strength. A material that flows significantly during loading can result with stresses in individual grains that represent the strength of the individual orientation of the grain. For example, if a cubic material has a very soft slip system on (1 0 0) planes, then grains with (1 0 0) parallel to the maximum stress will support higher values of σ_1 than other grains. Thus, the elastic strain for these grains will be greater than others. Such information would be incorrectly interpreted by using the above equation.

The results for MgO include effects of anisotropic strength. Analysis of the stress field using Eq. (4) during compression yielded a tensional stress in excess of 2 GPa parallel to the cylinder axis, yet the sample shortened considerably during this phase of the experiment. Uchida et al. (1998) observed the same strain distribution for MgO, but concluded that it was due to a change in the sign of anisotropy at about 3 GPa. This conclusion is contradicted by the elastic moduli data of Chen et al. (1998a) who found very little change in elastic anisotropy to much higher pressure. The specific observation is that the (2 0 0) diffraction peak indicated a larger strain than the other peaks. This could be due to tension along the diffraction direction (high stress axis) or higher stress on grains with (1 0 0) parallel to the compression axis. The latter results if the easy slip systems are on (1 0 0) planes. Taken together with the preferred orientation, this information indicates that the low temperature flow in MgO occurred on (1 0 0) planes and that the flow directions varied between the middle and the ends of the MgO sample in this study.

On heating, the situation changed for this MgO sample. Applying Eq. (4) to the diffraction data indicated a compressive stress parallel to the shortening axis of about 1 GPa at 600°C. The most stable results were obtained using only the two strongest peaks in the diffraction pattern (2 0 0) and (2 2 0). Heating to 1100°C resulted in a decrease in the differential stress to about 0.5 GPa. In both cases, the diffraction patterns were taken about 20 min after the temperature increase when strain rate was about $10^{-5}\,s^{-1}$. Frost and Ashby (1982) estimated the differential stress to overcome obstacles to be about 1.2 GPa for MgO. Their model would suggest that obstacles should limit flow at 600°C but that power law creep may be effective by 1100°C with a differential stress of about 0.4–0.5 GPa for a strain rate of $10^{-5}\,s^{-1}$. Thus, for the elevated temperature measurements of flow in MgO, the macro-scale stress–strain rate relation that is measured for a reference material is in excellent agreement with existing models.

Clearly there are uncertainties with the stress measurements in the macro-scale stress analysis. The approach outlined here determines this stress from diffraction data obtained at a fixed position relative to the stress axis. If the entire Debye ring can be defined, then it is much more straightforward to define the stress. Weidner (1998) demonstrate that the ellipticity of the Debye ring can yield a more robust measure of the differential stress and has fewer of the ambiguities discussed here. Our DIA geometry does not allow measurement of the full Debye ring, however, other guide-blocks and future strategies may.

3. Mantle minerals

We have carried out micro-scale stress measurements on several mantle minerals including olivine, wadsleyite, ringwoodite, majorite-rich garnet, and magnesium silicate perovskite. While many of these studies are still in progress to define the flow laws and deformation mechanisms, we will describe the relative properties here. The effect of water on the observed strength has already been presented (Chen et al., 1998b) and would not be further discussed below.

One useful measure of the mineral rheology is the 1000 s strength, that is the deviatoric stress supported by the sample upon holding temperature constant for 1000 s. In our studies, this stress is relatively independent of the time–temperature path that precede the data point as long as the stress exceeded saturation. From a flow law viewpoint, this is the stress for a strain rate of about $10^{-7}\,s^{-1}$. The 1000 s strength is illustrated in Fig. 13 for these minerals as a function of temperature. All minerals exhibit a strength of several GPa at low temperatures and a thermally induced softening at higher temperatures which is probably the result of a thermally activated process. The temperature for this process is lowest for olivine for which it is about 500°C. Wadsleyite and majorite soften at about 700°C, and ringwoodite softens at a slightly higher temperature. We did not observe a significant thermal softening of perovskite in the experimental range, which extended to 800°C.

The point of thermal softening is still not in the power law creep regime, because the effective n is well above 5 for all of these materials. Thus, we conclude

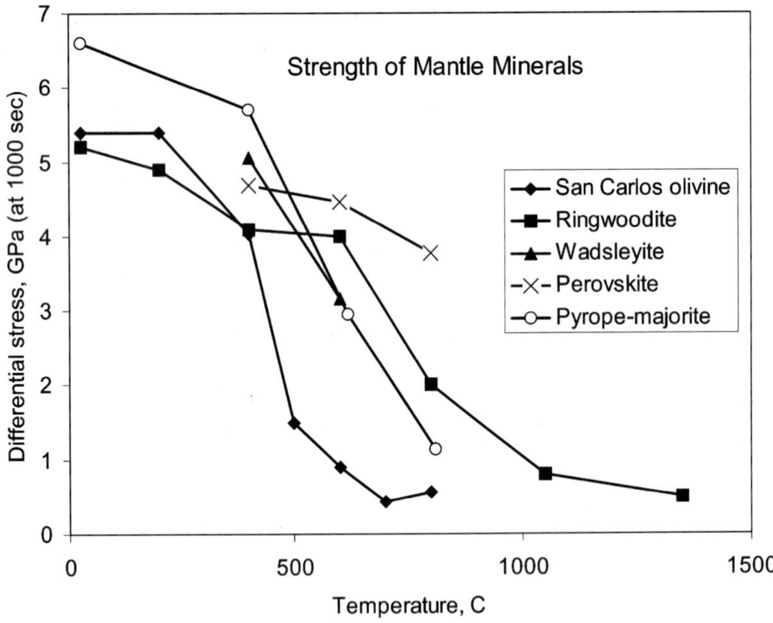

Fig. 13. The 1000 s strength of mantle minerals as a function of temperature. This is the measured differential stress that can be supported during a relaxation experiment 1000 s after the initial load was applied.

that this region generally corresponds to glide or to a transition region between glide and climb. Temperatures above the softening region may well correspond to power law creep.

The picture that emerges includes three thermally defined regions. The low temperature region exhibits a strength of several GPa, which is relatively temperature independent. Strain rate depends strongly on stress. For stress higher than the 1000 s strength, strain rate becomes very fast, yet for lower stresses, the strain rate is quite slow. Expressed as a power law flow law, the effective value of n is often above 20. The softening region is characterized by a strong temperature dependence of strength, with an effective n still greater than that expected in the normal power law creep regime. The high temperature region demonstrates a reduced temperature dependence of strength and less dependence of strain rate on stress. Values of n are often in the range of 3–5.

The thermal softening region has important rheological characteristics that may have a bearing on the deep earthquake process. Ogawa (1987) and Hobbs and Ord (1988) demonstrate that a plastic instability mechanism for deep earthquakes requires a rheology with a strong stress and temperature dependence. This is precisely the characteristic of the softening region that is identified by our data. At lower temperatures, the stress dependence of the strain rate is high, but the temperature dependence is low. Thus, softening due to flow-induced heating will not be so significant. At higher temperatures, the stress dependence diminishes, reducing the effect of stress concentrations at the 'crack' tip.

Elevated temperatures will eventually cause seismicity to cease as there is insufficient strength to support seismogenic stresses for sufficient times. Holt (1995) and Bevis (1988) estimate a strain rate of $10^{-15}\,s^{-1}$ for the flow process that results from deep earthquakes. If creep becomes sufficiently active that it exceeds this flow rate, then we would expect that deep seismicity would shut off since plastic flow will drain stress from the region more efficiently than seismic activity. Using the power law creep law of Hirth and Kohlstedt (1996), we estimate that for 100 MPa

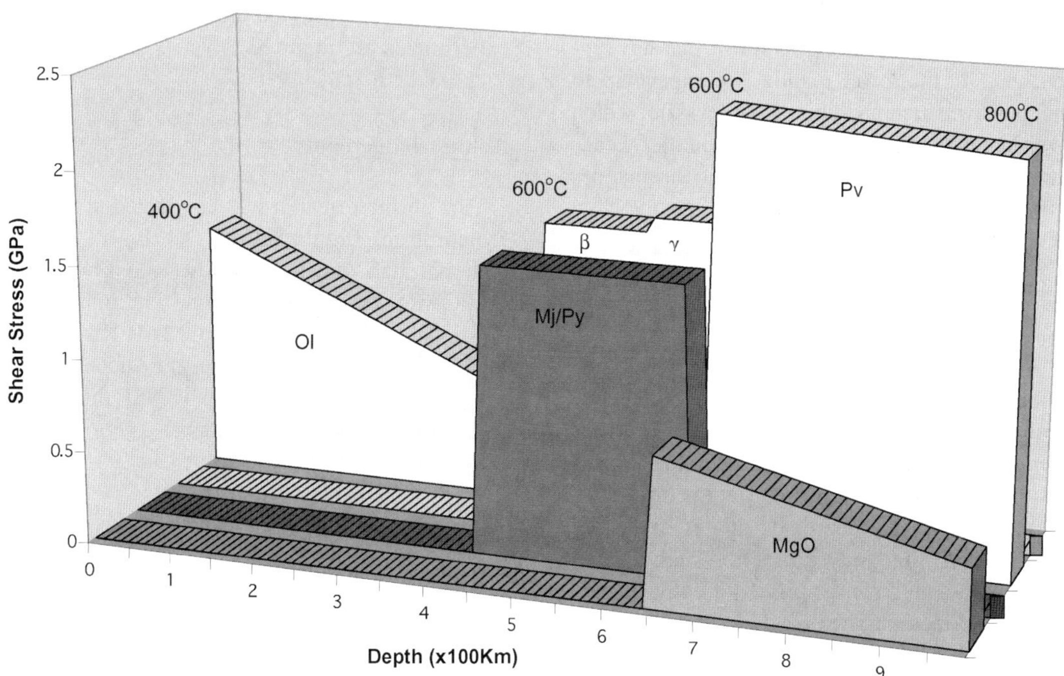

Fig. 14. Stress capacity of the subduction zone. The 1000 s strength is projected along a geotherm in the cold center of a subducting slab. Phase transformations give rise to sudden increases in strength at the appropriate depths.

of shear stress, creep will shut down the seismic process at 600°C in wet olivine (their preferred oceanic lithosphere model) and at 700°C for dry olivine. Thus, we expect that such temperatures represent a maximum temperature for seismicity above 400 km where olivine is the stable mineral.

A plastic instability origin of deep earthquakes is consistent with many of the depth variations in seismic character. The decrease in seismicity from 350 to 400 km could reflect the approach of the temperature to 600°C in the bulk of the slab by this depth. The increase in seismicity below 400 km depth reflects the higher temperature for thermal softening of the high pressure phases. Cessation of seismicity at the perovskite boundary may be a result of the strong, temperature independent nature of perovskite.

These measurements also provide important constraints on the rheology of the slab as is required for dynamic models of subduction. Fig. 14 illustrates the stress capacity of the subduction region along the cold center of the slab. Again, the 1000 s strength is illustrated. Phase transitions mark strong increases in strength as is illustrated here, with the perovskite region being extremely strong. We do not account for secondary phases. In the upper mantle and transition zone, they should be as strong as those that are represented here. In the lower mantle, we expect up to 15% magnesiowustite in addition to perovskite, which might weaken the slab. However, the amount of weakening will depend on the distribution of the magnesiowustite. Further experiments will define the flow laws of these materials.

4. Conclusion

Credible stress–strain rate relations are defined by synchrotron-based measurements. Micro-scale stress measurements are based on relating deviatoric stress to stress heterogeneities. The stress field is created as a powdered sample is compressed and is expressed as a broadening of the X-ray diffraction peak. Relaxation of the stress field can be observed with time resolved X-ray diffraction. The time dependence enables an empirical flow law to be defined. Literature data agree with the empirical flow laws for spinel, corundum, and olivine in the temperature regime reported here. Generally, these measurements have been limited to deviatoric stresses greater than about 100 MPa. This has allowed analyses of only a small region of parameter space where the defining flow regime is the high-temperature power law creep regime. However, the lower temperature regimes probably control flow in the cold part of a subducting slab.

Preliminary data are presented for most phases that exist within a subducting slab. Higher pressure phases tend to be stronger at elevated temperatures. In particular, thermal activation of the weakening process occurs at higher temperatures for the higher pressure phases. Association of the seismogenic zone with the temperature regime where strength is a strong function of temperature suggests that earthquakes in high pressure phases occur at higher temperatures than those in olivine. In fact, olivine probably becomes too weak to support earthquakes at temperatures in excess of 600–700°C. Thus, the decrease in seismicity with depth to 400 km may result from the weakening of olivine in the slab at these depths. The increase in seismicity deeper than 400 km would then result from the increased strength of the higher pressure phases. The disappearance of earthquakes in the perovskite field may reflect the loss of a plastic instability mechanism associated with the rheology of perovskite. More experiments are required to verify this possibility.

Acknowledgements

Part of this research was conducted at the National Synchrotron Light Source X17B of Brookhaven National Lab and at the Advanced Photon Source, GSECARS beamline, Argonne National Lab. Jerry Hastings, Yanbin Wang, and Mark Rivers, at these facilities were instrumental in the success of this research. Support was provided by NSF through CHiPR and EAR9909266.

References

Ando, K., Oishi, Y., 1974. J. Chem. Phys. 61, 625.

Bevis, M., 1988. Seismic slip and down-dip strain rates in Wadati–Benioff zones. Science 240, 1317–1319.

Chen, G., Liebermann, R.C., Weidner, D.J., 1998a. Single-crystal MgO to 8 GPa and 1600 K. Science 280, 1913–1916.

Chen, J., Inoue, T., Weidner, D.J., Wu, Y., Vaughan, M.T., 1998b. Strength and water weakening of mantle minerals, olivine, wadsleyite, and ringwoodite. Geophys. Res. Lett. 25, 575–578 and 1103–1104.

Choi, D.M., 1965. PhD thesis. North Carolina State University at Raleigh.

Evans, B., Goetze, C., 1979. The temperature variation of hardness of olivine and its implication for polycrystalline yield stress. J. Geophys. Res. 84, 5505–5524.

Frost, H.J., Ashby, M.F., 1982. Deformation-Mechanism Maps, The Plasticity and Creep of Metals and Ceramics. Pergamon Press, Oxford, 166 p.

Goetze, C., 1978. The mechanisms of creep in olivine. Philos. Trans. R. Soc. London A 288, 99–119.

Hirth, G., Kohlstedt, D.L., 1996. Water in the oceanic upper mantle implications for rheology, melt extraction and the evolution of the lithosphere. Earth Planet. Sci. Lett. 144, 93–108.

Hobbs, B.E., Ord, A., 1988. Plastic instabilities: implications for the origin of intermediate and deep earthquakes. J. Geophys. Res. 93, 10521–10540.

Holt, W.E., 1995. Flow fields within the Tonga slab determined form the moment tensors of deep earthquakes. Geophys. Res. Lett. 22, 989–992.

Karato, S.I., Rubie, D.C., 1997. Toward an experimental study of deep mantle rheology a new multi-anvil sample assembly for deformation studies under high pressures and temperatures. J. Geophys. Res. 102, 20111–20122.

Kinsland, G.L., Bassett, W., 1977. Strength of MgO and NaCl polycrystals to confining pressures of 250 Kbar at 25°C. J. Appl. Phys. 48, 978–985.

Mao, H.K. et al., 1998. Elasticity and rheology of iron above 200 GPa and the nature of the earth's inner core, Nature, 396, 741–743.

Meade, C., Jeanloz, R., 1990. The strength of mantle silicates at high pressures and room temperature: implications for the viscosity of the mantle. Nature 348, 533–535.

Ogawa, M., 1987. Shear instability in a viscoelastic material as the cause of deep focus earthquakes. J. Geophys. Res. 92, 13801–13810.

Phakey, P., Dollinger, G., Christie, J., 1972. Transmission electron microscopy of experimentally deformed olivine crystals in flow and fracture of rocks. In: Heard, H.C., Borg, I.Y., Carter, N.L., Raleigh, C.B. (Eds.), Flow and Fracture of Rocks. Geophysical Monograph. American Geophysical Union, Washington, DC.

Singh, A.K., 1993. The lattice strains in a specimen (cubic symmetry) compressed non-hydrostatically in an opposed anvil device. J. Appl. Phys. 73, 4278–4286.

Singh, A.K., Balasingh, C., 1994. The lattice strains in a specimen (hexagonal system) compressed non-hydrostatically in an opposed anvil high pressure setup. J. Appl. Phys. 75 (10), 4956–4962.

Singh, A.K., Kennedy, G.C., 1974. Uniaxial stress component in tungsten carbide anvil high-pressure X-ray cameras. J. Appl. Phys. 45 (11), 4686–4691.

Uchida, T., Yagi, T., Katsuya, O., Funamori, N., 1998. Analysis of powder X-ray diffraction data obtained under uniaxial stress field. Rev. High Pressure Sci. Technol. 7, 269–271.

Vaughan, M.T. et al., 1998. T-cup: a new high-pressure apparatus for X-ray studies. Rev. High Pressure Sci. Technol. 7, 1520–1522.

Weidner, D.J., 1998. Rheological studies at high pressure. In: Ribbe, P.H. (Ed.), Reviews in Mineralogy. I. Ultrahigh-pressure Mineralogy: Physics and Chemistry of the Earth's Deep Interior. Mineralogical Society of America, Washington, DC, pp. 493–524.

Weidner, D.J., et al., 1992a. Large volume high pressure research using the wiggler port at NSLS. High Pressure Res. 8, 617–623.

Weidner, D.J., et al., 1992b. Characterization of stress, pressure, and temperature in SAM85: a DIA type high pressure apparatus. In: Syono, Y., Manghnani, M.H. (Eds.), High-pressure Research: Application to Earth and Planetary Sciences. Terra Scientific Publishing Company and American Geophysical Union, Tokyo and Washington, DC, pp. 13–17.

Weidner, D.J., Wang, Y., Vaughan, M.T., 1994. Deviatoric stress measurements at high pressure and temperature. In: Schmidt, S.C., Shaner, J.W., Samara, G.A., Ross, M. (Eds.), High-pressure Science and Technology. Proceedings of the 1993 AIRAPT Conference, pp. 1025–1028.

Weidner, D.J., Wang, Y.B., Chen, G., Ando, J., 1998. Rheology measurements at high pressure and temperature. In: Manghnani, M.H., Yagi, T. (Eds.), Properties of Earth and Planetary Materials at High Pressure and Temperature. Geophysical Monograph. American Geophysical Union, Washington, DC, pp. 473–480.

Westbrook, J.H., 1966. Rev. Hautes Temper. et Refract. 3, 47.

ELSEVIER

Physics of the Earth and Planetary Interiors 127 (2001) 83–108

PHYSICS OF THE EARTH AND PLANETARY INTERIORS

www.elsevier.com/locate/pepi

Rheological structure and deformation of subducted slabs in the mantle transition zone: implications for mantle circulation and deep earthquakes

Shun-ichiro Karato [a,*], Michael R. Riedel [b], David A. Yuen [a,c]

[a] *Department of Geology and Geophysics, University of Minnesota, Minneapolis, MN 55455, USA*
[b] *GeoForschungs Zentrum, Telegrafenberg, Potsdam D-14473, Germany*
[c] *Minnesota Supercomputer Institute, University of Minnesota, Minneapolis, MN 55415-1227, USA*

Received 28 March 2000; accepted 2 October 2000

Abstract

Rheological structure of subducted slabs of oceanic lithosphere in the mantle transition zone is investigated based on mineral physics observations incorporating grain-size, stress, temperature and pressure dependence of rheology. It is shown that the rheological structure of slabs depends strongly on subduction parameters through temperature that controls the grain-size of spinel (ringwoodite) and the magnitude of forces acting on a slab. We use a theoretical model of grain-size evolution associated with the olivine–spinel transformation, plastic flow laws of olivine and spinel combined with a thermal model of subducting slab in which the effect of latent heat release is incorporated. Three deformation mechanisms for olivine and spinel (diffusional creep, power-law (dislocation) creep and the Peierls mechanism) are considered. Due to the large variation in temperature, stress and grain-size, a subducting slab is shown to have a complicated rheological structure which varies both laterally and with depth. A cold slab in the deep transition zone is characterized by a weak, fine-grained spinel region surrounded by narrow but strong regions. The flexural rigidity and the curvature of a slab are calculated using a new formulation in which the effects of stress-dependent rheology is incorporated in a self-consistent fashion. Although uncertainties in both the transformation kinetics and the rheology of high pressure phases are still large, the general trend of dependence of slab flexural rigidity and the curvature on the subduction parameters is well constrained. Slabs with very low thermal parameters (warm slabs) are weak, but slabs with large thermal parameters (cold slabs) are also weak due to small spinel grain-size and large external force (bending moment). Slabs with intermediate thermal parameters will have a relatively large flexural rigidity and could penetrate into the lower mantle without much deformation. Thus the 660 km discontinuity may work as a rheological filter for mantle convection. This prediction provides a natural explanation for a paradoxical observation that significant deformation of slabs is observed exclusively in the western Pacific where temperatures of the slabs are considered to be low.

Our slab rheology models also have important implications for deep earthquakes. Overall rheological weakening of slabs in the deep transition zone results in high rates of deformation under relatively low temperatures providing a favorable environment for thermal runaway instability (adiabatic shear instability). Our model predicts heterogeneous energy dissipation as a result of heterogeneous rheology: energy dissipation in deep, cold slabs is concentrated in high strength regions surrounding a weak, fine-grained spinel core. The regions of high energy dissipation are prone to thermal runaway instability and are likely

*Corresponding author. Present address: Department of Geology and Geophysics, Yale University, New Haven, CT 06520-8109; Tel.: +1 203-432-3164; fax: +1-203-432-3134.
E-mail address: karato@yale.edu (S. Karato).

PII: S0031-9201(01)00223-0

to be seismogenic. The width of this seismogenic region is well constrained by our model and is predicted to be ~40–60 km which is in excellent agreement with seismological observations. Other features of deep earthquakes including low seismic efficiency and low aftershock activities can also be explained by the thermal instability model.

Keywords: Rheology; Bending moment; 660 km discontinuity; Olivine–spinel transformation; Grain-size; Deep earthquakes; Shear instability; Mantle convection

1. Introduction

Deformation of subducted slabs consumes a large amount of energy and therefore subduction is considered to be the most important and violent of the processes that control the dynamics of the solid Earth (e.g. Forsyth and Uyeda, 1975; Chapple and Tullis, 1977). Presently, a number of problems related to subduction remain unsolved particularly those concerning deep processes. They include the cause of regional variation in the nature of interaction between slabs and the transition zone that has an important influence on the style of mantle convection and the origin of deep earthquakes. A better understanding of these problems requires the detailed understanding of rheological properties of minerals under deep mantle conditions. However, in most previous models of slab deformation, highly simplified rheological properties such as homogenous rheology across a slab and/or Newtonian rheology sometimes with a weak temperature dependence were assumed (Zhong and Gurnis, 1995; Christensen, 1996; Houseman and Gubbins, 1997; Schmeling et al., 1999).

Complications of mantle convection caused by the slab-transition zone interactions have been emphasized by numerical modeling studies that demonstrated the important effects of the 660 km discontinuity on the style of mantle convection (Christensen and Yuen, 1984; Honda et al., 1993; Tackley et al., 1993; Davies, 1995). These studies also showed that the nature of slab-transition zone interaction is highly sensitive to subtle differences in material properties and/or the Rayleigh number (Christensen and Yuen, 1984, 1985; King and Ita, 1995; Davies, 1995), and therefore one might expect a variety of slab-transition zone interactions in the present Earth. In fact, in some portions of the western Pacific, slabs appear to be deflected horizontally above the 660 km discontinuity and there is no clear image of slabs in most of the lower mantle (van der Hilst et al., 1991; Fukao et al., 1992; Ding and Grand, 1994; Bijwaard et al., 1998). In contrast, in most of the eastern Pacific (and in the Tethys), slabs appear to penetrate into the lower mantle without much deformation (Grand, 1994; van der Hilst et al., 1997). This is a rather paradoxical observation, because slabs in the eastern Pacific are expected to be warm and hence should deform more easily than the relatively cold ones in the western Pacific.

Another puzzling observation is the occurrence of deep earthquakes which occur under temperature–pressure conditions where brittle failure is unlikely to occur. Various models of deep earthquakes have been proposed including a thermal runaway instability (adiabatic instability) model (Griggs and Baker, 1969; Ogawa, 1987; Hobbs and Ord, 1988), a transformational faulting model (Kirby, 1987; Green and Burnley, 1989; Kirby et al., 1991, 1996; Green, 1994; Green and Houston, 1995) and instability caused by amorphization of hydrous minerals (Meade and Jeanloz, 1991). However, none of the existing models provide satisfactory explanations for important seismological observations such as the geometry of faults, seismic efficiency and the frequency of aftershocks. Thus, the paradox of deep earthquakes has not been solved despite the recent claim by Green (1994).

The main purpose of this study is to provide a better rheological model of deep slabs in the transition zone to answer the above questions using mineral physics data in comparison to geodynamic observations. A simpler model has been published (Riedel and Karato, 1997) in which only the role of grain-size reduction was considered. However, this earlier treatment contained several limitations including the formulation of slab deformation assuming a constant strain-rate and the large uncertainties in rheological parameters available at that time. In addition, comparisons with geodynamic observations were not made in detail. In this paper, we have extended this earlier work by including the use of a new self-consistent formulation of slab deformation and the use of new better-constrained rheological data that have recently become available. We first summarize geophysical observations on slab deformation and seek some correlation between slab deformation and slab thermal parameters. We then present a new method of calculating the "flexural rigidity" of a slab in a self-consistent fashion. We consider both the effects of phase transformations to modify the grain-size of newly formed spinel and of the stress-dependent deformation mechanisms in silicates. We then compare the calculated variation of flexural rigidity with subduction parameters and geodynamic observations related to slab deformation. Some implications of the present model for deep earthquakes will also be discussed including the geometry of faults, seismic efficiency and the frequency of aftershocks.

2. Geophysical observations of slab deformation

We use the geometry of Wadati–Benioff zones based on the zones of high seismic activity (Isacks and Barazangi, 1977; Jarrard, 1986; Giardini and Woodhouse, 1984; Giardini, 1992; Lundgren and Giardini, 1992; Sugi et al., 1989; Engdahl et al., 1998) and the results of seismic tomography (van der Hilst et al., 1991; Fukao et al., 1992; Ding and Grand, 1994; Grand, 1994; van der Hilst, 1995; Engdahl et al., 1995; Sakurai, 1996; van der Hilst et al., 1997; Bijwaard et al., 1998) to define the deformation of slabs. These two methods are complementary. The Wadati–Benioff zones represent zones of high seismicity and can be correlated with cold descending lithospheric slabs. Errors in focal positions after relocation are ~10 km or less and therefore maps of focal positions provide high resolution three-dimensional images of seismogenic regions (slabs) (Engdahl et al., 1998). However, these data are limited to the portions of slabs that have high seismicity and hence no constraints can be placed from the morphology of Wadati–Benioff zones where deformation is aseismic, such as below the 660 km discontinuity. Seismic tomography can provide insights into slab morphology, including aseismic regions, although its resolution is much worse than that of the Wadati–Benioff zones.

Fig. 1 shows two-dimensional morphologies of a section of some subducted slabs. Although, in reality, slabs deform in three-dimensions (Yamaoka et al., 1986), we choose several portions of slabs that show approximately two-dimensional deformation to facilitate comparison with our two-dimensional model. Because we focus on slab deformation in the transition zone, only the geometry of Wadati–Benioff zones below 200 km depth is considered. Several points can be noted. (1) Most of the slabs above ~400 km depth are nearly straight (or planar) implying a high rigidity. (2) In several slabs (Izu–Bonin, Banda and Tonga, etc.), significant deformation occurs near the bottom of the transition zone. (3) However, slabs in South America and Honshu (not shown here) are planar close to the 660 km discontinuity, suggesting relatively high strength. (4) The Tonga slab shows significant deformation not only toward the bottom of the transition zone, but also around ~400 km depth. Deformation at this depth is concentrated at ~21°S–178°E and is highly three-dimensional.

The deformation that occurs at around 400 km depth in the Tonga slab is anomalous for several reasons: (i) the focal mechanisms of deep earthquakes in that region are significantly different from those of surrounding regions (Giardini, 1992; Giardini and Woodhouse, 1984); (ii) this region corresponds to the extension of the Louiville ridge which has thicker than normal oceanic crust (Vogt et al., 1976); and (iii) the deformation of this region may be related to the opening of the Lau basin to its north-west (Isacks and Barazangi, 1977). Therefore we focus on the first three of the above points in this paper. We also observe significant regional variations in slab deformation. For example, pronounced deformation near the bottom of the transition zone is well documented in the Izu–Bonin and Banda subduction zones but not in the Mariana slab. Although the data are limited, the deep seismic zones under South America show much less deformation (Lundgren and Giardini, 1992). Therefore there is significant variation in slab deformation behavior. We will return to this point in the next section where observations from seismic tomography are summarized.

One of the most remarkable observations from high resolution tomography (van der Hilst et al., 1991; Fukao et al., 1992; Ding and Grand, 1994; Grand, 1994; van der Hilst et al., 1997; Bijwaard et al., 1998) is that there is a variety of styles of slab deformation:

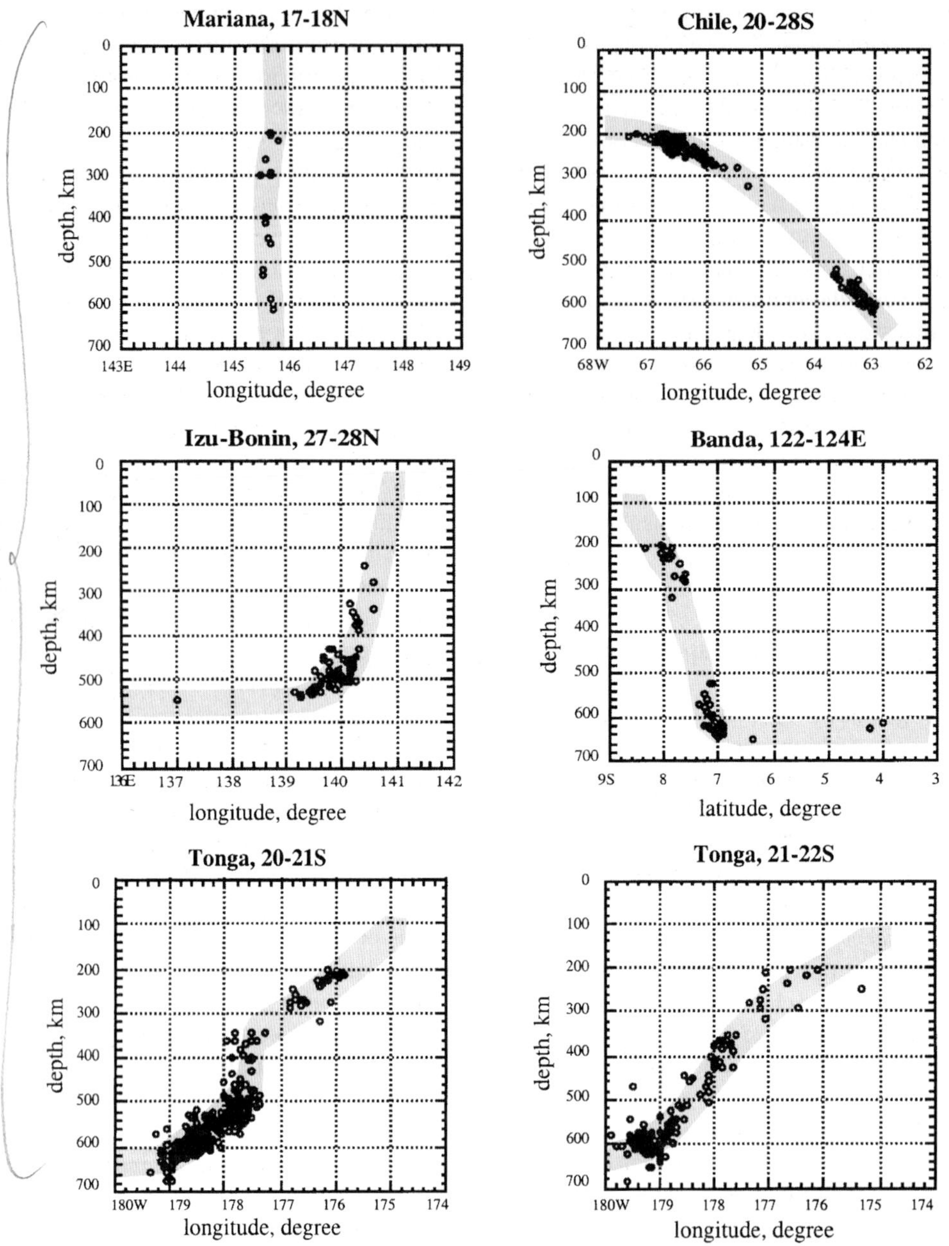

Fig. 1. Morphology of slabs (below 200 km) as inferred from distribution of earthquakes and from seismic tomography. Refined source locations by Engdahl et al. (1998) and tomographic results (van der Hilst et al., 1991; Fukao et al., 1992; van der Hilst, 1995; Sakurai, 1996; Bijwaard et al., 1998) are used. Each dot indicates a hypocenter (with a precision better than 10 km). Morphology of slabs in seismogenic regions is well constrained by hypocenter distribution. In general, slabs show little evidence for deformation from ~200 to ~400 km depth. Morphology in aseismic regions can only be inferred from tomographic images. Hatched regions show high velocity regions corresponding to subducted slabs inferred from seismic tomography. The shape and thickness of these regions are only poorly constrained. However, horizontal deflection of high velocity anomalies at around 600 km depth in the Izu–Bonin, Banda and Tonga is a robust feature of all high resolution tomographic studies. In contrast, no evidence for significant deformation is found in slabs in Americas and Mariana. Note a sharp change in slab morphology with a small change in latitude (i.e. three-dimensional morphology) in Tonga, which is presumably caused by the subduction of a buoyant ridge (Vogt et al., 1976) or by the opening of the Lau basin (Isacks and Barazangi, 1977).

in some regions such as Izu–Bonin, Banda and southern Kurile, the slabs show significant deflection at the 660 km discontinuity where they lie nearly horizontally. The best example is the Izu–Bonin slab which lies horizontally over a distance of more than 2000 km. In these areas, images of slabs in the lower mantle are poorly resolved although fast regions near the bottom of the mantle suggest that slabs do reach the bottom eventually. The lack of clear images of slabs in the lower mantle may indicate that the sinking velocity in the lower mantle is fast presumably because subduction into this region occurs only after the accumulation of a large amount of material on top of the 660 km discontinuity ("flushing events"; Honda et al., 1993; Tackley et al., 1993). In other regions, including beneath Americas and Tethys, continuous images of slabs can be observed more or less following the trend in the upper mantle, suggesting much less deformation of slabs in the transition zone.

2.1. Classification of slab deformation

The high resolution images of the Wadati–Benioff zones and the relatively low resolution images from seismic tomography both strongly suggest that there are two classes of slabs: (1) (type A) slabs that penetrate directly into the lower mantle without much deflection in the transition zone. Slabs under the Americas, north Kurile, Java and Kermadec belong to this category. The slab in the Mariana (and perhaps in the Kermadec) subduction zone appears to penetrate deep into the lower mantle, but the slab image becomes broad in the lower mantle suggesting significant deformation in this region; and (2) (type B) slabs that are significantly deflected above the 660 km discontinuity. The slab in the Izu–Bonin subduction zone is the best example. The Tonga slab is another example, although the horizontal extent of the slab deflection is not as large as that in Izu–Bonin. The New Guinea, south Kurile and Japan slabs are also examples of this type. The degree of deformation of slabs in the Americas is difficult to estimate because of poor resolution. Regional tomography by Engdahl et al. (1995) suggests some deformation but the geometry is poorly constrained. It is obvious, however, that the degree of slab deformation near the 660 km discontinuity in these subduction zones is significantly less than under the western Pacific. A wider variety of deformation behavior has been observed from laboratory experiments (Griffths et al., 1996; Guillou-Frottier et al., 1996), numerical modeling (Christensen, 1996) and Wadati–Benioff zone morphology (Giardini and Woodhouse, 1984; Jarrard, 1986; Giardini, 1992; Lundgren and Giardini, 1992). However, we adopt this simple classification, because the main issue in this paper is the deformation of slabs at around the 660 km discontinuity; in addition the limited resolution of current seismic tomography does not justify further sophistication.

Styles of slab deformation may be related to the various parameters of subduction. First, the age and velocity of subducting lithosphere will control deformation behavior through their effects on temperature which affects both slab strength and the buoyancy forces. Wortel and Vlaar (1988) suggested a slab thermal parameter defined by $\Phi = tv_{sub}$ (t: age of lithosphere at the time of subduction, v_{sub}: velocity of subduction) to characterize the minimum temperature of slabs. Second, trench migration velocity (Olbertz et al., 1997), v_m, may influence the deformation behavior of slabs as suggested by laboratory studies (Griffths et al., 1996; Guillou-Frottier et al., 1996) and by numerical modeling (Christensen, 1996). We estimated the slab thermal parameters and trench migration velocities, using a compilation of current subduction parameters (Olbertz, 1997) and subduction history (Lithgow-Bertelloni and Richards, 1998). Values of v_m and v_{sub} and morphology of slabs according to the above classification are summarized in Table 1.

Fig. 2 summarizes the correlation among slab deformation and subduction parameters. Note that in estimating the trench migration velocity and the slab thermal parameter, we choose the data corresponding to the slab properties from the surface to around 660 km depth. This is done because the geometry of slabs near the 660 km discontinuity in the transition zone is considered to be determined by the properties of slabs in that region of the mantle and the kinematic boundary condition, such as the trench migration velocity, during the period when a slab interacts with the 660 km discontinuity. For example, for a slab in the Izu–Bonin subduction zone, the latter corresponds to a period from ~10 million years ago to the present.

Several points can be noted from Fig. 2. First, despite an earlier claim (van der Hilst and Seno, 1993),

Table 1
Classification of slabs and subduction parameters[a]

Name	Type	v_m (cm per year)[b]	v_{sub} (cm per year)[c]	Age (million years)	Φ (km)[d]
North Kurile	A	1	7.3	120	8760
South Kurile	B	1	8.7	130	11300
Northeast Japan	B	0.7	9.6	130	12480
Izu–Bonin	B	3.7	8.7	146	12700
Mariana	A	1.2	9.5	155	14725
Java	A	0.5	7.7	96–134	7400–10300
Banda	B	–	8.1	150	12150
New Guinea	B	1.0	7.0	–	–
Tonga	B	4.2	9.9	113	11200
Kermadec	A	1	8.9	98	8720
South Chile	A	2.8	4.4	26	1140
Central Chile	A	3.1	4.4	48	2110
North Chile	A	3.4	4.5	82	3690
Peru	A	3.1	3.7	45	1670
Central America	A	1.4	5.7	23	1311
North America	A	1.3	–	0–5	–

[a] Data of v_m, v_{sub} are from the compile by Olbertz (1997) and Lithgow-Bertelloni and Richards (1998). Classification of slab deformation is based on the shape of the Wadati–Benioff zone from deep seismicity (Engdahl et al., 1998) and the results of seismic tomography (van der Hilst et al., 1991; Fukao et al., 1992; van der Hilst, 1995; Sakurai, 1996; Bijwaard et al., 1998).
[b] v_m: trench migration velocity.
[c] v_{sub}: velocity of subduction.
[d] Φ: slab thermal parameter (age × v_{sub}).

the local variation in slab deformation behavior does not show a clear relation with the trench migration velocity. Sharp changes in slab geometry observed in the Izu–Bonin–Mariana, northern to southern Kurile and in the Tonga–Kermadec subduction zones do not correlate with the variation in the trench migration velocity. In the Izu–Bonin–Mariana subduction zone where the most dramatic change in the style of slab deformation occurs, the trench migration velocities have remained similar in each region from ~10 million years ago to the present, although there was a significant variation between 30 and 17 million years ago due to the opening of the Shikoku basin (van der Hilst and Seno, 1993). Similarly, the change in the style of subduction from the northern to southern Kurile (Fukao et al., 1992; Ding and Grand, 1994) does not have an obvious relation to trench migration. Note also that in the Tonga–Kermadec trench, very significant trench migration is observed from ~50 million years ago to the present (Lithgow-Bertelloni and Richards, 1998). Yet, no significant slab deflection is observed in the Kermadec subduction zone. Furthermore, in the Americas where trench migration, similar to that of some of the western Pacific subduction zones occurs, evidence for slab deformation is less than that in the western Pacific. Thus, the correlation between trench migration and slab deflection appears rather indirect and weak. Second, there is a broad correlation between the slab thermal parameters and slab deformation: all the slabs with significant deformation around the 660 km discontinuity are in the western Pacific and have high slab thermal parameters, which indicate low temperatures and/or high subduction velocities.

The rather dramatic local variations observed in slab geometry suggest that deformation of a slab is controlled to a large extent by the local force balance and/or local tectonic history (kinematic boundary conditions). Thus, factors controlling slab deformation may be sought in some local parameters rather than global circulation, although global circulation certainly has an important effect on slab geometry (e.g. Hager and O'Connell, 1978). Based on the weak correlation between the observed trench migration and slab deformation, we consider that the role of trench migration in slab deformation is secondary. In contrast, the difference in the style of deformation between slabs with large and small thermal parameters seems more fundamental.

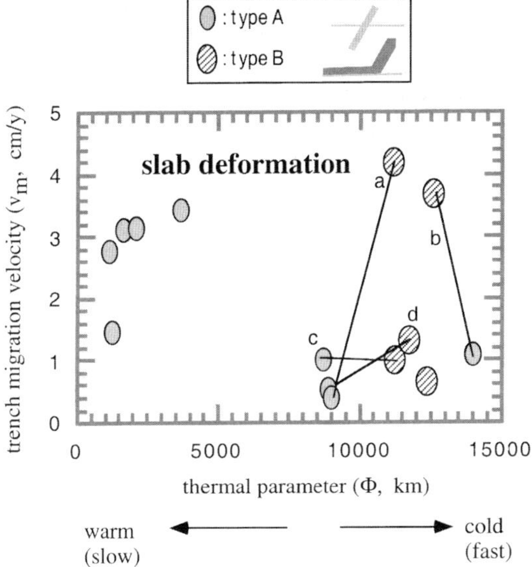

Fig. 2. Correlation of style of slab deformation with the slab thermal parameter Φ and trench migration velocity v_m. Slab geometry is classified into two categories (types A and B). The data are summarized in Table 1. Ages and kinematic data are the average values from the surface to the 660 km discontinuity. Portions of slabs that are connected are indicated by a tie line. Note a general correlation that type A behavior (direct penetration of slabs into the lower mantle without much deformation) occurs mostly for slabs with small thermal parameters and type B behavior (significant deformation above the 660 km discontinuity) occurs mostly for slabs with large thermal parameters. Correlation of deformation behavior with trench migration velocity suggested by previous studies (van der Hilst and Seno, 1993) appears secondary. An exception is the slab at the Mariana trench, which has the largest thermal parameter, yet penetrates into the lower mantle without much deformation above the 660 km discontinuity. The reason for this exception may be its very steep dip angle (~90°) which results in a small bending moment ((a) Tonga–Kermadec; (b) Izu–Mariana; (c) Java–Banda; (d) Kurile–Japan).

Other sets of observations on slab deformation or rheology include: (i) the geoid anomalies around the subduction zones of western Pacific (Moresi and Gurnis, 1996; Zhong and Davies, 1999); (ii) the magnitude of stress drops of deep earthquakes (Kanamori and Anderson, 1975; Sugi et al., 1989; Kikuchi and Kanamori, 1994); and (iii) a sudden steeping of slab dip angle at the seismicity cutoff (Castle and Creager, 1998). The analysis of geoid anomalies in the western Pacific suggests that slabs there are weak and only little stress is transmitted along the slabs. In addition, stress drops associated with deep earthquakes in the Tonga subduction zone (Sugi et al., 1989) are ~80 MPa, which are higher than those associated with the shallow earthquakes (Kanamori and Anderson, 1975), but not as high as would be expected if the slab stress is determined by viscosity that depends only on temperature and pressure. Note also, however, the stress drop associated with the Bolivia deep earthquake of 1994 is significantly higher (~110 MPa) than those of Tonga (Kikuchi and Kanamori, 1994). All of these observations, taken together, strongly suggest that deep slabs, particularly those in the western Pacific, are rather weak. Similarly, the results of numerical modeling of slab deformation are consistent with the observations of slab morphology and focal mechanisms of deep earthquakes only when a rather weak temperature-dependence of rheology is assumed (e.g. Zhong and Gurnis, 1995).

These observations indicate that: (i) a slab with a large thermal parameter (low temperature and/or high subduction rate) is easier to deform than those with smaller thermal parameters; and (ii) a high rate of trench migration favors slab deformation (or flattening above the 660 km discontinuity) but its effect is rather indirect. Therefore we suggest that variation in the slab rheology through the variation in the slab thermal parameters may play an important role in controlling the style of slab deformation. However, the above observations imply a seemingly strange temperature dependence of slab strength: a cold slab appears to be weaker than a warm slab. To solve this paradox, we need to investigate the relation between slab rheology and its deformation and the dependence of slab rheology on temperature under conditions characteristic of the deep transition zone.

3. Slab deformation and rheology

3.1. *Deformation of subducted slabs*

Deformation of subducted slabs in the transition zone may occur for a number of reasons. First, when the trench migration velocity is significantly higher than the terminal velocity of a slab in the lower mantle (due to a high viscosity there), then slabs will be deformed above the 660 km discontinuity to lie horizontally above it *if they are weak* (Griffths et al., 1996;

Christensen, 1996; Olbertz et al., 1997). Second, slab deformation may occur due to the resistance forces exerted by the 660 km discontinuity, for example, the buoyancy forces caused by the phase transformation(s) that occurs at around the 660 km discontinuity (Tackley, 1997) or by an increase in viscosity (Gurnis and Hager, 1988). The presence of resistance forces deep in the transition zone has been documented by the focal mechanisms of deep earthquakes (e.g. Isacks and Molner, 1969). In both cases, the balance of the moment caused by the external forces acting at/or around the 660 km discontinuity and the bending moment due to the strength of the deforming slab materials determines the extent of slab deformation. For slabs in the deep transition zone, contribution from elastic stress is relatively small and therefore the moment balance equation is (Turcotte and Schubert, 1982)

$$M = \int_{-h/2}^{h/2} \sigma y \, dy = -4 \int_{-h/2}^{h/2} \eta \dot{\varepsilon} y \, dy \tag{1}$$

where h is slab thickness, σ and $\dot{\varepsilon}$ the longitudinal stress and strain-rate in a slab, y the distance from the center of the slab (Fig. 3), M the bending moment caused by the external forces and η the effective viscosity defined by $\eta = -\sigma/4\dot{\varepsilon}$ which generally depends on strain-rate (or stress), grain-size, temperature and pressure.

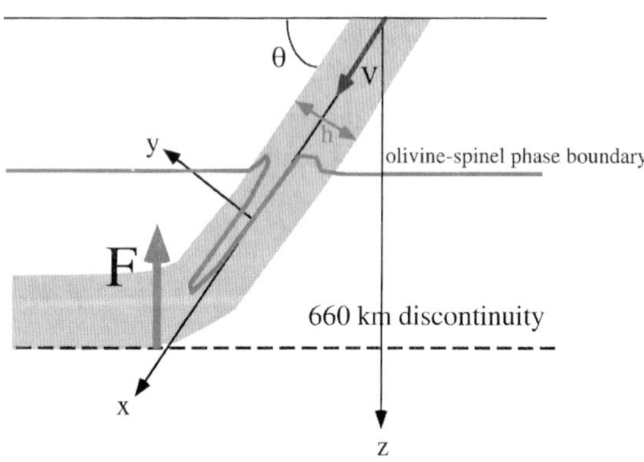

Fig. 3. Definition of the coordinate system. A slab encounters a barrier at the 660 km discontinuity which exerts a force in the vertical direction. The force due to this barrier provides a bending moment to the slab which tends to bend the slab. The degree of slab deformation depends on its rheological properties which depend on temperature, the magnitude of force (F) and subduction geometry such as the dip angle (θ).

For simplicity we assume that $\dot{\varepsilon} = \dot{\varepsilon}_0 y/h$, where $\dot{\varepsilon}_0$ is twice the strain-rate at the slab surface. Thus

$$M = \frac{\dot{\varepsilon}_0 D(\dot{\varepsilon}_0)}{h} \tag{2}$$

where we define D (flexural rigidity which has the dimensions of Nms) as

$$D \equiv 4h^3 \int_{-1/2}^{1/2} \eta(\dot{\varepsilon}) y'^2 \, dy' \tag{3}$$

For a case of uniform linear rheology, D is independent of $\dot{\varepsilon}_0$ and $D = h^3\eta/3$. However, for a more general case of non-linear and inhomogeneous rheology, Eq. (2) is an implicit equation that can be used to determine $\dot{\varepsilon}_0$ and hence D. Thus, Eq. (2) must be solved numerically by iteration using appropriate constitutive relations for slab materials. Note that the assumption of $\dot{\varepsilon} = \dot{\varepsilon}_0 y/h$ leads to an upper bound for a realistic value of D. More heterogeneous deformation would yield a smaller D. To solve Eq. (2) we need to know both the relevant rheological constitutive relation and $M(x)$. However, since the depth variation of rheological properties is significantly larger than that of $M(x)$ (for discussions on the depth variation of $M(x)$, see Bina, 1996, 1997; Yoshioka et al., 1997; Schmeling et al., 1999), we make a simplifying approximation that $M(x)$ is independent of x. Using this assumption, we can solve Eq. (2) to determine $D(x)$.

The forces acting on a slab at/or around the 660 km discontinuity include the forces due to the density anomalies caused by the deflection of the 660 km discontinuity (Tackley, 1997; Schmeling et al., 1999), forces due to the density anomalies caused by the buoyancy forces associated with subduction of a garnetite layer (Ringwood and Irifune, 1988; O'Neill and Jeanloz, 1994) and the resistance forces due to a high viscosity of the lower mantle. All of these forces increase with the velocity of subduction (and with the age of the lithosphere). For simplicity, we assume a linear relation between the moment M and the velocity of subduction. We choose bending moments of $M = 4 \times 10^{18}$ and 9×10^{18} N for velocities of 4 and 10 cm per year, respectively, which corresponds to a force of $F \sim 10^{13}\,\mathrm{N\,m^{-1}}$ (Davies, 1980). It must also be noted that the bending moment due to external forces depends on the dip angle: for a large dip angle the bending moment is small.

The curvature of a slab may also be calculated from the slab strain-rate calculated above. From the definition of curvature $1/R = \mathrm{d}^2 w/\mathrm{d}x^2$ (where R is the radius of curvature, w displacement normal to the surface of the slab), and the relation $\dot{\varepsilon} = -y(\mathrm{d}^3 w)/(\mathrm{d}^2 x\,\mathrm{d}t)$ (Turcotte and Schubert, 1982), one can write

$$\frac{\dot{\varepsilon}_0}{h} = \frac{\mathrm{d}(1/R)}{\mathrm{d}t} \approx v_{\mathrm{sub}} \frac{\mathrm{d}(1/R)}{\mathrm{d}x} \tag{4}$$

where v_{sub} is the velocity of subduction. Note that we assume a steady-state shape of the slab. Thus, given $\dot{\varepsilon}_0(x)$ one can integrate Eq. (4) to obtain the radius of curvature of the slab $R(x)$.

The flexural rigidity, D, is sensitive to the rheological properties of slab materials, which change with both depth and lateral position due to changes in temperature, stress, pressure and grain-size. Slab temperatures depend on a number of factors including the initial temperature profile before subduction which depends on the age of the slab, thermal conduction and the latent heat release associated with phase transformations. We use a model similar to that of McKenzie (1969) with the additional effects of latent heat release (Riedel and Karato, 1997). We consider slabs with different initial thicknesses from 60 to 100 km which correspond to ages of ~60–180 million years at the beginning of subduction. Here we show the results of detailed calculation for an initial thickness of 85 km with subduction velocities ranging from 4 to 10 cm per year. The range of parameters considered thus covers slab thermal parameters ranging from 2560 to 17,700 km.

3.2. Rheology of subducted slabs

3.2.1. Mineral physics considerations

To solve Eq. (2) with (3), we need to know the rheological properties of materials composing a slab. The rheological properties of subducting slabs can change both laterally and radially, because a number of parameters that affect rheology (temperature, pressure, stress and grain-size, etc.) can change both laterally and radially. In this paper, we will make the following simplifications. (1) We assume that the rheology of a subducting slab is controlled by its major component, $(Mg, Fe)_2SiO_4$, because this component is volumetrically dominant (e.g. Ringwood, 1991) and its rheology is usually softer than other components such as $(Mg, Fe)SiO_3$ (e.g. Karato, 1989a, 1997b). A thin strong garnetite layer may play an important role in the chemical evolution of slabs (Karato, 1997a), but its effects on the overall strength of slabs are not very large and will therefore be neglected in this paper. (2) Major phase transformations that may occur in $(Mg, Fe)_2SiO_4$ in the transition zone include olivine to wadsleyite, wadsleyite to ringwoodite (spinel) and olivine to ringwoodite. However, to simplify the analysis, we assume that only one transformation occurs in $(Mg, Fe)_2SiO_4$ in subducting slabs. We assume an equilibrium boundary corresponding to that of the olivine to wadsleyite transformation (e.g. Akaogi et al., 1989) and use the kinetic parameters for the olivine to spinel transformation from the experimental work by Rubie et al. (1990) on Ni_2SiO_4 to calculate the depth of transformation and the degree of grain-size reduction (for details, see Riedel and Karato, 1997). This admittedly crude approximation is made firstly because the behavior of wadsleyite in phase transformation appears similar to that of ringwoodite (Rubie and Ross, 1994) and secondly because detailed quantitative data on transformation kinetics are available only for the olivine to spinel transformation in Ni_2SiO_4 (Rubie et al., 1990; for a new data set on $(Mg, Fe)_2SiO_4$, see Mosenfelder et al., this volume). We believe that the essence of our model, i.e. semi-quantitative conclusions concerning the weakening of slabs and the effects of slab temperature on rheology, are not seriously affected by these simplifying assumptions.

Under these assumptions, we can characterize the rheology of a slab by knowing the rheology of olivine and ringwoodite, as well as knowing how the olivine to ringwoodite transformation affects rheology. We consider three deformation mechanisms, namely, power-law dislocation creep, diffusion creep and the Peierls mechanism. The rheology of olivine (Karato et al., 1986; Hirth and Kohlstedt, 1995a,b), wadsleyite (Kubo et al., 1998a; Mosenfelder et al., 2000; see however, Chen et al., 1998) and ringwoodite (Chen et al., 1998) is likely to be affected by water. However, we will consider rheology under water-free conditions, because oceanic lithosphere is considered to be depleted in water due to partial melting at mid-oceanic ridges (Karato, 1986; Hirth and Kohlstedt, 1996; Karato and Jung, 1998).

For olivine, we use the parameters determined by Evans and Goetze (1979), Karato et al. (1986),

Karato et al. (1993), Hirth and Kohlstedt (1995a,b) and Karato and Rubie (1997). The rheology of ringwoodite is less well constrained than that of olivine. However, a deformation mechanism map has now been constructed by the combination of direct mechanical tests on ringwoodite under transition zone conditions (Karato et al., 1998) and high resolution mechanical tests on analog spinels at lower pressures (Okamoto, 1989; Lawlis et al., 2001). In addition, the strength of ringwoodite (and wadsleyite) in the high stress, Peierls mechanism regime is constrained by the experimental data (Chen et al., 1998; Mosenfelder et al., 2000) together with the scaling law (Frost and Ashby, 1982). These data provide constraints on the degree of weakening caused by grain-size reduction resulting from the olivine–ringwoodite transformation.

A generic equation that describes the dependence of rheology on temperature, stress and pressure is given by

$$\dot{\varepsilon} = A\left(\frac{\sigma}{\mu}\right)^n \left(\frac{b}{d}\right)^m \exp\left[-g\frac{T_m(P)}{T}\left(1-\frac{\sigma}{\sigma_P}\right)^q\right] \tag{5}$$

where A is a constant, σ the stress, μ the shear modulus, n the stress exponent, m the grain-size exponent, b the length of the Burgers vector, d the grain-size, g a non-dimensional constant proportional to the activation enthalpy, T_m the melting temperature, T the temperature, P the pressure, σ_P the Peierls stress and q a constant that depends on the mechanism of dislocation glide (Frost and Ashby, 1982). One important assumption implicit in this formulation is that the pressure dependence of deformation can be parameterized through that of the melting temperature. The validity of this assumption has been evaluated by Weertman (1968), and Karato (1989a, 1997b, 1998) in particular showed that the melting temperature in this equation should be the melting temperature of constituent minerals as opposed to the solidus. We use the melting temperatures of major mantle minerals as compiled by Ohtani (1983) and Presnall and Walker (1993). Parameters in Eq. (5) used in this study are summarized in Table 2.

Important points to be emphasized are (1) due to the variation in stress in a slab, deformation mechanisms may change from low stress regions to high stress regions. In high stress regions, the Peierls mechanism will operate and this will limit the stress magnitude in these regions. (2) When grain-size reduction occurs due to a phase transformation, grain-size sensitive creep may operate. This will significantly reduce the strength when the degree of grain-size reduction is large. The degree of grain-size reduction depends on the temperature of a slab, as discussed below, and has

Table 2
Creep law parameters for olivine and spinel (ringwoodite)[a]

	A (s^{-1})	n	m	g	σ_P (GPa)	q
Olivine						
Peierls[b]	5.7×10^{11}	0	0	31	8.5	2
Power-law[c,d]	3.5×10^{22}	3.5	0	31	–	0
Diffusion[c,d]	8.7×10^{15}	1	2.5[e]	17	–	0
Spinel						
Peierls[f]	7×10^{11}	0	0	31	10	2
Power-law[g]	4×10^{22}	3.5	0	31	–	0
Diffusion[g]	2×10^{16}	1	2.5[e]	17	–	0

[a] A generic creep law formula of $\dot{\varepsilon} = A(\sigma/\mu)^n(b/d)^m \exp[-g(T_m(P)/T)(1-(\sigma/\sigma_P))^q]$ is assumed ($T_m(P)$: melting temperature).
[b] Evans and Goetze (1979).
[c] Karato et al. (1986), Hirth and Kohlstedt (1995b) and Karato and Rubie (1997).
[d] Karato et al. (1986), Karato et al. (1993) and Hirth and Kohlstedt (1995a).
[e] Grain-size exponent (m) is assumed to be 2.5 considering the uncertainties involved in extrapolation (see Karato et al., 1986; Karato and Wu, 1993).
[f] Estimated from Chen et al. (1998) and a scaling law by Frost and Ashby (1982).
[g] Estimated from Karato et al. (1998) and Lawlis et al. (2001).

an important influence on the temperature dependence of rheology in such regions.

The effects of phase transformations on rheology were reviewed by Karato (1997b) and include: (1) intrinsic effects of change in crystal structure and chemical bonding; (2) transient effects of internal stress/strain; (3) transient effects caused by the change in temperature due to latent heat; and (4) transient effects through the change in grain-size. As far as the olivine to spinel transformation is concerned, the first factor (effects of crystal structure change) is not very large (Karato, 1989a, 1997b; Lawlis et al., 2001). The second mechanism, namely, "transformational plasticity" was suggested by Gordon (1971), Sammis and Dein (1974), Poirier (1982) and Panasyuk and Hager (1998), but there has been no convincing experimental demonstration of this effect on silicate minerals or their analogs (Meike, 1993; Zhao et al., 1999). Paterson (1983) proposed that recovery would reduce this effect and this mechanism will not be effective at least in warm regions (say $T > 1300\,\mathrm{K}$). Recent experimental data on dislocation recovery in olivine under deep upper mantle conditions (Karato et al., 1993) support this notion. Therefore, in this paper, we consider the effects of crystal structure change, effects of latent heat release and the effects of grain-size reduction on rheology and ignore the effects of internal stress/strain.

Vaughan and Coe (1981) first suggested the possible importance of grain-size reduction associated with mantle phase transformations on mantle rheology. Rubie (1984) elaborated this idea and discussed the causes of regional variation in dynamics of deep slabs as related to the olivine–spinel transformation. However, none of these early works provided a quantitative estimate of these effects and, even qualitatively, the effects of temperature were not appropriately included because data on transformation kinetics and rheology of high pressure phase(s) was not available and a quantitative theory for transformation kinetics had not been developed. Riedel and Karato (1997) presented the first quantitative calculation of grain-size and its effects on rheology associated with the olivine to spinel transformation. The essence of the model is that the grain-size of a new phase (spinel) after the transformation is determined by impingement and therefore by the balance of nucleation and growth (Axe and Yamada, 1986). In general, the grain-size, d, of a new phase after the completion of a transformation and the time to complete the transformation, τ, are related to the kinetics of nucleation and growth as (Axe and Yamada, 1986; Riedel and Karato, 1996, 1997)

$$\tau \sim [Y(T, P)^{-k} I_k(T, P)]^{-1/(k+1)} \tag{6}$$

$$d \sim \left(\frac{Y(T, P)}{I_k(T, P)}\right)^{1/(k+1)} \tag{7}$$

respectively, where $Y(T, P)$ is the growth rate, $I_k(T, P)$ the nucleation rate and k a constant that depends on nucleation mechanism ($k = 2$ for inter-granular nucleation, $k = 3$ for intra-granular nucleation).

In a subducting slab, the characteristic time for phase transformation is largely controlled by the rate of subduction. It can be shown that under most conditions in subducting slabs (say $T > 900\,\mathrm{K}$ for the olivine to ringwoodite transformation), the characteristic time is given by $\tau \sim \Delta L/v_{\mathrm{sub}}$ (ΔL is the overshoot depth for transformation (an excess depth beyond the equilibrium depth needed for transformation), v_{sub} velocity of subduction, which is only weakly dependent on temperature (Riedel and Karato, 1997)). In this case, the temperature dependence of grain-size is controlled mostly by the temperature dependence of the growth rate. Consequently, the size of grains after the completion of a transformation tends to be small if the transformation occurs at low temperatures. Under these circumstances, the grain-size after the transformation depends on temperature as

$$d \sim Y\tau \propto \exp\left(-\frac{H_{\mathrm{g}}^*}{RT}\right) \tag{8}$$

where H_{g}^* is the effective activation enthalpy that controls grain-size, which is close to the activation enthalpy for growth.

The rheological flow laws summarized in Table 2 show that when the spinel grain-size is less than $\sim 100\,\mu\mathrm{m}$, then grain-size sensitive flow will dominate. In this case, the strain-rate depends on temperature as

$$\dot{\varepsilon} \propto \exp\left(\frac{mH_{\mathrm{g}}^* - gRT_{\mathrm{m}}(P)}{RT}\right) \tag{9}$$

where we assumed $q = 0$ for simplicity. For a typical range of parameters for mantle minerals, $mH_{\mathrm{g}}^* - gRT_{\mathrm{m}}(P) > 0$, which means that the effective activation enthalpy is negative and therefore the strain-rate

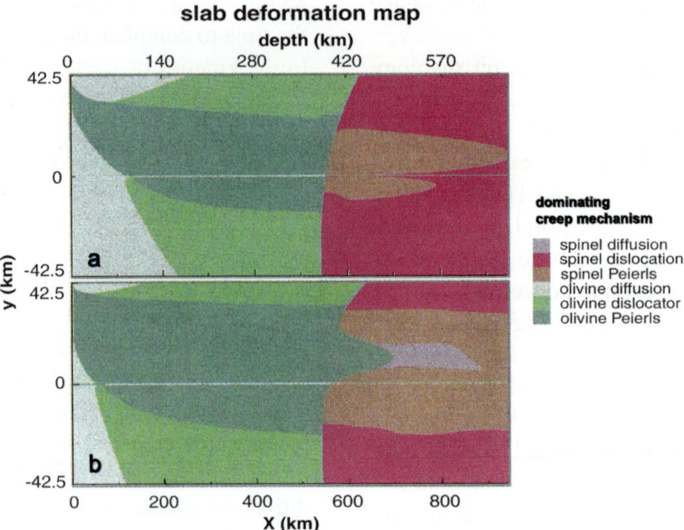

Fig. 4. Dominant deformation mechanisms in subducting slabs. Initial temperature distribution corresponds to that for a 100 million years oceanic lithosphere of 85 km thickness. The cases for subduction velocities of (a) 4 and (b) 10 cm per year are shown. Because stress, temperature, pressure and grain-size change significantly in space for a given slab, dominant mechanisms of deformation change in a complicated fashion. In high stress, low temperature regions, the Peierls mechanism dominates. In moderate stress, moderate to large grain-size regions, power-law creep dominates. Diffusion creep plays an important role in cold, fine-grain regions in the center of slabs after a phase transformation. Note that such a pattern also depends on the velocity of subduction, which controls the temperature distribution and the magnitude of stress.

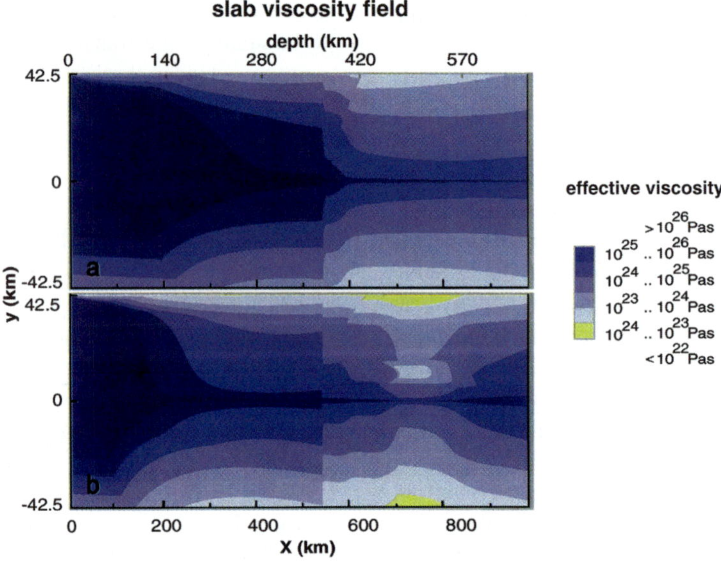

Fig. 5. Distribution of effective viscosity in subducting slabs corresponding to the cases shown in Fig. 4: (a) for $v_{sub} = 4$ cm per year and (b) for $v_{sub} = 10$ cm per year. Note that the center of the slab has high viscosities for 4 cm per year (a), but the central portions in the deep part of a slab for 10 cm per year (b) has low viscosity after the phase transformation because of small spinel grain-size. This low viscosity region recovers its strength due to progressive grain-growth at a greater depths.

will be *larger* for *lower* temperatures. This unusual behavior comes from the strong temperature dependence of grain-size after a phase transformation.

After the completion of a phase transformation, we use the grain-growth law to estimate the increase in grain-size. The kinetics of grain-growth have been investigated for olivine (Karato, 1989b), but a detailed data set on high pressure polymorphs is not available. In this study, we used preliminary data on grain-growth kinetics in wadsleyite studied at transition zone conditions ($T = 1473$–1873 K, $P = 16$ GPa; Karato, unpublished data) and assume that the kinetics of grain-growth in ringwoodite are the same as those in wadsleyite. We assume a canonical form of grain-growth kinetics (see Karato, 1989b)

$$d^2(t) - d^2(0) = k_0 \exp\left(\frac{-H^*_{gg}}{RT}\right) t \qquad (10)$$

where $d(t)$ is grain-size at time t, H^*_{gg} the activation enthalpy for grain-growth (~500 kJ/mol) and k_0 the pre-exponential factor (~9×10^3 m^2/s). When a phase transformation occurs at low temperatures, small grains will survive for a long period. In contrast, when a phase transformation occurs at high temperatures, grain-growth is so fast that small grain-size will not survive for a long time. Thus, a slab will develop and maintain a significant weak fine-grained

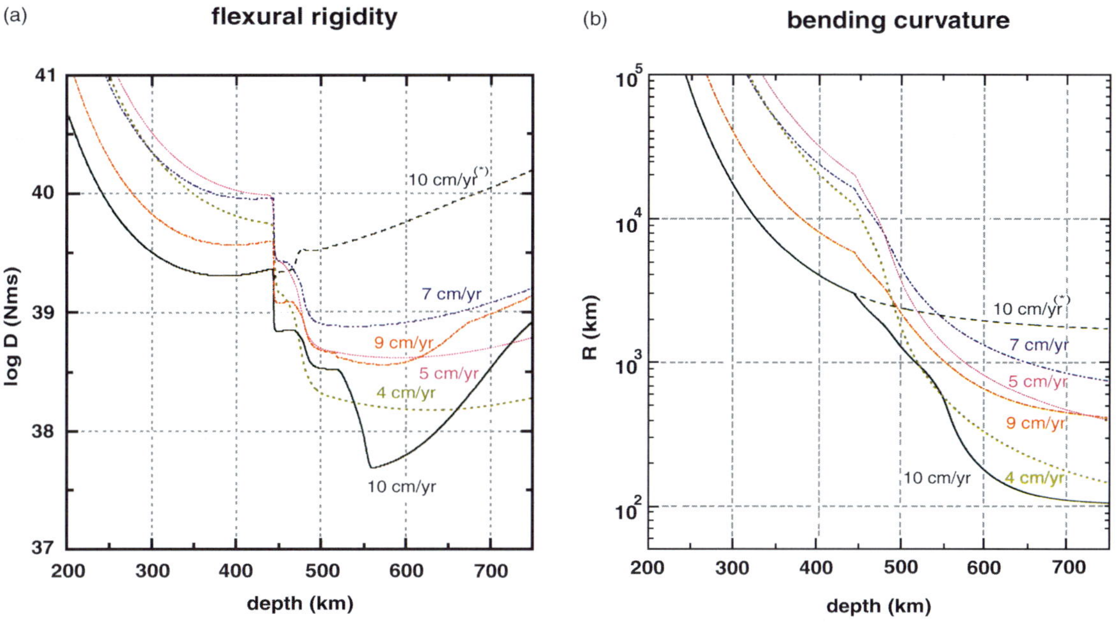

Fig. 6. (a) Flexural rigidity (D) vs. depth for slabs for various subduction velocities. An initial temperature distribution corresponding to oceanic lithosphere of 100 million years of age is used. The broken curve for 10 cm per year shows the results when the effects of latent heat release and grain-size reduction are ignored. Flexural rigidity generally decreases with depth due to the increase in temperature, and there is an abrupt decrease around 440 km due to the effect of latent heat release. In addition, the production of fine-grained spinel in the deep portions of fast slabs results in significant reduction in flexural rigidity. Velocity of subduction affects flexural rigidity through two different mechanisms. First, it affects flexural rigidity through the change in temperature (fast slabs are cold). Usually low temperatures result in high strength, but when the degree of grain-size reduction is large, strength will be low for low temperatures due to the large effect of temperature on grain-size. Second, velocity of subduction changes stress levels that also affect rheology. The second effect is important where the stress-sensitive Peierls mechanism dominates. The effects of grain-size reduction and of the stress magnitude on rheology result in a negative correlation between flexural rigidity and velocity of subduction. (b) Depth variation of radius of curvature. The broken curve for 10 cm per year shows the results when the effects of latent heat release and grain-size reduction are ignored. Steady-state deformation is assumed and the radius of curvature at 200 km is assumed to be infinite. The radius of curvature decreases significantly with depth due to rheological softening.

region only when a phase transformation occurs at low temperatures.

3.2.2. Slab rheology models

The calculation of flexural rigidity and rheological structure is made as follows. First, we assume a certain initial temperature profile for a slab corresponding to a certain age of subduction. A standard error function geothermal model of oceanic lithosphere is used (Turcotte and Schubert, 1982). We solve the equation for thermal conduction incorporating the equation for the kinetics of the olivine–spinel transformation that includes the effects of latent heat. This provides a temperature distribution (e.g. Daessler et al., 1996; Riedel and Karato, 1997). The solution of these equations provides nucleation and growth rates at each point in the slab, from which we estimate the grain-size using the scaling law described by Riedel and Karato (1996). Given temperature, pressure and grain-size, we calculate viscosity corresponding to an assumed moment M and initially assumed strain-rate $\dot{\varepsilon}_0$ for three different deformation mechanisms. We choose the one that gives the lowest viscosity. Then the right-hand side of Eq. (2) is used to calculate $\dot{\varepsilon}_0$. This calculation is repeated until the calculated value of $\dot{\varepsilon}_0$ agrees with the initial value within a fraction of 10^{-5}.

The results of such calculations are shown in Figs. 4–6. Fig. 4 shows the dominant mechanisms of deformation for slabs with two different temperature distributions. Dominant mechanisms of deformation change in a complicated way as a function of subduction parameter and space. The Peierls mechanism is a dominant deformation mechanism in the central portions of slabs where temperature is low and stress is high. However, in cold slabs, the central portion will be replaced with fine-grained ringwoodite after the olivine–ringwoodite phase transformation, and diffusional creep dominates in that portion thus significantly reducing the viscosity. In general, outer warm regions of slabs deform by dislocation creep. Effective viscosity varies laterally and also with the velocity of subduction (Fig. 5). Flexural rigidity D generally decreases and hence the radius of curvature decreases with depth as a result of the increase in temperature (Fig. 6a). Flexural rigidity (radius of curvature) decreases abruptly at ~440 km deep due to the effects of latent heat release. However, the radius of curvature remains large until a slab subducts deeper than ~500 km (Fig. 6b). The effect of grain-size reduction is large for cold fast slabs and is important particularly in regions deeper than ~500 km.

The flexural rigidity or the curvature in the deep transition zone has an important influence on the interaction of the slab with the 660 km discontinuity. Fig. 7a and b shows the variation of flexural rigidity (or the radius of curvature) at 600 km depth as a function of subduction velocity. The slab flexural rigidity (radius of curvature) first increases with subduction velocity, but there is a peak in D at ~7 cm per year

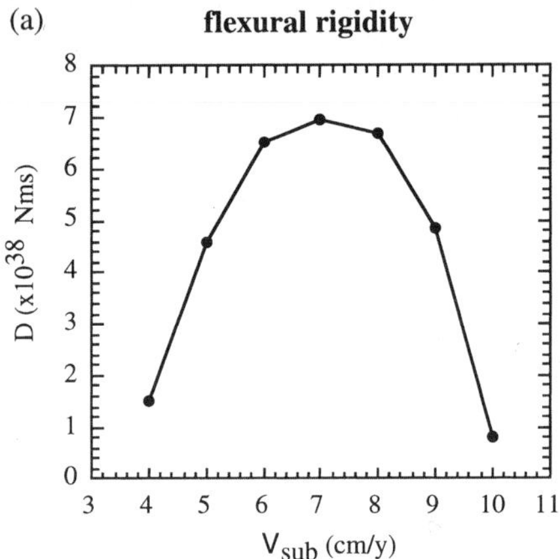

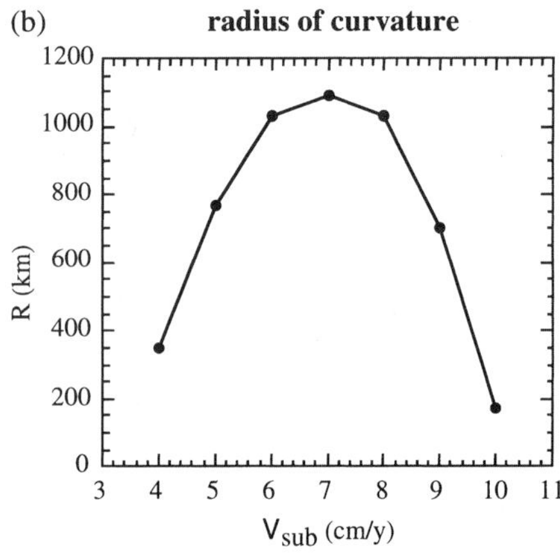

Fig. 7. The flexural rigidity D (a) and the radius of curvature (b) of subducted slabs at 600 km deep as a function of velocity of subduction. The rigidity and radius of curvature both increase with velocity in the range ~4–7 cm per year because of the decrease in temperature, but beyond ~8 cm per year, these parameters decrease with velocity of subduction mainly because of the effects of temperature on grain-size. Slabs are strongest at the intermediate velocities/temperatures.

and the flexural rigidity (radius of curvature) then decreases with subduction velocity at faster subduction velocities. Variation of the flexural rigidity (curvature) with the subduction velocity is partly due to the assumed increase in bending moment with subduction velocity. The sensitivity of slab rheology on the bending moment is discussed in Riedel et al. (1998) in detail. The bending moment can be modified by changing the force as well as the dip angle. A high dip angle near the 660 km discontinuity leads to a small bending moment.

We may note that the two trends shown in Figs. 5 and 7 can be seen clearly for the parameters (bending moment and initial thickness of the slab) that we choose. For other parameter sets, only one trend may be seen. For example, when one chooses much smaller initial thickness, then the slab temperatures are so high that the effects of phase transitions to reduce the slab strength are not seen clearly. The parameter set that we use here is chosen on the basis of geodynamical estimates (e.g. Davies, 1980) as well as mineral physics. The effects of change in parameters are explored in Riedel et al. (1998).

4. Discussion

4.1. Uncertainties in materials properties

4.1.1. Flow laws and grain-growth kinetics

Flow laws of olivine have now been well constrained at least under dry conditions (Karato et al., 1986, 1993; Hirth and Kohlstedt, 1995a,b, 1996). This includes the effects of pressure (Karato and Rubie, 1997) which are important under transition zone conditions. In contrast, flow laws for high pressure phases are still poorly constrained. The most critical points are the transition conditions that separates dislocation creep and grain-size sensitive creep and the flow law in the grain-size sensitive regime. However, due to an experimental study at transition zone conditions (Karato et al., 1998), we now have a constraint on the transition conditions between diffusion and dislocation creep in ringwoodite under laboratory conditions (the transition grain-size is $\sim 1\,\mu m$ at $\sigma/\mu \sim 10^{-2}$ and $T/T_m \sim 0.7$). To translate these conditions to geological conditions, we need to know flow law parameters particularly the stress exponents and grain-size exponents. Flow law parameters in the grain-size sensitive regime currently have some uncertainties. Vaughan and Coe (1981) fitted their data for Mg_2GeO_4 spinel with a non-linear grain-size sensitive creep law ($n = 2$ and $m = 2$). However, a later systematic study on creep in spinels (Okamoto, 1989) demonstrated that diffusion creep ($n = 1$ and $m = 2$–3) or power-law creep ($n = 3$–4 and $m = 0$) dominates under most conditions. Our detailed study on germanate spinels also supports this conclusion (Lawlis et al., 2001). Therefore we used the flow law parameters for power-law creep and diffusion creep in this study for estimating the transition conditions to grain-size sensitive creep and estimate the transition grain-size under geological conditions ($\sigma/\mu \sim 10^{-3}$, $T/T_m \sim 0.4$–0.7) as $\sim 10^2$–$10^3\,\mu m$. Similar results on transition conditions between dislocation and diffusion creep have also been obtained in wadsleyite (Lawlis et al., 1999). Likewise, Chen et al. (1998) and Mosenfelder et al. (2000) noted a similarity in rheology between wadsleyite and ringwoodite in the low temperature plasticity regime (Peierls mechanism).

Uncertainties in grain-growth kinetics are large at this stage. To see the effects of uncertainties on the rheology of slabs in our models, we calculated the grain-sizes using different activation enthalpies by changing k_0 and H^*_{gg} simultaneously to satisfy the experimental constraints. A change in H^*_{gg} from 500 to 400 kJ/mol leads to only minor changes for a critical temperature for grain-growth (~80 K). We therefore conclude that the uncertainties in grain-growth kinetics will not significantly affect the main conclusions of this paper.

4.1.2. Transformation kinetics: effects of intra-granular transformation

Recently Rubie and his colleagues reported new mechanisms of phase transformation. Kerschhofer et al. (1996, 1998, 2000) found that the olivine to wadsleyite and olivine to ringwoodite transformations can occur through intra-granular nucleation as well as inter-granular (grain boundary) nucleation. Under some conditions, new phases are formed inside of old grains as opposed to at grain-boundaries, usually associated with stacking faults or dislocations. Such a new mechanism could modify the conclusions reached above, although the details of its effects are unclear, because quantitative details of the kinetics of this mechanism have not been well constrained (some discussions on the kinetics of this process are given in Kerschhofer et al., 2000). We can make the following points.

1. In terms of the above analysis of grain-size evolution, the dependence of grain-size on temperature will be largely the same for both inter- and intra-granular nucleation mechanisms, because the temperature sensitivity of grain-size depends on the growth rate rather than the nucleation rate. Therefore one of the most important conclusions of our model, namely, the strong temperature dependence of grain-size after transformation will be valid even if this new mechanism operates.
2. The effects of intra-granular nucleation on rheology may be less dramatic than those of inter-granular nucleation, because the rheological effects of grain-size reduction depend strongly on the manner in which fine grained regions are connected (percolation transition). This percolation transition is easier for inter-granular nucleation than intra-granular nucleation (Riedel and Karato, 1996). Consequently, rheological weakening would occur at an earlier stage for inter-granular nucleation than for intra-granular nucleation.
3. Irrespective of the details, the operation of additional mechanisms will enhance transformation kinetics, so that the depth interval in which two phases coexist will be smaller.

In short, we consider that the recently-documented intra-granular nucleation mechanisms may modify the details of grain-size estimation, but the main conclusion on the temperature dependence of grain-size will not be modified significantly. However, the kinetic parameters relevant to intra-granular nucleation mechanisms have not been well constrained including the effects of stress (Mosenfelder et al., 2000). Therefore it is difficult to quantitatively discuss the effects of this mechanism at this stage.

4.1.3. Transformation kinetics: effects of transformation stress/strain

A first order phase transformation such as the olivine–spinel transformation is associated with a volume change. Consequently, plastic flow must occur to accommodate strain caused by the volume change. If untransformed material is completely surrounded by a strong material, then the volume change that occurs in that material results in a change in pressure that could prevent further transformation. Kubo et al. (1998b) and Mosenfelder et al. (2000) noted significant reduction in the growth rate of a new phase in the later stage of transformation and interpreted their results as caused by such an effect. However, the relevance of such observations on the phase transformation in a real slab is not clear, because the effect of plastic flow is dependent on the geometry of transforming materials. For a more general geometry of two phases, plastic flow to accommodate volumetric strain occurs mostly in a softer phase (olivine) (e.g. Greenwood and Johnson, 1965) and results in an increase in free energy of a weak phase. In such a case, the effects of plastic strain would *enhance* phase transformation rather than suppress it.

The effect of transformation stress/strain on the grain-size of newly formed spinel is not included in our model. However, a simple calculation shows that a change in free energy caused by plastic strain is usually significantly smaller than the difference in free energies between the two phases (olivine and spinel) and consequently the effects of transformation stress/strain on the grain-size of spinel will be small.

4.1.4. Effects of transformation strain-induced recrystallization

Rubie et al. (1998) also suggested that the strain-energy stored during a phase transformation might lead to grain-size reduction, a process analogous to dynamic recrystallization. Such a process is not considered in our model and could modify our conclusions. In principle, any additional mechanisms of grain-size reduction will further reduce the strength of subducting slabs. A key question is whether this new mechanism significantly modifies our main conclusion on the temperature dependence of rheology. Because the details of this process have not been analyzed, it is difficult to make any assessment of potential influence of this process at this stage. However, we can make some qualitative arguments. The size of grains formed by this process is determined by the competition between nucleation of "recrystallized" grains and their growth, similar to Eq. (7) (see Shimizu, 1998). The driving force for this type of recrystallization is dislocation energy which is proportional to dislocation density. The higher the dislocation density, the smaller is the recrystallized grain-size. If one uses the assumption that the dislocation density after the transformation is determined by the volume change due to "geometrically necessary dislocations" as proposed by Poirier (1982), then the driving force is nearly independent of temperature. In this case, the temperature sensitivity of grain-size will depend mostly on the growth rate and hence the grain-size will be smaller when transformation occurs at low temperatures. In short, the effects of temperature on grain-size due to this mechanism is at least qualitatively similar to that of our model.

4.2. Comparison with previous studies

4.2.1. Effects of trench migration

Most of the previous models for slab deformation considered that the variation in trench migration velocities control slab deformation (van der Hilst and Seno, 1993; Griffths et al., 1996; Guillou-Frottier et al., 1996; Olbertz et al., 1997). However, this kinematic explanation has two major problems. First, such a kinematic model would work only when slab deformation is dynamically possible. In fact, all the experimental (Kincaid and Olson, 1987; Griffths et al., 1996; Guillou-Frottier et al., 1996) and theoretical modeling of slab deformation (Christensen, 1996; Olbertz et al., 1997) assumed low slab viscosities to make deformation possible. However, if deformability of slabs can vary from one slab to the other, then the effect of such variation must also be examined in addition to the influence of kinematic boundary conditions.

Second, the observational basis for inferring the importance of trench migration is not strong (Fig. 2). There is no obvious reason to suggest a causal relation between trench migration and slab deflection from the seismological observation and plate kinematics as discussed before. The contrast between western Pacific slabs (slabs with large thermal parameters) and those in the Americas (slabs with small thermal parameters) appears more fundamental. It is possible that the rapid trench migration observed in some of the western Pacific subduction zones could be a *result* of weak slabs and that the rapid trench migration is not the *cause* of deformation.

4.2.2. Effects due to the variation in buoyancy forces

Variation in buoyancy forces that control slab deformation may also be a cause of variation in slab deformation behavior. In our model, this is partly incorporated by assuming that the bending moment acting on a slab depends on subduction velocity. Such an effect has been analyzed by Bina (1996, 1997), Marton et al. (1999) and Schmeling et al. (1999) in more detail (also see Daessler et al., 1996). The pronounced bend observed in some slabs in the middle part of the transition zone (e.g. Tonga, see Fig. 1) might be due to the variation in buoyancy forces in addition to a change in rheology.

4.2.3. Deformation by earthquake faulting

The rheological models considered here assume homogeneous deformation. However, there is evidence for localized deformation in deep slabs. Several analyses suggest that some fraction of the strain in subducted slabs develops by seismic faulting (Sugi et al., 1989; Holt, 1995). In this case, the assumption of homogeneous rheology is not valid in its simplest form. However, deep earthquakes can explain only a part of the total strain. This is true particularly for the deep portions of slabs beneath the seismicity cutoff where the presence of weak slabs is inferred (Castle and Creager, 1998). Therefore, weak rheology in the ductile regime must play a significant role in many slabs.

Weakening in the ductile regime, where deformation is more or less homogeneous, may also be important for the generation of deep earthquakes. Many mechanisms of deep earthquake faulting are based on some form of rheological weakening (Hobbs and Ord, 1988; Green and Burnley, 1989; Kirby et al., 1991). In addition, the complicated rheological structures of deep slabs predicted by the present model suggest that deep earthquakes may originate in the deep slabs where high stress concentration occurs outside the regions of weak fine-grained spinel. This point will be discussed in more detail in Section 4.4.

4.2.4. Preexisting weak zones?

The presence of zones of weakness such as fossil fault planes might influence the deformability of slabs (Zhong and Gurnis, 1995; Silver et al., 1995). Such fossil fault zones may survive deep into the transition zone particularly in the cold portions of slabs. Fossil fault zones may be weaker than other regions due to a small grain-size and/or to higher content of hydrous minerals (or higher content of hydrogen in nominally anhydrous minerals). Deformation along these zones may also contribute to slab deformation. However, it is not straightforward to explain the observed variations in the style of slab deformation by this mechanism, which would predict a positive correlation between seismicity at trenches and slab deformation in the transition zone. Such a correlation is not consistent with the observations (Engdahl et al., 1998).

4.3. Explanation of the variability of slab deformation behavior

How can we explain the diversity of slab deformation behavior by our model? First, our model provides a natural explanation for a puzzling observation that slabs with a large thermal parameter (i.e. low temperatures and/or fast subduction velocities) tend to be more prone to deformation. In our model, this occurs because of the strong temperature dependence of grain-size after the olivine–spinel transformation in addition to the higher stresses (bending moment) caused by a larger force. Obviously, if the slab temperature is very high, viscosity is too low and negative buoyancy forces are too small for penetration through the 660 km discontinuity. Over a reasonable range of slab temperatures, slabs with moderately high temperatures will have high enough strength to resist deformation near the 660 km discontinuity, whereas slabs with colder temperatures are weak enough to be significantly deformed. Thus, there will be a temperature range (a "window") in which slabs will be strong enough to penetrate through the 660 km discontinuity without much deformation (see Fig. 7). The 660 km discontinuity will act as a filter for slabs for rheological reasons. This provides a possible explanation for the observed correlation shown in Fig. 2. Among the slabs that are weak enough, the style of deformation will also depend on other factors including trench migration velocity and geometrical factors such as the dip angle.

4.4. Implications for deep earthquakes

Mechanisms of deep earthquakes have been reviewed in several recent papers (e.g. Frohlich, 1989; Kirby et al., 1991, 1996; Green, 1994; Green and Houston, 1995; Karato, 1997b). Among others, the transformation faulting model has attracted much attention recently because experimental support for that model has been reported (Kirby, 1987; Green and Burnley, 1989; Kirby et al., 1991; Green et al., 1990; Tingle et al., 1993; Green and Zhou, 1996). Some of the seismological observations including the depth variation in deep earthquake activity (Frohlich, 1989) and spatial distribution of deep earthquakes (Wiens et al., 1993) appear to be consistent with this model if metastable transformations are assumed. However, some fundamental problems remain with this model (Karato, 1997b). The most serious difficulty with the transformational faulting model is that it predicts much smaller fault dimensions (10–20 km) for deep earthquakes than are observed (40–60 km) (Kikuchi and Kanamori, 1994; Wiens et al., 1994; Tibi et al., 1999). It is also uncertain if a metastable olivine wedge exists or not, particularly in relatively warm slabs such as those beneath South America. In addition, the recent observation of repeating deep earthquakes (Wiens and Snider, 1999) is difficult to reconcile with the transformational faulting model.

Furthermore, the processes of instability associated with phase transformations are not well understood. Most of the experimental studies on transformation faulting are on analog materials at low pressures (Green and Burnley, 1989; Kirby et al., 1991; Green and Zhou, 1996) where the magnitude of differential stresses is comparable to the confining pressure. Green et al. (1990) reported the occurrence of transformation faulting associated with the olivine–wadsleyite transformation in their experiments under the transition zone conditions. However, later studies in which deformation during pressurization is minimized by an improved sample assembly failed to reproduce transformation faulting under almost exactly the same conditions (Dupas-Bruzek et al., 1998; Karato et al., 1998). Furthermore, theoretical understanding of transformation faulting is inadequate. Most of the models are at best qualitative (Kirby, 1987; Green and Burnley, 1989; Kirby et al., 1991; Green, 1994), and a published quantitative model (Liu, 1997) contains an inappropriate use of the Levy-von Mises formulation of non-linear rheology in treating the interaction of internal stress and deformation which leads to a significant overestimate of this interaction. Therefore, the only robust prediction from this model is that the fault plane is limited to the region where the phase transformation occurs and therefore the fault width should be less than 10–20 km (the expected width of the metastable olivine wedge). This prediction is obviously in contradiction to the observations of much grater fault widths, making this mechanism unlikely to be important.

Another model was proposed by Meade and Jeanloz (1991) who reported acoustic emissions associated with the dehydration and amorphization of serpentine under shallow and deep mantle pressures, respectively. They suggested that dehydration may cause intermediate earthquakes and amorphization could cause deep earthquakes. But this model for deep earthquakes has several difficulties. First and most important is that the recent high pressure, temperature experimental study on serpentine showed that the presumed amorphization does not occur under deep slab conditions (Irifune et al., 1996). Second, the physical mechanisms of instability have not been identified and therefore the scaling argument presented by Meade and Jeanloz (1991) is at best speculative. Therefore we consider that this model does not have strong support from mineral physics.

An alternative model for deep earthquakes is the thermal runaway instability (adiabatic instability) model originally proposed by Griggs and Baker (1969) and later elaborated by Ogawa (1987) and Hobbs and Ord (1988). Unlike the mechanisms discussed above, the physics of thermal runaway instability (adiabatic instability) is well understood both theoretically (Backofen, 1972; Poirier, 1980; Fressengeas and Molinari, 1987; Estrin and Kubin, 1988; Hobbs and Ord, 1988) and experimentally (e.g. Basinski, 1957, 1960; Backofen, 1972; Rogers, 1979; Staker, 1981). In addition, it has recently been demonstrated that the likely temperature dependence of thermal diffusivity under low temperature conditions significantly enhances thermal runaway instability particularly in spinel-rich regions (Branlund et al., 2000). Also, shear-induced melting in mantle rocks was documented in naturally-deformed mantle rocks from the Ivrea zone, northern Italy (Obata and Karato, 1995) and melting associated with the 1994 Bolivian deep earthquake was suggested (Kanamori et al., 1998). Therefore the thermal runaway model is worth special attention as a plausible cause of deep earthquakes. In this section, we will first show that thermal runaway is likely to occur in deep slabs, and that this model, when combined with our model of rheological structures, provides a natural explanation for a large width of faults associated with deep earthquakes. We will then discuss some additional implications of this model.

The basic physics of thermal runaway instability is that when deformation of a material occurs rapidly enough in comparison with the time scale of thermal diffusion, heat is accumulated in the regions of high strain that further enhances the deformation; there is therefore a positive feedback between deformation-induced heating and deformation, leading to thermal runaway (Fressengeas and Molinari, 1987; Estrin and Kubin, 1988). When the stiffness of the surrounding system is not too high, then this leads to a runaway instability causing earthquakes (Hobbs and Ord, 1988). Therefore the key issues in thermal runaway are (i) the rate of deformation must be fast and (ii) the degree of thermal feedback must be large.

We use the model proposed by Hobbs and Ord (1988) for the onset of instability, but use a more realistic rheology to evaluate the plausibility of the thermal runaway instability model. For power-law creep, the condition for thermal runaway (adiabatic instability) is given by

$$\Theta > 1 \tag{11}$$

with

$$\Theta \equiv \frac{H^{*}\sigma\varepsilon}{nhRT^{2}\rho C_{P}}\beta \tag{12}$$

where H^{*} is the activation enthalpy for deformation, n the stress exponent (Eq. (5)), h a parameter that is related to the hardening effect ($h \sim 0.1$), ε strain, ρ density, C_P specific heat and β a parameter that measures the efficiency of conversion of mechanical energy to heat defined by

$$\beta = 1 - \exp\left(-\frac{\dot{\varepsilon}L^{2}}{\kappa}\right) \tag{13}$$

where L is the size of the region and κ thermal diffusivity. A similar equation can be derived for a highly non-linear mechanism such as the Peierls mechanism (Kameyama et al., 1999). Eqs. (11)–(13) mean that both low temperature and a high strain-rate are needed for thermal runaway instability. Such a condition is

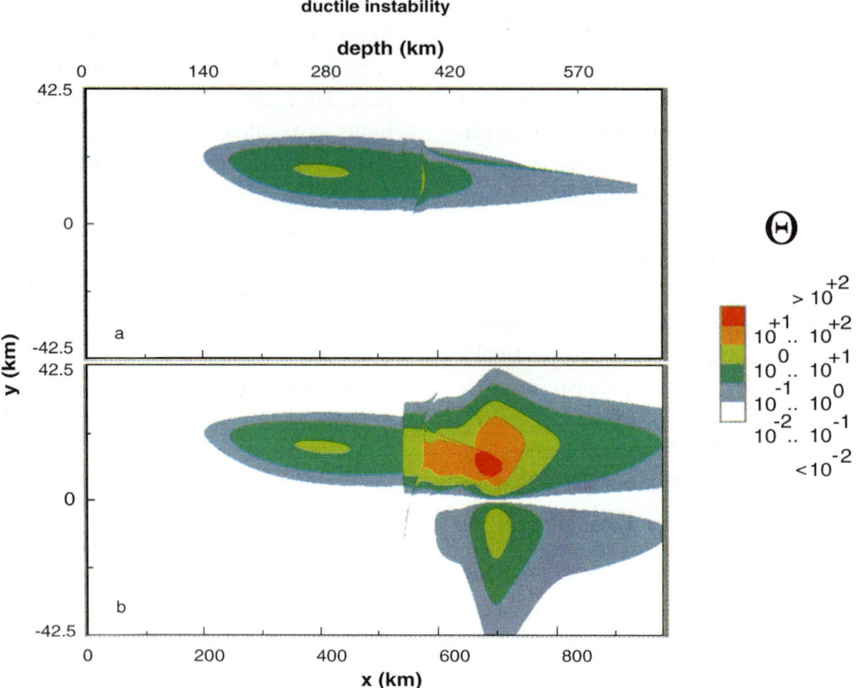

Fig. 8. Distribution of the instability parameter Θ in subducted slabs with $v_{sub} = 10$ cm per year (a) for a case where the effects of latent heat release and grain-size reduction are neglected (corresponding to a broken curve in Fig. 6) and (b) for a case where the effects of latent heat release and grain-size reduction are included. When this parameter exceeds ~1, thermal runaway instability occurs. For a cold slab (10 cm per year), Θ in high strength regions surrounding a weak, fine-grained spinel core in the deep slab exceeds the critical value for thermal runaway instability. Deep earthquakes likely nucleate in these regions. In this case one expects a double seismic zone as seen by Wiens et al. (1993). The width of the seismogenic regions is estimated to be ~40–60 km. Note also that the instability criterion is also satisfied in the shallow portions, providing explanation for intermediate earthquakes.

not usually met because high strain-rates are usually associated with high temperatures. However, because of the overall weakening of slabs (low flexural rigidity) at low temperature conditions, these conditions are likely to be met in deep slabs. The validity of instability criterion (11) is largely confirmed by finite amplitude calculations such as by Kameyama et al. (1999).

Fig. 8 shows plots of Θ. When this parameter becomes larger than ~1, thermal runaway instability will occur. Three points must be noted. First, in the deep cold slab, regions surrounding the weak, fine-grained spinel core likely satisfy this instability criterion. Second, the instability conditions are met also at the shallower depths and can therefore explain the occurrence of intermediate depth earthquakes. Third, the effects of grain-size reduction and the latent heat release associated with the olivine–spinel phase transformation significantly enhances deep earthquake activity (compare Fig. 8a and b).

The model of thermal runaway for deep earthquakes combined with our slab rheology model has some interesting features. First, our model provides a natural explanation for large dimensions of faults observed to be associated with deep earthquakes. In this model, earthquake faulting likely occurs in high strength regions when the strain-rate is high. In cold slabs, these would be the regions surrounding a weak, fine-grained spinel core (Fig. 8b). Thus, our model provides a natural explanation for the observed double seismic zones of deep earthquakes (Wiens et al., 1993) and gives a robust estimate of the width of the seismogenic region to be 40–60 km. This predicted fault width is in excellent agreement with seismological observations of the width of deep earthquake fault zones (Kikuchi and Kanamori, 1994; Wiens et al., 1994). It should be noted that the width of the fault zone for the 1994 Fiji earthquake is determined from the observations of aftershocks. Therefore the faults must have *nucleated* in a wide seismogenic zone rather than propagating to the observed dimensions. Note that thermal instability could also occur in relatively warm slabs if a large enough moment is applied (not shown here). In this case, the presence of a metastable olivine wedge is not required for deep earthquakes. The seismogenic regions would correspond to high strength regions in relatively cold portions of slabs. The width of the seismogenic zone depends on the magnitude of the bending moment and is estimated to be 30–60 km. In contrast to our model, the transformation faulting model proposed previously (Kirby, 1987; Green and Burnley, 1989; Kirby et al., 1991) clearly fails to account for this important observation and is therefore no longer a tenable model for deep earthquakes. In addition, a different mechanism is needed to explain intermediate earthquakes in the transformation faulting mode, which also makes this model less attractive.

How does this thermal instability model explain some of the other characteristics of deep earthquakes? The depth variation of seismicity is characterized by a continuous decrease down to ~400 km followed by a peak in seismicity at around 550–600 km and finally there is a sharp cutoff at ~680 km (e.g. Frohlich, 1989). The present model provides a clear explanation for the peak at around 550–600 km. Our model predicts that the strain-rates should be high in this depth region because of the low flexural rigidity. A high stress concentration in high strength regions surrounding a weak, fine-grained spinel core enhances the thermal runaway instability in this depth region. A large magnitude of latent heat release in the deep portions of slabs will also enhance thermal runaway instabilities (Bina, 1998). Intermediate depth earthquakes that occur from ~200 to ~400 km depth region may also be attributed to thermal runaway instability, because warm regions of slabs in this depth region could satisfy the necessary condition for thermal runaway instability (Hobbs and Ord, 1988; Fig. 8). A clear shutoff of seismicity at a depth of ~680 km is most likely due to the loss of strength of slabs in the lower mantle due to superplasticity (Ito and Sato, 1990; Karato et al., 1995) and/or to the absence of strong forces in this depth region.

Kanamori et al. (1998), Estabrook (1999) and Wiens (this volume) showed that deep earthquakes are characterized by low seismic efficiency and slow rupture velocity, particularly those that occur in warm slabs. Kanamori et al. (1998) argued that a thermal instability model provides a natural explanation for the low seismic efficiency: thermal runaway instability involves significant plastic flow and therefore some of the stored energy is dissipated as heat.

Another characteristic of deep earthquakes is the low frequency of aftershocks, particularly those with small magnitude (e.g. Frohlich, 1989). Wiens and Gilbert (1996) (see Wiens, this volume) also showed that the aftershock frequency tends to be lower for warmer slabs particularly for those with small magnitudes (which leads to lower b values in the Gutenberg–Richter relation for warm slabs). Other correlations between fault characteristics and slab temperature include those seismic efficiency and the nature of the source–time function (Estabrook, 1999). The thermal runaway instability model provides a natural explanation for these observations. Thermal runaway instability is scale-dependent because it is sensitive to the time scale of deformation relative to the time scale of thermal diffusion which depends on the length scale of deformation (see Eq. (13)). Consequently, small scale instabilities that would result in earthquakes with small magnitudes are difficult to realize through this process (for the same reason, laboratory studies of adiabatic instability is difficult; one needs an exceedingly high strain-rate to produce an adiabatic instability in a small sample; Kitamura et al., 1992).

5. Concluding remarks

In most laboratory experiments or numerical modeling of convection or slab deformation, a simple rheology for Earth materials, such as constant viscosity or a very weakly temperature dependent viscosity, has been assumed. This is firstly due to computational difficulties in incorporating more "realistic" rheologies but also because if one uses such "realistic" rheologies (strongly temperature-dependent rheology), one would not see much deformation of slabs (Davies, 1995). These studies have led to the notion that it is mainly the kinematic boundary conditions that control the variation in deformation behavior of slabs and the importance of the rheology of slabs has not been well appreciated. We have shown from both theoretical and observational backgrounds that realistic slab rheology must be considered as an important parameter that plays a critical role in slab deformation when the nature of slab-transition zone interaction is investigated. Consideration of realistic rheology is also critical in understanding the mechanisms of deep earthquakes.

The key features of our model are a strong dependence of the grain-size of spinel on temperature and a highly stress-dependent rheology through power-law creep and the Peierls mechanism. We used a theoretical model to estimate grain-size and experimental data on deformation of olivine and spinel (ringwoodite). Although many of the rheological parameters for high pressure minerals are poorly constrained, these qualitative features are well documented in material science (e.g. Frost and Ashby, 1982) and seem robust. In fact, recent direct high-pressure deformation experiments of silicate spinel (ringwoodite) demonstrate the important role of grain-size sensitive creep at small grain-sizes (Karato et al., 1998).

We have focused on slab deformation above the 660 km discontinuity. This is firstly because the deformability of slabs in that region controls their fate upon collision with the 660 km discontinuity, and secondly because much less is known about the rheology and the kinetics of phase transformations for lower mantle minerals. However, most results of seismic tomography show significant thickening of slabs in the lower mantle (van der Hilst et al., 1997; Bijwaard et al., 1998), suggesting that slabs there are weak. Effects of grain-size reduction, as investigated in this study, may also operate in the lower mantle (Ito and Sato, 1990; Kubo et al., 2000).

Finally, if some slabs are indeed weak in the deep transition zone, as our model suggests, then the role of the 660 km discontinuity to modify convection patterns must be significant. Such an effect would be more pronounced at high Rayleigh numbers (Christensen and Yuen, 1985), therefore the convection style could have evolved with time as mantle temperatures and hence Rayleigh number has changed (Breuer and Spohn, 1995; Honda, 1995). Thus, the nature of interaction of convection currents with the 660 km discontinuity is likely to vary both regionally and temporally. Rheology of mantle materials, particularly that of subducting slabs, appears to play an important role in controlling the complexity of mantle convection.

Acknowledgements

This research was supported by NSF (S.K., D.A.Y.), Alexander von Humboldt Stiftung (S.K.), the Deutsche Forschungsgemeinschaft (M.R.R.) and the European Science Foundation (M.R.R.). We thank Chuck Estabrook, Ljuba Kerschhofer, Carolina Lithgow-Bertelloni, Jed Mosenfelder, Dave Rubie, Tetsuzo Seno, Rob van der Hilst and Doug Wiens for helpful discussions. Comments by Craig Bina, Dave Rubie and an anonymous reviewer were helpful in improving the manuscript.

References

Akaogi, M., Ito, E., Navrotsky, A., 1989. Olivine-modified spinel–spinel transitions in the system Mg_2SiO_4–Fe_2SiO_4: calorimetric measurements, thermochemical calculation, and geophysical application. J. Geophys. Res. 94, 15671–15685.

Axe, J.D., Yamada, Y., 1986. Scaling relations for grain autocorrelation functions during nucleation and growth. Phys. Rev. B89, 10161–10177.

Backofen, W.A., 1972. Deformation Processing. Addison-Wesley, Reading, MA.

Basinski, Z.S., 1957. The instability of plastic deformation of metals at very low temperatures. Proc. R. Soc. London Ser. A 240, 229–242.

Basinski, Z.S., 1960. The instability of plastic deformation of metals at very low temperatures, II. Aust. J. Phys. 13, 354–358.

Bijwaard, H., Spakman, W., Engdahl, E.R., 1998. Closing the gap between regional and global travel time tomography. J. Geophys. Res. 103, 30055–30078.

Bina, C.R., 1996. Phase transition buoyancy contributions to stresses in subducting lithosphere. Geophys. Res. Lett. 23, 3563–3566.

Bina, C.R., 1997. Patterns of deep seismicity reflect buoyancy stresses due to phase transitions. Geophys. Res. Lett. 24, 3301–3304.

Bina, C.R., 1998. A note on latent heat release from disequilibrium phase transformations and deep seismogenesis. Earth Planet. Space 50, 1029–1034.

Branlund, J.M., Kameyama, M., Yuen, D.A., Kaneda, Y., 2000. Effects of temperature-dependent thermal diffusivity on shear instability in a viscoelastic zone: implications for faster ductile faulting and earthquakes in the spinel stability field. Earth Planet. Sci. Lett. 182, 171–185.

Breuer, D., Spohn, T., 1995. Possible flush instability in mantle convection at the Archean-Proterozoic transition. Nature 378, 608–610.

Castle, J.C., Creager, K.C., 1998. Topography of the 660 km seismic discontinuity beneath Izu–Bonin: implications for the tectonic history and slab deformation. J. Geophys. Res. 103, 12511–12527.

Chapple, W.M., Tullis, T.E., 1977. Evaluation of the forces that drive the plates. J. Geophys. Res. 82, 1967–1984.

Chen, J., Inoue, T., Weidner, D.J., Wu, Y., Vaughan, M.T., 1998. Strength and water weakening of mantle minerals, olivine, wadsleyite and ringwoodite. Geophys. Res. Lett. 25, 575–578.

Christensen, U.R., 1996. The influence of trench migration on slab penetration into the lower mantle. Earth Planet. Sci. Lett. 140, 27–39.

Christensen, U.R., Yuen, D.A., 1984. The interaction of a subducting lithospheric slab with a chemical or phase-boundary. J. Geophys. Res. 89, 4389–4402.

Christensen, U.R., Yuen, D.A., 1985. Layered convection induced by phase-transitions. J. Geophys. Res. 90, 291–300.

Daessler, R., Yuen, D.A., Karato, S., Riedel, M.R., 1996. Two-dimensional thermo-kinetic model for the olivine–spinel transition in subducting slabs. Phys. Earth Planet. Inter. 94, 217–239.

Davies, G.F., 1980. Mechanics of subducted lithosphere. J. Geophys. Res. 85, 6304–6318.

Davies, G.F., 1995. Penetration of plates and plumes through the mantle transition zone. Earth Planet. Sci. Lett. 133, 507–516.

Ding, X.-Y., Grand, S.P., 1994. Seismic structure of the deep Kurile subduction zone. J. Geophys. Res. 99, 23767–23786.

Dupas-Bruzek, C., Sharp, T.G., Rubie, D.C., Durham, W.B., 1998. Mechanisms of transformation and deformation in $Mg_{1.8}Fe_{0.2}SiO_4$ olivine and wadsleyite under non-hydrostatic stress. Phys. Earth Planet. Inter. 108, 33–48.

Engdahl, E.R., van der Hilst, R.D., Berrocal, J., 1995. Imaging of subducted lithosphere beneath South America. Geophys. Res. Lett. 22, 2317–2320.

Engdahl, E.R., van der Hilst, R.D., Buland, R.P., 1998. Global teleseismic earthquake relocation with improved travel times and procedures for depth determination. Bull. Seism. Soc. Am. 88, 722–743.

Estabrook, C.H., 1999. Body wave inversion of the 1970 and 1963 South American large deep-focus earthquakes. J. Geophys. Res. 104, 28751–28767.

Estrin, Y., Kubin, L.P., 1988. Plastic instabilities: classification and physical mechanisms. Res. Mechanics 23, 197–221.

Evans, B., Goetze, C., 1979. Temperature variation of hardness of olivine and its implication for polycrystalline yield stress. J. Geophys. Res. 84, 5505–5524.

Forsyth, D.W., Uyeda, S., 1975. On the relative importance of driving forces of plate motion. Geophys. J. R. Astr. Soc. 43, 163–200.

Fressengeas, C., Molinari, A., 1987. Instability and localization of plastic flow in shear at high strain rate. J. Mech. Phys. Solids 35, 185–211.

Frohlich, C., 1989. The nature of deep focus earthquakes. Ann. Rev. Earth Planet. Sci. 17, 227–254.

Frost, H.J., Ashby, M.F., 1982. Deformation Mechanism Maps. Pergamon Press, Oxford, p. 168.

Fukao, Y., Obayashi, M., Inoue, H., Nenbai, M., 1992. Subducting slabs stagnant in the mantle transition zone. J. Geophys. Res. 97, 4809–4822.

Giardini, D., 1992. Space-time distribution of deep seismic deformation in Tonga. Phys. Earth Planet. Inter. 74, 75–88.

Giardini, D., Woodhouse, J.H., 1984. Deep seismicity and modes of deformation in Tonga subduction zone. Nature 307, 505–509.

Gordon, R.B., 1971. Observations of crystal plasticity under pressure with application to the Earth's mantle. J. Geophys. Res. 76, 2413–2418.

Grand, S., 1994. Mantle shear structure beneath the Americas and surrounding oceans. J. Geophys. Res. 99, 11591–11621.

Green II, H.W., 1994. Solving the paradox of deep earthquakes. Sci. Am. 271, 64–71.

Green II, H.W., Burnley, P.C., 1989. A new self-organizing mechanism for deep focus earthquakes. Nature 341, 733–737.

Green II, H.W., Houston, H., 1995. The mechanics of deep earthquakes. Ann. Rev. Earth Planet. Sci. 23, 169–213.

Green II, H.W., Zhou, Y., 1996. Transformation-induced faulting requires an endothermic reaction and explains the cessation of earthquakes at the base of the mantle transition zone. Tectonophysics 256, 39–56.

Green II, H.W., Young, T.E., Walker, D., Scholtz, C., 1990. Anticrack-associated faulting at very high pressure in natural olivine. Nature 348, 720–722.

Greenwood, G.W., Johnson, R.H., 1965. Deformation of metals at small stresses during phase transformations. Proc. Roy. Soc. Ser A 283, 403–422.

Griffths, R.W., Hackney, R.I., van der Hilst, R.D., 1996. A laboratory investigation of effects of trench migration on the descent of subducted slabs. Earth Planet. Sci. Lett. 133, 1–17.

Griggs, D.T., Baker, D.W., 1969. The origin of deep focus earthquakes. In: Mark, H., Fernbach, S. (Eds.), Properties of Matter Under Unusual Conditions. Wiley/Intersciences, New York, pp. 23–42.

Guillou-Frottier, L., Buttles, J., Olson, P., 1996. Laboratory experiments on the structure of subducted lithosphere. Earth Planet. Sci. Lett. 133, 19–34.

Gurnis, M., Hager, B.H., 1988. Controls of the structure of subducted slabs. Nature 335, 317–321.

Hager, B.H., O'Connell, R.J., 1978. Subduction zone dip and flow driven by plate motions. Tectonophysics 50, 111–133.

Hirth, G., Kohlstedt, D.L., 1995a. Experimental constraints on the dynamics of the partially molten upper mantle: deformation in the diffusion creep regime. J. Geophys. Res. 100, 1981–2001.

Hirth, G., Kohlstedt, D.L., 1995b. Experimental constraints on the dynamics of the partially molten upper mantle. 2. Deformation in the dislocation creep regime. J. Geophys. Res. 100, 15441–15450.

Hirth, G., Kohlstedt, D.L., 1996. Water in the oceanic upper mantle: implications for rheology, melt extraction and the evolution of the lithosphere. Earth Planet. Sci. Lett. 144, 93–108.

Hobbs, B.E., Ord, A., 1988. Plastic instabilities: implications for the origin of intermediate and deep focus earthquakes. J. Geophys. Res. 93, 10521–10540.

Holt, W.E., 1995. Flow fields within the Tonga slab determined from the moment tensors of deep earthquakes. Geophys. Res. Lett. 22, 989–992.

Honda, S., 1995. A simple parameterized model of Earth's thermal history with the transition from layered to whole mantle convection. Earth Planet. Sci. Lett. 131, 357–370.

Honda, S., Yuen, D.A., Balachandar, S., Reuteler, D., 1993. Three-dimensional mantle dynamics with an endothermic phase transition. Science 259, 1308–1311.

Houseman, G.A., Gubbins, D., 1997. Deformation of subducted oceanic lithosphere. Geophys. J. Int. 131, 535–551.

Irifune, T., Kuroda, K., Funamori, N., Uchida, T., Yagi, T., Inoue, T., Miyajima, N., 1996. Amorphization of serpentine at high pressure and high temperature. Science 272, 1468–1470.

Isacks, B.L., Molner, P., 1969. Mantle earthquakes mechanisms and the sinking of the lithosphere. Nature 223, 1121–1124.

Isacks, B.L., Barazangi, M., 1977. Geometry of Benioff zones: lateral segmentation and downward bending of the subducted lithosphere. In: Talwani, M., Pitman III, W.C. (Eds.), Island Arcs, Deep Sea Trenches and Back-Arc Basins. American Geophysics Union, Washington, DC, pp. 99–114.

Ito, E., Sato, H., 1990. Aseismicity in the lower mantle by superplasticity in a subducting slab. Nature 351, 140–141.

Jarrard, R.D., 1986. Relations among subduction parameters. Rev. Geophys. 24, 217–284.

Kameyama, M., Yuen, D.A., Karato, S., 1999. Thermal–mechanical effects of low-temperature plasticity (the Peierls mechanism) on the deformation of a viscoelastic shear zone. Earth Planet. Sci. Lett. 168, 159–172.

Kanamori, H., Anderson, D.L., 1975. Theoretical basis of some empirical relations in seismology. Bull. Seism. Soc. Am. 65, 1073–1095.

Kanamori, H., Anderson, D.L., Heaton, T.H., 1998. Frictional melting during the rupture of the 1994 Bolivia earthquake. Science 279, 839–842.

Karato, S., 1986. Does partial melting reduce the creep strength of the upper mantle. Nature 319, 309–310.

Karato, S., 1989a. Plasticity-crystal structure systematics in dense oxides and its implications for the creep strength of the Earth's deep interior: a preliminary result. Phys. Earth Planet. Inter. 55, 234–240.

Karato, S., 1989b. Grain growth kinetics in olivine aggregates. Tectonophysics 155, 255–273.

Karato, S., 1997a. On the separation of crustal component from subducted oceanic lithosphere near the 660 km discontinuity. Phys. Earth Planet. Inter. 99, 103–111.

Karato, S., 1997b. Phase transformations and rheological properties of mantle minerals. In: Crossley, D. (Ed.), Earth's Deep Interior (Doornbus Volume). Gordon and Breach, pp. 223–272.

Karato, S., 1998. Effects of pressure on plastic deformation of polycrystalline solids: some geological applications. In: Wentzcovitch, R.M., Hemley, R.J., Nellis, W.J., Yu, P.Y. (Eds.), High Pressure Research in Materials Research. Materials Research Society, Warrendale, PA, pp. 3–14.

Karato, S., Rubie, D.C., 1997. Toward an experimental study of deep mantle rheology: a new multianvil sample assembly for deformation studies under high pressure and temperatures. J. Geophys. Res. 102, 20111–20122.

Karato, S., Jung, H., 1998. Water, partial melting and the origin of the seismic low velocity and high attenuation zone in the upper mantle. Earth Planet. Sci. Lett. 157, 193–207.

Karato, S., Wu, P., 1993. Rheology of the upper mantle: A synthesis. Science 260, 771–778.

Karato, S., Paterson, M.S., Fitz Gerald, J.D., 1986. Rheology of synthetic olivine aggregates: influence of grain size and water. J. Geophys. Res. 91, 8151–8176.

Karato, S., Rubie, D.C., Yan, H., O'Neill, St.H.C., 1993. Dislocation recovery in olivine under deep upper mantle conditions. J. Geophys. Res. 98, 9761–9768.

Karato, S., Zhang, S., Wenk, H.-R., 1995. Superplasticity in Earth's lower mantle: evidence from seismic anisotropy and rock physics. Science 270, 458–461.

Karato, S., Dupas-Bruzek, C., Rubie, D.C., 1998. Plastic deformation of silicate spinel under the transition zone conditions of the Earth. Nature 395, 266–269.

Kerschhofer, L., Sharp, T.G., Rubie, D.C., 1996. Intracrystalline transformation of olivine to wadsleyite and ringwoodite under subduction zone conditions. Science 274, 79–81.

Kerschhofer, L., Dupas, C., Liu, M., Sharp, T.G., Durham, W.B., Rubie, D.C., 1998. Polymorphic transformations between olivine, wadsleyite and ringwoodite: mechanisms of intracrystalline transformation and role of elastic strain. Mineral. Mag. 62, 617–638.

Kerschhofer, L., Rubie, D.C., Sharp, T.G., McConnell, J.D.C., Dupas-Bruzek, C., 2000. Kinetics of intracrystalline olivine–ringwoodite transformation. Phys. Earth Planet. Inter. 121, 59–76.

Kikuchi, M., Kanamori, H., 1994. The mechanism of the deep Bolivia earthquake of June 9, 1994. Geophys. Res. Lett. 22, 2341–2344.

Kincaid, C., Olson, P., 1987. An experimental study of subduction and slab migration. J. Geophys. Res. 92, 13832–13840.

King, S.D., Ita, J.J., 1995. Effect of slab rheology on mass transport across a phase transition boundary. J. Geophys. Res. 100, 20211–20222.

Kirby, S.H., 1987. Localized polymorphic phase transformations in high-pressure faults and applications to the physical mechanisms of deep earthquakes. J. Geophys. Res. 92, 13789–13800.

Kirby, S.H., Durham, W.B., Stern, L., 1991. Mantle phase changes and deep-earthquake faulting in subducting slabs. Science 252, 216–225.

Kirby, S.H., Stein, S., Okal, E.A., Rubie, D.C., 1996. Metastable mantle phase transformations and deep earthquakes in subducting oceanic lithosphere. Rev. Geophys. 34, 261–306.

Kitamura, M., Tsuchiyama, A., Watanabe, S., Syono, Y., Fukuoka, K., 1992. Shock recovery experiments on chondritic materials. In: Syono, Y., Manghnani, M.H. (Eds.), High-Pressure Research: Application to Earth and Planetary Sciences. Terra Science Publication, Tokyo, pp. 333–340.

Kubo, T., Ohtani, E., Kato, T., Shinmei, T., Fujino, K., 1998a. Effects of water on the α–β transformation kinetics in San Carlos olivine. Science 281, 85–87.

Kubo, T., Ohtani, E., Kato, T., Shinmei, T., Fujino, K., 1998b. Experimental investigation of the α–β transformation of San Carlos olivine single crystals. Phys. Chem. Mineral. 26, 1–6.

Kubo, T., Ohtani, E., Kato, T., Urakawa, S., Suzuki, A., Kanbe, Y., Funakoshi, K., Utsumi, W., Fujino, K., 2000.

Formation of metastable assemblages and mechanisms of the grain-size reduction in the postspinel transformation of Mg_2SiO_4. Geophys. Res. Lett. 27, 807–810.
Lawlis, J.D., Lee, K.-H., Frost, D., Rubie, D.C., Cordier, P., Karato, S., 1999. Deformation and fabric development in transition zone minerals. EOS 80, F975.
Lawlis, J.D., Zhang, Y., Karato, S., 2001. High-temperature creep in Ni_2GeO_4: A contribution to creep systematics in spinel. Chem. Mineral., in press.
Lithgow-Bertelloni, C., Richards, M.A., 1998. The dynamics of Cenozoic and Mesozoic plate motions. Rev. Geophys. 36, 27–78.
Liu, M., 1997. A constitutive model for olivine–spinel aggregates and its application to deep earthquake nucleation. J. Geophys. Res. 102, 5295–5312.
Lundgren, P.R., Giardini, D., 1992. Seismicity, shear failure and modes of deformation in deep subduction zones. Phys. Earth Planet. Inter. 74, 63–74.
Marton, F., Bina, C.R., Stein, S., Rubie, D.C., 1999. Effects of slab mineralogy on subduction rates. Geophys. Res. Lett. 26, 119–122.
McKenzie, D.P., 1969. Speculations on the consequences and causes of plate motions. Geophys. J. R. Astr. Soc. 18, 1–32.
Meade, C., Jeanloz, R., 1991. Deep-focus earthquakes and recycling of water into the Earth's mantle. Science 252, 68–72.
Meike, A., 1993. A critical review of investigations into transformational plasticity. In: Boland, J.N., Fitz Gerald, J.D. (Eds.), Defects and Processes in the Solid States: Geoscience Applications. Elsevier, Amsterdam, pp. 5–25.
Moresi, L., Gurnis, M., 1996. Constraints on the lateral strength of slabs from three-dimensional dynamic flow models. Earth Planet. Sci. Lett. 138, 15–28.
Mosenfelder, J.L., Connelly, J.A.D., Rubie, D.C., Liu, M., 2000. Strength of $(Mg, Fe)_2SiO_4$ wadsleyite determined by relaxation of transformational stress. Phys. Earth Planet. Inter. 120, 63–78.
Obata, M., Karato, S., 1995. Ultramafic pseudotachylyte from Balmuccia peridotite, Ivrea–Verbana zone, northern Italy. Tectonophysics 242, 313–328.
Ogawa, M., 1987. Shear instability in a viscoelastic material as the cause for deep focus earthquakes. J. Geophys. Res. 92, 13801–13810.
Ohtani, E., 1983. Melting temperature distribution and fractionation in the lower mantle. Phys. Earth Planet. Inter. 33, 12–25.
Okamoto, Y., 1989. Creep deformation mechanisms in oxides and deformation of spinel ferrites. In: Karato, S., Toriumi, M. (Eds.), Rheology of Solids and of the Earth. Oxford University Press, Oxford, pp. 83–104.
Olbertz, D., 1997. The long-term evolution of subduction zones: a modelling study. Ph.D. thesis, University of Utrecht, p. 152.
Olbertz, D., Wortel, M.J.R., Hansen, U., 1997. Trench migration and subduction zone geometry. Geophys. Res. Lett. 24, 221–224.
O'Neill, B., Jeanloz, R., 1994. $MgSiO_3$–$FeSiO_3$–Al_2O_3 in the Earth's lower mantle: perovskite and garnet at 1200 km depth. J. Geophys. Res. 99, 19901–19915.
Panasyuk, S.V., Hager, B.H., 1998. A model of transformational superplasticity in the upper mantle. Geophys. J. Int. 133, 741–755.
Paterson, M.S., 1983. Creep in transforming polycrystalline materials. Mech. Mater. 2, 103–109.
Poirier, J.-P., 1980. Shear localization and shear instability in materials in the ductile field. J. Struct. Geol. 2, 135–142.
Poirier, J.-P., 1982. On transformation plasticity. J. Geophys. Res. 87, 6791–6797.
Presnall, D.C., Walker, M.J., 1993. Melting of forsterite, Mg_2SiO_4, from 9.7 to 16.5 GPa. J. Geophys. Res. 98, 19777–19783.
Riedel, M.R., Karato, S., 1996. Microstructural development during nucleation and growth. Geophys. J. Int. 125, 397–414.
Riedel, M.R., Karato, S., 1997. Grain-size evolution in subducted oceanic lithosphere associated with the olivine–spinel transformation and its influence on rheology. Earth Planet. Sci. Lett. 148, 27–44.
Riedel, M.R., Karato, S., Yuen, D.A., 1998. Criticality of subducting Slabs. EOS, 79, F911.
Ringwood, A.E., 1991. Phase transformations and their bearings on the constitution and dynamics of the mantle. Geochim. Cosmochim. Acta 55, 2083–2110.
Ringwood, A.E., Irifune, T., 1988. Nature of the 650 km seismic discontinuity. Nature 331, 131–136.
Rogers, H.C., 1979. Adiabatic plastic deformation. Ann. Rev. Mater. Sci. 9, 283–311.
Rubie, D.C., 1984. The olivine–spinel transformation and the rheology of subducting lithosphere. Nature 308, 505–508.
Rubie, D.C., Ross II, C.R., 1994. Kinetics of the olivine–spinel transformation in subducting lithosphere: experimental constraints and implications for deep slab processes. Phys. Earth Planet. Inter. 86, 223–241.
Rubie, D.C., Tsuchida, Y., Yagi, T., Utsumi, W., Kikegawa, T., Shimomura, O., Brearley, A.J., 1990. An in situ X-ray diffraction study of the kinetics of the Ni_2SiO_4 olivine–spinel transformation. J. Geophys. Res. 95, 15829–15844.
Rubie, D.C., Mosenfelder, J.L., Kerschhofer, L., 1998. Mechanisms of grain size reduction and rheological weakening in subducting slabs. EOS 79, F911.
Sakurai, T., 1996. Whole mantle P-wave tomography and differential PP–P time measurement. M.Sc. thesis, University of Tokyo, p. 31.
Sammis, C.G., Dein, J.L., 1974. On the possibility of transformational superplasticity in the Earth's mantle. J. Geophys. Res. 79, 2961–2965.
Schmeling, H., Monz, R., Rubie, D.C., 1999. The influence of olivine metastability on the dynamics of subduction. Earth Planet. Sci. Lett. 165, 55–66.
Shimizu, I., 1998. Stress and temperature dependence of recrystallized grain size: a subgrain misorientation model. Geophys. Res. Lett. 25, 4237–4240.
Silver, P.G., Beck, S.L., Wallace, T.C., Meade, C., Myers, S.C., James, D.E., Kuehnel, R., 1995. Rupture characteristics of the deep Bolivia earthquakes of 9 June 1994 and the mechanism of deep-focus earthquakes. Science 268, 69–71.
Staker, M.R., 1981. The relation between adiabatic shear instability, strain and material properties. Acta Metall. 29, 683–689.
Sugi, N., Kikuchi, M., Fukao, Y., 1989. Mode of stress release within a subducting slab of lithosphere: implications of source mechanism of deep and intermediate-depth earthquakes. Phys. Earth Planet. Inter. 55, 106–125.

Tackley, P.J., 1997. Effects of phase transitions on three-dimensional mantle convection. In: Crossley, D.J. (Ed.), Earth's Deep Interior. Gordon and Breach Science Publisher, Amsterdam, pp. 273–335.

Tackley, P.J., Stevenson, D.J., Glatzmaier, G.A., Schubert, G., 1993. Effects of an endothermic phase-transition at 670 km depth in a spherical model of convection in the Earth's mantle. Nature 361, 699–704.

Tibi, R., Estabrook, C.H., Bock, G., 1999. The 1996 June 17 Flores Sea and 1994 March 9 Fiji-Tonga earthquakes: source processes and deep earthquake mechanisms. Geophys. J. Int. 128, 625–642.

Tingle, T.N., Green II, H.W., Scholtz, C.H., Koczynski, T.A., 1993. The rheology of faults triggered by the olivine–spinel transformation in Mg_2GeO_4 and its implications for the mechanism of deep focus earthquakes. J. Struct. Geol. 15, 1249–1256.

Turcotte, D.L., Schubert, G., 1982. Geodynamics. Wiley, New York.

van der Hilst, R.D., 1995. Complex morphology of subducted lithosphere in the mantle beneath the Tonga trench. Nature 374, 154–157.

van der Hilst, R.D., Seno, T., 1993. Effects of relative plate motion on the deep structure and penetration depth of slabs below the Izu–Bonin and Mariana arcs. Earth Planet. Sci. Lett. 120, 395–407.

van der Hilst, R.D., Engdahl, E.R., Spakman, W., Nolet, G., 1991. Tomographic imaging of subducted lithosphere below northwest Pacific island arcs. Nature 353, 37–43.

van der Hilst, R.D., Widiyantoro, S., Engdahl, E.R., 1997. Evidence for deep mantle circulation from global tomography. Nature 386, 578–584.

Vaughan, P.J., Coe, R.S., 1981. Creep mechanisms in Mg_2GeO_4: effects of a phase transition. J. Geophys. Res. 86, 389–404.

Vogt, P.R., Lowrie, A., Bracey, D.R., Hey, R.N., 1976. Subduction of aseismic oceanic ridges: effects on shape, seismicity and other characteristics of consuming plate boundaries. Special Paper 162. Geological Society of America, p. 59.

Weertman, J., 1968. Dislocation climb theory of steady-state creep. Trans. ASM 61, 681–694.

Wiens, D.A., Gilbert, H.J., 1996. Effect of slab temperature on deep-earthquake aftershock productivity and magnitude–frequency relations. Nature 372, 153–156.

Wiens, D.A., Snider, N.O., 1999. Repeating deep earthquakes: evidence for fault reactivation at great depth. EOS, Trans. Am. Geophys. Union 80, F667.

Wiens, D.A., McGuire, J.J., Shore, P.J., 1993. Evidence for transformational faulting from a deep double seismic zone in Tonga. Nature 364, 790–793.

Wiens, D.A., McGuire, J.J., Shore, P.J., Bevis, M.G., Draunidalo, K., Prasad, G., Helu, S.P., 1994. A deep earthquake aftershock sequence and implications for the rupture mechanism of deep earthquakes. Nature 372, 540–543.

Wortel, M.J.R., Vlaar, N.J., 1988. Subduction zone seismicity and the thermo-mechanical evolution of down-going lithosphere. PAGEOPH 128, 625–659.

Yamaoka, K., Fukao, Y., Kumazawa, M., 1986. Spherical shell tectonics: effect of sphericity and inextensibility on the geometry of the descending lithosphere. Rev. Geophys. 24, 27–53.

Yoshioka, S., Daessler, D., Yuen, D.A., 1997. Stress fields associated with metastable phase transitions in descending slabs and deep-focus earthquakes. Phys. Earth Planet. Inter. 104, 345–361.

Zhao, Y.-H., Lawlis, J.D., Lee, K.-H., Karato, S., 1999. Deformation of $(Mg, Ni)_2GeO_4$: the effects of the olivine–spinel transformation. EOS 80, 1027–1028.

Zhong, S.J., Gurnis, M., 1995. Mantle convection with plates and mobile, faulted plate margins. Science 267, 838–843.

Zhong, S., Davies, G.F., 1999. Effects of plate and slab viscosities on the geoid. Earth Planet. Sci. Lett. 170, 487–496.

ELSEVIER

Physics of the Earth and Planetary Interiors 127 (2001) 109–143

PHYSICS
OF THE EARTH
AND PLANETARY
INTERIORS

www.elsevier.com/locate/pepi

“Detached” deep earthquakes: are they really?

Emile A. Okal

Department of Geological Sciences, Northwestern University, Evanston, IL 60208, USA

Received 28 March 2000; accepted 27 July 2000

Abstract

We use primarily the generation of acoustic T waves into the ocean by deep seismic sources to investigate the propagation of high-frequency seismic energy from the bottom of subduction zones to the shoreline at the earth's surface. Conversion from shear waves to oceanic acoustic waves can be used as a proxy for the existence of a continuous slab featuring low anelastic attenuation. With the help of other techniques, such as the estimation of Q from S-to-P spectral amplitude ratios, we examine systematically a number of regions where earthquakes have been described as “detached”. We establish the mechanical continuity of the slab to the hypocenters of the 1990 Sakhalin and 1982 Bonin events, which occurred several hundred kilometers in front of the mainstream seismic zone. The study of the 1989 Paraguay shock is inconclusive, probably due to its much smaller size. The vertical continuity of the South American slab through its aseismic depth range is verified, and a similar situation probably exists in Java. Attenuation data suggests that the deep Spanish earthquakes occur within a vertically large segment of colder material, and a similar situation may exist in Colombia. The only clearly detached deep events with no mechanical connection to the surface make up the Vityaz cluster, under the North Fiji Basin. Based on a variety of geophysical evidence, the small deep earthquakes under New Zealand are likely to take place in a detached blob at least 350 km below the termination of mainstream seismicity. These results support a model integrating buoyancy forces over a long continuous slab as the source of the down-dip compressional stresses observed in large earthquakes at the bottom of the transition zone.

Keywords: Detached; Deep earthquakes; High-frequency seismic energy

1. Introduction and background

This paper studies a number of selected deep earthquakes, often described as “detached” because of the discontinuous character of seismicity at the bottom of the relevant subduction zones. Our principal tool of investigation are the T phases received at teleseismic distances across the ocean following these events; we are motivated by the fact that T waves can be channeled by the SOFAR low-velocity waveguide (Ewing et al., 1946) only at frequencies $f > 2.5$ Hz, and that their excitation, thus, expresses the ability for the earthquake to efficiently send high-frequency seismic energy, convertible into T waves, to the coastal area. We conclude that several such earthquakes can be better described as occurring in warped but continuous sections of the Wadati–Benioff zone (WBZ). We also show on several examples that T waves are indeed routinely recorded from large deep earthquakes at the bottom of subduction zones.

In a previous study concerned with the 1994 Bolivian earthquake (Okal and Talandier, 1997, 1998) (hereafter Paper I), we showed that the T phases it generated had to be converted from high-frequency S waves, which in turn required a path with low shear attenuation from the hypocenter to the water

E-mail address: emile@earth.nwu.edu (E.A. Okal).

0031-9201/01/$ – see front matter
PII: S0031-9201(01)00224-2

column, which could be explained only if a thermally and hence mechanically continuous slab extends upwards from the 1994 hypocenter. This observation was of particular interest since the Bolivian earthquake occurred in a region with no previous record of seismicity deeper than 280 km, between the Peruvian and south Bolivian–Argentinian deep clusters (Kirby et al., 1995). The combination of the 1994 seismicity (including aftershocks of the 9 June event), of several recent earthquakes, and of the efficient propagation of high-frequency regional S waves (with Q_μ estimated as at least 800 in Paper I), indicates that rather than being torn, the slab is merely warped along a jog linking the Peruvian and Argentinian segments, and remains continuous both laterally and vertically.

A conclusion of Paper I, substantiated by the examination of spectra of regional S waves, was that the existence of an $S \rightarrow T$ conversion from a deep earthquake at the bottom of a subduction zone can be viewed as a proxy for a mechanically continuous slab, an especially valuable result in the presence of a gap of activity in the WBZ. Similarly, a recent study by Mele (1998) in the Calabrian arc, has estimated $Q_\mu \approx 1000$ and calculated that a 25-km gap in the continuity of the slab would suffice to eradicate high-frequency (6 Hz) S waves, such resolution being of course much finer than that of even the best tomographic models. Armed with this technique, we focus in this paper on a number of regions where several earthquakes have been previously described as "detached" in the literature. Their geographical layout can take several forms which we now examine in detail.

1.1. Wadati–Benioff zones with depth gaps

Ever since the pioneering work of Benioff (1949), it was noticed that seismicity in several slabs is not downward continuous, prompting early authors to propose that the slab may be mechanically "broken", with earthquakes taking place in individual blobs of sinking lithosphere, (e.g., Isacks and Molnar, 1971; Wortel, 1984). The primary example of such a geometry is South America, where seismicity gaps exist between depths of 337 and 502 km in northern Argentina, and between 211 and 506 km in Peru–Brazil (Engdahl et al., 1998), while farther North, activity below 292 km is documented only in the form of three very large, very deep events in 1921, 1922 and 1970 (Okal and Bina, 1994), and of three similarly very deep but very small 1997 shocks scattered between the 1921–1922 hypocentral area and the northern end of the Peru–Brazil deep cluster (Okal and Bina, 2001). The occurrence of the great 1994 Bolivian earthquake, its aftershocks, and several recent earthquakes (14 March 1995; 28 November 1997) indicated that while the slab itself must be continuous horizontally between the foci of abundant deep seismicity in Peru–Bolivia and Argentina, it does feature a vertical gap in seismicity in central Bolivia from 280 to 566 km depth. Another example of a gap in seismicity with depth is the Java WBZ, West of 115°E, where no earthquakes are known between 338 and 470 km, (e.g., Kirby et al., 1996).

1.2. Outboard earthquakes

Lundgren and Giardini (1994) have reviewed a number of cases of earthquakes occurring several hundred km in front of the general trend of the WBZ in the relevant subduction zone. The most prominent examples are the 1990 Sakhalin, 1982 Bonin Islands and 1989 Paraguay events. Under the South Fiji Basin, West of Tonga, a number of narrow fingers of seismicity extending as much as 700 km in front of the WBZ have also been documented (Okal and Kirby, 1998; Brudzinski and Chen, 1998).

1.3. The Vityaz deep cluster

Under the North Fiji Basin, Okal and Kirby (1998) have analyzed in detail a large cluster of frequent, relatively small earthquakes at depths of 570–660 km. They concluded that these events take place in a piece of slab severed from a deactivated subduction system and lying recumbent on the bottom of the transition zone.

1.4. The small deep earthquakes under New Zealand

The bottom limit of the WBZ rises regularly from 650 km under the Kermadec Islands at 30°S to 260 km under Cook Strait at 41°S. It is probable that this process is controlled by the reduction in thermal parameter Φ (Kostoglodov, 1989; Kirby et al.,

1991) resulting from a slower convergence rate as one moves southwards closer to the pole of rotation of the Pacific–Australian plate system (DeMets et al., 1990). However, a few earthquakes have been documented at depths of 570–622 km under the North Island of New Zealand at 39°S (Adams, 1963; Adams and Ferris, 1976), where mainstream seismic activity stops at 250 km. Additional events in 1991–1998 confirm the existence of this intriguing seismic cluster. Incidentally, it should be emphasized that this situation is not unique among subduction zones, and that individual events are occasionally reported at depths greater than that of cessation of abundant seismicity at the tip of WBZs. A striking example is the South Sandwich system, where seismicity is confined to the upper 200 km of slab, but where rare events can occur down to ~300 km (the latest one on 5 October 1997; $h = 273$ km; $m_b = 6.3$). However, we do not address such cases of detached earthquakes at intermediate depths in the present paper.

1.5. The deep Spanish earthquakes

The origin of the large 1954 shock at 627 km, studied in detail by Chung and Kanamori (1976), remains to this day a largely unresolved puzzle; only three events are known in its vicinity, in 1973, 1990, and 1993. Intermediate seismicity is known to the SSW, down to ~135 km depth (Mezcua and Rueda, 1997).

In all above geometries, the question arises whether the deep seismicity occurs in blobs of sinking lithosphere actually detached from the main slab, or rather is merely a result of a change in geometrical or thermomechanical properties controlling the existence or the release of ambient stresses in an otherwise physically continuous slab. The answer to this question is of course of great importance for our understanding of the dynamics of the subduction process.

2. Previous approaches

Three main lines of evidence have previously been used to explore the nature of gaps in deep seismicity: first, experiments in seismic tomography have been used to infer thermal continuity for the South American slab (Engdahl et al., 1995), to suggest at least partial deflection of the latter's southern segment, as well as of the Izu–Bonin slab (van der Hilst et al., 1993), and to image a flat lying extension of the Tonga slab under the South Fiji Basin (van der Hilst, 1995), where it hosts the seismic fingers described by Okal and Kirby (1998). Also, in the case of the 1982 "detached" event in the Bonin Islands, Okino et al. (1989) used travel-time residuals at Japanese stations to argue for the presence of fast material in the event's immediate vicinity.

Second, Isacks and Molnar (1971) and Huang et al. (1998) (among others) have noted the coherence of focal geometries of events below the depth gap in South America, notably in Argentina, where the mechanisms express down-dip compression, arguing for the mechanical continuity of the slab through its aseismic segment. The latter was explained by Engebretson and Kirby (1992) as involving an age discontinuity in the downgoing lithosphere. Conversely, Okal and Kirby (1998), noting the wide variety of focal mechanisms in the Vityaz deep cluster, have argued for its being mechanically unrelated to any shallower structures. Lundgren and Giardini (1994) also presented evidence that outboard events have mechanisms differing from those of neighboring events in the mainstream WBZ.

Finally, a number of regional studies have identified high-frequency S waves in various subduction zones, some of them with seismicity gaps. In particular, in South America, Sacks (1969), and later Isacks and Barazangi (1973), Snoke et al. (1974a) and James and Snoke (1990) used this technique to propose the mechanical continuity of the slab through the aseismic depth gap, although Snoke et al. (1974b) also proposed an underside reflection on the subducting slab as an alternate explanation for the late phases. Mooney (1970) in New Zealand, Barazangi et al. (1972) in Tonga, and Mele (1998) in the Calabrian arc, also used similar techniques. van der Hilst and Snieder (1996) modeled the propagation of high-frequency P waves through a three-dimensional model of lateral heterogeneity under New Zealand and concluded that their observation almost certainly requires the continuity of the slab. Our approach will be conceptually similar, building on the same principle, i.e. that high-frequency S waves need a cold, continuous medium to propagate from the bottom of the WBZ, but will use a different observational strategy, and be global in scope.

3. Dataset and methodology

In this general framework, we investigate in the present study the propagation of high-frequency S waves up slab from a number of deep earthquakes located in targeted subduction zones. We focus on the process of generation of T waves in the oceanic column, complemented occasionally by the direct spectral analysis of regional S waves. We refer to Paper I for a description of the general techniques used.

3.1. Dataset

In the present study, we rely on the following datasets:

- for recent (post-1989) events, IRIS continuous broad-band channels;
- for a few events in 1994–1996, continuous broad-band records of the Micronesian Seismic Experiment (Richardson, 1998);
- since ∼1996, continuous hydrophone channels of the Wake Hydrophone array, available from the Prototype International Data Center;
- from ∼1975 to 1989, GDSN short-period channels (usually available through the IRIS Data Center), in general, the recording of these channels was triggered, and only P (and occasionally S) windows are available, which can constitute a very significant handicap for the study of T waves;
- analog (paper) continuous records at the French Polynesia seismic array, available since 1962, with

Table 1
Seismic events used in the T wave study

Code	Name	Island			Coordinates		Distance to closest conversion point[a]
		Name	Chain	Nature[b]	°N	°E	
French Polynesia							
PMO	Pomariorio	Rangiroa	Tuamotu	a	−15.017	−147.906	50 m
TPT	Tiputa	Rangiroa	Tuamotu	a	−14.984	−147.619	50 m
REAO	Reao	Reao	Tuamotu	a	−18.51	−136.40	50 m
MEH	Mehetia	Mehetia	Society	h	−17.875	−148.066	200 m
PPT	Pamatai	Tahiti	Society	h	−17.569	−149.574	8 km
TET	Tetiaroa	Tetiaroa	Society	a	−16.996	−149.586	50 m
RKT	Rikitea	Mangareva	Gambier	e	−23.118	−134.972	12 km
TBI	Tubuai	Tubuai	Austral	h	−23.349	−149.461	8 km
Hawaii							
KAA	Kalahiki	Hawaii	Hawaii	h	19.266	−155.871	7 km
HUL	Heiheiahulu	Hawaii	Hawaii	h	19.419	−154.979	7 km
KIP	Kipapa	Oahu	Hawaii	h	21.420	−158.020	22 km
Other							
RPN	Rapa Nui	Easter		h	−27.127	−109.334	8 km
KOS	Kosrae	Kosrae	Caroline	h	5.324	163.009	6 km
TKK	Moen	Chuuk	Caroline	e	7.447	151.887	19 km
NAU	Nauru	Nauru		u	−0.509	166.932	200 m
RAR	Rarotonga	Rarotonga	Cook	h	−21.210	−159.770	10 km
AFI	Afiamalu	Opulu	Samoa	h	−13.910	−171.780	18 km
GUMO	Guam	Guam	Mariana	b	13.588	144.866	12 km
COCO	Cocos Island	Cocos	Keeling	a	−12.190	96.835	2 km
WK30	Wake Hydrophone				19.410	167.499	
SYP	Santa Ynez Peak, California				34.527	−119.978	82 km
SNCC	San Nicolas Island, California				33.248	−119.524	73 km
SCZ	Santa Cruz, California				36.60	−121.40	58 km
PET	Petropavlovsk–Kamchatskiy, Russia				53.017	158.650	36 km
HOPE	Hope, South Georgia				−54.824	−36.488	58 km

[a] This is the minimum distance to a conversion point, but depending on the geometry of arrival, the actual distance used to estimate a correction may be larger.

[b] a: atoll; h: high island (volcanic); e: eroded high island inside large lagoon; u: uplifted atoll; b: uplifted fore-arc basement.

high-frequency (T wave) channels (Talandier and Kuster, 1976) available since the mid-1970s;
- analog (paper and develocorder film) continuous records of the Hawaii Volcano Observatory (HVO) short-period array;
- analog (paper) continuous records of the Caltech short-period network in southern California;
- analog (micro-film) continuous short-period records of the WWSSN, they suffer from the often mediocre gains used at ocean island stations.

It should be emphasized that signal processing methods, such as high-pass filtering and spectrogram analysis can be applied routinely only to digital datasets. In the case of analog records, we were occasionally able to hand-digitize time series for further processing, after enlarging the record several times using a magnifying photocopier. The quality of the resulting time series remains low, and they can be used only at the lower end of the frequency window of interest here ($f \leq 3$ Hz).

Whenever possible, we use in this study T wave receiving stations located on atolls characterized by steep reefs optimizing the acoustic-to-seismic conversion on the receiver side (Talandier and Okal, 1996). Table 1 lists parameters for stations used in the present work, and Table 2 lists events studied with the T wave technique.

3.2. Station corrections

As detailed in Paper I, and based upon the work of Talandier and Okal (1998), we introduce station corrections to compensate for the faster propagation inside the insular or continental structure after conversion back into seismic energy on the receiver side. These corrections are unnecessary in the case of atoll sites such as the Rangiroa stations in French Polynesia, or for very small high islands whose dimensions can be neglected (Mehetia, Pitcairn).

3.3. Example and methodology

As intriguing as this situation may be, given that their sources are de facto removed from the oceanic water mass, the excitation of T phases by deep earthquakes is the rule rather than the exception. Fig. 1 shows a typical example of T wave recorded at station RPN, (Easter Island) following the earthquake of 21 July 1994 under the Primorye province of eastern Russia. This event is discussed in more detail in Appendix A, but we use this record here to describe the methodology of the present study.

The two T wave arrivals are easily extracted by filtering in Fig. 1b, and further resolved by the spectrogram in frame (c). The T phase is clearly composed of two arrivals, with maxima separated by 90 s, at 21:09:14 and 21:10:44 GMT, respectively. We use a station correction of 4 s to account for 11 km of on-land propagation at Easter Island following acoustic-to-seismic conversion (Talandier and Okal, 1998) for final times of 21:09:18 and 21:10:48, with a precision estimated at ± 5 s. In Fig. 2, which is conceptually similar to Figs. 3–5 of Paper I,[1] we study the residual (defined as the difference between the observed arrival time of the T phase and the arrival time computed from a modeled conversion), as a function of the latitude of the point of conversion along the Japan–Kuril shoreline. The fundamental result from Fig. 2 is that no $P \rightarrow T$ conversion anywhere along the shore can explain the second arrival, which remains always more than 1 min late. On the other hand, it is easily explained by an $S \rightarrow T$ conversion taking place at 41.5°N, 141.9°E, which corresponds to a concave bight in the 1200 m isobath at the Hokkaido corner (Fig. 3). Note that the first arrival can be explained by a $P \rightarrow T$ conversion at essentially the same location (it could also, conceivably, be explained by an $S \rightarrow T$ conversion off Cape Erimo). We conclude that the two pulses in the T phase correspond to $P \rightarrow T$ and $S \rightarrow T$ conversions, respectively, generated by a scatterer at the Hokkaido corner.

The $S \rightarrow T$ conversion is observed in Fig. 1 to be of larger amplitude than the $P \rightarrow T$ one; this is found to be a common occurrence, which can be justified along the following arguments: first, as is well known from elementary seismic theory, (e.g., Okal, 1992), the generation of S waves by a double-couple is $(\alpha/\beta)^3$ times more efficient (5.2 times in a Poisson solid) than that of P waves. This results in teleseis-

[1] Due to a production error, Figs. 3 and 4 were permuted in (Okal and Talandier, 1997). The figures were reprinted correctly in (Okal and Talandier, 1998).

Table 2
Seismic events used in the T wave study

Date (D M (J) Y)	Origin time (GMT)	Hypocenter			Published CMT solution	
		Latitude (°N)	Longitude (°E)	Depth (km)	M_0 (10^{27} dyn cm)	ϕ, δ, λ (°)
Okhotsk–Sakhalin–Primorye						
30 AUG (242) 1970	17:46:09.0	52.38	151.60	645	1.1	229, 78, 284
5 SEP (248) 1970	07:52:27.9	52.10	150.99	561	0.07	4, 79, 281
29 JAN (029) 1971	21:58:05.4	51.73	150.95	524	0.25	31, 73, 255
10 JUL (192) 1976	11:37:12.8	47.36	145.72	421	0.02	64, 81, 263
21 DEC (355) 1975	10:54:17.7	51.84	151.75	545	0.21	44, 77, 265
21 JUN (172) 1978	11:10:38.2	47.98	149.01	403	0.07	185, 85, 116
1 FEB (032) 1984	07:28:27.8	49.10	146.31	581	0.04	231, 85, 83
18 MAY (138) 1987	03:07:34.7	49.12	147.39	552	0.17	50, 83, 270
12 MAY (132) 1990[a]	04:50:08.7	49.04	141.85	605	0.82	172, 29, 210
21 JUL (202) 1994	18:36:31.7	42.34	132.87	471	1.1	64, 34, 178
Izu–Bonin–Marianas						
31 JAN (031) 1973	20:55:53.1	28.18	138.86	506	0.25	327, 79, 308
13 MAY (133) 1977	11:13:31.2	28.12	139.73	440	0.06	342, 72, 286
18 MAY (138) 1979	20:18:01.1	23.94	142.66	581	0.06	57, 47, 267
4 JAN (004) 1982	06:05:01.3	17.92	145.46	595	0.07	167, 58, 271
4 JUL (185) 1982[a]	01:20:07.6	27.92	136.48	552	0.12	80, 80, 248
6 MAR (066) 1984	02:17:21.1	29.60	139.110	446	1.4	332, 88, 290
5 AUG (217) 1990	01:34:57.5	29.48	137.500	520	0.06	296, 87, 322
23 AUG (235) 1995	07:06:02.6	18.88	145.30	599	0.46	136, 42, 242
16 MAR (076) 1996	22:04:06.2	28.98	138.94	477	0.11	68, 23, 197
Argentina–Paraguay						
21 DEC (355) 1983	12:05:06.3	−28.19	−63.17	602	0.27	202, 14, 115
28 FEB (059) 1989[a]	13:01;57.6	−23.11	−61.47	569	0.07	194, 24, 290
23 JUN (174) 1991	21:22:28.9	−26.80	−63.35	558	0.86	162, 68, 256
29 APR (119) 1994	07:11:30.3	−28.51	−63.22	566	0.25	169, 71, 260
10 MAY (130) 1994	06:36:28.4	−28.50	−63.10	600	0.28	256, 10, 183
19 AUG (231) 1994	10:02:51.8	−26.72	−63.42	563	0.06	159, 73, 253
Bolivia						
9 JUN (160) 1994	00:33:16.2	−13.841	−67.553	631	26	302, 10, 300
Colombia						
31 JUL (212) 1970	17:08:05.2	−1.46	−72.56	651	21	148, 58, 261
Vityaz						
13 APR (103) 1995	02:34:38.0	−13.45	170.43	637	0.021	312, 45, 287
New Zealand						
14 SEP (257) 1991	14:14:42.0	−39.180	174.434	602		

[a] The three outboard events studied by Lundgren and Giardini (1994).

mic S classically having larger amplitudes than P at typical long periods (10 s and above). At the high frequencies used in this study ($\omega > 20$ rad/s), teleseismic S waves are generally eradicated by anelastic attenuation, but propagation over a short distance (≈1000 km) up a cold slab with high Q_μ (≈800) can result in S waves of greater amplitude than P at the conversion point.

In addition, the radiation pattern of S waves up the slab can be more favorable that than of P for the particular focal mechanism involved. Also, Talandier and Okal (1998) have shown in the idealized case of a planar shore dipping at a constant angle that $S \rightarrow T$ conversions can be favored over $P \rightarrow T$ for a wide range of combinations of shore dips and ray incidences.

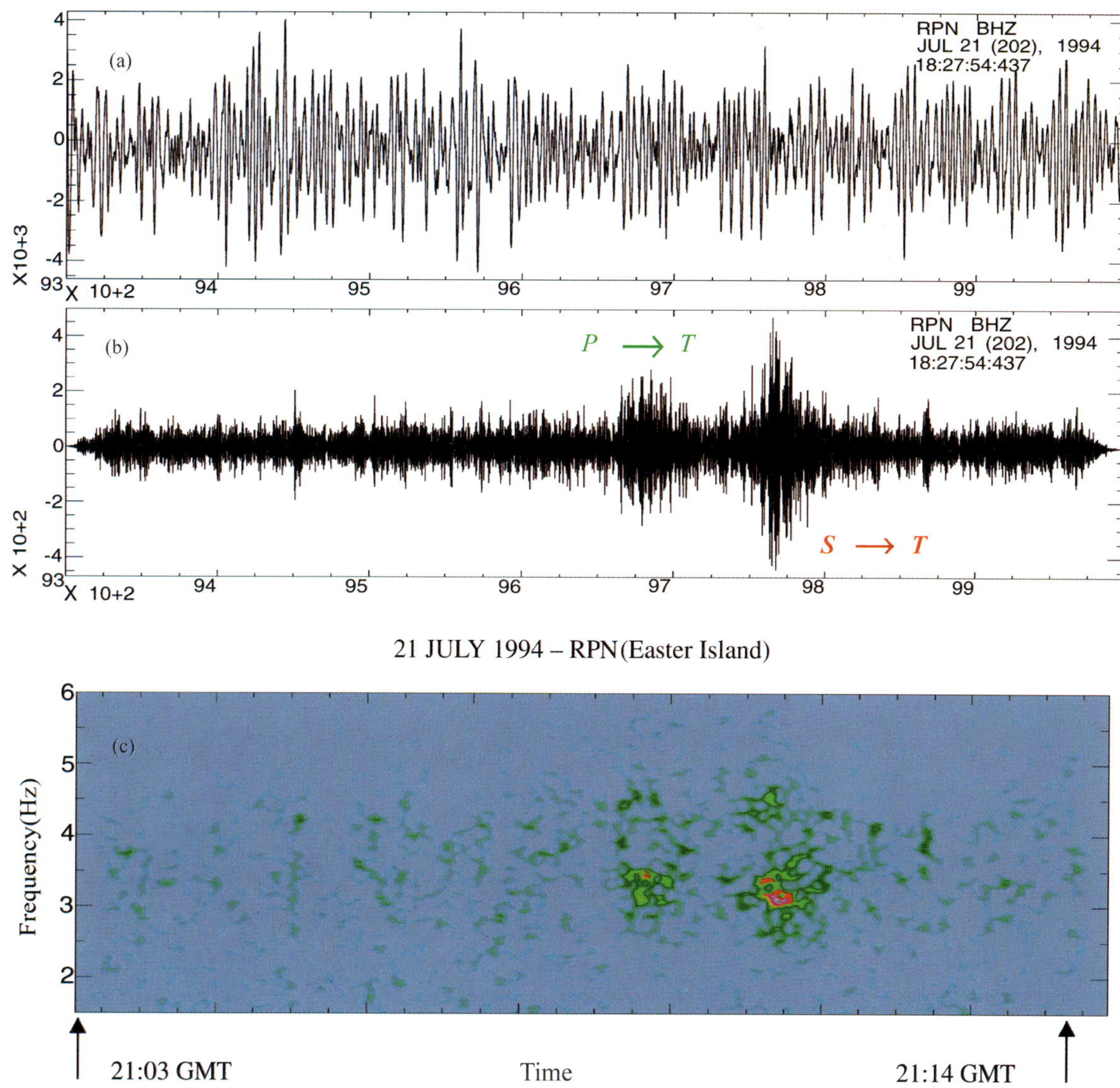

Fig. 1. Typical example of a teleseismic record of T wave from a deep earthquake (Primorye event of 21 July 1994 recorded at Easter Island (RPN)). The epicentral distance is 127.8°, and the oceanic path of the acoustic wave 13,450 km. (a) Original broad-band record. (b) High-pass filtered record ($f \geq 1.5$ Hz). Note the two strong arrivals interpreted as $P \to T$ and $S \to T$ conversions at the source. (c) Spectrogram of the record, in the frequency window 1.5–6 Hz.

Finally, the exact location and mechanism of the seismic-to-acoustic conversion has to be controlled by the morphology of the shoreline at the depth of the SOFAR channel, on a scale comparable to a few acoustic wavelengths, typically 1 km. For example, the presence of bays, coves or bights at the SOFAR depths will affect strongly and in a different manner the $P \to T$ and $S \to T$ conversions. In the absence of its precise knowledge, the combination of all above factors can easily justify $S \to T$ conversions with stronger amplitudes than $P \to T$ ones.

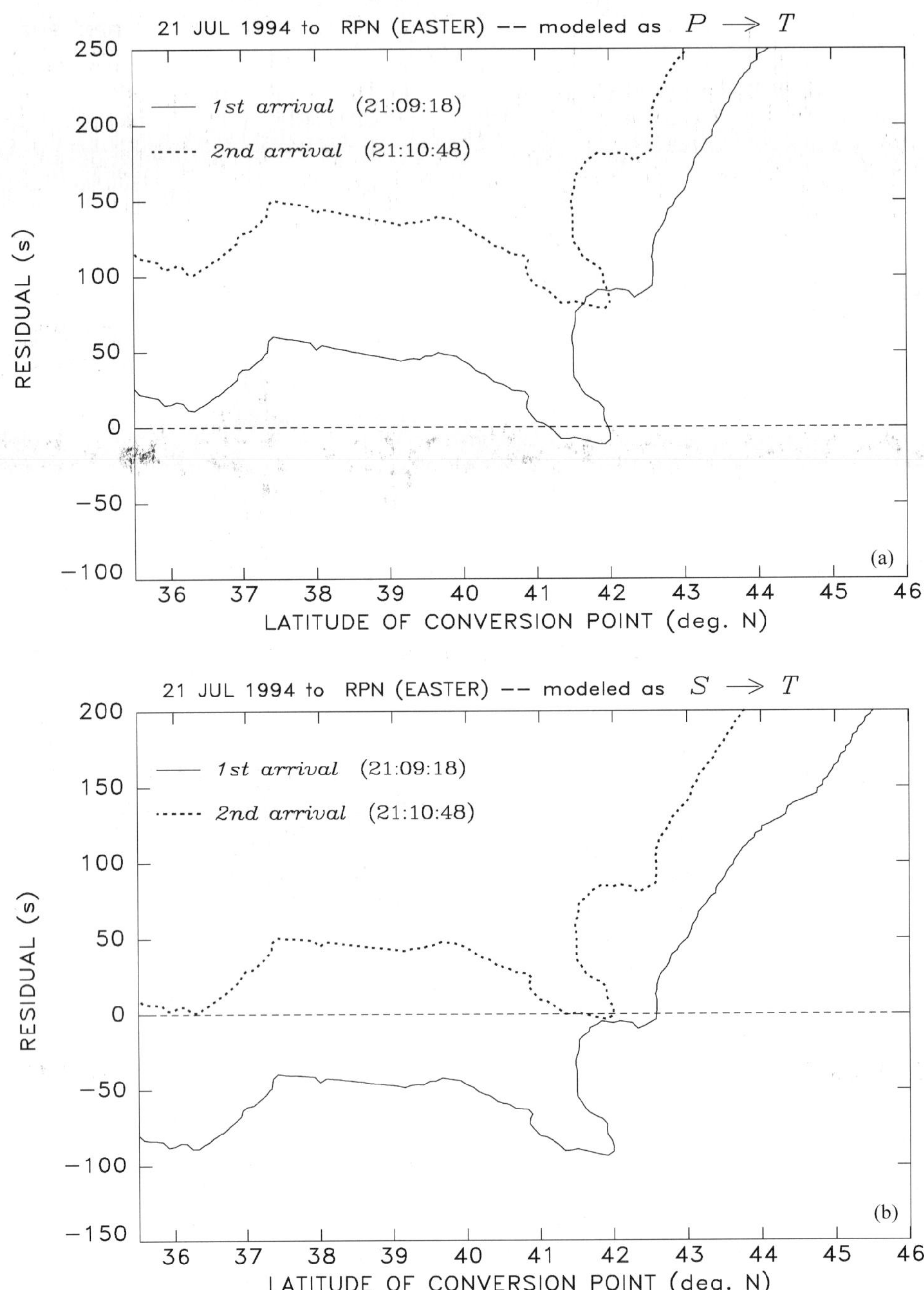

Fig. 2. Residual (observed minus computed) travel-times for the first (solid line) and second (dashes) arrivals at RPN, plotted as a function of the latitude of a hypothetical point of seismic-to-acoustic conversion, on the 1200 m isobath along the Japan–Kuril trench (see Fig. 3). The top frame assumes initial seismic propagation as a P wave, the bottom one as an S wave.

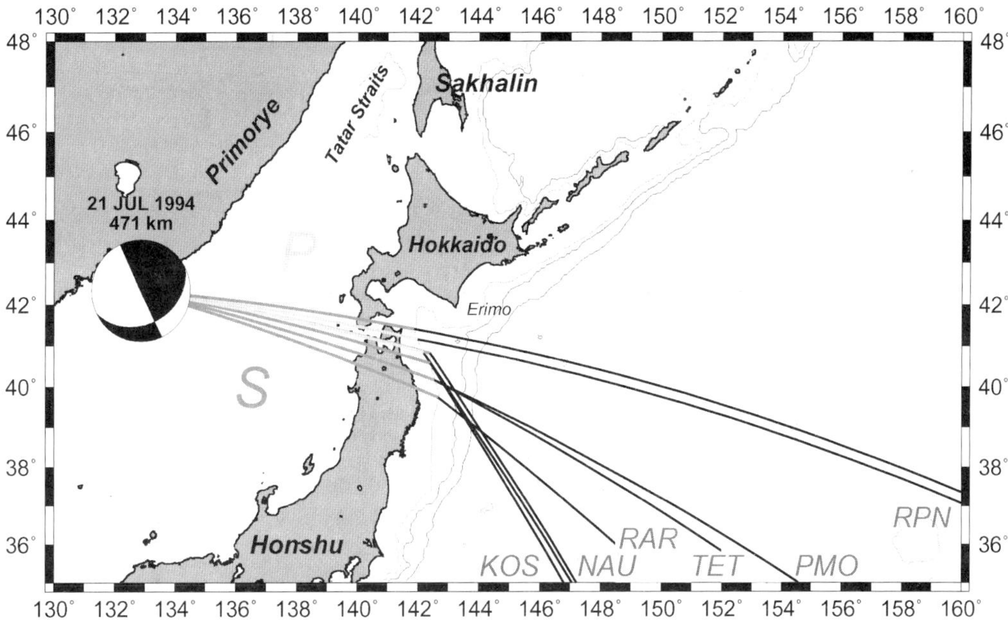

Fig. 3. Map of the epicentral and conversion area for the deep event of 21 July 1994 in Primorye. The seismic paths are plotted as solid (P) or dotted (S) lines. All arrivals can be explained by conversion of either P or S to an acoustic wave at the Hokkaido corner.

4. Regional investigations

In this section, we investigate systematically the characteristics of T waves generated by deep events in subduction zones characterized by seismicity gaps and/or detached events. We discuss in the main text only the principal events, in particular the so-called "detached" earthquakes. All details regarding additional shocks are given in Appendix A.

4.1. *Kuril–Sakhalin subduction zone*

In this region, we are of course motivated by the large deep earthquake under Sakhalin on 12 May 1990. The WBZ is well defined to the East, but deep earthquakes were largely undocumented West of 148°E ($h > 600$ km) or 146.5°E ($h > 550$ km). In this respect, the event occurred significantly North of the perceived extent of the WBZ, and was widely described as "isolated" (Lundgren and Giardini, 1994). However, in an independent study involving relocation of historical earthquakes, we have identified a string of hypocenters extending continuously from the Hokkaido corner at (45.5°N, 137°E; 310 km), under the Tatar Straits down to the 1990 Sakhalin event (Okal et al., 1995; Huang et al., 1998).

4.1.1. *Sakhalin, 12 May 1990; 49.0°N, 141.8°E; 605 km; $M_0 = 8.2 \times 10^{26}$ dyn cm*

We obtained excellent paper records of T arrivals in Polynesia. We use stations PMO (Pomariorio, Rangiroa), MEH (Mehetia, Society) and TBI (Tubuai, Austral). Fig. 4 shows the record at PMO. It is obviously characterized by two main arrivals, with maxima at 06:38:12 and 06:39:38 GMT, respectively. No correction is needed at MEH and PMO, since the stations are less than 500 m from the point of acoustic-to-seismic conversion. A 2 s correction was performed at TBI.

Unfortunately, we were unable to find any other record of T waves from that event. This is due to several circumstances. In 1990, the demise of the WWSSN was complete, but the IRIS network was not yet operating at full strength, and in particular few broad-band stations were continuously recording, hence no records are available at RAR, RPN,

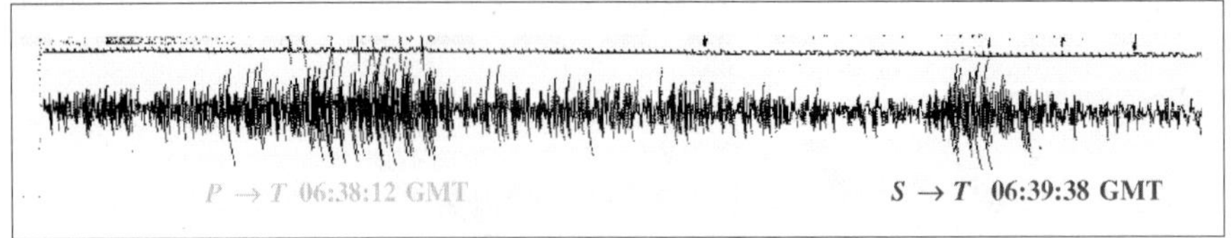

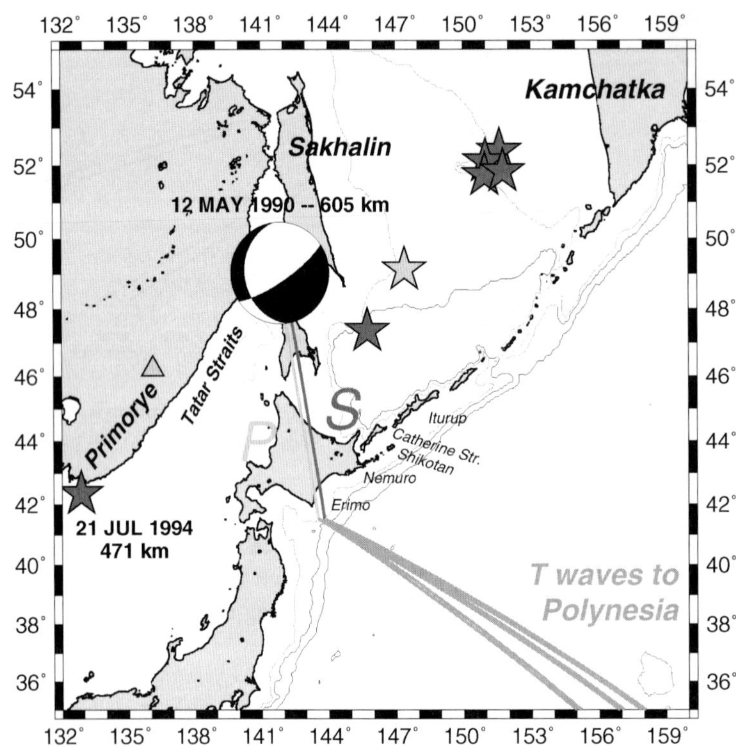

Fig. 4. T waves recorded in Polynesia from the isolated deep earthquake of 12 May 1990. Top: original paper record at PMO. Note the two puffs of high-frequency activity, separated by ~90 s, which compose the T phase. Bottom: map of the Primorye–Sakhalin–Okhotsk area showing the mechanism of conversion at Cape Erimo, Hokkaido. The stars are other deep events generating T phases into the Pacific Ocean, including the large 1970 Okhotsk Sea earthquake. The event at extreme left is the Primorye earthquake described in Figs. 1–3. The triangle is a small 1997 event at 416 km, confirming the warped geometry of the slab.

AFI. While the broad-band channel at KIP (Kipapa, Hawaii) is continuously available, the station is masked by Kauai, and no T wave arrival could be identified; similarly, a systematic search of the HVO records failed to turn up a T phase. Finally, propagation to California coastal stations is blocked by the Aleutian arc.

Fig. 5 shows that whereas the first arrivals (solid lines) at stations PMO, MEH, and TBI are readily interpreted by a $P \rightarrow T$ conversion, the second maxima in the T waves (dotted lines) occur too late to correspond to conversion from a P wave anywhere along the coastline, from 37 to 50°N (conversion at 42°N; 142°E is not acceptable, because at this location, the offshore direction faces the source rather than the receiver). On the other hand, these arrivals, as well as the maximum in T at TET, are readily explained by $S \rightarrow T$ conversion at the southern tip of Hokkaido,

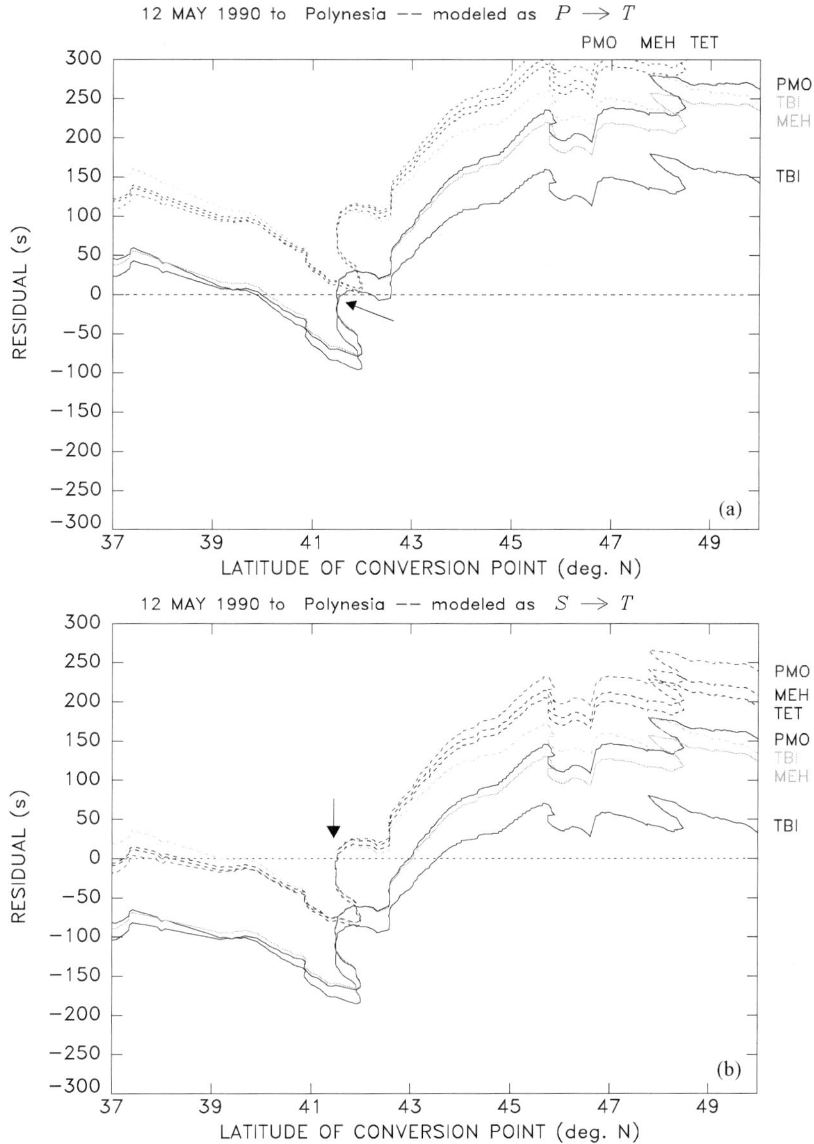

Fig. 5. Same as Fig. 2 for the 1990 Sakhalin event.

in the vicinity of Cape Erimo, at 41.5°N and 143.5°E, essentially the same location as for the $P \rightarrow T$ conversion.

Even though the unavailability of digital data prevents formal spectrogram analysis, the mere propagation of the T wave in the Pacific SOFAR channel requires frequencies of at least 3 Hz, and, thus, our observations indicate the possibility of propagating high-frequency S waves from the Sakhalin deep focus to the Hokkaido corner. We interpret this as evidence for mechanical continuity of the slab in this area.

T waves from additional large deep shocks in the Sea of Okhotsk were studied systematically; all details are given in Appendix A. The emerging pattern is that of the routine generation of T waves from both P and S body waves at the Kuril trench for eastern

Sea of Okhotsk events, and on the southern shores of Hokkaido for the western ones. We note that this conversion is efficient — events as small as 10^{25} dyn cm routinely generate T waves detectable in Polynesia.

In this respect, the 1990 Sakhalin earthquake does not exhibit any singularity in its generation of T waves, as compared with events both in the Sea of Okhotsk to the East (Fig. 1) and in the Sea of Japan to the West (Fig. 3). The slab is vertically mechanically continuous with the lithosphere subducted at the Hokkaido corner, as also documented by the seismic "finger" reaching the 1990 hypocenter (Huang et al., 1998) (we failed to find any new events (1994–1998) which would have updated the seismicity along the finger).

Rather, the northerly position of the 1990 Sakhalin deep shock indicates that the slab is warped, with the subducting angle significantly shallower in the western part of the Sea of Okhotsk than in its eastern part, an interpretation in line with the model of Glennon and Chen (1993), and with the tomographic results of van der Hilst et al. (1991). Finally, it is also borne out by the occurrence of a moderately deep shock under Primorye on 1 October 1997 ($h = 416$ km; $m_b = 5.2$; triangle in Fig. 4), which constrains the WBZ to a northerly location. Faint T waves were recorded from this event at the Wake hydrophones, but were of an amplitude too small for a meaningful study.

4.2. *Bonin–Marianas arc*

4.2.1. *4 July 1982; 27.9°N, 136.5°E; 552 km; $M_0 = 1.2 \times 10^{26}$ dyn cm*

This "isolated" event, mentioned by Okino et al. (1989), was discussed in detail by Lundgren and Giardini (1994). Strong T waves were detected at PMO, where two arrivals are separated by 68 s. No WWSSN or GDSN records could be found for the expected time windows of T waves, but a strong T phase with maximum at 02:34:05 was found at KAA, on the western coast of the Big Island of Hawaii, and a weak but undeniable record is also present at SYP at the western end of the southern California network.

Because the Izu–Mariana arc is composed of discrete island and seamount structures, few of which penetrate the SOFAR channel, the latitude sampling in Fig. 6(c and d) is discontinuous. Even so, the interpretation of the records is somewhat ambiguous. Most arrivals can be interpreted as conversions either at the northern end of the Volcano Islands (Kita Iwo Jima) or at a site on the Bonin Island group. The exception is the SYP record, which can only be interpreted as an $S \to T$ conversion on Haha Jima, on the southern tip of the Bonin group. Note, however, that the path to SYP involves a 32 s receiver side correction (accounting for 86 km of travel in continental structure), which is bound to be imprecise, given the complex geometry of the shoreline.

The most important observation from Fig. 6(c and d) is that it is impossible to account for the second, stronger arrival at PMO (dotted line in Fig. 6c) by invoking a $P \to T$ conversion at any of the shallow structures available along the arc. Rather, it is readily interpreted as an $S \to T$ conversion. Here again, this requires the propagation of strong S waves at frequencies greater than 3 Hz, proving that the slab must be mechanically continuous from the hypocenter to the ocean. The 1982 earthquake cannot have occurred in a detached blob of subducted lithosphere; this result also upholds van der Hilst et al.s' (1991) tomographic section, in which the Bonin slab sags westward above the 670 km discontinuity.

4.2.2. *T waves from other events*

Other deep events in the Bonin–Mariana WBZ were studied systematically, with full details reported in Appendix A. A remarkable result of this study is that while some travel times can be explained by conversion on the Volcano Islands, none require it; on the other hand, the Wake hydrophone signals (and the possible arrival at SCZ) from the 1996 event (purple star in Fig. 6a) can be explained only by conversion in the Bonin group. In this framework, the latter becomes the only proven scatterer of P and S energy into the SOFAR channel.

This can be understood by noting that unlike the presently active Volcano Islands, the Bonin group is an uplifted fragment of fore-arc basement whose volcanics are at least 40-million year-old, (e.g., Umino, 1985; Taylor et al., 1994). In the fully decoupled Bonin–Mariana subduction zone (Uyeda and Kanamori, 1979), there are few shallow slopes permitting conversion of seismic energy into the SOFAR channel, and in practice, only islands or large seamounts are adequate structures. However, in the geometry of that steeply dipping slab, the seismic rays will arrive vertically to the island and, thus, if the

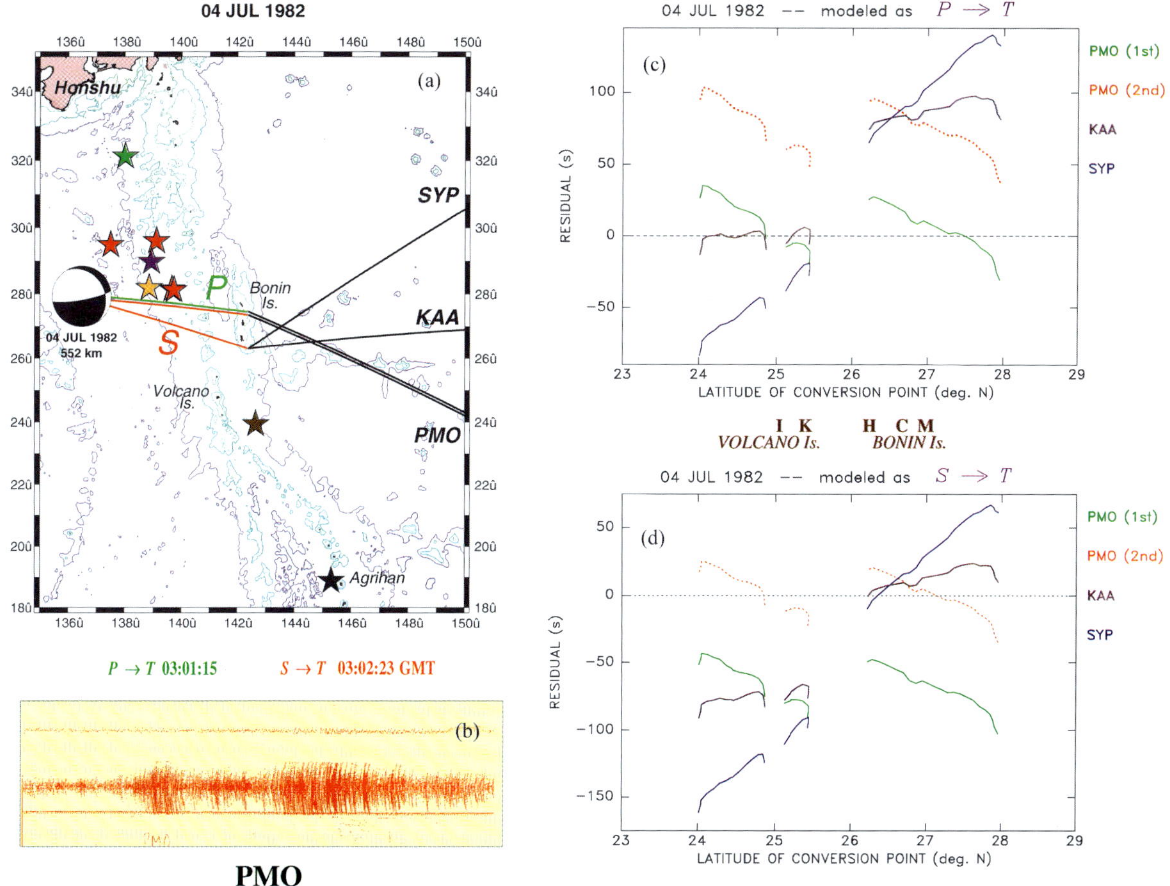

Fig. 6. Composite figure for the 1982 "detached" Bonin earthquake. (a) Location map; ray paths travelled as P waves shown in green, as S waves in red. In addition, red stars show the additional events studied in Appendix A, for which $S \rightarrow T$ phases were observed; other colors are explained in Appendix A. (b) T wavetrain recorded at PMO on the T wave channel. The times of maximum amplitude of the two arrivals are given. (c and d) Modeling as $P \rightarrow T$ and $S \rightarrow T$ conversions. The latitudinal bands of the two island groups are given in brown, with initials indicating the position of the islands mentioned in the text: I, Iwo Jima; K, Kita Iwo Jima; H, Haha Jima; C, Chichi Jima; and M, Muko Jima.

site is volcanic, travel through its magmatic system, characterized by strong anelastic attenuation. As a result, the high frequencies necessary for channeling into the SOFAR will be eradicated, and conversion to an acoustic wave impossible. The Bonin Islands, however, are offset laterally 120 km from the magmatic body under the active arc and have cooled off since 40 million year, to the extent that they can provide a high-Q path from the bottom of the slab to the 1200 m isobath, capable of delivering high-frequency seismic energy into the SOFAR channel. This situation is of course in contrast to more traditional, shallower-dipping and more strongly coupled subduction systems featuring a well-developed fore-arc structure, long known to offer high-Q reception of seismic energy from the bottom of the slab (Utsu, 1971).

We, thus, explain that the Bonin Islands are the only locale along the Bonin–Mariana WBZ capable of efficiently generating T waves from deep earthquakes in a fashion similar to the Kuril shoreline (Section 4.1) or the coast of South America (Paper I and Section 4.3). Note the 1995 shock to the South (black star in Fig. 6a), which did not generate T phases into the

Pacific Basin, despite having the largest seismic moment of the regional group studied, and being in the immediate vicinity of an island (the volcanically active Agrihan). On the contrary, the last event studied, event farther South, did generate a T wave, through conversion on Guam, which is not an active member of a volcanic arc, but rather a limestone-capped uplifted segment of fore-arc basement, estimated to be at least Early Miocene in age (Tracey et al., 1964). Finally, this pattern is also upheld in the Izu region to the North, as documented by the 1993 earthquake, for which conversions are not observed at the nearby volcanic islands and seamounts (e.g. Hachichojima), but rather off the continental structure of southeastern Honshu (see Appendix A).

4.2.3. Other detached events

In the general vicinity of the 1982 shock, we were able to identify two outlying earthquakes, on 23 June 1988 ($m_b = 4.5$) and 12 September 1997 ($m_b = 4.1$), which qualify as "detached" in the sense that they are clearly located in front of the mainstream WBZ (Fig. 7). We relocated these events using the formalism of Wysession et al. (1991). In particular, we attempted unsuccessfully to force the events into the WBZ, by arbitrarily deleting stations. In addition, we performed Monte Carlo relocations after injecting random noise into the dataset, using $\sigma_G = 1.5$ s as the standard deviation of the Gaussian noise, a generous value for such modern events. As shown in Fig. 7b, the resulting error ellipses do not reach the WBZ, and we must conclude that the events are indeed located outside the main body of WBZ.

Another potential candidate, on 9 October 1963, was relocated with its Monte Carlo ellipse reaching the WBZ, and a historical shock on 20 April 1933, listed by the ISS at 20.5°N, 140°E (477 km), is probably an intermediate depth earthquake under the Sea of Okhotsk.

4.2.4. Discussion

We regard the identification of the 1988 and 1997 outliers as a very important result, in that it establishes the continuity of seismogenic material from the WBZ to the 1982 hypocenter. Together with the detection of the $S \rightarrow T$ conversion from the latter, this establishes the mechanical continuity of a segment of slab extending to the location of the 1982 shock. This is in general agreement with regional tomographic models such as Van der Hilst et al.s' (1991) and Fukao et al. (1992), which show a zone of fast P wave velocities extending West of the Bonin arc at the relevant depth (550 km), thus suggesting that the slab sags and stagnates above the lower mantle. This feature is, however, absent from their models for the southern part of the Philippine Basin, West of the Mariana arc. More recently, and based on a comparison between tomographic inversions of P- and S-travel-times, Widiyantoro et al. (1999) have suggested that the stagnating and subducting portions of the Izu–Bonin slab may have different signatures, the latter being unseen in the S tomography. A model compatible with the high-Q path required by our documented $S \rightarrow T$ conversions would have to involve a change of mineralogy increasing the Poisson ratio of the stagnant material, while at the same time keeping a low attenuation.

There is no clear explanation as to why outboard seismicity takes place in front of the WBZ only at the latitude of the 1982 shock and of its two small outlying companions. Of course, the latter observation may be an artifact of a short time sampling of seismological observations. We note the intriguing coincidence of this feature with the presence of the Bonin Islands group to the East, which constitute the only emerged uplifted fore-arc along the whole Izu–Bonin subduction system. One can only speculate that there may exist a common geodynamic agent explaining the apparent upwards deflection of both systems, with obviously very different vertical scales.

4.3. South America

We are motivated in this region by the 1989 earthquake in Paraguay, the third of the main "detached" events reported in the literature (Lundgren and Giardini, 1994).

4.3.1. Paraguay, 28 Feb 1989; 23.11°S, 61.47°W; 569 km; $M_0 = 7.2 \times 10^{25}$ dyn cm

Unfortunately, we were unable to document a consistent set of T waves at Pacific stations which could be associated with this event. There are two T wave signals in Polynesia: a very weak one, legible only on the T wave channel, at RKT (Gambier) at 14:24:47 GMT, the other one at TPT, detectable on the regular

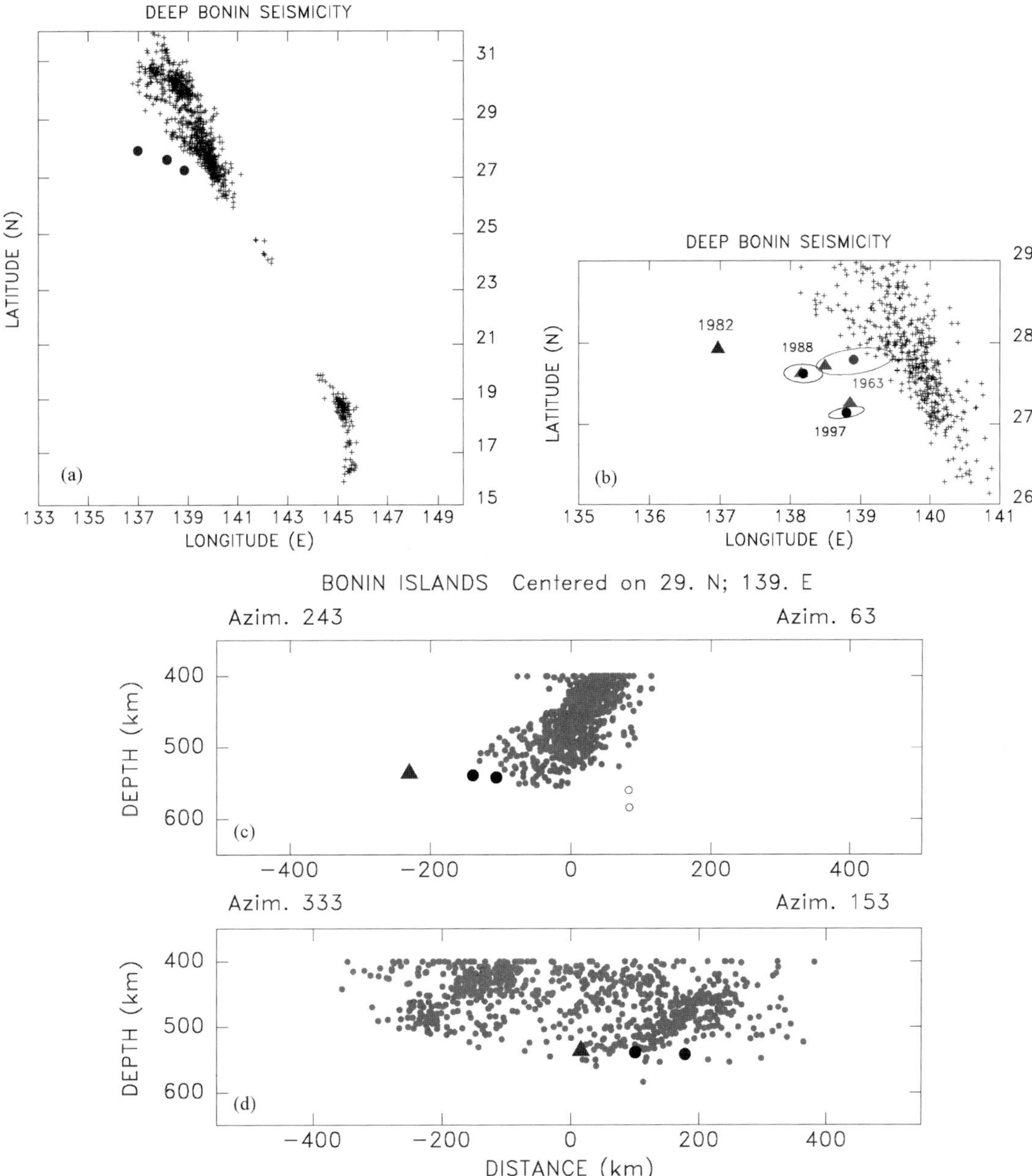

Fig. 7. (a) Deep seismicity of the Izu–Bonin WBZ, showing the three outlying events (1982, 1988 and 1997) as large dots. The '+' signs are unrelocated NEIC earthquakes ($h = 400$ km) for the period 1964–1998. (b) Relocation of the outlying events. The triangles show the original NEIC locations of the 1988 and 1997 events; the dots are our relocations, with associated error ellipses ($\sigma_G = 1.5$ s). Also shown in gray is the 1963 earthquake, listed as outlying in the NEIC catalogue, but relocating to the mainstream WBZ. The ISC estimates of the uncertainty on the 1982 epicenter are ≤ 1.5 km, less than the size of the symbol. (c) Cross-section of the dataset in panel (a) along the azimuth 63°, perpendicular to the strike of the WBZ; only events North of 25°N are included; we verified that the two deepest events, shown as open circles, are poorly located shocks which are not genuine outliers. (d) Cross-section along azimuth 153°, parallel to the strike of the WBZ.

short-period channel at 14:29:19 GMT. These times indicate that the two signals cannot share a common source on the coast of South America. Indeed, they can be reconciled with a small earthquake at 13:19:36 GMT, off the coast of Guerrero, Mexico, a region for which station TPT is particularly sensitive to T waves. A systematic examination of both paper and develocorder archives at HVO failed to turn up a legible T phase at the coastal station HUL. No IRIS/GDSN digital data are available at appropriate combinations of time windows and sampling rates; similarly we could not find digital S wave records at coastal South American stations to investigate the possible presence of high-frequency S waves, using the spectral ratio techniques of Paper I.

We conclude that no T waves from the 1989 Paraguay earthquake were recorded in the Pacific Basin. Given the adequate performance of the Polynesian stations, and despite the absence of continuous broad-band digital records at that date, we believe that the event did not send measurable T phases into the ocean. Noting that its moment is smaller

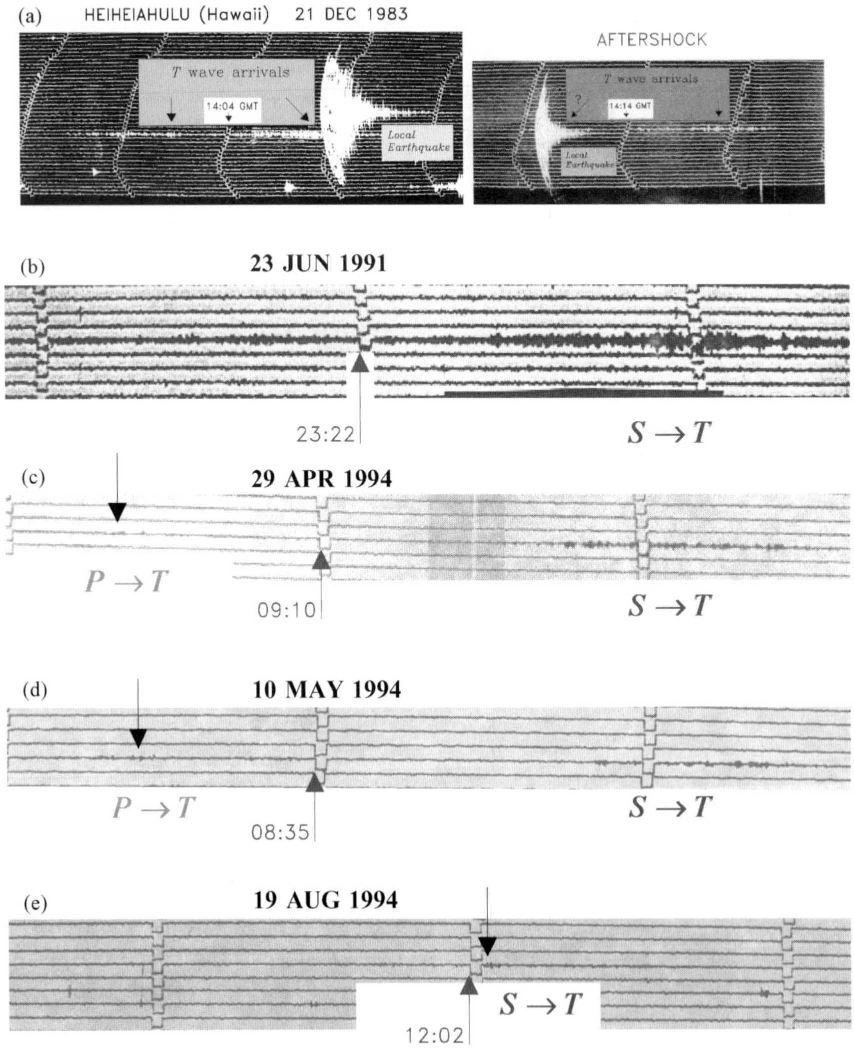

Fig. 8. Comparison of T waves received at station HUL from deep events in northern Argentina. This figure illustrates a threshold of $\sim 10^{26}$ dyn cm for their detection.

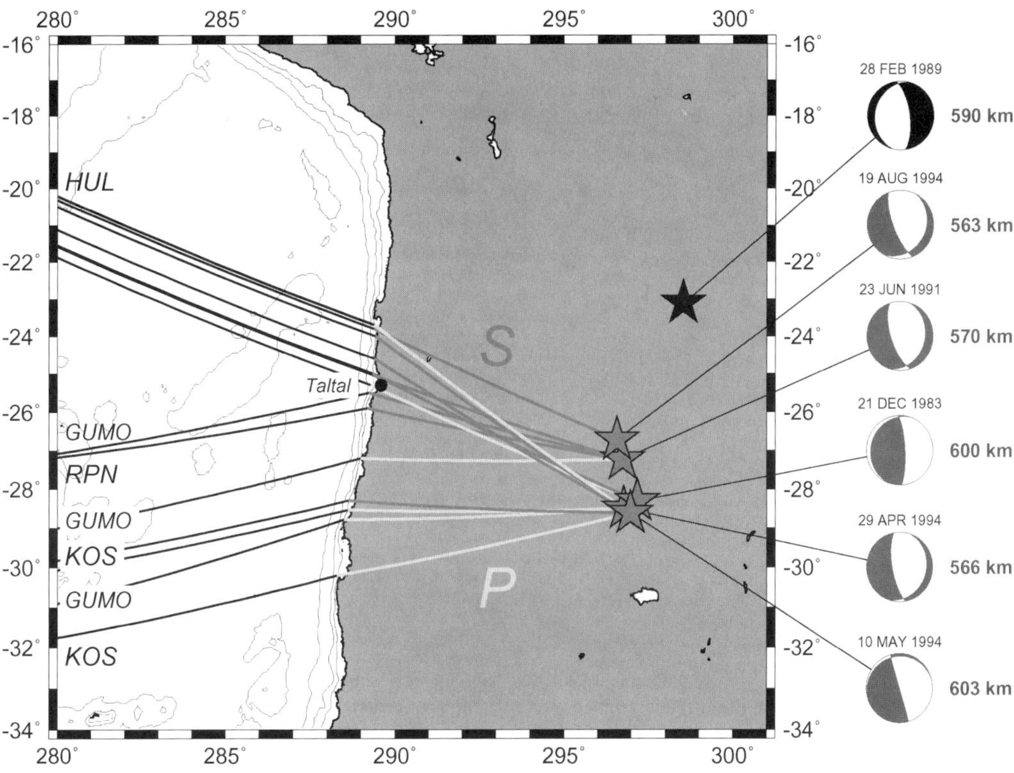

Fig. 9. Map of northern Argentina illustrating the various conversions documented for the events used in the threshold study. The 1989 Paraguay earthquake, for which no T waves were detected, is shown as the black star.

than that of the 1982 Bonin Island and a *fortiori* 1990 Sakhalin "detached" events, we decided to investigate the threshold of detectability of T waves from deep Argentine earthquakes. Specifically, we focus on five events listed in Table 2, and investigate their T waves, primarily at the Hawaiian station HUL. Fig. 8 compares the T waveshapes recorded at HUL, and Fig. 9 verifies that the conversion location is fundamentally similar for all observed T waves. Further details are given in Appendix A. This experiment suggests a threshold of detectability for T waves from deep earthquakes in northern Argentina recorded at HUL of between $M_0 = 2 \times 10^{26}$ and 5.6×10^{25} dyn cm. A practical threshold may then be 10^{26} dyn cm, but unfortunately, no CMT solution exists to fill in the moment gap and refine this estimate. We note that this threshold is significantly higher than in other subduction zones, such as Kuril and Bonin–Mariana.

Of course, the final amplitude of the T wave is controlled in part by a number of unknown factors, such as small scale lateral heterogeneity at the source and conversion point, and high-frequency behavior of the earthquake's source time function, which may not be exactly repeated between events. Thus, an exact correlation between the observability of T waves and seismic moment (or magnitude m_b) would not necessarily be expected. In this sense, our experiment is designed to provide no more than a general order of magnitude of the moment threshold under which T waves from a given region are not expected to emerge from background noise at a given station.

In this context, the absence of T waves from the 1989 Paraguay earthquake may simply express the small size of the event. However, the actual detachment of a blob of seismogenic material cannot be ruled out.

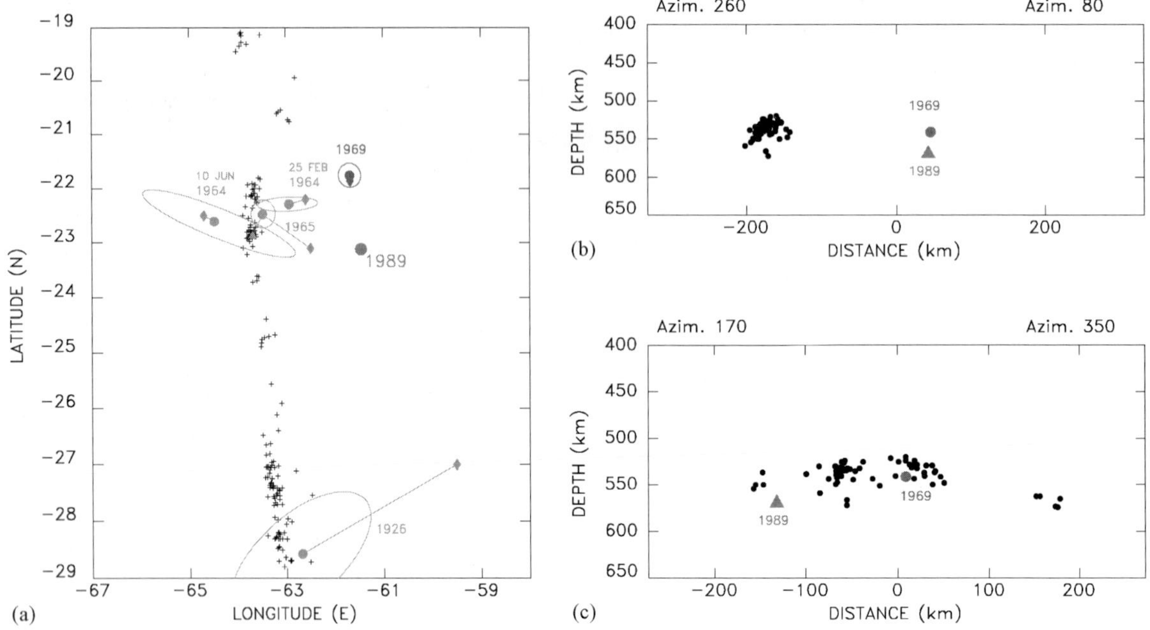

Fig. 10. Outlying deep seismicity in northern Argentina and Paraguay. (a) Map view: the '+' signs are unrelocated NEIC foci ($h \geq 400$ km) for the period 1964–1998. The diamonds identify the original NEIC (or G-R) epicenters of outliers; the inverted triangles are the ISS/ISC locations, when different. The solid dots are our relocations, with error ellipses, as described in detail in Appendix A. The ISC estimates of the uncertainty on 1989 epicenter are ±3.5 km, less than the size of the symbol. (b) Cross-section of the seismicity between latitudes 21 and 24°S, perpendicular to the strike of the WBZ. (c) Cross-section of the seismicity between 20 and 24°S, along the strike of the WBZ.

4.3.2. Other detached events

Again in this region, we investigated systematically the possible existence of additional outboard deep events. The NEIC catalogue (1964–1998) shows three such shocks in the immediate vicinity of the 1989 Paraguay earthquake, on 25 February 1964, 18 February 1965 and 15 April 1969 (Fig. 10). In addition, a historical earthquake in 1926 is given an ISS location 370 km East of the active WBZ, and a lone 1975 earthquake is listed by the NEIC and the ISC 500 km South of the termination of deep seismicity in South America. Finally, a small event on 10 June 1964 is given an "inboard" deep location, West of the WBZ. We relocated all these shocks, using the techniques of Wysession et al. (1991), with all details given in Appendix A. Only the 1969 earthquake is a genuine outboard event. Its final location, 150 km North of the 1989 epicenter, and at a comparable depth (550 km), suggests a similar origin for the two events. The tomographic model of Engdahl et al. (1995) (their Fig. 3f) suggests that around 23°S, the slab sags horizontally before partially penetrating the lower mantle, a behavior not repeated along their "e" profile at 20°S. This could explain that the only known outlying events are at that latitude. The exact mechanism for the distortion of the slab remains unclear. One can simply note (Fig. 10) that the 1989 event takes place ahead of a lateral gap in deep seismicity, which could argue for a local warping of the slab, similar to a curtain fold; this is not the case, however, for the 1969 earthquake.

4.3.3. Bolivia and Colombia

We recall here the study of the 1994 Bolivian deep earthquake in Paper I, which proved that the slab is continuous, both laterally and vertically in that region.

We also refer to Paper I for an analysis of the lone T wave, detected at Reao, from the 1970 Colombian earthquake, which we showed, was generated by a $P \rightarrow T$ conversion along the great circle path from the epicenter. Apart from traces on the analog record at RKT, and despite a systematic search of WWSSN and other databases, we failed to identify T phases from that event at any other Pacific receiver location. A search for T waves from the small 1997 earthquakes was similarly unsuccessful, so that no definite conclusion can be reached as to the nature of the material separating the three deep hypocenters (1921, 1922, 1970) from the intermediate depth WBZ.

The best argument for mechanical continuity would be the continuous subduction of the Farallon, later Nazca, plate for the past 50 million years; the absence of an $S \rightarrow T$ conversion from the 1970 earthquake can be ascribed to an unfavorable radiation coefficient at the source ($R^{SV} = 0.07$); note also that the P axes for the two historical events are only 16 and 26° away from the one in 1970, and that this common direction would correspond to down-dip compressional stress release in the model of a continuous slab. As for the 1997 earthquakes, they may be simply too small ($m_b \leq 4.8$) to excite detectable T phases. On the other hand, the tomographic results of Engdahl et al. (1995) fail to image the slab through the transition zone, and Grand's (1994) tomographic model is inconclusive.

4.4. Java

The deep seismicity under the Sunda arc has been described in detail by Kirby et al. (1996). Deep earthquakes, absent from Sumatra, are present East of the Sunda Straits at 107°E, but a gap in seismicity exists West of 115°E, between 338 and 470 km. This gap ends significantly shallower than in Argentina. We have verified through relocation (Wysession et al., 1991) the depth of the limiting event (1 June 1997, $h = 470 \pm 17$ km). Based on tomographic studies, Widiyantoro et al. (1997) have proposed necking of the slab under western Java, in a region which would grossly coincide with the seismicity gap. We sought to analyze T waves from deep shocks from this zone; unfortunately, the eastern part of the gap is masked from many sites by Western Australia; in addition, the instrumentation of the Indian Ocean is only very recent, so that we could study only one event, on 19 January 1997.

4.4.1. Java, 19 January 1997; 5.03°S, 108.40°E; 651 km; $M_0 = 1.7 \times 10^{24}$ dyn cm

T wave arrivals could be identified at two stations: Cocos Island and Hope (South Georgia). The records at COCO are characterized by two strong arrivals, which can be interpreted as $P \rightarrow T$ and $S \rightarrow T$ conversions, from a scatterer south of Cape Ujungkulong, at 7.1°S; 105.6°E. This interpretation is, however, non-unique since other points along the coast of Java would also be adequate converters, without the need to invoke $S \rightarrow T$ conversions (Fig. 11). Furthermore, the T waves recorded at COCO are characterized by very low frequencies (down to 1.7 Hz), which should not propagate in a standard Pacific SOFAR channel, but may be explained if the channel is less well defined, i.e. extends over a greater depth range and features faster axial velocities, estimated locally at 1489 m/s (Levitus et al., 1994). Such low frequencies then significantly diminish the power of the method as a proxy for the continuity of high-Q material along the S fragment of the path.

The HOPE record also shows two arrivals, at 04:52:48 and 04:54:51 GMT, the second one being much stronger (Fig. 11(e and f)). Unfortunately, travel-time corrections to Hope are practically impossible to assess given that the 12,700 km great-circle path grazes the Kerguelen Plateau and Antarctica, and samples extreme southern latitudes where the morphology of the SOFAR channel is expected to be strongly altered, with minimum velocities being probably lower but largely uncharted (Levitus et al., 1994). This in turn would suggest off great-circle propagation and quite possibly reflections.

As for the use of spectral properties of S waves, we could not find digital stations in Java sufficiently removed from active volcanic structures to compute a meaningful estimate of Q_μ.

In order to obtain some quantitative assessment of the slab's continuity, we investigated the distribution of deep focal mechanisms both in the zone featuring the seismic gap (West of 115°E) and in the portion with continuous seismicity with depth, defined here as between 115 and 121°E (we do not include mechanisms farther East into the Banda Sea, where the regime of subduction is significantly different). We consider all 47 published CMT solutions deeper than 500 km, including those for 1957–1976 inverted by Huang et al. (1997, 1998), 15 of which belong to the Eastern group, and 32 to the Western one. For each group, we calculate a composite focal mechanism, by summing the moment tensors and solving for the best-fitting double-couple (without weighting the solutions according to moment, which in essence would

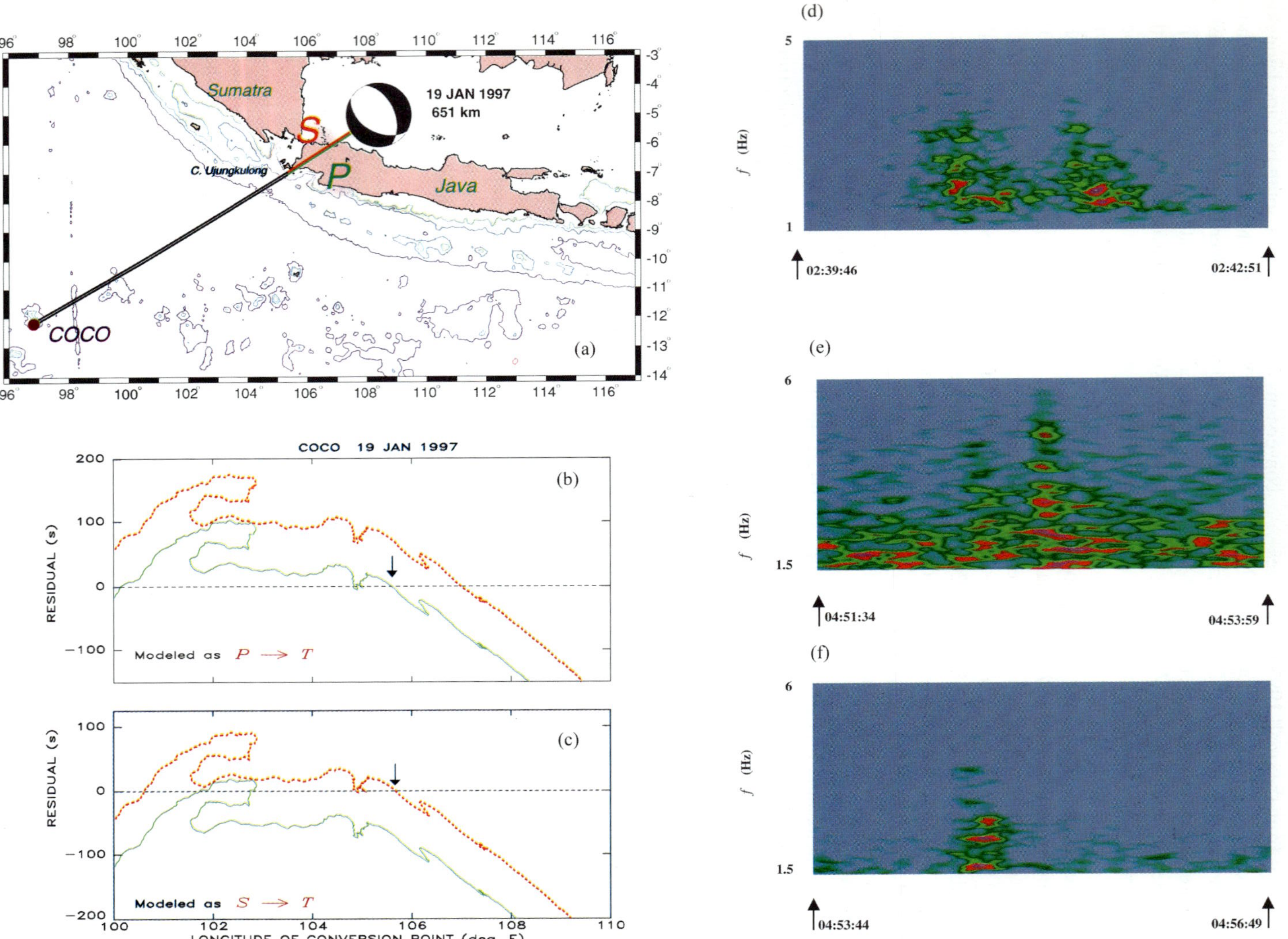

Fig. 11. The deep Java earthquake of 19 January 1997. (a) Map of the proposed paths to Coco Island (COCO). (b and c) Modeling of COCO arrivals as conversions at the Sunda shoreline. (d) Spectrograms of the COCO record. Note the significantly low frequency of the *T* phase. (e and f) Spectrograms of records obtained at Hope, South Georgia. The scale for spectral amplitudes is 50 times smaller for (e) than for (f).

keep only the contribution of the largest earthquake in the dataset). The results show that the stress release in the Western group remains coherent, and essentially expresses the release of quasi-vertical compressional stress, as it does in the Eastern group. The average P axes East and West of the divide dip 85° at azimuth N189°E, and 77° at azimuth N273°E, respectively. These two directions are separated in space by only 13°. Furthermore, the average best double-couples are separated by a rotation of 28° about a steeply (67°) dipping axis, in the formalism of Kagan (1991). This indicates that the predominant down-dip compressional stress believed to control the strain release of the deepest earthquakes is transmitted in a coherent fashion through the seismic gap.

4.5. New Zealand

We investigate here the cluster of deep New Zealand earthquakes, originally discovered by Adams (1963). Table 3 is a list of apparently detached events, updated to June 1999, on the basis of the latest available catalogues. Since the 1998 event has not yet been processed by the ISC, we relocated it based on the dataset available from the EDR files. The essential point is that the depth of the event is perfectly controlled by regional stations, at a value of 600 ± 10 km. Thus, a total of 10 deep New Zealand earthquakes are now confirmed from teleseismic records. None of the six recent shocks were large enough to be processed by the Harvard CMT project, and so, no modern focal mechanism is available. Adams (1963) proposed a thrust mechanism for the 1960 event, based on world-wide first motion readings. Fig. 12 shows a map view and a cross-section along strike of the relevant seismicity.

T waves are routinely recorded from those and other deep Kermadec events at Pacific stations, such as TBI, PPT, RPN, and even KIP, often showing complex wavetrains, suggesting multipathing. However, the interpretation of their arrival times is made difficult, if not outright impossible, by the presence of numerous shallow structures, including the Chatham Rise, which provide a large selection of potential converters. Consequently, the identification of possible $S \rightarrow T$ converted phases at distant receivers is not a realistic means of investigating the structure of the slab.

Rather, we use here the spectral ratio technique described in Paper I, based on three-component broad-band data at station South Kariro (SNZO). IRIS records are only available for the three events of 8 July 1992, 8 April 1994 and 4 July 1998. The records of the small event of 1992 have poor signal-to-noise ratios, and only the other two events could be studied (Figs. 13 and 14). Over the exact same frequency range (1–3 Hz), we obtain an average estimate of $Q_\mu = 300$, which is very significantly lower than found in Paper I for Bolivia–northern Chile, where a continuous path had been documented by $S \rightarrow T$ conversions. By contrast with this situation, we surmise that the path from the deep seismic cluster to SNZO is not made of continuous cold-slab material. This value of Q_μ is only an estimate, but we note that the results for the two events are surprisingly consistent, despite different waveshapes, suggestive of different focal geometries. We propose that this value represents a weighted average of Q_μ^{-1} along

Table 3
Deep earthquakes under New Zealand

Date (D M (J) Y)	Origin time (GMT)	Epicenter		Depth (km)	Magnitude	Reference
		°N	°E			
24 MAR (083) 1953	03:37:13.0	−38.9	174.5	570	5 M_L	Adams (1963)
23 MAR (083) 1960	01:32:18.0	−39.05	174.87	607	6.25 M_L	Adams (1963)
23 MAR (083) 1960	01:36:35.7	−39.10	175.07	612	6.2 M_L	Adams (1963)
7 FEB (038) 1975	15:19:43.0	−39.27	174.26	582	4.9 M_L	Adams and Ferris (1976)
14 SEP (257) 1991	14:14:42.0	−39.180	174.434	602	5.0 m_b	ISC on line Bulletin
26 FEB (057) 1992	00:05:33.7	−38.896	174.693	602	4.0 m_b	ISC on line Bulletin
8 JUL (190) 1992	12:25:25.0	−39.116	174.362	622	4.9 M_L	ISC on line Bulletin
5 MAY (125) 1993	03:51:19.2	−38.903	174.440	577	4.6 M_L	ISC on line Bulletin
8 APR (098) 1994	04:06:54.8	−39.194	174.455	602	3.8 m_b	ISC on line Bulletin
4 JUL (185) 1998	07:47:47.5	−39.16	174.510	600	4.3 m_b	This study

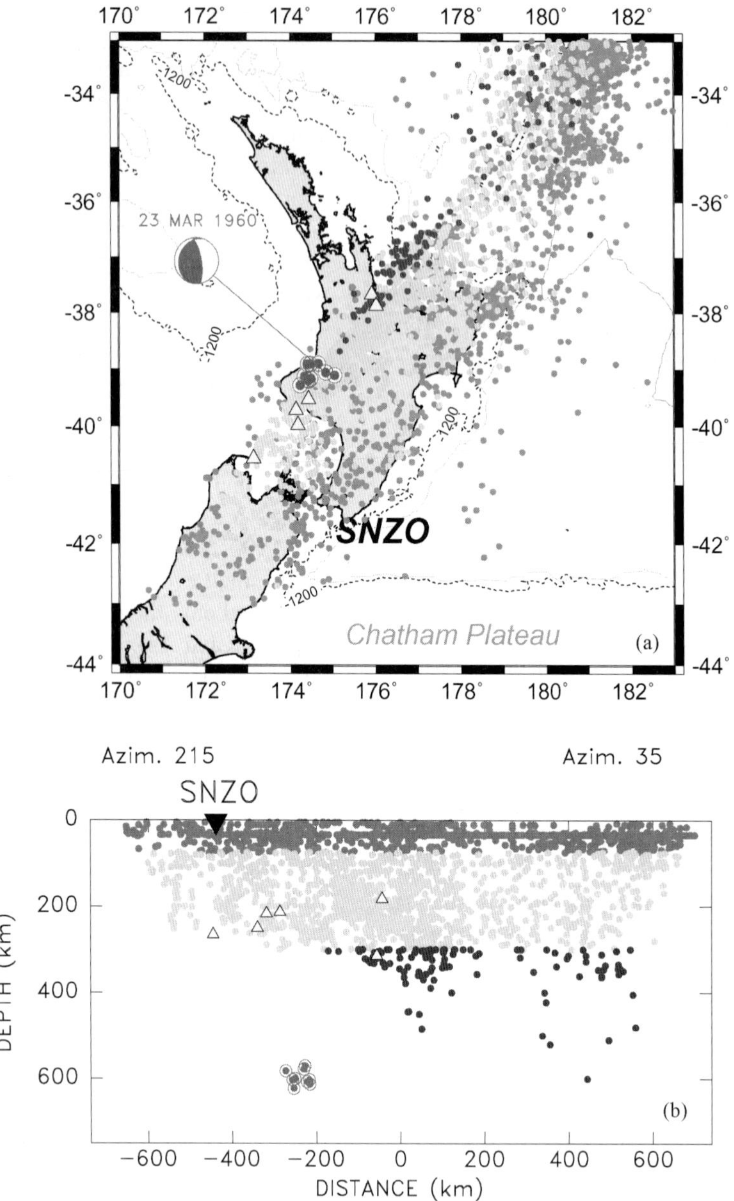

Fig. 12. (a) Map of New Zealand seismicity emphasizing the deep cluster. The events are coded according to depth ($h \leq 75$ km: small, light gray dots; $75 < h \leq 300$ km: medium gray dots; $h > 300$ km: black dots). The deep events listed in Table 3 are shown as larger bull's eye symbols. Focal mechanism is after Adams (1963). The triangles are shallower events also investigated. (b) Along-strike cross-section of the seismicity. Both frames use unrelocated NEIC hypocenters.

a path composed of one-third cold, seismogenic, slab (~200 km with $Q_\mu \approx 800$), and two-thirds background mantle (~400 km with $Q_\mu \approx 250$, a possible value for the bottom of the transition zone (Okal and Jo, 1990)), all these numbers being tentative.

Finally, we note that Adams and Ferris (1976) had proposed the identification of a possible structural

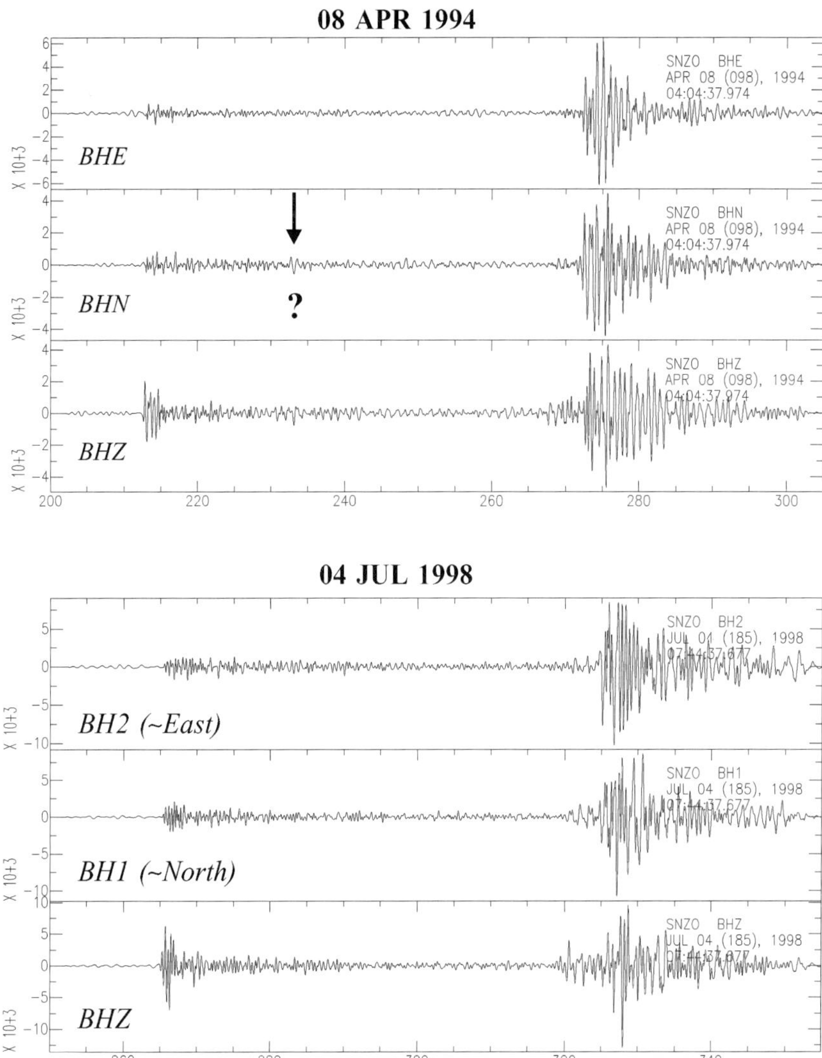

Fig. 13. Three-component P and S waveshapes at South Karori (SNZO) for the two deep New Zealand earthquakes of 8 April 1994 (top) and 4 July 1998 (bottom). For the former, the arrow identifies a possible converted phase, similar to those discussed by Adams and Ferris (1976).

transition, based on intermediate arrivals observed between P and S at New Zealand stations from events in the deep cluster, which they interpreted as $P \rightarrow S$ conversions. Fig. 13 shows that such an arrival may be present on the SNZO record of the 1994 event. Being most prominent on the NS component (naturally polarized as SV), it could represent a $P \rightarrow S$ conversion, in contrast to the $S \rightarrow P$ phase claimed by Adams and Ferris (1976). Its timing, 20 s after P, would correspond to conversion two-third of the way to the station, or at ~200 km, which is approximately the local depth extent of mainstream seismicity. This interpretation is, however, non-unique since the conversion could for example occur as part of an internal reflection inside a mechanically continuous slab. Indeed, we have found arrivals intermediate

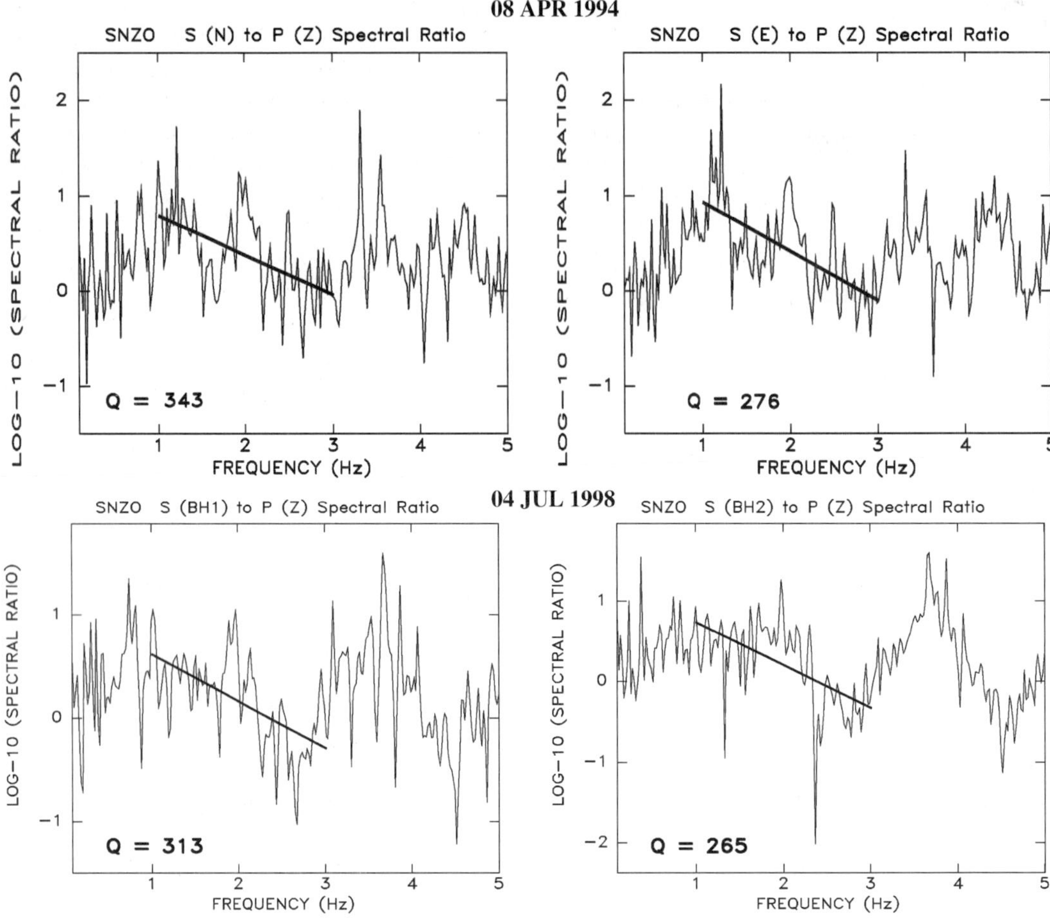

Fig. 14. Estimation of Q_μ on the quasi-vertical path from the New Zealand deep shocks to station SNZO. Each figure shows the variation of the spectral ratio $X_S(\omega)/X_P(\omega)$, for two events and two sets of components: at left North–South (essentially SV); at right East–West (SH). The straight segments provide the regressed best linear fit to the logarithmic decay of the ratio with frequency between 1 and 3 Hz.

between P and S on vertical records at SNZO of shallower, mainstream events (shown as triangles in Fig. 12). Such phases would more precisely match the description given in Adams and Ferris (1976), but would violate their model because of the continuity of the slab to those shallow sources. Finally, the $P \rightarrow S$ conversion reported here is absent for the 1998 event, but this could be due to a different focal mechanism.

To a large extent, the body of data available from the few and small deep New Zealand earthquakes is still inconclusive. However, the following arguments would favor a mechanical separation between the deep cluster and the mainstream WBZ: the relatively low value of Q on the near-vertical path to SNZO, the tentative $P \rightarrow S$ converted phase on the 1994 record, and the down-dip tensional character of the mechanism ($\phi = 350°$; $\delta = 71°$; $\lambda = 79°$) determined by Adams (1963) for the 1960 earthquake. In particular, the first two observations are generally consistent with the mechanical termination of the slab at the depth of cessation of mainstream seismicity. Finally, we note that the tomographic results of van der Hilst and Snieder (1996) would suggest the existence of a body of slow wavespeed between depths of 250 and 500 km at 40°S (their Fig. 12e).

4.6. Vityaz

The Vityaz deep cluster, under the North Fiji Basin, was studied in detail by Okal and Kirby (1998). Since this study was completed, 34 more deep earthquakes (spanning July 1996–June 1999) have been given preliminary locations in the cluster. While the precise depths of these events have not been determined, their epicenters generally fit the several groups identified by Okal and Kirby (1998); the two available new CMT solutions are of small moment (1×10^{24} and 2×10^{24} dyn cm); one mechanism is normal faulting, the other strike-slip. Finally, an intermediate shock given at 200 km under the North Fiji Basin, on 14 November 1997, is most probably a shallow crustal earthquake in the North Fiji Basin: the dataset of 10 stations is found to have no depth resolution.

Despite the availability of good records at KIP, PET, GUMO, SNCC and TKK, we failed to identify T phases from the largest recent deep Vityaz event, on 13 April 1995. This suggests that high-frequency energy is not transmitted to the earth's surface North of the Vityaz trench, where adequate converters would exist in the form of numerous shallow bathymetric structures. This supports Okal and Kirby's (1998) conclusion, namely that the Vityaz seismic cluster resides in a severed piece of lithospheric slab, orphaned from the Pacific plate after the reorganization of subduction along the Tonga and Vanuatu systems, and having lain recumbent at the bottom of the transition zone ever since.

4.7. Spain

In this relatively complex tectonic province, the main unanswered question remains the origin of the deep seismicity. Any integrated tectonic model should also explain a growing body of geophysical data at shallower depths, in particular intermediate depth seismicity.

A large scale tomographic experiment by Blanco and Spakman (1993) has imaged a possibly continuous slab, extending in a generally SW–NE direction, between depths of 200 and 670 km. Its structure, with P wavespeed anomalies of $+1.5\%$, would be compatible with subduction of oceanic lithosphere, and its size much larger than the 100 km scale, which Grimison and Chen (1986) argued was the maximum width of a paleo-ocean separating Europe and Africa from which to drain such material. Blanco and Spakman (1993) also note that the mapped anomaly does not extend to the surface, and thus speculate that the imaged slab may have been detached, possibly as early as the Miocene, whereas Grimison and Chen (1986) suggest sinking over only a few million years.

The focal mechanism of the 1954 earthquake was worked out by Chung and Kanamori (1976), Udías et al. (1976) and Buforn et al. (1991). Their solutions are generally similar, with the azimuth of dip of the P axis varying from N87°E to N61°E, only in fair agreement with a down-dip compression in the geometry of Blanco and Spakman (1993), which would require a more southerly azimuth. The small events of 1973 and 1990 were also worked out by Buforn et al. (1991, 1997), who found mechanisms close, but not exactly similar, to that in 1954.

At shallower depths, we refer to the recent relocations by Seber et al. (1996) and Mezcua and Rueda (1997), who mapped the main body of intermediate depth seismicity as spanning a 140 km long segment oriented N10°E across the Sea of Alborán, from Spain to the northern coast of Morocco. This places its northern end 50 km southeast of the epicenters of the deep shocks (Fig. 15a). Seber et al. (1996) also document strong seismic attenuation in a lower crust–upper mantle body extending at depths of 20–60 km under the Sea of Alborán, and underlain by a more rigid and seismically active mantle structure which they interpret as a fragment of continental (Spanish) lithosphere, delaminated below asthenospheric material under the Sea, that they relate to Neogene volcanism in the Sea of Alborán (Bellon, 1981). On the other hand, Lonergan and White (1997) question the generation of melt in this geometry, and Morales et al. (1999) present a model supported by local P and S wave nseismic tomography and gravity data, suggesting active continental subduction under the Sea of Alborán.

In this general framework, the question remains of the exact source of seismogenic material for the deep Spanish earthquakes, of its history, and of its relationship, if any, to the subducting or delaminating segments carrying the present intermediate seismicity. It should be emphasized that the relative geographical offset between the intermediate and deep seismicity is not compatible with a single subducting structure along the geometry outlined by Blanco and Spak-

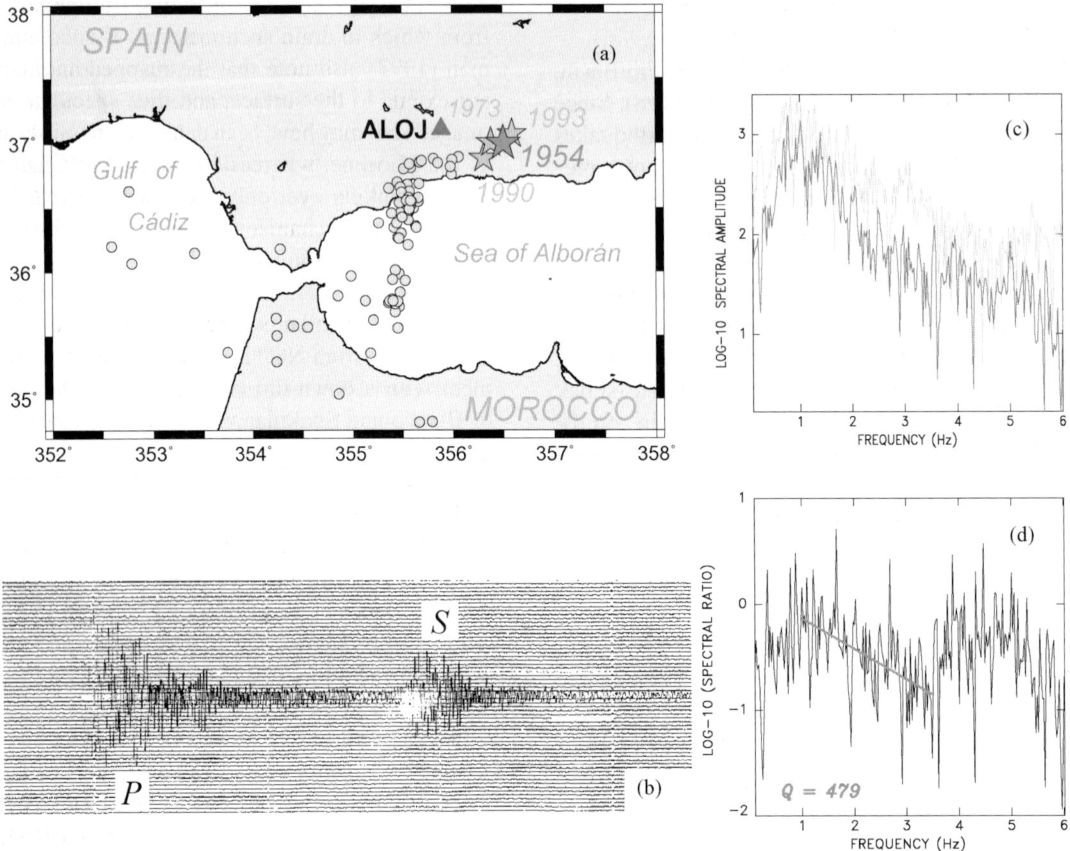

Fig. 15. (a) Map of southern Spain and northern Morocco, showing the epicenters of the deep Spanish earthquakes (black star: 1954 event; white stars: subsequent smaller shocks). The solid dots are unrelocated NEIC epicenters of intermediate depth earthquakes ($h \leq 70$ km) for the period 1964–1998. Also shown (triangle) is station ALOJ used in the attenuation analysis. (b) Original analog record at ALOJ of the deep shock of 8 March 1990 (courtesy of Dr. J. Morales); tick marks are minutes. The epicentral distance is 44 km, and the focal depth 637 km. (c) Spectral amplitude of the P (gray) and S (black) wavetrains obtained after hand-digitizing (b). (d) P-to-S spectral ratio at ALOJ, used to estimate Q_μ between 1 and 3 Hz.

man (1993); rather, such an association would require subduction with a strike of ~N10°E, which would be more in line with the suggestion (Royden, 1993; Lonergan and White, 1997) of the westward rollback of a short eastward-dipping subduction zone. In such a geometry, a cold slab continuous from the deep foci to the surface could provide efficient channeling of high frequency energy, with the possibility of a conversion into T waves at the eastern shores of the Gulf of Cádiz. We therefore attempted to identify T phases received from the deep Spanish events.

We could not find any T phase records from any of the four deep Spanish shocks. The dataset inspected included the 1954 short-period record at Bermuda, the available WWSSN collection of Atlantic stations for the 1973 earthquake, and all available IRIS time series for the 1990 and 1993 earthquakes.

This negative search remains, however, inconclusive, given on the one hand the low quality of instrumentation for the large 1954 shock, and on the other hand the extremely small size ($m_b \leq 4.1$) of the more recent earthquakes.

We then studied spectral amplitude ratios for a vertical path from the deep 1990 event to a station in Andalucía. An analog record adequate for hand-digitizing and processing was obtained at station

ALOJ, for which the quasi-vertical path would be expected to sample the structure imaged by Blanco and Spakman (1993). We obtain $Q_\mu = 479$, a value intermediate between that of a typical upper mantle average (200) and the much higher values (800–1000) characteristic of propagation through actively subducting cold slabs (Paper I; Mele, 1998). Being higher than the typical mantle average, it requires sustained propagation through cold material, and also a path avoiding the asthenosphere. A tentative model involving a 440 km path in (Blanco and Spakman, 1993) structure, assumed to have $Q_\mu = 700$ (this lower value reflecting possible reheating after detachment), would require $Q_\mu = 283$ for the complementary segment from 200 km depth to the surface, which incidentally samples the zone of intermediate depth seismicity, and remains significantly North of the attenuating body, as mapped in Fig. 2 of Seber et al. (1996). The estimate $Q_\mu = 283$ would be in the range of values for continental lithosphere, given in particular the uncertainty on the thermal structure of a subducted or delaminated fragment (Mitchell, 1995).

It is clear that these results remain very tentative, and that further experimentation, hopefully with digital data acquired during future deep Spanish earthquakes, is warranted.

5. Conclusion

Having examined in detail "detached" or isolated deep earthquakes in eight different environments, we can draw the following conclusions.

1. T phases are routinely generated by the deepest earthquakes, as documented for example in our studies of the Sea of Okhotsk, northern Argentina or the Izu–Bonin systems. The threshold of detection at teleseismic distances varies significantly with the relevant subduction zone; we have found it to be particularly low in Java and much higher in South America. This variation is probably related to the morphology of the converting slope for each subduction zone.
2. In two cases out of three (Sakhalin and Bonin) of detached earthquakes occurring outboard of well-documented WBZs with abundant seismicity, we offer proof of mechanical continuity with the subducting slab. The evidence includes $S \rightarrow T$ conversions, and a string of seismic activity at lower magnitudes. In the third case (Paraguay), we fail to document T phases, but we take note that the earthquake falls below the threshold of detectability from nearby deep shocks; we document an even smaller outboard event 145 km to the North. In general, our results are also compatible with tomographic images, and suggest that the slab undergoes warping, resulting in an offset of the seismicity, rather than tearing and detachment of a lump. In Sakhalin, this is probably related to the cusp in subduction at the Hokkaido corner; in Bonin, we can only make the intriguing observation that the fold in the slab seems to correlate geographically with the presence of the uplifted Bonin Islands at the same latitude along the subduction zone. In Paraguay, we can offer no insight into the mechanism of warping.
3. In subduction zones where a depth gap in seismicity is observed, the presence of $S \rightarrow T$ conversions proves that the slab is indeed mechanically continuous through the gap, the latter expressing only a change in the seismogenic character of the material, rather than the physical separation of a deep blob. In Argentina and Bolivia, this is upheld by the consistency of focal mechanisms, and by tomographic imaging. The temporary loss of seismogenic potential (between 337 and 502 km in Argentina) can be ascribed to an age discontinuity in the subducting material, in the model of Engebretson and Kirby (1992). In Java, where tomography suggests necking of the slab, our T wave results support a continuous slab, without requiring it, the best evidence for mechanical continuity being the coherence of focal mechanisms for the deep events below the gap.
4. The cases of Colombia and Spain are somewhat comparable regarding the extreme isolation of the seismicity, mostly expressed by very large shocks, but differ in relation to small events, present in Spain at the same location as the large ones, but apparently offset to the South in Colombia. Also, Colombia is part of a well-documented major subducting system, whereas Spain is not; on the other hand, a large fast slab is imaged by tomography under Spain, but is not detected under Colombia. In both cases, an argument can be made for

mechanical continuity based on the geometry of stress release: Bina (1997) has modeled downward compressional stresses at the bottom of sinking slabs from the integral of buoyancy forces resulting from the retardation of phase transformations in the cold interior of slabs. A continuous slab, extending several hundred kilometers up from the seismic foci, would provide a consistent domain over which these forces can be integrated in a coherent fashion, whereas a small blob extending only 100 km or less around the seismogenic zone, would not. Thus, it is probable that both Colombia and Spain have a mechanically and thermally continuous slab, since they both feature consistent down-dip compressional stresses. Additional evidence supporting this model exists but its nature is different: in Spain, it comes from tomography and high Q values; in Colombia, mainly from the history of subduction of the Pacific lithosphere and from the geometry of stress release. Both interpretations leave substantial questions unanswered: in Spain, the origin of the material, as discussed by Grimison and Chen (1986), and the orientation of the stress with respect to the slab imaged by tomography; in Colombia, the absence of any detectable tomographic signal, and of low-level seismicity around the 1970 focus (Okal and Bina, 1994, 2001).

5. On the other hand, the Vityaz cluster, where abundant seismicity occurs at relatively low moment levels, and with incoherent geometries of stress release, is most easily explained as a severed piece of lithosphere lying recumbent on the top of the lower mantle, as proposed by Okal and Kirby (1998). The failure to detect any T waves from these deep earthquakes would support this model, but a major difficulty resides in the thermal state of the severed fragment which is expected to be cooling too fast to preserve metastable olivine as a candidate seismogenic material (Van Ark et al., 1999).
6. This leaves the case of the deep New Zealand earthquakes as perhaps the most intriguing and enigmatic detached events. The extreme complexity of the local coastlines makes it impossible to interpret their T waves. A significant body of evidence consisting of the lone available focal mechanism, tomographic results, relatively low Q values, and the tentative observation of converted phases, would support the concept of a mechanically detached lump of seismogenic material, but its origin and past history remain elusive.

Acknowledgements

Digital data used in the present study were obtained from the IRIS, GEOSCOPE and pIDC data centers. I am grateful to Paul Okubo at HVO and Olivier Hyvernaud at PPT for access to their respective analog archives. Numerous visits to the Caltech and Lamont–Doherty collections are also acknowledged. Dr. José Morales kindly provided the Andalusian records used in Section 4.7. Discussions with Steve Kirby, Craig Bina and Ray Russo are gratefully acknowledged. I thank Paul Lundgren and Rob van der Hilst for constructive comments on a previous version of the paper, and Renata Dmowska for a personalized international library loan. This study was supported in part by the National Science Foundation under grants EAR-93-16396 and EAR-97-06152, and by the Department of Defense, under grant DTRA-01-00-C0065. Many figures were drafted using the GMT software (Wessel and Smith, 1991).

Appendix A

We present here a detailed discussion of various events and records used in this study.

A.1. Kuril–Sakhalin subduction zone

A.1.1. Sea of Okhotsk, 30 August 1970; 52.4°N, 151.69°E; 645 km; $M_0 = 1.1 \times 10^{27}$ dyn cm

This deep event, studied by, among others, Strelitz (1980) and Sasatani (1980), and more recently, Huang and Okal (1998), is one of the largest recorded at depth under the Sea of Okhotsk. We document T waves at four short-period stations in the Pacific Basin: KIP, AFI and RAR (WWSSN), and PMO (Rangiroa). Note that "T wave channels" were not yet operational in Polynesia in 1970. Records at RAR and AFI are arguably faint, but their spectral characteristics establish them beyond any possible doubt as high-frequency waves without counterparts in the time series for that day, and thus as T phases. That they are at all legible

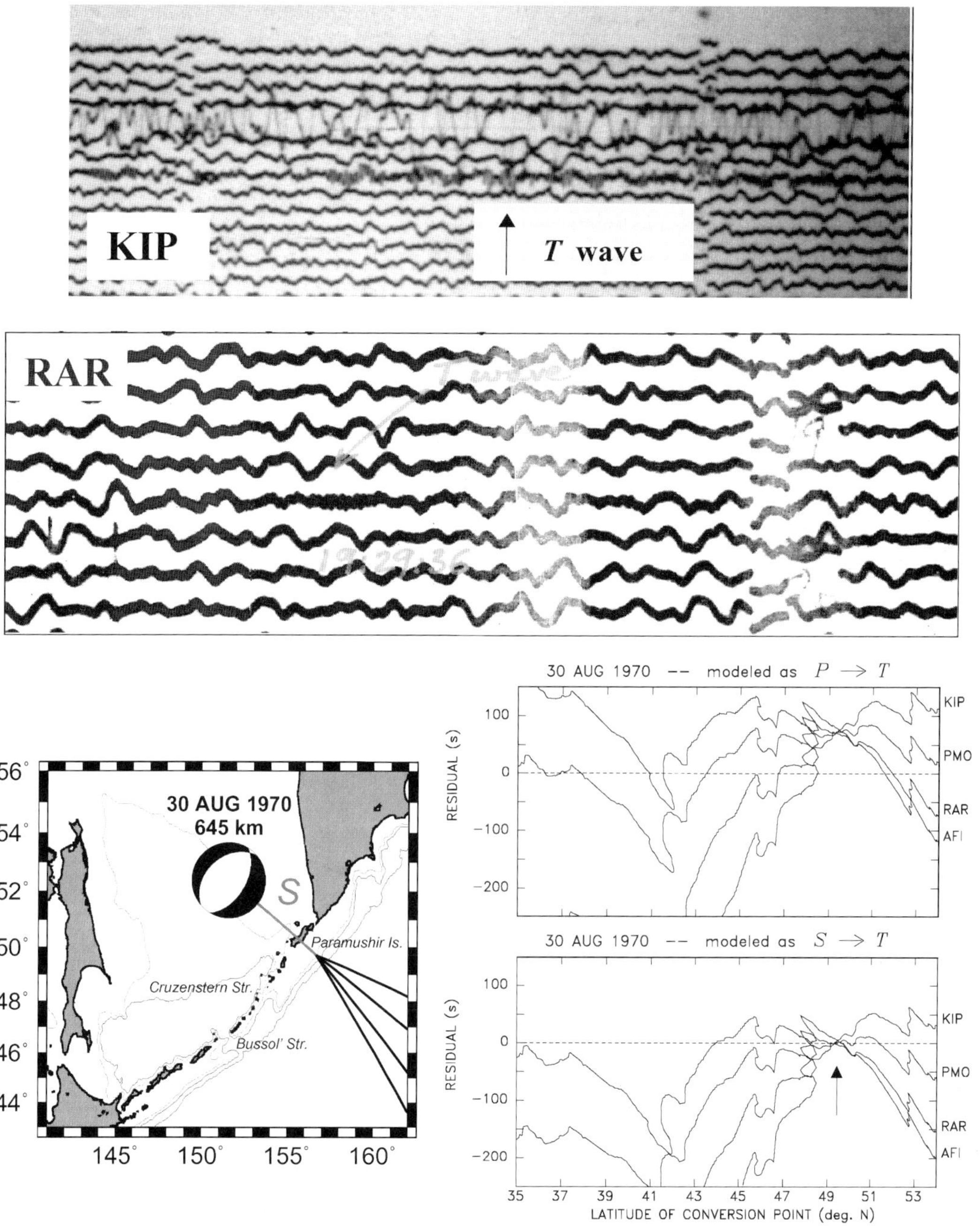

Fig. 16. Examples of T waves recorded on short-period channels of the WWSSN following the large 1970 Sea of Okhotsk deep shock. Top: record at Kipapa, Oahu. The duration of the window is 97 s and the original magnification of the record 12,500. The T wave is clearly apparent three traces below (i.e. 45 mn after) the prominent body waves. Bottom: record at Rarotonga. The duration is 56 s. Note the unmistakingly high-frequency character of the arrival, which emerges from the noise level despite a particularly mediocre gain at that station (originally 6250 at 1 Hz). The geometry of the conversion is shown on the bottom left map (similar to Fig. 3) and its nature as $S \rightarrow T$ established in the diagrams at bottom right (similar to Fig. 2).

on WWSSN records written at the mediocre magnifications used (6250 at RAR; Fig. 16) suggests a very strong acoustic wave. The record at KIP is clearer, with an estimated peak-to-peak amplitude of ground motion of 7 μm, a remarkable figure for a wave having travelled at least 40 km inside the structure of Oahu before reaching the station.

While in principle, it would be possible to interpret each arrival as resulting from a $P \rightarrow T$ conversion at an individual point of the Kuril coast (ranging from 47°N for PMO to 52°N for RAR), it is possible to model all four arrivals by diffraction of an S wave incident at 49.5°N, 156.3°E, off Paramushir Island (arrow in Fig. 16). This model is supported by a very favorable radiation of SV energy by the source towards the conversion point ($R^{SV} = 0.90$).

A.1.2. Sea of Okhotsk, 5 September 1970; 52.1°N, 150.99°E; 561 km; $M_0 = 6.4 \times 10^{25}$ dyn cm

For this aftershock of the previous event, we could find only one T phase, at PMO. The arrival time requires a much more southerly $S \rightarrow T$ conversion, at 46.5°N, 151.5°E, off the Bussol' Straits, which is easily explained by the difference in focal mechanism between the two shocks (Huang et al., 1997).

A.1.3. Sea of Okhotsk, 29 January 1971; 51.73°N, 150.95°E; 524 km; $M_0 = 2.5 \times 10^{26}$ dyn cm

We obtained two clear arrivals at KIP, separated by 108 s, and one at RAR. The record at PMO was noisy and could not be used. The T phase at KIP is interpreted as converted to P_n on the northern coast of Kauai, and corrected for the subsequent 199 km to the station. The travel-times are consistent with a common $S \rightarrow T$ scatterer to KIP and RAR at 46.8°N, 152.3°E, near the Bussol' Straits, and a conversion farther North, off Cruzenstern Straits (48.3°N, 153.9°E) for the $P \rightarrow T$ phase at KIP.

A.1.4. Sea of Okhotsk, 21 December 1975; 51.84°N, 151.75°E; 545 km; $M_0 = 2.1 \times 10^{26}$ dyn cm

We read two arrivals at PMO, and a clear one at KIP. They are interpreted as $P \rightarrow T$ and $S \rightarrow T$ conversions to PMO at Bussol' Straits (46.6°N, 151.5°E). The same $S \rightarrow T$ scatterer can be invoked on the path to KIP if an allowance is made for conversion back to P_n on the northern shore of Kauai.

A.1.5. Sea of Okhotsk, 21 June 1978; 47.98°N, 149.01°E; 402 km; $M_0 = 6.5 \times 10^{25}$ dyn cm

We obtained T phase doublets at both PMO and PPT, separated by 81 and 79 s, respectively. They are readily interpreted as conversions from a $P \rightarrow T$ scatterer at 45°N, 149.7°E (off Iturup), and an $S \rightarrow T$ one farther South off the Catherine Straits, at 44°N, 148°E.

A.1.6. Sea of Okhotsk, 10 July 1976; 47.36°N, 145.72°E; 421 km; $M_0 = 1.9 \times 10^25$ dyn cm

The record at PMO shows two arrivals of fair quality, which can be interpreted as $P \rightarrow T$ and $S \rightarrow T$ conversions, respectively, off Shikotan Island (43.5°N, 147°E) and Nemuro Peninsula (42.9°N; 145.7°E).

A.1.7. Sea of Okhotsk, 1 February 1984; 49.10°N, 146.31°E; 581 km; $M_0 = 3.6 \times 10^{25}$ dyn cm

A double arrival at PMO is readily interpreted as $P \rightarrow T$ and $S \rightarrow T$ conversions South of Nemuro (42.6°N, 145°E). The record at TBI could not be easily interpreted.

A.1.8. Sea of Okhotsk, 18 May 1987; 49.12°N, 147.39°E; 552 km; $M_0 = 1.7 \times 10^{26}$ dyn cm

This earthquake generated exceptionally long T wavetrains at PMO. The first arrival is readily identified as a $P \rightarrow T$ conversion at Bussol' Straits, but the interpretation of the later arrivals requires multipathing through many conversion points along the southern Kuril arc. Because we cannot assertively demonstrate the occurrence of an $S \rightarrow T$ conversion, we show this event as an open star in Fig. 4.

A.1.9. Primorye, 21 July 1994; 42.3°N, 132.9°E; 471 km; $M_0 = 1.1 \times 10^{27}$ dyn cm

Finally, we complete here the investigation of this strong event under the Primorye province of eastern Russia, whose record at RPN was used in the introduction to illustrate our methodology. Most other stations in the Pacific Basin recorded T phases. We identified $S \rightarrow T$ conversions at KOS, NAU, RAR, TET, PMO and RPN, with NAU and RPN showing earlier $P \rightarrow T$ phases. The conversion points are located on a 150 km stretch of the northernmost shore of Honshu (Fig. 3).

A.2. Bonin–Marianas arc

We start with a discussion of events located at a similar latitude to the 1982 "detached" earthquake, but in the mainstream WBZ.

A.2.1. Bonin Islands, 3 May 1991; 28.14°N, 139.65°E; 453 km; $M_0 = 1.3 \times 10^{26}$ dyn cm

The T wave channel at PMO saturated, but the regular seismic channel recorded a superb wavepacket, composed of two arrivals, generally similar to those from the 1982 event shown in Fig. 6. They are easily interpreted as $P \to T$ and $S \to T$ conversions at either Iwo Jima or Muka Jima.

Despite significant background noise at Easter Island (RPN), spectrogram techniques extract a T arrival at 04:48:35 GMT, in the 3–4 Hz frequency band. This signal is interpreted as an $S \to T$ conversion at Haha Jima, reflected on the northwestern Hawaiian Islands. The exact location of the reflector cannot be further constrained.

A.2.2. Bonin Islands, 13 May 1977; 28.12°N, 139.73°E; 440 km; $M_0 = 6.1 \times 10^{25}$ dyn cm

Results are similar to those of the previous shock: a double arrival at PMO interpreted as $P \to T$ and $S \to T$ conversions at either Iwo Jima or Chichi Jima.

A.2.3. Bonin Islands, 31 January 1973; 28.18°N, 138.86°E; 506 km; $M_0 = 2.5 \times 10^{26}$ dyn cm

We could obtain only one record, at PMO, on the regular seismic channel. A prominent arrival at 22:36:45, corresponding to $P \to T$ conversion at Iwo Jima or Chichi Jima, is followed by sustained high-frequency energy, from which no distinctive second arrival can be extracted. This event is shown as a yellow star in Fig. 6.

- To the North of the Bonin Islands, we examine the following events:
  - *Izu trench, 6 March 1984; 29.60°N, 139.11°E; 446 km; $M_0 = 1.4 \times 10^{27}$ dyn cm*

    This earthquake (the largest deep CMT solution in the Bonin–Mariana arc) generated spectacular T phases in Polynesia, which saturated the T wave channel at PMO for 10 min. The regular channels allow the identification of the strongest arrival at 03:59:35, with a precursor triggering saturation of the T wave channel at 03:58:12. These two times are readily interpreted as $P \to T$ and $S \to T$ paths converted either on the northern Bonin or northern Volcano Islands. Similarly, a powerful arrival at RKT is well matched by a $P \to T$ conversion at Muko Jima, the northernmost Bonin Island.

    A very intense T phase is present at TBI, even though the station is masked by the western Marshall Islands. The arrival time of the maximum amplitude, 04:12:35 GMT, could correspond to a reflection on Johnston Atoll (16.7°N; 169.5°W) for an $S \to T$ conversion at Haha Jima.
  - *Bonin Islands, 16 March 1996; 28.98°N, 138.94°E; 477 km; $M_0 = 1.1 \times 10^{26}$ dyn cm*

    Despite the size of this event, we could not identify T wave signals above noise level at any of the following stations: KIP, RAR, SNCC, PPT. A strong signal at SCZ, at 23:44:57, could be interpreted as a T wave converted from S at Haha Jima at the southern end of the Bonin Islands (26.3°N; 142.4°E). This interpretation remains speculative in the absence of corroborative signals in California (e.g. SNCC). We obtained a distinct doublet at WK30, readily interpreted as $P \to T$ and $S \to T$ converted at Muko Jima; these arrivals cannot be explained by conversion at the Volcano group. This event is shown as the purple star in Fig. 6.
  - *Bonin Islands, 5 August 1990; 29.48°N, 137.50°E; 520 km; $M_0 = 5.7 \times 10^{25}$ dyn cm*

    We obtained a small but distinctive doublet on the T wave channel at PMO, interpreted as $P \to T$ and $S \to T$ conversions at either Iwo Jima or Muko Jima. No other signal could be convincingly read at other stations. We show this event as a yellow star in Fig. 6.
  - *Izu Islands, 11 October 1993; 32.12°N, 138.02°E; 364 km; $M_0 = 2.5 \times 10^{26}$ dyn cm*

    This large event, about half-way between the Bonin Islands and Japan, produced a singular pattern of T waves. The PMO record in Polynesia shows a series of arrivals, the first one being the familiar $S \to T$ conversion at Haha Jima, but the most intense one, at 17:40:29 GMT, being too late to fit conversion anywhere along the islands or shallow seamounts of the Izu–Bonin arc. Rather, it can be successfully modeled by a $P \rightarrow T$ conversion off the southeastern extremity of Honshu. Similarly, the main doublet of arrivals at RAR is best explained by $P \rightarrow T$ conversions on the coasts of Iwase (southern Honshu) and farther West off Murotozaki (Shikoku). This event is shown as the green star in Fig. 6.

- To the South of the Bonin Islands, we investigate the following events:
  - *Mariana Islands, 18 May 1979; 23.94°N, 142.66°E; 581 km;* $M_0 = 5.5 \times 10^{25}$ *dyn cm*

    A very weak T wave is readable on the T wave channel at PMO, at 21:58:18, and is readily interpreted as a $P \rightarrow T$ conversion, at either Iwo Jima or Haha Jima. No distinctive second arrival can be identified in the subsequent part of the record. We show this event as a brown star in Fig. 6.
  - *Mariana Islands, 23 August 1995; 18.88°N, 145.30°E; 599 km;* $M_0 = 4.6 \times 10^{26}$ *dyn cm*

    We could not find any definitive T wave record from this large deep event, despite examining data from TBI, PPT, KIP, RAR. We show this event as a black star in Fig. 6.
  - *Mariana Islands, 4 January 1982; 17.92°N, 145.46°E; 595 km;* $M_0 = 6.7 \times 10^{25}$ *dyn cm*

    A strong T phase is observed at PMO on the T wave channel, followed by several subsequent arrivals. The first signal corresponds to a $P \rightarrow T$ conversion on the eastern shore of Guam (13.6°N; 145.1°E). The latter arrivals could not be modeled assertively.

A.3. South America

A.3.1. Argentina, 23 June 1991; 26.8°S, 63.3°W; 558 km; $M_0 = 8.6 \times 10^{26}$ *dyn cm*

This is the largest deep Argentinian event with an available CMT solution. T waves were recorded in Hawaii (HUL), Easter (RPN), and Guam (GUMO). Significant volcanic activity at the Hollister Ridge during the relevant time window (Talandier and Okal, 1996) prevents the use of Polynesian stations. The HUL signal does not lend itself well to a clear separation into $P \rightarrow T$ and $S \rightarrow T$ arrivals (Fig. 8b). Based on comparisons with the other sources in the same region, we interpret the maximum in amplitude as an $S \rightarrow T$ conversion near Taltal, Chile (25.0°S; 70.6°W). The GUMO arrival is composed of a strong wavepacket lasting approximately 150 s, with maximum amplitude at 00:26:03 GMT on 24 June, with a smaller precursor at 00:23:47. We add a 4 s station correction at the receiver. These times then fit a $P \rightarrow T$ conversion at 27.2°S, 71.0°W, and an $S \rightarrow T$ one near Taltal (25.5°S). Finally, the RPN record shows a somewhat tentative arrival at 22:08:29 GMT. After applying a 3 s correction, this time fits an $S \rightarrow T$ conversion at 25.9°S; 70.8°W. Thus, most of the T phases recorded from this event would evolve from scatterers located along the bight offshore from Taltal, and activated by the S wave, 191 s after the origin time of the earthquake.

A.3.2. Argentina, 21 December 1983; 28.2°S, 63.2°W; 602 km; $M_0 = 2.7 \times 10^{26}$ *dyn cm*

This large event occurred at 12:05:06 GMT and was followed by an aftershock (no CMT available) at 12:15:07 GMT. Both shocks were well recorded at HUL on the southern shore of the Big Island of Hawaii, despite interference with local seismicity. Arrival times for the first event are easily interpreted as $P \rightarrow T$ and $S \rightarrow T$ conversions at 25.3 and 25.0°S, respectively. For the aftershock, the arrival corresponds to the $S \rightarrow T$ conversion; the expected $P \rightarrow T$ conversion is obscured by a local earthquake (Fig. 8a).

A.3.3. Argentina, 29 April 1994; 28.51°S, 63.22°W; 566 km; $M_0 = 2.5 \times 10^{26}$ *dyn cm*

The $S \rightarrow T$ conversion is well recorded at HUL, while the $P \rightarrow T$ one is much fainter (Fig. 8c). Scatterers are modeled at 23.7°S (P) and 24.0°S (S). No definitive T phases from this event were identified elsewhere in the Pacific Basin.

A.3.4. Argentina, 10 May 1994; 28.50°S, 63.10°W; 600 km; $M_0 = 2.8 \times 10^{26}$ *dyn cm*

The HUL record, shown in Fig. 8d, shows a faint $P \rightarrow T$ conversion and a better developed $S \rightarrow T$ one. The scatterers are located to the North at 23.7°S (P) and 24.6°S (S). In addition, we obtained a clear doublet at KOS, with scatterers at 30.2°S (P) and 28.3°S (S).

A.3.5. Argentina, 19 August 1994; 26.72°S, 63.42°W; 563 km; $M_0 = 5.6 \times 10^{25}$ dyn cm

With this event, we reach what we regard as the limit of detectability at HUL (Fig. 8e). By analogy with the previous events, we believe that the faint signal at 12:02:03 GMT corresponds to an $S \to T$ conversion, which would then involve a scatterer at 23.8°S, 70.6°W. We could not extract definite signals at other Pacific sites.

A.3.6. Paraguay, 15 April 1969; 21.86°S, 61.68°W; 541 km; $m_b = 4.0$

We relocated this small earthquake just 10 km to the North; Monte Carlo relocations with a generous noise ($\sigma_G = 1.5$ s) are not sufficient to bring it into the mainstream WBZ, while keeping its depth robust (521–574 km). We conclude that the earthquake is indeed an outboard outlier. It remains much too small for a study of any possible T waves.

A.3.7. Argentina, 8 February 1965; 23.1°S, 62.5°W; 468 km; $m_b = 4.7$

The above NEIC location stands out, while the ISC location is in the mainstream WBZ. We also relocate the earthquake in the WBZ, and at a significantly greater depth (22.47°S; 63.49°W; 547 km), a hypocenter close but not identical to that of a foreshock 2 min earlier.

A.3.8. Argentina, 25 February 1964; 22.2°S, 62.6°W; 470 km; $m_b = 4.1$

The NEIC location is outboard, but the dataset has poor longitudinal resolution. The Monte Carlo ellipse drawn with very little noise ($\sigma_G = 1$ s) intersects the mainstream WBZ. Our preferred location (22.29°S; 62.94°W) also involves a greater depth (506 km).

A.3.9. Argentina, 10 June 1964; 22.5°S, 64.7°W; 480 km; no magnitude reported

The above NEIC location would make this an "inboard outlier" to the West of the WBZ. While our relocation converges on the same solution, the dataset has poor resolution in the ESE–WNW direction, and the 1 s ellipse intersects the mainstream WBZ.

A.3.10. Argentina, 9 February 1926; 27°S, 59.5°W; 540 km; $M_{PAS} = 6$

The above ISS location would qualify as an outboard hypocenter, although Gutenberg and Richter's (1954) solution (28°S; 62°W) lies within the WBZ. We relocate the event even farther West (28.57°S; 62.68°W), at the very bottom of the active WBZ (626 km).

A.3.11. Chile, 18 September 1975; 34.15°S, 68.90°W; 429 km; no magnitude reported

If its location were confirmed, this small earthquake would represent an event detached both vertically and horizontally, in a section of the Andean subduction zone where no deep seismicity is otherwise observed. We relocate the earthquake even farther South (35.46°S; 67.67°W; 465 km), but note that the dataset has only 5 P times, with the Monte Carlo ellipsoid ($\sigma_G = 1.5$ s) intersecting the active WBZ at intermediate depths.

References

Adams, R.D., 1963. Source characteristics of some deep New Zealand earthquakes. N.Z. J. Geol. Geophys. 6, 209–220.

Adams, R.D., Ferris, B.G., 1976. A further earthquake at exceptional depth between New Zealand. N.Z. J. Geol. Geophys. 9, 269–273.

Barazangi, M., Isacks, B., Oliver, J., 1972. Propagation of seismic waves through and beneath the lithosphere that descends under the Tonga Island arc. J. Geophys. Res. 77, 952–958.

Bellon, H., 1981. Chronologie radiométrique (K–Ar) des manifestations magmatiques autour de la Méditerraneé occidentale entre 33 et 1 M a. In: Wezel, F.C. (Ed.), Sedimentary Basins of Mediterranean Margins, Tecnoprint, Bologna, pp. 341–360.

Benioff, H., 1949. Seismic evidence for the fault origin of oceanic deeps. Geol. Soc. Amer. Bull. 60, 1837–1856.

Bina, C.R., 1997. Patterns of deep seismicity reflect buoyancy stresses due to phase transitions. Geophys. Res. Lett. 24, 3301–3304.

Blanco, M.J., Spakman, W., 1993. The P wave velocity structure of the mantle below the Iberian Peninsula: evidence for subducted lithosphere below southern Spain. Tectonophysics 221, 13–34.

Brudzinski, M.R., Chen, W.-P., 1998. Anomalous seismicity disassociated with high P wave speeds in the transition zone beneath the Tonga back-arc, Eos. Trans. Am. Geophys. Union, 79 (45) F633 (abstract).

Buforn, E., Udías, A., Mezcua, J., Madariaga, R., 1991. A deep earthquake under south Spain, 8 March 1990. Bull. Seismol. Soc. Am. 81, 1403–1407.

Buforn, E., Coca, P., Udías, A., Lasa, C., 1997. Source mechanism of intermediate and deep earthquakes in southern Spain. J. Seismol. 1, 113–130.

Chung, W.-Y., Kanamori, H., 1976. Source process and tectonic implication of the Spanish deep-focus earthquake of March 29, 1954. Phys. Earth Planet. Int. 13, 85–96.

DeMets, D.C., Gordon, R.G., Argus, D.F., Stein, S., 1990. Current plate motions. Geophys. J. Intl. 101, 425–478.

Engdahl, E.R., van der Hilst, R., Berrocal, J., 1995. Imaging of subducted lithosphere beneath South America. Geophys. Res. Lett. 22, 2317–2320.

Engdahl, E.R., van der Hilst, R., Buland, R., 1998. Global teleseismic earthquake relocation with improved travel-times and procedures for depth determination. Bull. Seismol. Soc. Am. 88, 722–743.

Engebretson, D.C., Kirby, S.H., 1992. Deep Nazca slab seismicity: why is it so anomalous? Eos. Trans. Am. Geophys. Union 73 (43) 379 (abstract).

Ewing, W.M., Woollard, G.P., Vine, A.C., Worzel, J.L., 1946. Recent results in submarine geophysics. Geol. Soc. Am. Bull. 57, 909–934.

Fukao, Y., Obayashi, M., Inoue, H., Nenbai, M., 1992. Subducting slabs stagnant in the mantle transition zone. J. Geophys. Res. 97, 4809–4822 .

Glennon, M.A., Chen, W.-P., 1993. Systematics of deep-focus earthquakes along the Kuril–Kamchatka arc and their implications on mantle dynamics. J. Geophys. Res. 98, 735–769.

Grand, S.P., 1994. Mantle shear structure beneath the Americas and surrounding oceans. J. Geophys. Res. 99, 11591–11621.

Grimison, N.L., Chen, W.-P., 1986. The Azores–Gibraltar plate boundary: focal mechanisms, depths of earthquakes, and their tectonic implications. J. Geophys. Res. 91, 2029–2047.

Gutenberg, B., Richter, C.F., 1954. Seismicity of the Earth and Associated Phenomena. Princeton University Press, Princeton, NJ, 310 p.

Huang, W.-C., Okal, E.A., 1998. Centroid moment tensor solutions for deep earthquakes predating the digital era: discussion and inferences. Phys. Earth Planet. Inter. 106, 191–218.

Huang, W.-C., Okal, E.A., Ekström, G., Salganik, M.P., 1997. Centroid moment tensor solutions for deep earthquakes predating the digital era: the WWSSN dataset (1962–1976). Phys. Earth Planet. Int. 99, 121–129.

Huang, W.-C., Okal, E.A., Ekström, G., Salganik, M.P., 1998. Centroid moment tensor solutions for deep earthquakes predating the digital era: the historical dataset (1907–1961). Phys. Earth Planet. Int. 6, 181–190.

Isacks, B.L., Barazangi, M., 1973. High-frequency shear waves guided by a continuous lithosphere descending beneath western South America. Geophys. J. Roy. Astron. Soc. 33, 129–139.

Isacks, B.L., Molnar, P., 1971. Distribution of stresses in the descending lithosphere from a global survey of focal mechanism solutions of mantle earthquakes. Rev. Geophys. Space Phys. 9, 103–174.

Kagan, Y.Y., 1991. 3-D rotation of double-couple earthquake sources. Geoph. J. Intl. 106, 709–716.

Kirby, S.H., Durham, W.B., Stern, L.A., 1991. Mantle phase changes and deep-earthquake faulting in subducting lithosphere. Science 252, 216–225.

Kirby, S.H., Okal, E.A., Engdahl, E.R., 1995. The 9 June 1994 great Bolivian deep earthquake: an exceptional deep earthquake in an extraordinary subduction zone. Geophys. Res. Lett. 22, 2233–2236.

Kirby, S.H., Stein, S., Okal, E.A., Rubie, D., 1996. Deep earthquakes and metastable mantle phase transformations in subducting oceanic lithosphere. Rev. Geophys. Space Phys. 34, 261–306.

Kostoglodov, V.V., 1989. Maximal'naya glubina zemletryasenii i fazovye prevrashcheniya v litosfere, pograzhayushchejsya v mantiyu. In: Magnitskii, V.A. (Ed.), Fizika i vnutrennee stroienie zemli. Nauka, Moskva, pp. 52–57 (in Russian).

Levitus, S., Boyer, T.P., Antonov, J., Burgent, R., Conkright, M.E., 1994. World Ocean Atlas 1994, NOAA/NESDIS, Silver Spring, MD.

Lonergan, L., White, N., 1997. Origin of the Betic–Rif mountain belt. Tectonics 16, 504–522.

Lundgren, P.R., Giardini, D., 1994. Isolated deep earthquakes and the fate of subduction in the mantle. J. Geophys. Res. 99, 15833–15842.

Mele, G., 1998. High-frequency wave propagation from mantle earthquakes in the Tyrrhenian Sea: constraints for the geometry of the South Tyrrhenian subduction zone. Geophys. Res. Lett. 25, 2877–2880.

Mezcua, J., Rueda, J., 1997. Seismological evidence for a delamination process in the lithosphere under the Alborán Sea. Geophys. J. Intl. 129, F1–F8.

Mitchell, B.J., 1995. Anelastic structure and evolution of the continental crust and upper mantle from seismic surface wave attenuation. Rev. Geophys. 33, 441–462.

Mooney, H.M., 1970. Upper mantle inhomogeneity beneath New Zealand: seismic evidence. J. Geophys. Res. 75, 285–309.

Morales, J., Serrano, I., Jabaloy, A., Galindo-Zaldivar, J., Zhao, D., Torcal, F., Vidal, F., Gonzálaz-Lodeiro, F., 1999. Active continental subduction beneath the Betic Cordillera and the Alborán Sea. Geology 27, 735–738.

Okal, E.A., 1992. A student's guide to teleseismic body-wave amplitudes. Seismol. Res. Lett. 63, 169–180.

Okal, E.A., Bina, C.R., 1994. The deep earthquakes of 1921–1922 in northern Peru. Phys. Earth Planet. Int. 87, 33–54.

Okal, E.A., Bina, C.R., 2001. The deep earthquakes of 1997 in Western Brazil. Bull. Seismol. Soc. Amer., 91, 161–164.

Okal, E.A., Jo, B.-G., 1990. Q measurements for phase X overtones. Pure Appl. Geophys. 132, 331–362.

Okal, E.A., Kirby, S.H., 1998. Deep earthquakes beneath the North and South Fiji Basins, SW Pacific: Earth's most intense deep seismicity in stagnant slabs. Phys. Earth Planet. Int. 109, 25–63.

Okal, E.A., Talandier, J., 1997. T waves from the great 1994 Bolivian deep earthquake in relation to channeling of S wave energy up the slab. J. Geophys. Res. 102, 27421–27437.

Okal, E.A., Talandier, J., 1998. Correction to "T waves from the great 1994 Bolivian deep earthquake in relation to channeling of S wave energy up the slab". J. Geophys. Res. 103, 2793–2794.

Okal, E.A., Engdahl, E.R., Kirby, S.H., Huang, W.-C., 1995. Earthquake relocations in the Kuril slab: was the 1990 Sakhalin event not isolated after all? Eos. Trans. Am. Geophys. Union 76 (17), S199, (abstract).

Okino, K., Ando, M., Kaneshima, S., Hirahara, K., 1989. The horizontally lying slab. Geophys. Res. Lett. 16, 1059–1062.

Richardson, W.P., 1998. Surface-wave tomography of the Ontong–Java plateau: seismic probing of the world's largest igneous province. Ph.D. Thesis. Northwestern University, Evanston, IL, 191 p.

Royden, L.H., 1993. Evolution of retreating subduction boundaries formed during continental collision. Tectonics 12, 628–638.

Sacks, I.S., 1969. Distribution of Absorption of Shear Waves in South America, and Its Tectonic Significance. Vol. 67, Yearbook Carnegie Institute Washington, pp. 339–344.

Sasatani, T., 1980. Source parameters and rupture mechanism of deep-focus earthquakes. J. Fac. Sci. Hokkaido Univ. Ser. 7 Geophys. 6, 301–384.

Seber, D., Barazangi, M., Ibenbrahim, A., Demnati, A., 1996. Geophysical evidence for lithospheric delamination beneath the Alborán Sea and Rif–Betic mountains. Nature 379, 785–790.

Snoke, J.A., Sacks, I.S., Okada, H., 1974a. Empirical models for anomalous high-frequency arrivals from deep-focus earthquakes in South America. Geophys. J. Roy. Astron. Soc. 37, 133–139.

Snoke, J.A., Sacks, I.S., Okada, H., 1974b. A model not requiring continuous lithosphere for anomalous high-frequency arrivals from deep-focus South American earthquakes. Phys. Earth Planet. Int. 9, 199–206.

Strelitz, R.A., 1980. The fate of downgoing slab: a study of the moment tensors from body waves of complex deep-focus earthquakes. Phys. Earth Planet. Int. 21, 83–96.

Talandier, J., Kuster, G.T., 1976. Seismicity and submarine volcanic activity in French Polynesia. J. Geophys. Res. 81, 936–948.

Talandier, J., Okal, E.A., 1996. "Monochromatic T waves from underwater volcanoes in the Pacific Ocean: ringing witnesses to geyser processes". Bull. Seismol. Soc. Am. 86, 1529–1544.

Talandier, J., Okal, E.A., 1998. On the mechanism of conversion of seismic waves to and from T waves in the vicinity of island shores. Bull. Seismol. Soc. Am. 88, 621–632.

Taylor, R.N., Nesbitt, R.W., Vidal, P., Harmon, R.S., Auvray, B., Croudace, I.W., 1994. Mineralogy, chemistry and genesis of the Boninite series volcanics, Chichijima, Bonin Islands, Japan. J. Petrol. 35, 577–617.

Tracey Jr., J.I., Schlanger, S.O., Stark, J.T., Doan, D.B., May, H.G., 1964. General geology of Guam. US Geol. Surv. Prof. Paper 403-A, A1–A104.

Udías, A., López-Arroyo, A., Mezcua, J., 1976. Seismotectonics of the Azores–Alborán region. Tectonophysics 31, 259–289.

Umino, S., 1985. Volcanic geology of Chichijima, the Bonin Islands (Ogasawara Islands). J. Geol. Soc. Jpn. 91, 505–523.

Utsu, T., 1971. Seismological evidence for anomalous structures of island arcs with special reference to the Japanese region. Rev. Geophys. Space Phys. 9, 839–890.

Uyeda, S., Kanamori, H., 1979. Back-arc opening and the mode of subduction. J. Geophys. Res. 84, 1049–1061.

Van Ark, E., Marton, F., Stein, S., Bina, C., Rubie, D.C., 1999. Persistence of metastable olivine in detached slabs as a possible cause of deep earthquakes. In: Proceedings of the 1999 Alfred-Wegener Conference. Vol. 7, Terra Nostra, 1999, p. 106 (abstract).

van der Hilst, R., 1995. Complex morphology of subducted lithosphere in the mantle beneath the Tonga trench. Nature 374, 154–157.

van der Hilst, R.D., Snieder, R., 1996. High-frequency precursors to P wave arrivals in New Zealand: implications for slab structure. J. Geophys. Res. 101, 8473–8488.

van der Hilst, R.D., Engdahl, E.R., Spakman, W., Nolet, G., 1991. Tomographic imaging of subducted lithosphere below northwest Pacific island arcs. Nature 353, 37–43.

van der Hilst, R., Engdahl, E.R., Spakman, W., 1993. Tomographic inversion of P and pP data for aspherical mantle structure below the northwest Pacific region. Geophys. J. Intl. 115, 264–302.

Wessel, P., Smith, W.H.F., 1991. Free software helps map and display data. Eos. Trans. Am. Union 72, 441; 445–446.

Widiyantoro, S., Kennett, B.L.N., van der Hilst, R.D., 1997. Mantle structure beneath Indonesia inferred from high-resolution tomographic imaging. Geophys J. Intl. 130, 167–182.

Widiyantoro, S., Kennett, B.L.N., van der Hilst, R.D., 1999. Seismic tomography with P and S data reveals lateral variations in the rigidity of deep slabs. Earth Planet. Sci. Lett. 173, 91–100.

Wortel, M.J.R., 1984. Spatial and temporal variations in the Andean subduction zone. J. Geol. Soc. London 141, 783–791.

Wysession, M.E., Okal, E.A., Miller, K.L., 1991. Intraplate seismicity of the Pacific Basin, 1913–1988. Pure Appl. Geophys. 135, 261–359.

ELSEVIER

Physics of the Earth and Planetary Interiors 127 (2001) 145–163

PHYSICS OF THE EARTH AND PLANETARY INTERIORS

www.elsevier.com/locate/pepi

Seismological constraints on the mechanism of deep earthquakes: temperature dependence of deep earthquake source properties

Douglas A. Wiens*

Department of Earth and Planetary Sciences, Washington University, 1 Brookings Dr., St. Louis, MO 63130, USA

Received 25 May 2000; accepted 26 November 2000

Abstract

Seismological observations are of vital importance for understanding the mechanism of deep earthquakes. Most of the seismological observables for earthquakes deeper than 300 km are similar to shallow earthquakes. Deep earthquakes clearly represent shear failure on a planar surface as shown by their double couple mechanisms. Deep earthquake aftershock sequences show temporal decay rates and magnitude–frequency relations (b-values) that are similar to shallow earthquakes, with the aftershocks occurring preferentially along the mainshock fault planes. In addition, deep earthquake rupture velocities are similar to those of shallow earthquakes. However, there are also some observations that are clearly distinctive relative to shallow earthquakes. Stress drops of deep earthquakes show a large variation but are larger, on average, than shallow earthquakes. Different deep seismic zones show very different b-values, in contrast to shallow earthquakes, which show similar b-values worldwide. In addition, deep earthquakes show fewer aftershocks than shallow earthquakes.

A variety of observations suggest that deep earthquakes are highly sensitive to the temperature of the slab. Both deep earthquake b-values and the rate of deep earthquake aftershock occurrence are inversely correlated with the temperature of the deep slabs, suggesting that these factors are temperature-controlled to an extent much greater than with shallow earthquakes. Large deep earthquakes in warm slabs show slower rupture velocities, larger stress drops, and lower seismic efficiencies than similar earthquakes in cold slabs. The width of deep seismic zones is also probably temperature controlled, but deep earthquake rupture can propagate outside the normal limits of Benioff zone seismicity. Simple thermal models for the Tonga slab near the 1994 deep earthquake suggest that the temperature at the rupture termination point was at least 200°C warmer than the temperature that limits smaller earthquakes in the slab.

These observations can be used to evaluate physical models for deep earthquakes, including brittle slip along fluid-weakened faults, transformational faulting, and thermal (and perhaps melt lubricated) shear instabilities. It is difficult to explain the large fault widths of some deep earthquakes using a fluid weakening model, since hydrated materials are expected in only a narrow depth zone at the top of the slab, and the fault planes do not have the expected orientations for reactivated faults. The lateral extent of the largest deep earthquakes cast doubt on the transformational faulting model, in which events should be confined within a narrow metastable wedge. Seismological studies have also failed to find evidence for the existence of metastable olivine in slabs. The temperature dependence of deep earthquakes argues in favor of a temperature-activated phenomenon, such as thermal shear instabilities and perhaps fault zone melting.

Keywords: Seismology; Deep earthquake; Magnitude–frequency relation

1. Introduction

The mechanism and tectonic interpretation of deep earthquakes remains one of the major questions in

* Tel.: +1-314-935-6517; fax: +1-314-935-7361.
E-mail address: doug@mantle.wustl.edu (D.A. Wiens).

0031-9201/01/$ – see front matter
PII: S0031-9201(01)00225-4

seismology and rock mechanics. It has long been recognized that the occurrence of deep earthquakes is problematic, since brittle failure should be prohibited by confining pressure at great depth (Leith and Sharpe, 1936). In addition, deep earthquakes show interesting differences with shallow brittle faulting earthquakes, including much lower aftershock production rates (Page, 1968; Frohlich, 1989; Wiens et al., 1997), a dependence of source duration and rise time on earthquake depth (Houston and Williams, 1991; Vidale and Houston, 1993; Houston et al., 1998), and a dependence of magnitude–frequency relations and aftershock productivity on slab temperature (Wiens and Gilbert, 1996).

In recent years, a variety of mechanisms have been proposed to explain deep earthquake faulting. High fluid pressure resulting from possible dehydration reactions have been proposed to reduce the effective confining pressure and allow movement on pre-existing faults (Meade and Jeanloz, 1991; Silver et al., 1995). Transformational faulting (anticracks), in which the transformation from metastable α-olivine to a denser β-phase is accompanied by localization of strain into thin shear zones, has been suggested based on laboratory experiments (Kirby, 1987; Green and Burnley, 1989; Burnley et al., 1991; Kirby et al., 1991, 1996b; Green and Houston, 1995; Marone and Liu, 1997). It has also been proposed that shear failure propagates as ductile faulting or plastic instabilities (Ogawa, 1987; Hobbs and Ord, 1988) rather than as brittle failure, or that rupture propagation is facilitated by melt production at the crack tip (Kanamori et al., 1998). All of these proposals involve large extrapolation from laboratory experiments or calculations to the depths and scale lengths of deep earthquakes, and therefore remain speculative.

The major constraints on the mechanism and interpretation of deep earthquakes are provided by seismological observations, in combination with mineral physics experiments and geodynamic calculations. In this paper, I review the basic seismological observations concerning deep earthquakes, and discuss their implications for our understanding of the mechanism of deep earthquakes and the geodynamics of subducting slabs. The focus of this paper will be earthquakes deeper than 300 km; intermediate depth earthquakes (70–300 km) show many characteristics in common with deep earthquakes but it is unclear whether they result from similar processes (Meade and Jeanloz, 1991; Abers, 1996; Kirby et al., 1996a).

2. Comparison of deep and shallow earthquakes

Our knowledge of earthquakes is based primarily on observations of shallow earthquakes, which can be intensively studied using close-in seismological and geodetic instrumentation and direct field observations. Since good models have been developed for shallow earthquakes based on these observations, I will first review the characteristics of deep earthquakes in comparison with shallow earthquakes before discussing specific models of deep earthquake occurrence.

2.1. Seismic source geometry

The seismic source parameters of deep earthquakes are generally indistinguishable from the source parameters of typical shallow earthquakes. In the past, it was thought that deep earthquake sources might deviate systematically from the double couple source that is indicative of shear movement along a plane, and which is observed for nearly all shallow earthquakes. Dziewonski and Gilbert (1974) proposed that some deep earthquakes showed significant isotropic components, indicating a volume change. Randall and Knopoff (1970) proposed that deep earthquakes showed a significant compensated linear vector dipole (CLVD) component, which can be visualized as motion away from (or toward) the source along a polar axis, with a corresponding opposite motion along an equatorial band (Frohlich, 1994). These anomalous seismic sources were thought to be indicative of the causative mechanism of deep earthquakes, such as phase changes.

It is now clear that deep earthquakes generally show negligible isotropic components (Fig. 1a) (Hara et al., 1996; Kawakatsu, 1996; Russakoff et al., 1997). Deep earthquakes do show significant CLVD components, but in approximately the same proportions observed for shallow earthquakes (Fig. 1b) (Kuge and Kawakatsu, 1993). The CLVD components could be due to the effects of compound rupture, where slip occurs along several faults of different orientation during the event (Kuge and Kawakatsu, 1993; Frohlich, 1994). These results indicate that deep earthquakes

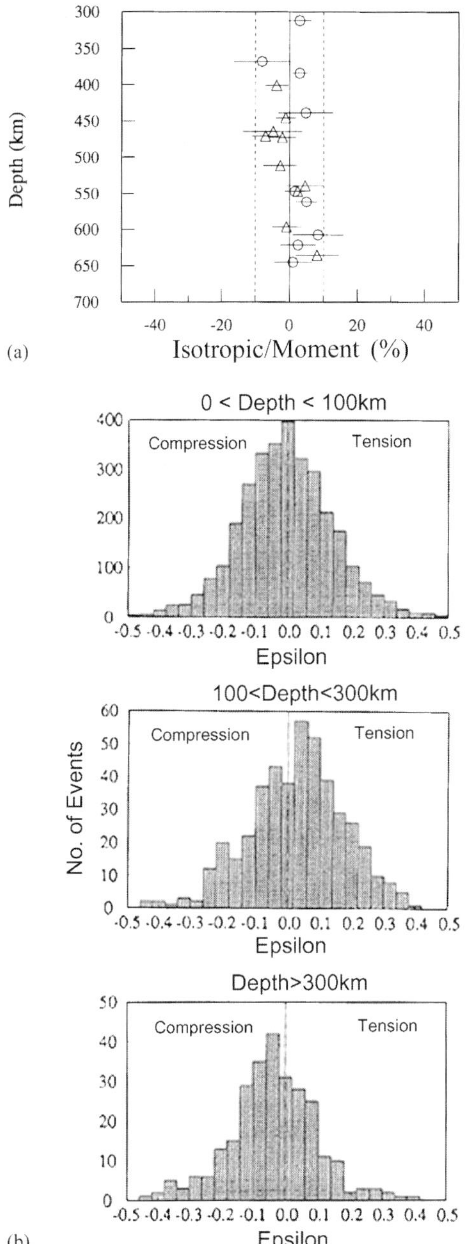

Fig. 1. (a) The isotropic component (%) of the moment tensor for 19 deep earthquakes, with earthquake depth plotted along the vertical axis (Kawakatsu, 1996). None of the earthquakes show a resolvable isotropic (volumetric) component. (b) Histograms showing the deviatoric non-double couple component for Harvard centroid moment tensor solutions (Kuge and Kawakatsu, 1993). Epsilon is 0 for a pure double couple and ±0.5 for a pure CLVD source. There is no significant increase in non-double couple component with depth.

must represent largely shear motion along a planar surface; any volumetric component resulting from a phase transformation must be much smaller than the shear slip.

2.2. Rupture characteristics

Seismological source studies allow seismologists to determine a variety of rupture characteristics for deep earthquakes, including average rupture velocity, static stress drop, and maximum seismic efficiency. The static stress drop represents the difference between the stress on the fault before and after the earthquake, and is proportional to the ratio of the fault displacement to the fault width. The maximum seismic efficiency is the ratio of the energy radiated elastically to the total energy release (Kikuchi, 1992). The difference between radiated and total energy is lost to friction along the fault. Due to the absence of near-field geodetic and strong ground motion measurements, these parameters are more uncertain for deep earthquakes than for well-studied shallow events. The uncertainties for deep events are a function of the spatial and temporal separation of the rupture events, so the rupture characteristics are better resolved for the larger events. For the largest deep earthquakes, the uncertainty in stress drop and seismic efficiency is probably a factor of 2, whereas the average rupture velocity can be somewhat better resolved.

Table 1 presents a synthesis of the best estimates of the rupture parameters of all deep earthquakes of $M_W = 7.5$ and greater since 1960. Details about how these parameters were determined are given in Appendix A, along with Table 2 listing the actual derived parameters from individual prior studies.

Deep earthquakes show a wide variety of rupture velocities, stress drops, and seismic efficiencies. Average rupture velocities range from 1.7 to 4.1 km/s. These rupture velocities are indistinguishable from rupture velocities for large shallow earthquakes. For example, in the tabulation of Geller (1976), rupture velocities range from 1.4 to 4.0 km/s, with most in the range of 2–3 km/s. As rupture velocity is thought to be limited by the shear wave velocity (Johnson and Scholz, 1976), it may be more appropriate to compare the ratio of rupture velocity to shear velocity. Since shear velocity is higher for deep earthquakes, this ratio is lower for the deep events. However, any differences

Table 1
Source parameters of the largest deep earthquakes

Date	Location	M_w	Depth (km)	Thermal parameters (km)	Aftershocks $m_b \geq 4.5$	Mo (10^{20} Nm)	Radiated energy (10^{14} J)	Rupture velocity (km/s)	Stress drop (MPa)	Rupture diameter (km)	Duration (s)	Scaled duration[a] (s)	Maximum efficiency
9 June 94	Bolivia	8.3	637	1800	3	26	290	1.7	44	60 × 45	40	40	0.061
31 July 70	Colombia	8.0	651	1800	0	14	23	2.0	38	100 × 20	60	72	0.010
17 June 96	Flores Sea	7.8	587	9400	14	7.3	180	4.0	16	90 × 25	23	35	0.370
29 September 73	Japan Sea	7.7	575	5600	0	5	10	3.3	10	60 × 40	24	41	0.046
15 August 63	Bolivia	7.7	543	1800	0	3.9	12	2.0	22	60 × 20	34	64	0.033
9 November 63	Brazil	7.6	600	1800	2	3.5	16	3.0	27	40 × 25	20	39	0.041
9 March 94	Tonga-Fiji	7.6	564	11800	29	3.1	24	4.1	15	45 × 30	14	28	0.120

[a] Duration of source time function divided by $M_0^{1/3}$ to remove the effect of event size and scaled to the duration of the Bolivia event.

are still minor compared to variations within both the shallow earthquake and deep earthquake populations.

Static stress drops for the deep events range from 10 to 44 MPa, somewhat greater than the stress drops of 3–10 MPa typically found for shallow events (Kanamori and Anderson, 1975). Thus, deep earthquakes show greater static stress drops than shallow earthquakes on average, as proposed by several previous studies (Chung and Kanamori, 1980; Fukao and Kikuchi, 1987). However, some deep earthquakes, such as the 1973 Japan Sea, 1994 Tonga–Fiji, and 1996 Flores Sea events show stress drops lower than 20 MPa and thus within the range shown by shallow earthquakes. Higher static stress drops can also explain previous suggestions that deep earthquakes show smaller rise times (Houston and Williams, 1991) and shorter source durations (Vidale and Houston, 1993). It should be noted that some reports of uniform stress drop with depth (Houston and Williams, 1991) are based on the Orowan stress drop (Vassiliou and Kanamori, 1982), a dynamic stress drop (equal to the static stress drop times the seismic efficiency) that assumes conditions unlikely to be fulfilled for deep earthquakes.

Deep earthquakes also show a wide range of maximum seismic efficiencies, ranging from 0.01 to 0.31. Thus, some deep earthquakes are quite efficient at converting available energy into seismic radiation, whereas other events loose more energy to friction and other dissipative processes. The seismic efficiencies of shallow earthquakes are not well constrained, but range from 0.04 to 0.22 in the study of Kikuchi (1992), and from 0.1 to 1.0, if the average apparent stresses for various tectonic settings given in Choy and Boatwright (1995) are converted to maximum seismic efficiencies. So, the seismic efficiencies of most deep earthquakes fall within the range of shallow earthquakes, but there is evidence that some deep earthquakes show unusually low seismic efficiencies, suggesting a more dissipative process (Kanamori et al., 1998).

2.3. Aftershock sequences

The smaller number of aftershocks observed for deep earthquakes may be the most obvious difference between shallow and deep earthquakes, and has been discussed by many investigators (Page, 1968; Kagan and Knopoff, 1980; Frohlich, 1987). Until recently, little was known about deep earthquake aftershocks, since significant deep aftershock sequences were unknown. However, the identification of several substantial deep aftershock sequences (Wiens et al., 1994; Myers et al., 1995; Wiens, 1998; Wu and Chen, 1999) has recently lead to a better understanding of deep earthquake aftershock characteristics.

Deep earthquake aftershock sequences generally show aftershock productivity rates of one to three orders of magnitude less than shallow earthquakes of equivalent size. A survey of aftershock sequences from moderate-sized deep earthquakes in the Tonga subduction zone suggests that they show an order of magnitude less aftershock activity than sequences from similar-sized mainshocks in California (Wiens et al., 1997). The 9 March 1994 Tonga deep earthquake showed the strongest known deep aftershock sequence, with 144 aftershocks (Wiens et al., 1994; Wiens and McGuire, 2000). This sequence is equiv-

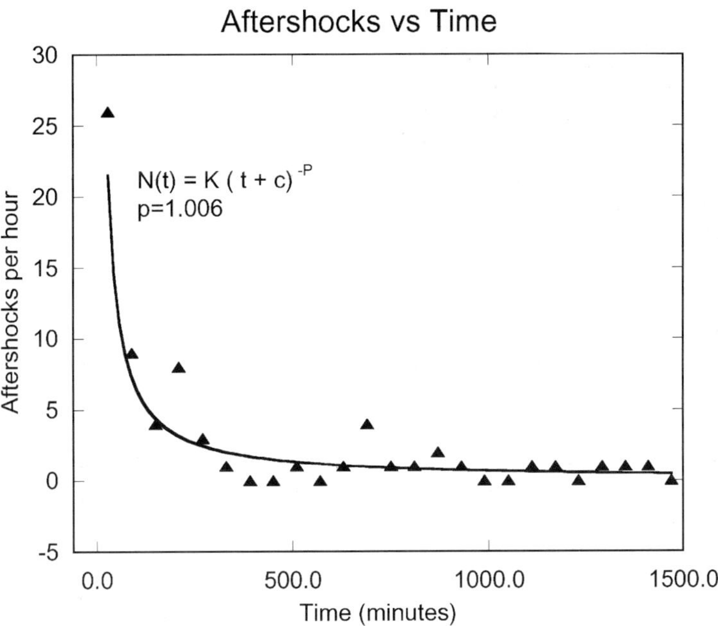

Fig. 2. Temporal decay of the aftershock occurrence rate for aftershocks with $m_b \geq 3.8$ during the first day following the 9 March 1994 Tonga deep earthquake. The line shows the fit obtained from a maximum likelihood analysis of the entire sequence.

alent to a rather weak aftershock sequence for a shallow earthquake of the same size. The stronger deep aftershock sequences generally show b-values and temporal decay rates that are similar to shallow earthquakes (Wiens et al., 1994; Nyffenegger and Frohlich, 2000; Wiens and McGuire, 2000). Fig. 2 shows that the temporal decay characteristics of the 9 March 1994 Tonga earthquake are well fit by a power law decay with an exponent (P-value) of 1.0, which is identical to typical shallow earthquakes (Wiens and McGuire, 2000). The similarity of deep and shallow earthquake P-values may be significant, since prior studies of shallow earthquakes suggest that P-values are indicative of the physical characteristics of the source region (Kisslinger, 1996).

The aftershock productivity of deep earthquakes also shows a strong variation between different deep seismic zones (Wiens and Gilbert, 1996). Deep earthquakes in Tonga and the Marianas generally show relatively strong aftershock sequences, whereas deep earthquakes in the Japan, Kurile, and South American subduction zones show few aftershocks relative to their seismic moment. For example, the two largest deep earthquakes known have occurred in South America, and both show few aftershocks (the 1994 Bolivia event ($M_W = 8.3$) shows only three aftershocks with $m_b \geq 4.5$, and the 1970 Colombia event ($M_W = 8.0$) had no known aftershocks). In contrast, the 1994 Tonga event ($M_W = 7.6$) showed 29 aftershocks with $m_b \geq 4.5$, and even a much smaller event such as the 1995 Mariana deep earthquake ($M_W = 7.0$) showed nine aftershocks with $m_b \geq 4.5$ (Wu and Chen, 1999). The systematic variation in deep earthquake aftershock production as a function of source region (Wiens and Gilbert, 1996) seems unlike shallow earthquakes, where aftershock productivity has been characterized as a simple function of mainshock fault area (Yamanaka and Shimazaki, 1990).

Aftershocks of deep earthquakes appear to concentrate along the mainshock fault plane (Wiens et al., 1994), as do aftershocks of shallow earthquakes. For shallow earthquakes Kisslinger (1996), classifies aftershocks as class 1, located along the rupture zone of the mainshock, class 2, located along the mainshock fault but not within the rupture zone, and class 3, which may be located anywhere in the vicinity. The majority of shallow aftershocks are class 1, particularly if they occur within the first 24–48 hours after

Fig. 3. Three-dimensional view of the best-located aftershocks of the 9 March 1994 Tonga deep earthquake. Ellipsoids denote the 95% confidence region of each earthquake location, with the white ellipsoid denoting the mainshock and the green ellipsoids denoting the best located aftershocks. The planar surface that best fits the aftershock locations is also shown; this plane is in good agreement with the mainshock focal mechanism.

the mainshock. Although past studies have suggested that deep earthquakes show aftershocks mostly of class 3 (Willemann and Frohlich, 1987), recent large deep earthquakes suggest that most aftershocks are concentrated near or along the mainshock rupture zone (classes 1 or 2). These deep earthquakes include the 1994 Tonga (Wiens et al., 1994; McGuire et al., 1997) and 1996 Flores Sea (Wiens, 1998; Tinker et al., 1998; Tibi et al., 1999) events. Fig. 3 illustrates the concentration of aftershocks along the fault plane of the 9 March 1994 Tonga deep earthquake. The 1994 Bolivia event shows fewer aftershocks along the mainshock fault plane, but the aftershocks are smaller and difficult to locate precisely (Myers et al., 1995).

The 1994 Tonga deep earthquake also showed an aftershock zone expansion pattern (Wiens and McGuire, 2000) similar to those observed for shallow earthquakes (Tajima and Kanamori, 1985). Aftershocks within the first 10 h were located mostly in regions of substantial mainshock moment release. This aftershock area expanded by about a factor of 2 over the next 24 h.

2.4. Magnitude–frequency relations

The slopes of deep earthquake magnitude–frequency relations (b-values) show a remarkable variation between different deep seismic zones (Giardini, 1988; Frohlich and Davis, 1993; Okal and Kirby, 1995; Wiens and Gilbert, 1996). Japan, the Kuriles, and South America are characterized by b-values (calculated from M_W) of 0.4–0.6 (Wiens and Gilbert, 1996),

(a)

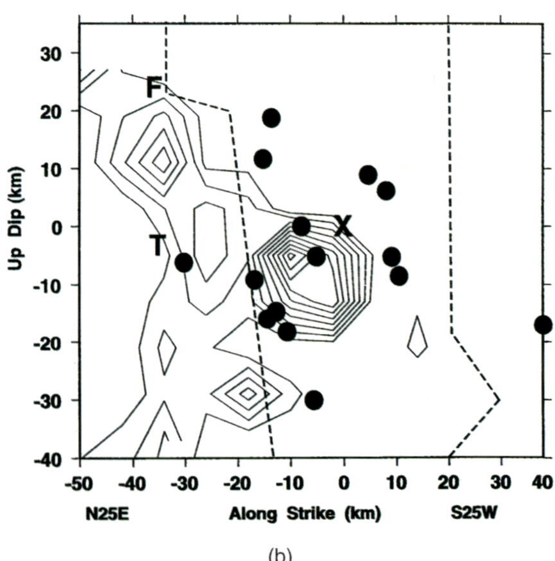

(b)

which is anomalously low relative to the typical values of 0.8–1.1 found for shallow events in all parts of the world (Frohlich and Davis, 1993). This indicates that large deep earthquakes are more numerous relative to small events than would be expected. In contrast, the b-value found for the Tonga deep subduction zone (1.28 ± 0.02) is higher than found for any shallow tectonic region, indicating a dearth of large earthquakes relative to the global average (Wiens and Gilbert, 1996). This behavior is different from shallow earthquakes, which show relatively constant b-values (Frohlich and Davis, 1993; Rundle, 1989), and where changes in b-values with increasing magnitude result from tectonic limitations on the width of the fault (Pacheco et al., 1992; Okal and Romanowicz, 1994).

3. Sensitivity of deep earthquakes to slab temperature

Some understanding that deep earthquakes are sensitive to the temperature of the slab has existed since early in the plate tectonics era. The maximum depth of intermediate and deep earthquakes is related to the temperature of the slab, as determined by the age of the subducting lithosphere and the convergence velocity (Molnar et al., 1979; Brodholt and Stein, 1988; Wortel and Vlaar, 1988). Recent evidence suggests that a variety of other deep earthquake statistical and rupture parameters are also a function of slab temperature. These observations may hold important clues about the mechanisms producing deep earthquakes.

3.1. Temperature limits on seismic nucleation and rupture

If the occurrence of deep earthquakes is temperature limited, slab thermal models suggest that the width of the seismically active zone should be controlled by the temperature range of faulting. Recent results show that the rupture and aftershock zone widths of the largest deep earthquakes are larger than the Benioff zone defined by smaller slab earthquakes (Wiens et al., 1994; McGuire et al., 1997). Fig. 4a shows the aftershock zone of the 9 March 1994 Tonga deep earthquake along with background seismicity. Two aftershocks occur 10–15 km beyond the limits of the Benioff zone as defined by small earthquakes. Fig. 4b shows a map of the slip distribution along the fault of the same earthquake, as determined by body waveform inversion, including regional broadband waveforms from the Southwest Pacific Seismic Experiment. This shows that the rupture also extended about 15 km beyond the nucleation limit of smaller slab earthquakes.

Slab thermal models allow some quantification of the different temperature limits for deep earthquake rupture and nucleation. Fig. 5 shows a thermal model of the Tonga subduction zone calculated using a finite difference method (Toksoz et al., 1973). Assuming that the center of the active seismic zone represents the coldest part of the slab, the features of the 1994 Tonga rupture can be mapped onto the thermal profiles. Unfortunately, uncertainties in many of the model parameters, such as the initial temperature structure of the Tonga slab, the subduction velocity, and the thermal conductivity (Hauck et al., 1999) are such that the absolute temperature structure is not

Fig. 4. (a) A cross-sectional view of the aftershock zone of the 9 March 1994 Tonga earthquake, viewed horizontally along the strike of the slab (N50°W) from the northwest (McGuire et al., 1997). Blue ellipsoids denote background seismicity, the white ellipsoid denotes the hypocenter of the 9 March 1994 Tonga event, the yellow ellipsoid denotes a foreshock, green ellipsoids denote aftershocks, and the red ellipsoid denotes the rupture termination point as determined from directivity analysis. The view is 90 km across, and earthquake depths range from 520 to 610 km. This figure shows that the background seismicity in this region forms an indistinct double seismic zone, and that the rupture propagated into the previously aseismic region. At least two aftershocks also initiated in the aseismic region. (b) A view of the seismic slip as a function of position along the fault plane of the 1994 Tonga deep earthquake, as determined by body waveform inversion of 47 P- and SH-phases (McGuire et al., 1997). The slip is shown with a contour interval of 0.5 m. The view is from the northwest, perpendicular to the fault plane, and is directly comparable with (a). The location of the edges of the background seismicity are plotted as dotted lines, the hypocenter is denoted as 'X', well-located aftershocks as dots, a foreshock is denoted as 'F', and the location of the end of the rupture as determined by the directivity analysis is plotted as 'T'. The rupture initiated in the interior of the previously seismically active zone and proceeded to the northeast, extending about 15–20 km beyond the edge of the previously active seismic zone (vertical dotted lines). Aftershocks are concentrated in the active slab core in regions of low seismic moment release adjacent to the first subevent.

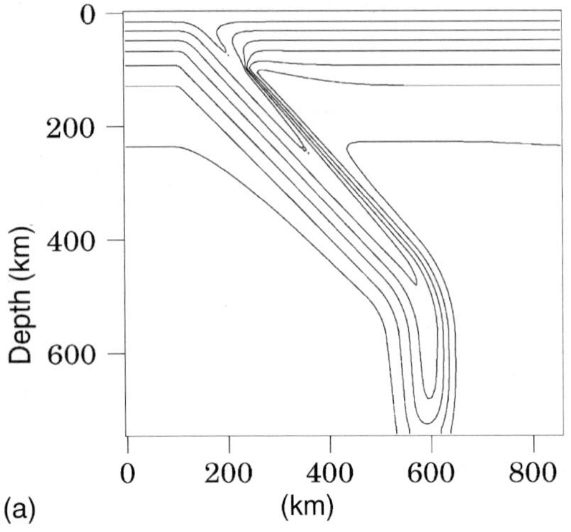

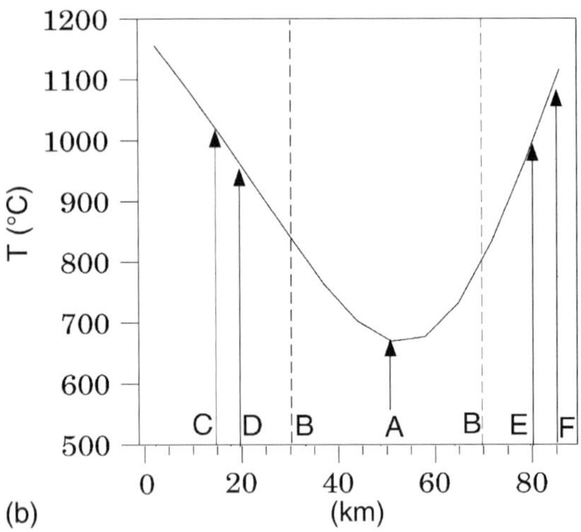

Fig. 5. (a) Thermal model of the Tonga slab, calculated with a finite difference method (Toksoz et al., 1973). The model was calculated assuming a convergence rate of 10 cm per year relative to the mantle. (b) Horizontal profile of temperatures through the slab model at the depth of the 1994 Tonga earthquake. Lateral locations of various features are shown, assuming that the seismic zone is centered on the coldest part of the slab. A denotes the epicenter location; B the edges of the background Wadati–Benioff zone seismicity, C the end of the rupture, D and E denote the outlying aftershocks to the northeast and southwest, respectively, and F denotes the position of the top of former oceanic crust in the thermal model. Although there is considerable uncertainty about the thermal model for the lower Tonga slab, the temperature difference of about 200°C between the limit of the background seismicity and the termination rupture is relatively robust. This suggests that deep earthquake ruptures can propagate through material that is at least 200°C warmer than the limiting temperatures for earthquake nucleation.

well constrained. However, by testing a wide variety of parameters for the thermal models, McGuire et al. (1997) determined that the temperature difference between the rupture termination point and the edge of the background seismicity is relatively robust, with differences of at least 200°C.

Thus, large earthquake ruptures can propagate through generally aseismic material until the rupture reaches temperatures that are 200–300°C warmer than the normal limiting temperature of deep seismicity in the slab. This suggests that slabs may be composed of a cold core, where seismic rupture initiates and small earthquakes occur, and a thermal "halo" of warmer material, which can sustain rupture, but no smaller earthquakes, and where aftershocks are generated only with difficulty during transient processes following large earthquakes.

3.2. Magnitude–frequency relations

The slope of the magnitude–frequency relation (b-value) is relatively constant within a given deep seismic zone but varies greatly between different zones (Giardini, 1988; Frohlich and Davis, 1993; Okal and Kirby, 1995; Wiens and McGuire, 1995). The variation in b-values between different deep seismic zones is related to the thermal properties of the subducting slabs, and seems to be a function of the slab thermal parameter (Wiens and Gilbert, 1996). The thermal parameter, the product of the slab vertical descent rate and the age of the subducting lithosphere, is a simple way of estimating the overall temperature structure of the deep slab. The use of the thermal parameter to compare temperatures at depth in the slab is justified theoretically if heating of the subducted lithosphere occurs by conduction and if the lithospheric temperature structure is given by a halfspace cooling model prior to subduction (Molnar et al., 1979). Larger thermal parameters correspond to cooler slab temperatures at depth. In this study, we use slab thermal parameters computed using the data given in Wiens and Gilbert (1996).

Fig. 6 shows b-values determined by a least squares fit to 1977–1995 M_w values for each deep seismic zone. The b-values show an obvious dependence on slab thermal parameter. Deep seismic zones in relatively warm slabs such as Peru–Chile show low b-values, indicating more large earthquakes relative

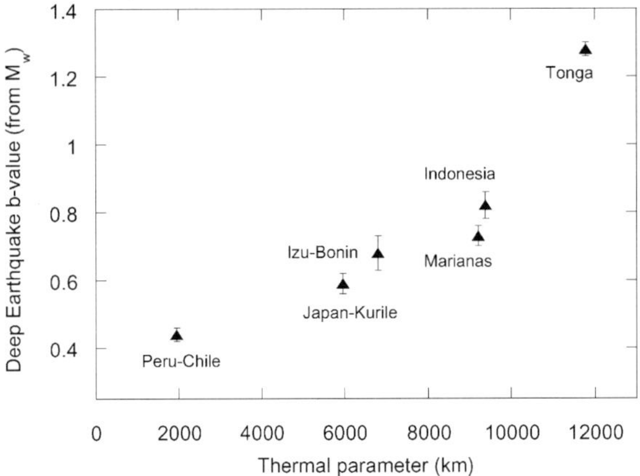

Fig. 6. The slope of the magnitude–frequency relation (b-value) for various deep seismic zones as a function of slab thermal parameter (Wiens and Gilbert, 1996). Cold deep seismic zones show a much higher b-value than warmer slabs, demonstrating that the statistical behavior of deep earthquakes is temperature dependent.

to small earthquakes. Deep earthquakes in colder slabs, such as Tonga, show much higher b-values, indicating more small earthquakes. This suggests that the statistics of deep earthquake occurrence is fundamentally temperature sensitive in a way that shallow earthquakes are not.

3.3. Aftershock productivity

Although there is substantial variability between events, the aftershock productivity of deep earthquakes when normalized by the mainshock size is correlated with the thermal parameter (Fig. 7) (Wiens and Gilbert, 1996). Fig. 7a tabulates aftershocks of all deep earthquakes 1958–1996 with $M_W \geq 7.0$ as a function of slab thermal parameter. Since the detection levels of small aftershocks varies with time and location, only aftershocks with $m_b \geq 4.5$ are tabulated. The number of aftershocks are normalized to the number expected for a $M_W = 8.3$ event, assuming that the number of aftershocks scales linearly with the seismic moment (Wiens and McGuire, 1995), as is observed in Japan and California (Yamanaka and Shimazaki, 1990). Fig. 7b tabulates all aftershocks of deep earthquakes with $M_W \geq 6.0$ from 1990 to 1996.

Both datasets suggest that the maximum number of aftershocks increases with the thermal parameter.

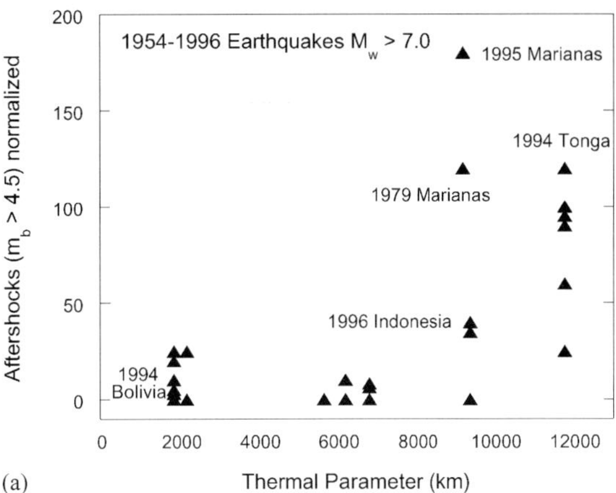

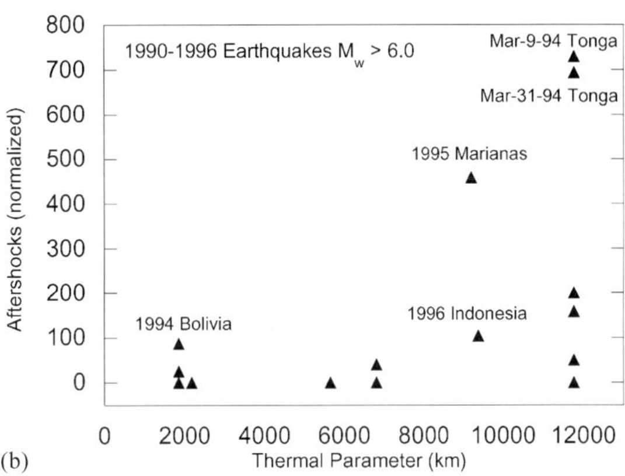

Fig. 7. (a) Number of normalized aftershocks ($m_b \geq 4.5$) for large 1958–1996 deep earthquakes as a function of subduction zone thermal parameter (Wiens and Gilbert, 1996). Larger thermal parameters correspond to cooler slabs. The number of aftershocks is normalized to correct for different mainshock sizes, assuming that the logarithm of the number of aftershocks is proportional to M_W, as observed for shallow earthquakes. (b) Total number of normalized aftershocks recorded for 1990–1996 deep earthquakes as a function of subduction-zone thermal parameter. Deep earthquakes in cold slabs show a much higher aftershock rate for both datasets.

Earthquakes with large numbers of aftershocks, such as the 1994 Tonga (Wiens and McGuire, 2000), the 1996 Flores Sea (Wiens, 1998), and 1995 Mariana (Wu and Chen, 1999) events are found only in relatively cold slabs with high thermal parameters. In contrast, earthquakes in warm slabs, such as the 1970 Colombia and 1994 Bolivia events, show few aftershocks relative to the mainshock moment. This observation is also correlated with the b-value of the deep seismic zone, since zones showing low *b*-values also show few aftershocks, and vice-versa.

Deep earthquakes that produce many aftershocks are generally located in regions that show many smaller earthquakes, such as northern Tonga. Conversely, deep earthquakes producing few aftershocks are located in more aseismic regions, as measured by the number of small earthquakes. It is interesting to note that the four largest deep earthquakes since 1954 all occurred in regions showing essentially no smaller earthquakes at all, and all produced very few aftershocks (Wiens and McGuire, 1995). This phenomenon can be successfully described by the rate-and-state friction approach to aftershocks proposed by Dieterich (1994). In this model, a stress step such as an earthquake mainshock gives rise to a given seismicity rate change relative to the background rate. Therefore, mainshocks in highly seismic areas will produce many aftershocks, whereas similar sized mainshocks in largely aseismic areas will produce few aftershocks, in agreement with the deep earthquake data. However, this model does not explain the difference in aftershock rates between deep and shallow earthquakes, since some deep seismic zones show higher seismicity rates than shallow faults, yet produce one to three orders of magnitude fewer aftershocks.

Thus, the correlation of *b*-values and aftershock productivity with slab temperature may result from a similar phenomenon, the tendency for small earthquakes to be inhibited in warmer slabs. This phenomenon results in lower overall seismicity rates, low *b*-values, and fewer aftershocks in warmer slabs relative to cold slabs. Both the *b*-value and aftershock observations suggest that the statistics and occurrence of deep earthquakes are controlled by the slab temperature in a way that is fundamentally different from shallow earthquakes.

3.4. Earthquake rupture characteristics

Recent data indicates that the deep earthquake rupture characteristics may also be a function of the slab thermal structure (Table 1). Fig. 8 shows plots of rupture velocity, stress drop, scaled duration, and maximum seismic efficiency for the largest deep earthquakes (Table 1) as a function of slab thermal parameter. All of these parameters seem to show some dependence on the thermal parameter of the subduction zone. Earthquakes in warm slabs such as South America are characterized by slow rupture velocity, high stress drop, longer durations, and lower seismic efficiencies. Colder slabs, such as Tonga and Indonesia, are characterized by faster rupture velocities, lower stress drops, shorter durations, and higher seismic efficiencies. Since the low thermal parameter dataset is dominated by four large South American earthquakes, it could be that these differences arise through some other unique characteristic of the South American slab relative to western Pacific slabs, such as tectonics or composition. However, I choose to interpret these differences as resulting from thermal structure, in accord with the *b*-value and aftershock observations.

In general, the source parameters and aftershock characteristics of colder slabs are relatively similar to shallow earthquakes, whereas the observations from warmer slabs are quite different. For example, rupture velocities for the Tonga and Flores Sea events are about 75% of the shear velocity, which is similar to the average for shallow earthquakes (Geller, 1976). Conversely, rupture velocities for warmer slabs range from 30 to 55% of the shear velocity, or much lower than generally found for shallow earthquakes. The stress drops for deep earthquakes in cold slabs (10–16 MPa) are similar to the average stress drops for shallow intraplate earthquakes (~10 MPa, Kanamori and Anderson, 1975), whereas stress drops for deep earthquakes in warm slabs are much greater (22–44 MPa). Similarly, deep earthquake efficiency, aftershock production, and *b*-values in cold slabs are more similar to shallow earthquakes than these parameters in warm slabs.

Overall, this survey of the source and aftershock parameters suggests that slab temperature has a major effect on the characteristics of deep earthquakes. This is an important clue in understanding the mechanism of deep earthquakes.

4. Proposed physical mechanisms of deep earthquakes

It has long been recognized that at depths greater than about 50 km, the stresses required for brittle faulting exceed any feasible strength of the material (Griggs and Handin, 1960; Leith and Sharpe, 1936). A variety of mechanisms to enable shear failure at

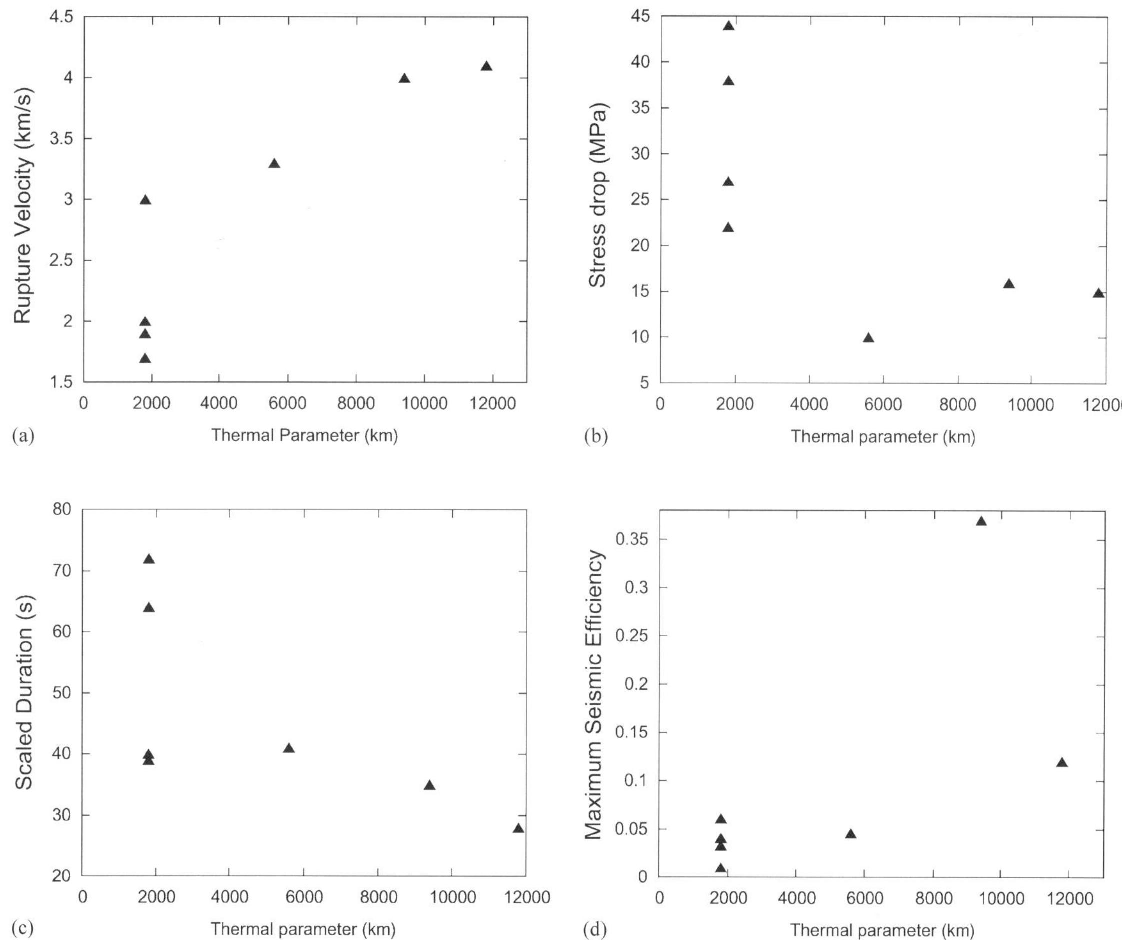

Fig. 8. (a) Average rupture velocity of large deep earthquakes as a function of slab thermal parameter, (b) static stress drop as a function of slab thermal parameter, (c) source time function duration, normalized by $M_0^{1/3}$ to correct for different seismic moments, as a function of slab thermal parameter and (d) maximum seismic efficiency as a function of thermal parameter. All source values are given in Table 1.

great depth have been proposed over the years, and are covered in several review papers (Frohlich, 1989; Green and Houston, 1995; Kirby et al., 1996b). Here, I briefly discuss the major proposed mechanisms and the evidence in light of recent research results.

4.1. Pore pressure and dehydration embrittlement

Earthquakes occurring under high confining pressures are commonly thought to result from high pore pressures, which reduce the effective normal stress on the fault (Hubbert and Rubey, 1959). It is unlikely that initial pore fluids can survive to great depths. However, water may be transported to depth as hydrous phases within a subducting slab (Thompson, 1992; Navrotsky and Bose, 1995; Bose and Navrotsky, 1998). Dehydration reactions may then liberate free fluids within regions of the slab containing hydrous minerals, thus leading to brittle failure at much smaller shear stresses (Raleigh and Paterson, 1965). Meade and Jeanloz (1991) documented acoustic emissions due to dehydration reactions in serpentinite at pressures of 2–9 GPa, and emissions resulting from metastable amorphization reactions at greater pressures.

Many studies have suggested that intermediate depth earthquakes (75–300 km depth) result from dehydration embrittlement (Meade and Jeanloz, 1991; Green and Houston, 1995; Kirby et al., 1996a), since fluid phases are probably expelled from the slab at these depths (Tatsumi, 1989). However, the persistence of water and the existence of dehydration reactions at greater depth are less well established. Some deep dehydration reactions have negative ΔV (Bose and Navrotsky, 1998), such that the reaction produces a reduction in volume, which will generally prohibit dehydration embrittlement (Wong et al., 1997). Another difficulty is that hydrated phases are thought to exist largely in the upper few kilometers of the oceanic mantle (Meade and Jeanloz, 1991), whereas large deep earthquakes show much larger fault dimensions. This has lead to the proposal that shallow faults in the oceanic lithosphere represent weak zones along which water can circulate and form hydrated minerals. Deep earthquakes may thus represent dehydration embrittlement of these faults due to the concentration of hydrated minerals (Silver et al., 1995). However, the fault orientations of many deep earthquakes are not consistent with typical shallow fault orientations (McGuire et al., 1997; Jiao et al., 2000).

4.2. *Transformational faulting*

Early proposals linking deep earthquakes to phase transformations generally envisioned a process of sudden implosion caused by rapid transformation of a metastable lower-density phase (Bridgeman, 1945; Evison, 1967). As noted in the previous section, such volumetric mechanisms are inconsistent with the observation that deep earthquakes have double couple sources. However, Kirby (1987) suggested that deep earthquakes may result from shear dislocations associated with metastable phase transformations, and several high-pressure experimental studies soon found evidence of such phase transformations in both analogue and olivine systems (Green and Burnley, 1989; Green et al., 1990; Burnley et al., 1991).

The transformational faulting model suggests that the transition from α-olivine to β- or γ-phase is kinetically inhibited in the center of the coldest subducting slabs, resulting in a wedge of metastable α-olivine below the 410 km discontinuity (Sung and Burns, 1976; Rubie and Ross, 1994; Kirby et al., 1996b). Within this wedge, small lenses of transformed material (anticracks) form, and under a strong external stress field these lenses link up. The transformed material in the developing fault has little shear strength due to its fine grain size (Green and Burnley, 1989), and unlike brittle failure, faulting is not suppressed by confining pressure. The resulting failure results in the metastable transformation of the material in the fault zone to β- or γ-phase.

The transformational faulting model predicts that deep earthquakes should occur within a rather narrow, tapering metastable olivine wedge. The wedge should be largest for the coldest slabs, with thermal and kinetic modeling (Kirby et al., 1996b; Dassler and Yuen, 1996; Devaux et al., 1997; Hauck et al., 1999) as well as buoyancy considerations (Marton et al., 1999; Schmeling et al., 1999) suggesting widths of 5–25 km. In contrast, studies of the 1994 Tonga and Bolivia deep earthquakes suggest fault widths on the order of 40–50 km perpendicular to the slab (Wiens et al., 1994; Silver et al., 1995; McGuire et al., 1997). The total width of the aftershock zone of the 1994 Tonga event is about 65 km, demonstrating that earthquakes actually nucleate over a wide region. Both rupture propagation and aftershocks occurred outside the Benioff zone defined by smaller background seismicity, showing that the rupture zones and aftershocks of large deep events are not subject to the same spatial limitations that delineate small earthquakes in deep seismic zones.

There is also substantial doubt as to whether a metastable olivine wedge actually exists in deep slabs. Iidaka and Suetsugu (1992) found evidence of a metastable wedge in northern Izu–Bonin from seismic travel times. However, an experiment involving a better geometry of stations, including ocean bottom seismographs, in a colder slab (Tonga) yielded no conclusive evidence of a metastable wedge, and suggested that resolution of olivine metastability using travel times might be difficult (Koper et al., 1998). A finite difference modeling study of the Tonga slab suggested that a metastable wedge should produce prominent waveform distortion and later arriving phases due to guided energy propagating within the wedge; however, a search of the appropriate seismic records failed to find such effects (Koper and Wiens, 2000). Thus, there is little seismological evidence for olivine metastability in slabs, a primary requirement of the transformational faulting model.

4.3. Shear instabilities and melting

Griggs and Baker (1969), Ogawa (1987), and Karato (1997 and this issue) propose that deep earthquakes occur through a runaway thermal creep event along a shear zone. Rock creep induced by stress in a shear zone will produce heat, which conducts away very slowly. Since most creep processes are highly temperature sensitive, heating of the shear zone will lower the viscosity and promote a positive-feedback process in which further slip and heat is generated. Such a runaway creep event may result in melting along the fault zone. Hobbs and Ord (1988) suggested that below a critical temperature, mantle materials show strain rate softening behavior and may undergo catastrophic plastic shear. This model of plastic shear instabilities differs from the thermal shear instabilities in that the dominant control of shear zone development is strain rate hardening or softening rather than thermal effects.

Karato (this issue) suggests that the thermal runaway phenomenon would be favored in deep slabs due to rheological weakness caused by small grain size of the slab interior after the olivine to β-spinel transformation. Bina (1998) also notes that if olivine persists metastably below the depth for equilibrium transformation, latent heat release by the metastable transformation reaction will produce local superheating, which may favor the production of deep earthquakes by the thermal shear instability mechanism.

Kanamori et al. (1998) and Bouchon and Ihmle (1999) suggest that the slow rupture velocity, high stress drop, and low seismic efficiency indicate that runaway thermal creep and fault zone melting occurred during the 1994 Bolivia deep earthquake. They further show that the high stress drop would facilitate frictional heating, and that the low seismic efficiency might be indicative of significant energy dissipation due to heating and melt production during the faulting, leaving less energy available for seismic radiation.

Fig. 9 shows one conception of how the fault zone melting model might work, based on geologically observed fault zone melts (pseudo-tachylytes) (McKenzie and Brune, 1972; Spray, 1993; Obata and Karato, 1995). In this model, a zone of frictional melt production moves with the rupture front along the fault. Melt injection under high pressure may also occur ahead of the melt zone, perhaps facilitating crack propagation at the tip of the rupture. Melt production will reduce the friction along the fault, but this will then reduce the production of melt, such that an equilibrium between friction, melt production, and lubrication may allow the fault to propagate smoothly.

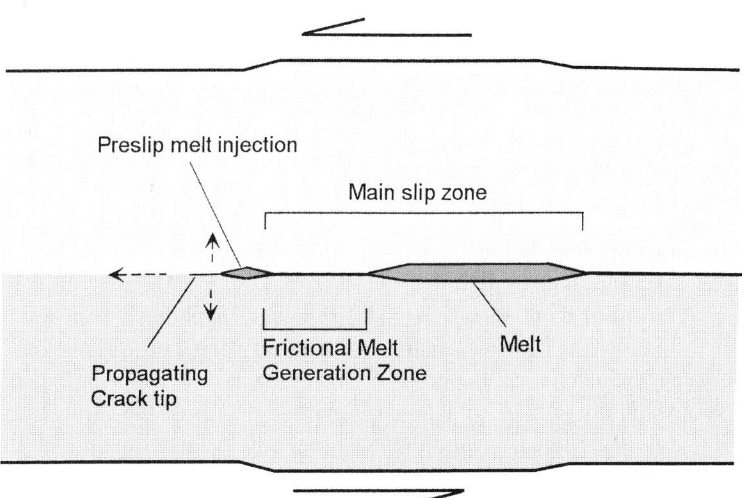

Fig. 9. Model for faulting with lubrication induced by frictional melting, originally developed to explain geological observations of pseudo-tachylytes in exhumed fault zones (after Spray, 1993). As the active slip zone moves along the fault, melting is induced by friction. If rupture propagation is slow, some melt may be injected ahead of the main slip zone, perhaps facilitating crack propagation.

Fig. 8 and Table 1 show that slow rupture, high stress drop, and low seismic efficiency seem to be characteristic of the largest deep earthquakes in warm slabs. According to the arguments presented in Kanamori et al. (1998), all of these earthquakes may involve fault zone melting, such that much of the elastic energy release is taken up in frictional heat production and melting, resulting in low seismic efficiency. In contrast, the earthquakes in colder slabs show faster rupture propagation, lower stress drop, and moderate seismic efficiency. In addition, deep earthquake magnitude–frequency relations and aftershock production show a dependence on slab temperature. Such a dependence of earthquake source properties on the temperature of the slab is to be expected if thermal shear instabilities and melting are the cause of deep earthquakes, since the precise mechanics of the thermal shear instability depend critically on the initial temperature of the rock.

One possibility, suggested by the different values of seismic efficiencies, is that deep earthquakes in warm slabs generally involve coseismic melting, whereas earthquakes in cold slabs represent thermal shear instabilities with little melting. The difference in rupture mechanics may result from the fact that the temperature in warmer slabs is closer to the melting point, and thus a highly dissipative thermal runaway involving melt is more readily attained. Magnitude–frequency relations would also be affected by slab temperature, since warm slabs would favor stress release through exceptionally large thermal runaway events, whereas colder slabs would show more typical *b*-values.

5. Conclusions

Deep earthquakes show many features that are similar to shallow earthquakes. Deep earthquakes represent shear failure on a planar surface, and show many similar statistical properties to shallow earthquakes, such as aftershock temporal decay characteristics. In addition, average rupture parameters are similar to shallow earthquakes, except that deep earthquakes tend to show higher stress drops.

However, some deep earthquake properties show a remarkable dependence on the source region, and vary systematically as a function of inferred slab thermal properties. Deep earthquake *b*-values and aftershock production rates are inversely correlated with slab temperature. In addition, deep earthquakes in warm slabs show slower rupture velocity, higher stress drop, and lower seismic efficiency relative to deep earthquakes in warm slabs. This dependence on temperature is unlike shallow brittle rupture.

Several models for the occurrence of deep earthquakes have been proposed. Dehydration embrittlement is a viable mechanism if it can be shown that hydrous minerals carry water to depths of 670 km. However, it is difficult to explain the large fault widths and orientations of some deep earthquakes, since hydrated phases are expected in only a narrow depth zone at the top of the slab, and the earthquakes do not have the expected orientations for reactivated shallow faults. In addition, dehydration mechanisms provide no explanation for the dependence of deep earthquake properties on slab temperature.

The transformational faulting model also fails to predict the large fault widths observed for deep earthquakes. Deep earthquakes rupture and aftershock nucleation occur throughout a wide region of the slab, in contrast to the narrow expected seismogenic zone within the metastable olivine wedge. There is little evidence that a large region of metastable olivine actually exists in slabs.

The shear instability model suggests that earthquake properties should be fundamentally sensitive to the temperature of the rock. In this way, it is qualitatively consistent with observations that magnitude–frequency relations, aftershocks, and rupture parameters vary systematically with slab temperature. Deep earthquakes in warm slabs show low seismic efficiencies, suggesting that they loose more energy to dissipative processes such as frictional heating and melting, perhaps as a result of the initially high temperatures of the rock in these regions. The thermal shear instability model is currently the most promising hypothesis concerning the physical mechanism of deep earthquakes, but further work on both observations of deep earthquake source parameters and the details of the shear instability mechanism are needed to evaluate this possibility.

Acknowledgements

I thank Patrick Shore, Jeffrey McGuire, Hersh Gilbert, and Keith Koper for use of figures and assistance. I also thank Alan Rubin, Craig Bina, and Rob van der Hilst for helpful reviews, and Rigobert Tibi for comments on an earlier draft. Many of the results described here resulted from field studies, which used instruments provided by the PASSCAL program of Incorporated Research Institutions in Seismology (IRIS), and teleseismic data were obtained from the IRIS Data Management Center. This research was supported by the National Science Foundation under Grants EAR9418942 and EAR9805309.

Appendix A. Seismic source parameters of deep earthquakes

The seismic source parameters shown in table one were derived from numerous separate source studies performed by various investigators. Estimates of the source parameters obtained by these previous studies are tabulated in Table 2, and the references are listed in Table 3. Many of the parameters shown in the table are calculated somewhat differently by the various studies. Therefore, the following procedure was used to estimate the parameters given in Table 1.

Values for radiated energy, rupture velocity, rupture dimensions, and rupture duration were taken from the median of the values given by the previous studies. Greater weight was given to some of the studies. For example, in establishing the rupture dimension of the 1994 Bolivia earthquake, later studies (Antolik et al., 1996; Ihmle, 1998) make a strong case that the actual dimensions of the fault was larger than determined in some of the initial studies. Therefore, the fault dimensions chosen here (60 km ×

Table 2
Estimates of deep earthquake source parameters from previous studies

Date	Location	Radiated energy (10^{14} J)	Rupture velocity (km/s)	Stress drop (MPa)	Rupture diameter (km)	Rupture duration (s)	Maximum efficiency	Reference (Table 3)
9 June 94	Bolivia	520	1.0	110	40 × 40	40	0.036	KK
			1–2.6		50 × 30	40		S
			1.0	114	50 × 20	40		LG
			2–3	283	40 × 30	40		G
			1–2		50 × 30	45		C
		150	1.5–4	51	50 × 50	47	0.026	E
			1.7	150	60 × 40	40		A
			1.8	70	90 × 90	40		I
		250		71			0.02	WR
		320						U
31 July 70	Colombia	21	1.3	7	100 × 60	50	0.016	FK
		25	2	39	100 × 20	60	0.01	E
17 June 96	Flores Sea	290	4.0	21	95 × 20	23	0.45	T
			3.5	7–20	90 × 25	23		W
			4.5		75 × 30	29		TW
		58	2–4	16	105 × 20	20	0.05	G, WR
			>4.5	10	120 × 40	18		A
		180						U
29 September 73	Japan Sea	9.6	3.3	10	60 × 40	24	0.012	FK
15 August 63	Bolivia	17	3.3	4	80 × 50	37	0.043	FK
		7	2.0	17	60 × 20	34	0.033	E
9 November 63	Brazil	16	3.0	27	40 × 25	20	0.026	FK
9 March 94	Tonga-Fiji	11		7			0.25	WR
		37	5.0	25	40 × 20	14	0.15	T
			4.1	13	45 × 30	14		M
			3.0	14	60 × 20	17		LG
			4.0–5.0	26	40 × 30	12		G
		24						U

Table 3
References to deep earthquake source studies

Abbreviation	Reference
A	Antolik et al. (1996), (1999)
C	Chen (1995)
E	Estabrook and Bock (1995); Estabrook (1999)
FK	Fukao and Kikuchi (1987); Kikuchi (1992)
G	Goes and Ritsema (1995); Goes et al. (1997)
I	Ihmle (1998)
KK	Kikuchi and Kanamori (1994); Kanamori et al. (1998)
LG	Lundgren and Giardini (1995)
M	McGuire et al. (1997)
S	Silver et al. (1995)
T	Tibi et al. (1999)
TW	Tinker et al. (1998)
U	USGS (Preliminary determination of epicenters)
WR	Winslow and Ruff (1999)
W	Wiens (1998)

45 km) are somewhat larger than the median of the studies.

Because static stress drop calculations vary somewhat between different studies, and to ensure that the values given in Table 1 are self-consistent, stress drops were recalculated in this study using

$$\Delta\sigma = 2.4 \times \left(\frac{M_0}{A^{3/2}}\right)$$

This formula assumes that the rupture can be approximated as a circular fault, where A represents the fault area calculated from the dimensions given in Table 1, and M_0 the seismic moment from the Harvard centroid moment tensor solutions (CMT) (Dziewonski et al., 1981). Thus, the static stress drop in Table 1 represents the average stress drop over the entire fault, and not the maximum stress drop associated with high moment release areas. Because the stress drop depends on the cube of the linear dimension of the fault, stress drop estimates have uncertainties of about a factor of 2.

The seismic efficiency is defined as the ratio of the radiated seismic energy to the total energy associated with the faulting event. It is difficult to estimate the actual seismic efficiency. However, with the assumption that the final stress after faulting is 0, it is possible to estimate the effective, or maximum seismic efficiency (Kikuchi, 1992) from

$$\eta = 2 \times \left(\frac{E_s}{M_0}\right)\left(\frac{\mu}{\Delta\sigma}\right)$$

where E_s is the radiated seismic energy and μ is the shear modulus. Assuming that the sliding friction is greater or equal to the final stress on the fault, this is the maximum possible seismic efficiency, since if the final stress on the fault is greater than 0, the seismic efficiency would be less than calculated with this method. Since the maximum seismic efficiency is calculated from the stress drop and the radiated seismic energy, both of which have considerable uncertainties, the derived values of are probably accurate to only a factor of 2.

References

Abers, G.A., 1996. Plate structure and the origin of double seismic zones. In: Bebout, G.E., Scholl, D.W., Kirby, S.H., Platt, J.P. (Eds.), Subduction Top to Bottom. Geophysics Monograph Series, AGU, Washington, DC, pp. 223–228.

Antolik, M., Dreger, D., Romanowicz, B., 1996. Finite fault source study of the great 1994 Bolivia deep earthquake. Geophys. Res. Lett. 23, 1589–1592.

Bina, C.R., 1998. A note on latent heat release from disequilibrium phase transformations and deep seismogenesis. Earth Planets Space 50, 1029–1034.

Bose, K., Navrotsky, A., 1998. Thermochemistry and phase equilibria of hydrous phases in the system $MgO–SiO_2–H_2O$: implications for volatile transport to the mantle ($MgO–SiO_2–H_2O$). J. Geophys. Res. 103, 9713–9720.

Bouchon, M., Ihmle, P.I., 1999. Stress drop and frictional heating during the 1994 Bolivia deep earthquake. Geophys. Res. Lett. 26 (23), 3521–3524.

Bridgeman, P.W., 1945. Polymorphic transitions and geologic phenomena. Am. J. Sci. 243A, 90–97.

Brodholt, J., Stein, S., 1988. Rheological control of Wadati–Benioff zone seismicity. Geophys. Res. Lett. 15, 1081–1084.

Burnley, P.C., Green, H.W., Prior, D.J., 1991. Faulting associated with the olivine to spinel transformation in Mg_2GeO_4 and its implications for deep focus earthquakes. J. Geophys. Res. 96, 425–443.

Choy, G.L., Boatwright, J.L., 1995. Global patterns of radiated seismic energy and apparent stress. J. Geophys. Res. 100, 18205–18228.

Chung, W.-Y., Kanamori, H., 1980. Variation of seismic source parameters and stress drops within a descending slab and its implications in plate mechanics. Phys. Earth Planet. Inter. 23, 134–159.

Dassler, R., Yuen, D.A., 1996. The metastable olivine wedge in fast subducting slabs: constraints from thermo-kinetic coupling. Earth Planet Sci. Lett. 137, 109–118.

Devaux, J.P., Schubert, G., Anderson, C., 1997. Formation of a metastable olivine wedge in a descending slab. J. Geophys. Res. 102, 24627–24638.

Dieterich, J., 1994. A constitutive law for rate of earthquake production and its application to earthquake clustering. J. Geophys. Res. 99, 2601–2618.

Dziewonski, A.M., Gilbert, F., 1974. Temporal variation of the seismic moment tensor and the evidence of precursive compression for two deep earthquakes. Nature 247, 185–188.

Dziewonski, A.M., Chou, T.-A., Woodhouse, J.H., 1981. Determination of earthquake source parameters from waveform data for studies of global and regional seismicity. J. Geophys. Res. 86, 2825–2852.

Evison, F.F., 1967. On the occurrence of volume change at the earthquake source. Bull. Seismol. Soc. Am. 57, 9–25.

Frohlich, C., 1987. Aftershocks and temporal clustering of deep earthquakes. J. Geophys. Res. 92, 13944–13956.

Frohlich, C., 1989. The nature of deep focus earthquakes. Ann. Rev. Earth Planet. Sci. 17, 227–254.

Frohlich, C., 1994. Earthquakes with non-double couple mechanisms. Science 264, 804–809.

Frohlich, C., Davis, S.D., 1993. Teleseismic *b*-values; or, much ado about 1.0. J. Geophys. Res. 98, 631–644.

Fukao, Y., Kikuchi, M., 1987. Source retrieval for mantle earthquakes by iterative deconvolution of long-period P-waves. Tectonophysics 144, 249–269.

Geller, R.J., 1976. Scaling relations for earthquake source parameters and magnitudes. Bull. Seismol. Soc. Am. 66, 1501–1523.

Giardini, D., 1988. Frequency distribution and quantification of deep earthquakes. J. Geophys. Res. 93, 2095–2105.

Green, H.W., Burnley, P.C., 1989. A new self-organizing mechanism for deep focus earthquakes. Nature 341, 733–737.

Green, H.W.I., Houston, H., 1995. The mechanics of deep earthquakes. Ann. Rev. Earth Planet. Sci. 24, 169–213.

Green, H.W., Young, T.E., Walker, D., Scholz, C.H., 1990. Anticrack-associated faulting at very high pressure in natural olivine. Nature 348, 720–722.

Griggs, D.T., Baker, D.W., 1969. The origin of deep focus earthquakes. In: Mark, H., Fernback, S. (Eds.), Properties of Matter under Unusual Conditions. Wiley, New York, pp. 23–42.

Griggs, D.T., Handin, J., 1960. Observations on fracture and hypothesis of earthquakes. Geol. Soc. Am. Mem. 79, 347–373.

Hara, T., Kuge, K., Kawakatsu, H., 1996. Determination of the isotropic component of deep focus earthquakes by inversion of normal-mode data. Geophys. J. Int. 127, 515–528.

Hauck, S.A., Phillips, R.J., Hofmeister, A.M., 1999. Variable conductivity: effects on the thermal structure of subducting slabs. Geophys. Res. Lett. 26 (21), 3257–3260.

Hobbs, B.E., Ord, A., 1988. Plastic instabilities: implications for the origin of intermediate and deep focus earthquakes. J. Geophys. Res. 93, 10521–10540.

Houston, H., Williams, Q., 1991. Fast rise times and the physical mechanism of deep earthquakes. Nature 352, 520–522.

Houston, H., Benz, H.M., Vidale, J.E., 1998. Time functions of deep earthquakes from broadband and short-period stacks. J. Geophys. Res. 103, 29895–29914.

Hubbert, M.K., Rubey, W.W., 1959. Role of fluid pressure in mechanics of overthrust faulting. 1. Mechanics of fluid-filled porous solids and its application to overthrust faulting. Geol. Soc. Am. Bull. 70, 115–166.

Ihmle, P.F.I., 1998. On the interpretation of subevents in teleseismic waveforms: the 1994 Bolivia deep earthquake revisited. J. Geophys. Res. 103, 17919–17932.

Iidaka, T., Suetsugu, D., 1992. Seismological evidence for metastable olivine inside a subducting slab. Nature 356, 593–595.

Jiao, W., Silver, P.G., Fei, Y., Prewitt, C.T., 2000. Do deep earthquakes occur on preexisting weak zones? An examination of the Tonga subduction slab. J. Geophys. Res. 105, 28125–28138.

Johnson, T.L., Scholz, C.H., 1976. Dynamic properties of stick-slip friction in rock. J. Geophys. Res. 81, 881–888.

Kagan, Y.Y., Knopoff, L., 1980. Dependence of seismicity on depth. Bull. Seismol. Soc. Am. 70, 1811–1822.

Kanamori, H., Anderson, D.L., 1975. Theoretical basis of some empirical relations in seismology. Bull. Seismol. Soc. Am. 65, 1073–1095.

Kanamori, H., Anderson, D.L., Heaton, T.H., 1998. Frictional melting during faulting. Science 279, 839.

Karato, S.-I., 1997. Phase transformations and rheological properties of mantle minerals. In: Crossleym D.J. (Ed.), Earth's Deep Interior. Gordon and Breach, Amsterdam, pp. 223–272.

Kawakatsu, H., 1996. Observability of the isotropic component of a moment tensor. Geophys. J. Int. 126, 525–544.

Kikuchi, M., 1992. Strain drop and apparent strain for large earthquakes. Tectonophysics 211, 107–113.

Kirby, S.H., 1987. Localized polymorphic phase transformations in high-pressure faults and applications to the physical mechanism of deep earthquakes. J. Geophys. Res. 92, 13789–13800.

Kirby, S.H., Durham, W.B., Stern, L.A., 1991. Mantle phase changes and deep earthquake faulting in subducting lithosphere. Science 252, 216–225.

Kirby, S., Engdahl, E.R., Denlinger, R., 1996a. Intermediate-depth intraslab earthquakes and arc volcanism as physical expressions of crustal and uppermost mantle metamorphism in subducting slabs (overview). In: Bebout, G.E., Scholl, D.W., Kirby, S.H., Platt, J.P. (Eds.), Subduction Top to Bottom. Geophysics Monograph Series, AGU, Washington, DC, pp. 195–214.

Kirby, S.H., Stein, S., Okal, E.A., Rubie, D., 1996b. Deep earthquakes and metastable mantle phase transformations in subducting oceanic lithosphere. Rev. Geophys. 34, 261–306.

Kisslinger, C., 1996. Aftershocks and fault-zone properties. Adv. Geophys. 38, 1–36.

Koper, K.D., Wiens, D.A., 2000. The waveguide effect of metastable olivine in slabs. Geophys. Res. Lett. 27, 573–576.

Koper, K.D., Wiens, D.A., Dorman, L.M., Hildebrand, J.A., Webb, S.C., 1998. Modeling the Tonga slab: can travel time data resolve a metastable olivine wedge? J. Geophys. Res. 103, 30079–30100.

Kuge, K., Kawakatsu, H., 1993. Significance of non-double couple components of deep and intermediate-depth earthquakes: implications from moment tensor inversions of long-period seismic waves. Phys. Earth Planet. Inter. 75, 243–266.

Leith, A., Sharpe, J.A., 1936. Deep focus earthquakes and their geological significance. J. Geol. 44, 877–917.

Marone, C., Liu, M., 1997. Transformation shear instability and the seismogenic zone for deep earthquakes. Geophys. Res. Lett. 24, 1887–1890.

Marton, F.C., Bina, C.R., Stein, S., Rubie, D.C., 1999. Effects of slab mineralogy on subduction rates. Geophys. Res. Lett. 26, 119–122.

McGuire, J.J., Wiens, D.A., Shore, P.J., Bevis, M.G., 1997. The 9 March 1994 Tonga deep earthquake: rupture outside the seismically active slab. J. Geophys. Res. 102, 15163–15182.

McKenzie, D., Brune, J.N., 1972. Melting on fault planes during large earthquakes. Geophys. J. Roy. Astron. Soc. 29, 65–78.

Meade, C., Jeanloz, R., 1991. Deep focus earthquakes and recycling of water into the Earth's mantle. Science 252, 68–72.

Molnar, P., Freedman, D., Shih, J.S.F., 1979. Lengths of intermediate and deep seismic zones and temperatures in downgoing slabs in the mantle. Geophys. J. Roy. Astron. Soc. 56, 41–54.

Myers, S.C., et al., 1995. Implications of spatial and temporal development of the aftershock sequence for the $M_w = 8.3$, 9 June 1994 Bolivian deep earthquake. Geophys. Res. Lett. 22, 2269–2272.

Navrotsky, A., Bose, K., 1995. Thermodynamic stability of hydrous silicates: some observations and implications for water in the Earth, Venus, and Mars. In: Farley, K. (Ed.), Processes of Deep Earth and Planetary Volatiles. American Institute of Physics, New York, pp. 221–228.

Nyffenegger, P., Frohlich, C., 2000. Aftershock occurrence rate decay properties for intermediate and deep earthquake sequences. Geophys. Res. Lett. 27, 1215–1218.

Obata, M., Karato, S., 1995. Ultramafic pseudo-tachylyte from Balmuccia peridotite, Ivrea-Verbania zone, northern Italy. Tectonophysics 242, 313–328.

Ogawa, M., 1987. Shear instability in a visco-elastic material as the cause of deep focus earthquakes. J. Geophys. Res. 92, 13801–13810.

Okal, E.A., Kirby, S.H., 1995. Frequency-moment distribution of deep earthquakes Implications for the seismogenic zone at the bottom of slabs. Phys. Earth Planet. Inter. 92, 169–187.

Okal, E.A., Romanowicz, B.A., 1994. On the variation of *b*-values with earthquake size. Phys. Earth Planet. Inter. 87, 55–76.

Pacheco, J.F., Scholz, C.H., Sykes, L.R., 1992. Changes in frequency-size relationship from small to large earthquakes. Nature 355, 71–73.

Page, R., 1968. Focal depths of aftershocks. J. Geophys. Res. 73, 3897–3903.

Raleigh, C.B., Paterson, M.S., 1965. Experimental deformation of serpentinite and its tectonic significance. J. Geophys. Res. 70, 3965–3985.

Randall, M.J., Knopoff, L., 1970. The mechanism at the focus of deep earthquakes. J. Geophys. Res. 75, 4965–4976.

Rubie, D.C., Ross II, C.R., 1994. Kinetics of the olivine–spinel transformation in subducting lithosphere: experimental constraints and implications for deep slab processes. Phys. Earth Planet. Inter. 86, 223–241.

Rundle, J.B., 1989. Derivation of the complete Gutenberg–Richter magnitude–frequency relation using the principle of scale invariance. J. Geophys. Res. 94, 12337–12342.

Russakoff, D., Ekstrom, G., Tromp, J., 1997. A new analysis of the great 1970 Columbia earthquake and its isotropic component. J. Geophys. Res. 102, 20423–20434.

Schmeling, H., Monz, R., Rubie, D.C., 1999. The influence of olivine metastability on the dynamics of subduction. Earth Planet. Sci. Lett. 165, 55–66.

Silver, P., et al., 1995. Rupture characteristics of the Bolivian deep earthquake of 9 June 1994 and the mechanism of deep focus earthquakes. Science 268, 69–73.

Spray, J.G., 1993. Viscosity determinations of some frictionally generated silicate melts: implications for fault zone rheology at high strain rates. J. Geophys. Res. 98, 8053–8068.

Sung, C.M., Burns, R.G., 1976. Kinetics of high-pressure phase transformations: implications to the evolution of the olivine to spinel transition in the downgoing lithosphere and its consequences on the dynamics of the mantle. Tectonophysics 31, 1–32.

Tajima, F., Kanamori, H., 1985. Global survey of aftershock area expansion patterns. Phys. Earth Planet. Inter. 40, 77–134.

Tatsumi, Y., 1989. Migration of fluid phases and genesis of basalt magmas in subduction zones. J. Geophys. Res. 94, 4697–4707.

Thompson, A.B., 1992. Water in the Earth's upper mantle. Nature 358, 295–302.

Tibi, R., Estabrook, C.H., Bock, G., 1999. The 17 June 1996 Flores Sea and 9 March 1994 Tonga–Fiji earthquakes: source processes and deep earthquake mechanisms. Geophys. J. Int. 138, 625–642.

Tinker, M.A., Beck, S.L., Jiao, W., Wallace, T.C., 1998. Mainshock and aftershock analysis of the 17 June 1996 deep Flores Sea earthquake sequence: implications for the mechanism of deep earthquakes and the tectonics of the Banda Sea. J. Geophys. Res. 103, 9987–10002.

Toksoz, M.N., Sleep, N.H., Smith, A.T., 1973. Evolution of the downgoing lithosphere and the mechanisms of deep focus earthquakes. Geophys. J. Roy. Astron. Soc. 35, 285–310.

Vassiliou, M.S., Kanamori, H., 1982. The energy release in earthquakes. Bull. Seismol. Soc. Am. 72, 371–387.

Vidale, J.E., Houston, H., 1993. The depth dependence of earthquake duration and implications for rupture mechanisms. Nature 365, 45–47.

Wiens, D.A., 1998. Source and aftershock properties of the 1996 Flores Sea deep earthquake. Geophys. Res. Lett. 25, 781–784.

Wiens, D.A., Gilbert, H.J., 1996. Effect of slab temperature on deep earthquake aftershock productivity and magnitude–frequency relations. Nature 384, 153–156.

Wiens, D.A., McGuire, J.J., 1995. The 1994 Bolivia and Tonga events: fundamentally different types of deep earthquakes. Geophys. Res. Lett. 22, 2245–2248.

Wiens, D.A., McGuire, J.J., 2000. Aftershocks of the 9 March 1994 Tonga earthquake: the strongest known deep aftershock sequence. J. Geophys. Res. 105, 19067–19083.

Wiens, D.A., et al., 1994. A deep earthquake aftershock sequence and implications for the rupture mechanism of deep earthquakes. Nature 372, 540–543.

Wiens, D.A., Gilbert, H.J., Hicks, B., Wysession, M.E., Shore, P.J., 1997. Aftershock sequences of moderate-sized intermediate and deep earthquakes in the Tonga subduction zone. Geophys. Res. Lett. 24, 2059–2062.

Willemann, R.J., Frohlich, C., 1987. Spatial patterns of aftershocks of deep focus earthquakes. J. Geophys. Res. 92, 13927–13943.

Wong, T., Ko, S., Olgaard, D.L., 1997. Generation and maintenance of pore pressure excess in a dehydrating system. 2. Theoretical analysis. J. Geophys. Res. 102, 841–852.

Wortel, M.J.R., Vlaar, N.J., 1988. Subduction zone seismicity and the thermo-mechanical evolution of downgoing lithosphere. Pageophysics 128, 625–659.

Wu, L.-R., Chen, W.-P., 1999. Anomalous aftershocks of deep earthquakes in Mariana. Geophys. Res. Lett. 26 (13), 1977–1980.

Yamanaka, Y., Shimazaki, K., 1990. Scaling relationship between the number of aftershocks and the size of the mainshock. J. Phys. Earth 38, 305–324.

ELSEVIER

Physics of the Earth and Planetary Interiors 127 (2001) 165–180

PHYSICS OF THE EARTH AND PLANETARY INTERIORS

www.elsevier.com/locate/pepi

Experimental constraints on the depth of olivine metastability in subducting lithosphere

Jed L. Mosenfelder [a,*], Frederic C. Marton [b], Charles R. Ross II [c], Ljuba Kerschhofer [d], David C. Rubie [a]

[a] *Bayerisches Geoinstitut, Universität Bayreuth, D-95440 Bayreuth, Germany*
[b] *Geophysical Laboratory and Center for High Pressure Research, Carnegie Institution of Washington, 5251 Broad Branch Rd., NW, Washington, DC 20015-1305, USA*
[c] *Department of Structural Biology, St. Jude Children's Research Hospital, 332 N. Lauderdale, Memphis, TN 38105, USA*
[d] *Siemens Business Services GmbH & Co. OHG, Otto-Hahn-Ring 6, D-81739 München, Germany*

Received 31 May 2000; accepted 16 March 2001

Abstract

The hypothesis that metastable olivine persists in some subducting slabs into the transition zone has wide implications for mantle dynamics and rheology. In order to evaluate this possibility we derive new thermo-kinetic subduction zone models to predict the extent of olivine metastability within the stability fields of its high-pressure polymorphs, wadsleyite and ringwoodite. Our updated models improve on previous work by incorporating experimental kinetic data on realistic mantle compositions ($(Mg, Fe)_2SiO_4$) rather than analogue systems. Furthermore, latent heat due to the transformation is fed back into both the kinetics and the thermal model. We also consider the effects of transformation stress on growth kinetics and the possibility of an intracrystalline transformation mechanism, previously thought to be important only at high shear stresses. Our models predict significantly smaller wedges of metastable olivine than previous work. In the case of Tonga, for example, where high values of lithospheric age (100–140 million years) and convergence rate (~14 cm per year) are most favorable for metastability, models considering only grain boundary nucleation and interface-controlled growth predict olivine metastability to ~600 km depth, in contrast to ~660 km predicted previously by Kirby et al. [Rev. Geophys. 34 (1996) 261]. When intracrystalline transformation is considered, the depth of metastability is further reduced by as much as 100 km, due to the large increase in the density of nucleation sites. Inhibition of growth by transformation stress can increase the depth interval over which the transformation takes place, but is unlikely to be a dominant factor, especially if the intracrystalline mechanism operates. These results indicate that the existence of metastable olivine at depths corresponding to those of the deepest earthquakes (~680 km) requires subduction of old lithosphere (>100 million years) and a high vertical subduction velocity ($\gtrsim$15 cm per year). Such conditions currently may only be achieved in the very northern part of the Tonga subduction zone, yet earthquakes occur at depths down to 680 km in other subduction zones in which lithospheric age and/or subduction velocity are relatively low (e.g. Indonesia and the Marianas). Therefore, mechanisms other than transformational faulting in metastable olivine must operate to cause the deepest earthquakes. © 2001 Elsevier Science B.V. All rights reserved.

Keywords: Olivine; Wadsleyite; Ringwoodite; Kinetics; Subduction; Deep earthquakes

*Corresponding author. Present address: Department of Geological and Planetary Sciences, California Institute of Technology, M/C 170-25, Pasadena, CA 91125, USA.
E-mail address: jed@gps.caltech.edu (J.L. Mosenfelder).

0031-9201/01/$ – see front matter
PII: S0031-9201(01)00226-6

1. Introduction

For over 20 years it has been hypothesized that high-pressure phase transformations can be kinetically inhibited in the interior of subducting lithosphere, where relatively low temperatures are maintained into the transition zone and lower mantle (Sung and Burns, 1976a,b). Particular attention has been paid to phase transformations involving olivine, the major constituent of the upper mantle. Under equilibrium conditions, α-(Mg, Fe)$_2$SiO$_4$ (olivine) transforms sequentially to β-(Mg, Fe)$_2$SiO$_4$ (wadsleyite) at a depth of about 410 km, then to γ-(Mg, Fe)$_2$SiO$_4$ (ringwoodite) [1] at a depth of about 520 km, and finally to (Mg, Fe)SiO$_3$ perovskite + (Mg, Fe)O (magnesiowüstite) at a depth of about 660 km.

Because the large density increases associated with these transformations affect the buoyancy of subducting slabs, kinetic hindrance will have a significant effect on the dynamics of subduction (Marton et al., 1999; Schmeling et al., 1999; Bina et al., 2001). Furthermore, the mechanical behavior of subducting slabs may be affected by high-pressure phase transformations in a number of ways. First, it has been proposed (Kirby, 1987; Green and Burnley, 1989) that a shear instability, postulated to occur during the transformation of metastable olivine to wadsleyite or ringwoodite, could be the cause of deep-focus earthquakes, although this hypothesis is currently under debate (Stein and Rubie, 1999; Karato et al., 2001). Second, regardless of whether this hypothesis is correct or not, the stress state in subducting slabs, which ultimately controls the distribution of deep-focus earthquakes, must be affected by buoyancy stresses associated with phase transformations (Goto et al., 1987; Bina, 1996, 1997; Yoshioka et al., 1997). Finally, it has been proposed that grain size reduction occurring during phase transformations leads to rheological weakening that strongly affects the mechanical properties of subducting slabs (Rubie, 1984; Riedel and Karato, 1997; Karato et al., 2001).

[1] In this paper, we use the mineralogical names olivine, wadsleyite and ringwoodite when referring to phases with the naturally-occurring (Mg, Fe)$_2$SiO$_4$ composition; for other compositions we refer to the phases as α, β, and γ.

Results of seismic studies concerning the presence of metastable olivine in subducting slabs are contradictory (e.g. Iidaka and Suetsugu, 1992; Collier and Hellfrich, 1997; Koper et al., 1998; Koper and Wiens, 2000). An alternative approach to determining the extent of olivine metastability entails thermal modeling combined with extrapolation of experimental kinetic data (Rubie and Ross, 1994; Däßler et al., 1996; Kirby et al., 1996; Devaux et al., 1997). These "thermo-kinetic" models have been based primarily on an empirical extrapolation (Rubie and Ross, 1994) of kinetic data from analogue systems (Mg$_2$GeO$_4$ and Ni$_2$SiO$_4$) to the mantle composition (Mg, Fe)$_2$SiO$_4$. These models suggest that growth kinetics control transformation rates deep in cold subducting slabs, rather than the poorly understood nucleation process, and that the persistence of a metastable olivine wedge is likely in rapidly subducting slabs of old lithosphere (e.g. Tonga). However, because the uncertainties in the above mentioned extrapolation could not be quantified, the reliability of this conclusion is uncertain. More recently, kinetic experiments have been performed on Mg$_2$SiO$_4$ (Kubo et al., 1998a; Kubo, 1999) and Mg$_{1.8}$Fe$_{0.2}$SiO$_4$ compositions (Kerschhofer et al., 1996; Kubo et al., 1998b,c; Liu et al., 1998; Mosenfelder et al., 2000) that are representative of the mantle. Here, we use the new growth rate data from these studies to develop more reliable thermo-kinetic models. We also take into account the effect of latent heat feedback into the *thermal model* as well as the kinetics of the transformation, in contrast to some previous studies (Rubie and Ross, 1994; Kirby et al., 1996). Recent experimental and theoretical studies have also highlighted several complexities in the transformation, including an intracrystalline reaction mechanism (Liu and Yund, 1995; Kerschhofer et al., 2000), the effects of stress and strain on nucleation and growth kinetics (Morris, 1992; Liu and Yund, 1995; Dupas-Bruzek et al., 1998; Kubo et al., 1998b,c; Liu et al., 1998; Mosenfelder et al., 2000), and the effects of trace amounts of hydrogen on the transformation (Young et al., 1993; Kubo et al., 1998c). Although the effects of all such complexities on kinetics in subduction zones are difficult to assess quantitatively, we present preliminary results on the effects of intracrystalline transformation and inhibition of growth by transformation stress on olivine metastability.

2. Reaction mechanisms

Extrapolation of experimental kinetic data (e.g. to lower temperatures) requires characterization of the mechanism of transformation because different mechanisms proceed at different rates (e.g. Rubie and Thompson, 1985). Here, we briefly review the mechanisms by which olivine is likely to transform to wadsleyite or ringwoodite in subducting slabs.

At the 410 km discontinuity in "normal" mantle, where olivine transforms to wadsleyite under near-equilibrium conditions in a two-phase stability field, growth of the new phase is likely to be diffusion-controlled, with long-range Fe–Mg interdiffusion being the rate controlling step (Rubie, 1993; Solomatov and Stevenson, 1994). In contrast, in subducting slabs, where the olivine–wadsleyite and olivine–ringwoodite transformations may occur at conditions far from equilibrium without any change in composition, the most likely mechanism is interface-controlled growth (Rubie and Ross, 1994; Kirby et al., 1996). Growth by this mechanism is controlled by the rate of diffusion of atoms across the migrating interface between the two phases. In contrast to diffusion-controlled growth, interface-controlled growth is theoretically expected to be linear with time (Turnbull, 1956). However, recent experimental studies have provided evidence for time-dependent growth rates, even when no compositional changes occur (Kubo et al., 1998b,c; Liu et al., 1998; Mosenfelder et al., 2000). This phenomenon is most likely associated with inhibition of growth by accumulation of elastic strain energy, as discussed below.

2.1. Grain boundary nucleation and interface-controlled growth

Numerous experimental studies have documented transformation in $(Mg,Fe)_2SiO_4$ compositions by incoherent nucleation on grain-boundaries and growth of the product phase, with the same composition as the reactant, presumably by the interface-controlled mechanism (Fig. 1a and b; Young et al., 1993; Brearley and Rubie, 1994; Kubo et al., 1998b,c; Mosenfelder et al., 2000). Nucleation occurs most easily on grain boundaries because of the high energy of these defects compared to intracrystalline defects such as dislocations and subgrain boundaries. When nucleation rates are high, the grain boundaries rapidly become saturated with grains of the new phase (Fig. 1b). The olivine grains then become separated by continuous rims of

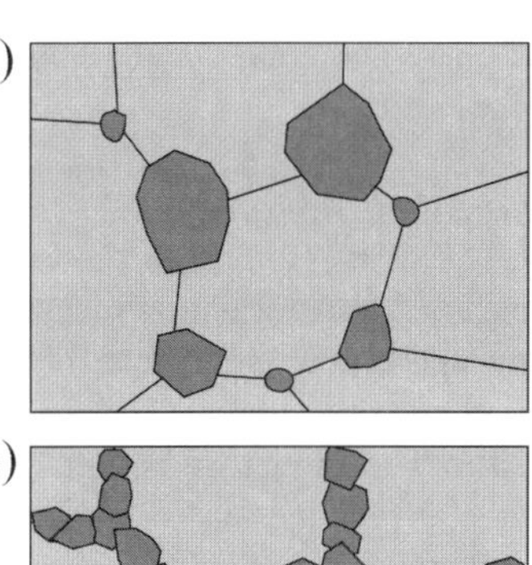

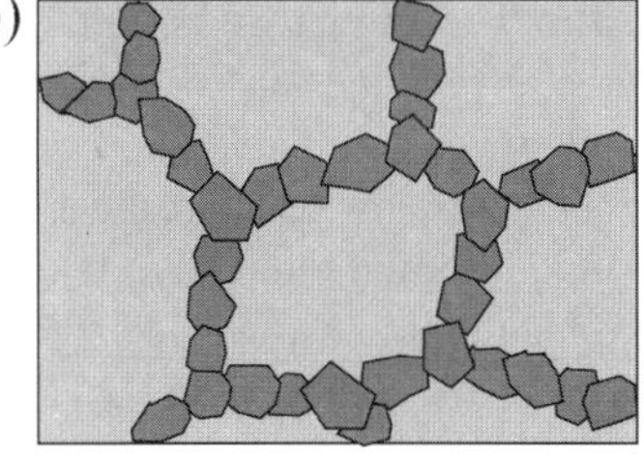

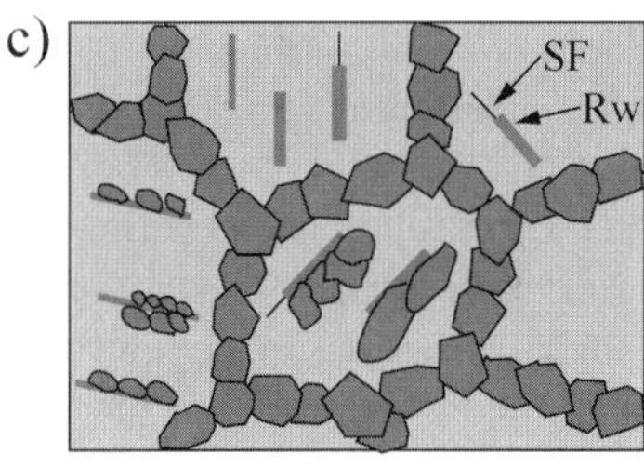

Fig. 1. Schematic cartoons illustrating reaction mechanisms, (a) grain-boundary nucleation and interface-controlled growth. Nucleation typically occurs first at grain corners (cf. Fig. 5a in Kerschhofer et al., 1998). When the nucleation rate is slow relative to the growth rate, isolated grains grow and well-defined reaction rims do not form; (b) grain-boundary nucleation and interface-controlled growth under conditions where nucleation rates are high relative to growth rates, resulting in the formation of continuous rims of the product phase around olivine grains; (c) in addition to grain-boundary nucleation and growth, various stages of intracrystalline transformation are shown. Semi-coherent ringwoodite (Rw) platelets nucleate on (0 1 0) stacking faults (SF) in olivine. Incoherent grains subsequently nucleate on the ringwoodite platelets. The increase in the density of sites available for incoherent nucleation (due to the growth of ringwoodite platelets) effectively increases the grain-boundary area per unit volume, causing faster overall transformation. Note that the size of the ringwoodite platelets is greatly exaggerated in this cartoon; in reality they are typically a few hundred nanometers long and up to a few tens of nanometers wide and can only be observed using TEM (see Kerschhofer et al., 2000).

the product phase and are gradually consumed by the growth of these rims.

2.2. Intracrystalline transformation

A "martensitic" transformation mechanism involving coherent nucleation and growth of γ-phase in α was proposed theoretically by Hornstra (1960) and Poirier (1981). The model of Poirier (1981) entails shear along (1 0 0) planes in α, producing stacking faults with the oxygen sublattice of γ, coincident with movement of cations (by "synchroshear") to complete the γ structure. Poirier's model predicts a specific crystallographic relationship between the phases, with $(1\,0\,0)_\alpha // \{1\,1\,1\}_\gamma$ and $[0\,0\,1]_\alpha // \langle\bar{1}\,1\,1\rangle_\gamma$. This topotaxial relationship has been observed in experimental studies on a variety of compositions in which thin lamellae or platelets of γ-phase formed in the interior of α crystals (see Kerschhofer et al., 2000, for references).

Until recently most of the experiments in which intracrystalline transformation was observed were performed under conditions of high differential stress (e.g. $\geq$1 GPa); the mechanism was therefore, considered to be unimportant in subduction zones, where stresses are likely to be lower (Rubie and Ross, 1994). However, recent experimental studies show that intracrystalline transformation can occur under conditions of low to moderate differential stress, provided the equilibrium phase boundary is overstepped by several GPa (Burnley, 1995; Kerschhofer et al., 1996; Martinez et al., 1997). This observation is also predicted by theoretical calculations (Liu and Yund, 1995).

Kerschhofer et al. (2000) studied intracrystalline transformation in $Mg_{1.8}Fe_{0.2}SiO_4$ olivine at 18–20 GPa and 1173–1373 K and observed four sequential stages. (1) Formation of $(1\,0\,0)_\alpha$ stacking faults in olivine; (2) coherent nucleation of thin ringwoodite platelets on these stacking faults, with the orientation relationship noted above; (3) slow semi-coherent growth of the platelets and (4) incoherent nucleation of ringwoodite and/or wadsleyite grains at the interfaces between the ringwoodite platelets and the host olivine (Fig. 1c). The conditions required to form the $(1\,0\,0)_\alpha$ stacking faults are not well understood; experimentally, temperatures $\geq$1173 K are required and there is no evidence that high stresses are necessary. The rates of processes (2) and (3) are thermally activated and are likely controlled by Si–O bond breaking (Kerschhofer et al., 2000). An important conclusion of this work is that even if the contribution to the overall transformation by coherent nucleation and growth is small, the increase in nucleation sites available for incoherent nucleation and growth can greatly enhance overall transformation rates.

2.3. Interactions between deformation and reaction kinetics

Although externally-applied stress can enhance transformation kinetics, for example by increasing the density of nucleation sites (e.g. Dupas-Bruzek et al., 1998) or promoting anisotropic growth (Green et al., 1992; Sharp et al., 1994; Dupas-Bruzek et al., 1998), such effects have not been quantified. In contrast, recent experimental studies have highlighted and quantified the effects of internal stresses in inhibiting growth (Kerschhofer et al., 1996; Kubo et al., 1998b,c; Liu et al., 1998; Mosenfelder et al., 2000). The volume decreases of the olivine to wadsleyite or ringwoodite transformations (~6 and 8%, respectively) result in internal "transformation stress" and the development of elastic strain energy. The effect of transformation stress on growth kinetics depends on the morphology of the grains of the product phase (Liu et al., 1998). When nucleation rates are low such that transformation involves the growth of isolated grains (Fig. 1a), the elastic strain energy does not depend on the size of the grains and the growth rate remains constant with time. On the other hand, when nucleation rates are high such that continuous rims of the product phase develop early in the transformation (Fig. 1b), the elastic strain energy increases continuously as the rims increase in width. In this case, growth eventually becomes inhibited by the accumulated strain energy and the rate of reaction is then controlled by the rate of stress relaxation in the rim (e.g. Morris, 1992). In general, the inhibition of growth by transformation stress will widen the depth interval over which transformation occurs. However, growth rates during intracrystalline transformation are unlikely to be affected by transformation stress because this mechanism results in isolated clusters of grains of the product phase rather than continuous rims (Fig. 1c).

2.4. Effects of hydrogen

Olivine can incorporate substantial amounts of hydrogen (up to 0.12 wt.% H_2O at high pressures close to its stability limit; Kohlstedt et al., 1996). The high-pressure polymorphs of olivine can dissolve considerably greater amounts of hydrogen into their structures, up to 2–3 wt.% H_2O (e.g. Kohlstedt et al., 1996). Growth kinetics may be affected by hydrogen in at least two ways. First, the presence of hydrogen is likely to enhance diffusion rates by lowering the activation energy needed to break Si–O bonds (e.g. Rubie, 1986). Secondly, trace amounts of hydrogen may promote mechanical weakening, thus relaxing the transformation stress that would otherwise inhibit growth (e.g. Kubo et al., 1998c).

Based on available data it is difficult to distinguish the relative importance of these two mechanisms for growth enhancement. Kubo et al. (1998c) studied the effect of varying OH contents on the growth kinetics of wadsleyite reaction rims that formed around single crystals of San Carlos olivine at 1303 K and 13.5 GPa. The difference in dislocation microstructures as well as growth rates between nominally dry samples (with ~0.02 wt.% H_2O in wadsleyite) and water-added samples (~0.05 wt.% H_2O) suggests that enhanced stress relaxation was an important effect (see also Mosenfelder et al., 2000). This result contrasts strongly with the conclusion of Chen et al. (1998), based on in situ stress measurements, that hydrogen has little effect on the strength of wadsleyite (in contrast to the well known hydrolytic weakening effect in olivine). The effects of water partitioning between the phases are also unknown. Preferential partitioning of water into wadsleyite (Young et al., 1993) may promote hydrolytic weakening during the transformation, but may also result in depletion of hydrogen at the interface between the phases, thus slowing the growth rate. More work is needed to address this question and to resolve the contradictory evidence for and against hydrolytic weakening in the high-pressure phases.

3. Experimental kinetic data

Early, systematic experimental kinetic studies were performed at relatively low pressures (up to 3.7 GPa) on the analogue materials Ni_2SiO_4 and Mg_2GeO_4 (see Rubie and Ross, 1994, for a review). Following the development of the multi-anvil apparatus, kinetics have now been investigated in $(Mg,Fe)_2SiO_4$ mantle compositions at 13.5–20 GPa and temperatures up to 1673 K using both traditional quench techniques and in situ X-ray diffraction. These recent data provide better constraints on kinetics in subducting slabs than the previous empirical extrapolation of Rubie and Ross (1994), which was based primarily on results of the low-pressure analogue experiments.

Incoherent grain-boundary nucleation and interface-controlled growth has been inferred as the dominant mechanism in experimental studies up to 17 GPa, whereas intracrystalline transformation becomes important at higher pressures. The starting material for all the studies considered here consisted of hot-pressed, polycrystalline olivine and/or olivine single crystals that remained intact during pressurization. Previous studies have shown the importance of hot pressing starting materials in kinetic studies in order to eliminate defects introduced by grinding powders and porosity reduction during pressurization (e.g. Brearley et al., 1992).

Here, we review experimental data on growth kinetics in Mg_2SiO_4 and $Mg_{1.8}Fe_{0.2}SiO_4$ compositions and use these data to derive a new rate equation for thermo-kinetic modeling. Because nucleation kinetics are still poorly understood, we assume that grain boundary nucleation is a rapid process during deep subduction of cold slabs (cf. Rubie and Ross, 1994; Kirby et al., 1996). This hypothesis is supported by the observation that the presence of clinoenstatite in peridotites enhances nucleation kinetics (Sharp and Rubie, 1995) and the inference that transformation strain energy enhances nucleation on grain boundaries (Rubie and Champness, 1987; Brearley et al., 1992).

3.1. Interface-controlled growth kinetics

Brearley et al. (1992) performed experiments at 15 GPa and a nominal temperature of 1173 K in which wadsleyite and ringwoodite formed from Mg_2SiO_4 olivine. The growth rate of wadsleyite (Table 1; see Rubie and Ross, 1994) was determined using TEM, an accurate technique for measuring the small grain size of the product phases (50–500 nm). Rubie and Ross (1994) re-estimated the temperature of these experiments to be 1233 K because Pt–Pt13% Rh ther-

Table 1
Experimentally determined growth rates for grain-boundary nucleated wadsleyite and ringwoodite[a]

Composition	Starting material	Product[b]	T (K)	P (GPa)	ΔG_{rxn} (J mol^{-1})	$\dot{x}$ (m s^{-1})	Reference
Mg_2SiO_4	Polycrystalline	Wd	1233 (50)	15.0 (0.5)	−5473 (1366)	9.2 (2.0) × 10^{-11}*	Brearley et al. (1992)
Mg_2SiO_4	Polycrystalline	Wd	1173 (20)	13.0 (0.2)	−1231 (567)	3.3 (1.7) × 10^{-10}	Kubo et al. (1998a)
Mg_2SiO_4	Polycrystalline	Wd	1158 (20)	14.1 (0.2)	−3921 (547)	7.1 (3.7) × 10^{-10}	Kubo et al. (1998a)
Mg_2SiO_4	Polycrystalline	Wd	1148 (20)	14.9 (0.2)	−5929 (535)	1.5 (0.8) × 10^{-9}	Kubo et al. (1998a)
Mg_2SiO_4	Polycrystalline	Wd	1173 (20)	14.9 (0.2)	−5805 (538)	2.4 (0.9) × 10^{-11}*	Kubo, 1999
Mg_2SiO_4	Polycrystalline	Wd	1258 (20)	15.8 (0.2)	−7136 (539)	3.7 (1.4) × 10^{-10}*	Kubo, 1999
$Mg_{1.8}Fe_{0.2}SiO_4$	Polycrystalline	Wd/Rw	1173 (50)	18.0 (0.5)	−14981 (1963)	1.8 (0.6) × 10^{-10}*	Liu et al. (1998)
$Mg_{1.8}Fe_{0.2}SiO_4$	Polycrystalline	Wd/Rw	1173 (50)	18.0 (0.5)	−14981 (1963)	4.8 (2.6) × 10^{-10}*	Liu et al. (1998)
$Mg_{1.8}Fe_{0.2}SiO_4$	Polycrystalline	Wd/Rw	1273 (50)	18.0 (0.5)	−14351 (2002)	9.4 (1.6) × 10^{-10}*	Liu et al. (1998)
$Mg_{1.8}Fe_{0.2}SiO_4$	Polycrystalline	Wd/Rw	1273 (50)	18.0 (0.5)	−14351 (2002)	1.4 (0.3) × 10^{-9}*	Liu et al. (1998) (for reinterpretation of data, see text)
$Mg_{1.8}Fe_{0.2}SiO_4$	Polycrystalline	Wd/Rw	1373 (50)	18.0 (0.5)	−13697 (2036)	2.7 (0.5) × 10^{-8}*	Liu et al. (1998)
$Mg_{1.8}Fe_{0.2}SiO_4$	Polycrystalline	Wd/Rw	1373 (50)	18.0 (0.5)	−13697 (2036)	2.4 (0.6) × 10^{-8}*	Liu et al. (1998) (for reinterpretation of data, see text)
$Mg_{1.8}Fe_{0.2}SiO_4$	Single crystal	Wd	1303 (50)	13.5 (0.5)	−1022 (1489)	4.7 (2.0) × 10^{-8}	Kubo et al. (1998c)
$Mg_{1.8}Fe_{0.2}SiO_4$	Single crystal	Wd	1303 (50)	13.5 (0.5)	−1022 (1489)	2.8 (0.7) × 10^{-8}	Kubo et al. (1998b)
$Mg_{1.8}Fe_{0.2}SiO_4$	Single crystal	Wd	1403 (50)	14.0 (0.5)	−1877 (1486)	5.5 (0.6) × 10^{-8}	Kubo et al. (1998b)
$Mg_{1.8}Fe_{0.2}SiO_4$	Single crystal	Wd	1503 (50)	14.0 (0.5)	−1472 (1503)	8.8 (2.5) × 10^{-8}	Kubo et al. (1998b)
$Mg_{1.8}Fe_{0.2}SiO_4$	Single crystal	Wd	1603 (50)	15.0 (0.5)	−3652 (1477)	1.2 (0.1) × 10^{-7}	Kubo et al. (1998b)
$Mg_{1.8}Fe_{0.2}SiO_4$	Single crystal	Wd	1373 (50)	16.0 (0.5)	−6973 (1405)	1.9 (0.6) × 10^{-8}	Mosenfelder et al. (2000)
$Mg_{1.8}Fe_{0.2}SiO_4$	Single crystal	Wd	1373 (50)	17.0 (0.5)	−9404 (1371)	3.0 (0.7) × 10^{-8}	Mosenfelder et al. (2000)
$Mg_{1.8}Fe_{0.2}SiO_4$	Single crystal	Wd/Rw	1173 (50)	18.0 (0.5)	−14981 (1963)	2.2 (1.1) × 10^{-10}	Liu et al. (1998)
$Mg_{1.8}Fe_{0.2}SiO_4$	Single crystal	Wd/Rw	1273 (50)	18.0 (0.5)	−14351 (2002)	1.2 (0.1) × 10^{-9}	Liu et al. (1998)
$Mg_{1.8}Fe_{0.2}SiO_4$	Single crystal	Wd/Rw	1373 (50)	18.0 (0.5)	−13697 (2036)	1.0 (0.2) × 10^{-8}	Liu et al. (1998)
$Mg_{1.8}Fe_{0.2}SiO_4$	Single crystal	Wd/Rw	1373 (50)	19.0 (0.5)	−17047 (1987)	5.5 (0.6) × 10^{-9}	Liu et al. (1998)
$Mg_{1.8}Fe_{0.2}SiO_4$	Single crystal	Wd/Rw	1373 (50)	20.0 (0.5)	−20353 (1937)	3.3 (0.6) × 10^{-9}	Liu et al. (1998)

[a] Estimated uncertainties are listed in parentheses. Growth rates used for fitting (see Table 2) are marked with asterisks.
[b] Wd: wadsleyite; Rw: ringwoodite.

mocouples (type R) were used, which are believed to suffer a greater effect of pressure on emf (Ohtani et al., 1982; Walter et al., 1995). Because emf pressure corrections are highly uncertain, we estimate a temperature uncertainty of ±50 K for these experiments.

Kubo et al. (1998a) and Kubo (1999) conducted in situ X-ray diffraction experiments on the olivine to wadsleyite transformation in Mg_2SiO_4 to measure the extent of transformation as a function of time. The growth rate of wadsleyite was estimated by fitting the data to the following equation (Cahn, 1956):

$$\xi = 1 - \exp[-2S\dot{x}t] \quad (1)$$

where ξ is the volume fraction of the new phase, S the grain boundary area per unit volume (approximated by $3.35/d$, where d is the olivine grain size), $\dot{x}$ the growth rate, and t the time. This equation is only valid for transformation following nucleation site saturation but can be used in this case because the shape of the transformation–time curves indicates that site saturation occurred early in the experiments (Kubo et al., 1998a; Kubo, 1999). The experiments were conducted using two different types of pressure media, comprised either of boron–epoxy or MgO–ZrO_2 mixtures. Growth rates in experiments performed in the two pressure media at nearly the same temperature differ by approximately one-order of magnitude (Table 1). The discrepancy is likely caused by catalysis of the transformation by hydrogen derived by devolatilization of the boron–epoxy medium as well as insufficient hot-pressing of the starting material (Kubo et al., 1998a).

Growth rates of wadsleyite and ringwoodite during transformation from $Mg_{1.8}Fe_{0.2}SiO_4$ San Carlos olivine have been studied by Kerschhofer et al. (1996, 1998) and Liu et al. (1998) employing two types of starting materials: (1) hot-pressed olivine with a bimodal grain size consisting of a small proportion of 40 μm grains and a larger proportion of 10 μm grains; (2) a composite starting material consisting of a 500 μm × 600 μm × 700 μm olivine single crystal contained in a fine-grained matrix as described under (1). In both case, the samples were hot-pressed for 3 h in the olivine-stability field, at 1473 K and 11 GPa, prior to transformation. Reaction in the fine-grained polycrystalline olivine proceeded by incoherent grain boundary nucleation and interface-controlled growth, whereas the single crystals transformed by a combination of grain boundary and intracrystalline nucleation and growth. Both wadsleyite and ringwoodite formed at the conditions of the experiments with similar growth rates. Growth rates for the grain boundary nucleation and growth process were determined by measuring the width of reaction rims around both the single crystals and larger olivine grains in the matrix, using optical and transmission electron microscopy (Table 1; see Liu et al., 1998, for details). Liu et al. (1998) assumed that nucleation effectively occurred instantaneously at the start of the experiments and growth rates were estimated by dividing the rim thickness by the experimental duration. Examination of the data for rims around the finer-grained olivine crystals, however, suggests that there was an incubation period for nucleation in the experiments at 1173 and 1273 K (Fig. 2). Therefore, we have re-analyzed the results at these two temperatures by fitting straight lines through the data points (Fig. 2) to determine the growth rate (the intercept on the time axis thus defines the "incubation period"). At 1373 K, there is no apparent nucleation barrier and the growth rate is determined by regressing a line through the data, forced to fit through the origin. We believe this new estimation of the growth rates to be more accurate than that of Liu et al. (1998), but we consider both possibilities in our overall analysis (see below).

Kubo et al. (1998b,c) studied the transformation of 1 mm^3 single crystals of $Mg_{1.8}Fe_{0.2}SiO_4$ olivine to wadsleyite. The single crystals were contained

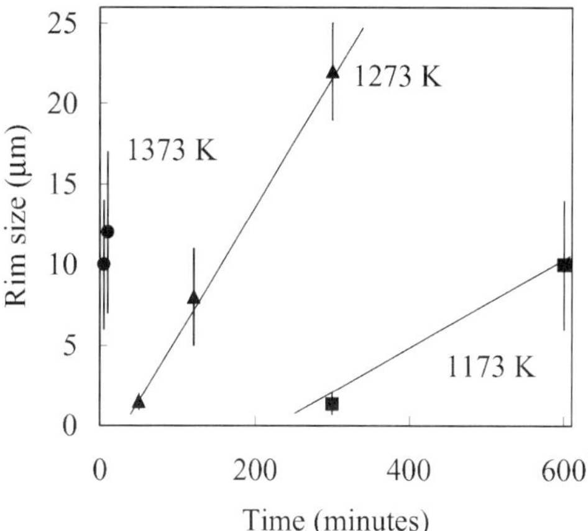

Fig. 2. Growth rates of wadsleyite/ringwoodite at 18 GPa in fine-grained $Mg_{1.8}Fe_{0.2}SiO_4$ polycrystalline olivine (Kerschhofer et al., 1996, 1998; Liu et al., 1998) at 1173 K, squares; 1273 K, triangles and 1373 K, circles. Linear fits to the data at 1173 and 1273 K suggest an incubation period for nucleation at those temperatures.

in a matrix of NaCl (to provide a quasi-hydrostatic pressure environment) and the experiments were performed under both "nominally anhydrous" (Kubo et al., 1998b) and "wet" (Kubo et al., 1998c) conditions, by adding brucite as a water (and hydrogen) source. The FTIR measurements showed that wadsleyite reaction rims that developed around the olivine crystals under "dry" conditions contained approximately 0.02 wt.% H_2O, after quenching. In the wet experiments, the wadsleyite rims contained between 0.05 and 0.21 wt.% H_2O, depending on the amount of brucite added. Growth rates in the "dry" experiments and "wet" experiments with 0.05 wt.% H_2O decreased with time, consistent with inhibition of growth by transformation stress as discussed in Section 2.3. Furthermore, reaction rims were observed in "zero-time" experiments (i.e. experiments in which temperature was increased to the desired value and then immediately quenched). By correcting for these two phenomena, the initial growth rate (i.e. prior to when transformation stress becomes significant) can be estimated by subtracting the "zero time" growth distance from the growth distance of the next shortest duration experiments (Table 1).

Mosenfelder et al. (2000) studied the transformation to wadsleyite using 500 μm diameter spheres of San

Carlos olivine. Experiments were conducted at 1373 K and 16 or 17 GPa, using quasi-hydrostatic pressure media (NaCl or Au). The growth rate of wadsleyite reaction rims that formed around the spherical single crystals also decreased with time due to the effect of transformation stress. Initial growth rates can be estimated from reaction rim measurements in the shortest duration experiments (15 min) at 16 and 17 GPa (Table 1). Small amounts of hydrogen may have infiltrated into these samples during transformation, as in the experiments of Kubo et al. (1998b,c). However, hydrogen could not be detected in the recovered samples (Mosenfelder et al., 2000).

For the purpose of fitting a rate equation to the growth-rate data, in order to model transformation kinetics in subduction zones, we make two assumptions. First, we assume that growth kinetics in the Mg_2SiO_4 and $Mg_{1.8}Fe_{0.2}SiO_4$ systems are similar. Although normalization of the data by homologous temperatures could be employed (cf. Rubie and Ross, 1994), we consider the utility of this approach to be limited by the proximity of melting temperatures in the two compositions (~100 K difference; Bowen and Schairer, 1935) and uncertainties in the growth rate data. Furthermore, although the presence of Fe substantially changes the point defect chemistry of olivine, it is less likely to be an important factor in interface-controlled growth where reaction is rate-limited by the breaking of Si–O bonds. The second assumption is that growth rate parameters are similar for wadsleyite and ringwoodite. This assumption was made previously by Rubie and Ross (1994), based on observations of similar growth distances in the experiments of Brearley et al. (1992). It may also be justified by the similarity in crystal structures between the two high-pressure phases, which implies that rates of interface diffusion should be similar in both cases.

The presence of relatively large amounts of hydrogen in the experiments of Kubo et al. (1998a,b,c) complicates comparison with the other experimental studies discussed here. Furthermore, estimates of initial growth rates in the experiments on olivine single crystals (Kubo et al., 1998b,c; Liu et al., 1998; Mosenfelder et al., 2000) may have been affected by transformation stress. Therefore, in order to derive activation energy parameters for growth, we exclude these data and use only the data point of Brearley et al. (1992), the two in situ measurements (under "dry" conditions) of Kubo (1999), and the three measurements of growth rates in fine-grained olivine presented in Liu et al. (1998), with and without the alternate interpretation regarding delayed nucleation as discussed above. The data have been fitted to the following equation for interface-controlled growth (Turnbull, 1956):

$$\dot{x} = k_0 T \exp\left\{-\frac{\Delta H_a + PV^*}{RT}\right\}\left\{1 - \exp\frac{\Delta G_{rxn}}{RT}\right\} \quad (2)$$

where k_0 is a constant, T the absolute temperature, ΔH_a the activation enthalpy for growth, P the pressure, V^* an activation volume, R the gas constant, and ΔG_{rxn} the free energy of reaction. For calculation of ΔG_{rxn}, we used the thermodynamic data of Akaogi et al. (1989). There are large uncertainties in the phase diagram and thermodynamic properties of the phases, even for the end-member compositions. For the Mg_2SiO_4 system, we took into account the redetermination of the α–β-phase boundary for Mg_2SiO_4 (Morishima et al., 1994) by adjusting the nominal pressure of the experiments (according to the new phase diagram) for the purpose of calculating ΔG_{rxn}.

Fig. 3 shows the fitted data on an Arrhenius plot, where $\dot{x}'$ is essentially the growth rate normalized to the free energy of the reaction (see Rubie and Ross, 1994). In order to fit Eq. (2) to these data, an estimate of the activation volume is required. The activation volume for processes controlled by diffusion across interphase boundaries is poorly constrained. Rubie and Ross (1994) fitted growth-rate data using a model suggested by O'Connell (1977) (Eq. (8) in Rubie and Ross, 1994) in which V^* varies with pressure, from approximately 4.3 $cm^3\,mol^{-1}$ at 14 GPa to 3.4 $cm^3\,mol^{-1}$ at 20 GPa. It has also been proposed that the activation volume for interface-controlled growth may be close to zero. Burnley (1995) estimated a value for V^* of $0 \pm 2\,cm^3\,mol^{-1}$ based on experimentally determined growth rates during the α–γ transformation in Mg_2GeO_4 at 1–2 and 15.5 GPa. However, because of the difficulty in comparing experiments conducted in two different types of apparatus, large temperature uncertainties, and assumptions about the nucleation behavior of the high-pressure phases, this estimate must be regarded with caution. More recently, Kubo et al. (1998a) constrained the value of V^* to lie between 0 and 4 $cm^3\,mol^{-1}$ on the basis of the in situ growth rate determinations

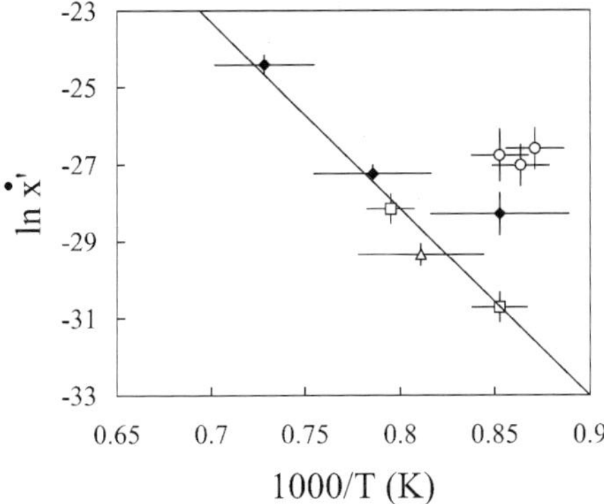

Fig. 3. Growth rate data selected for the determination of growth rate activation energy parameters. $\dot{x}'$ is defined as $\dot{x}/\{T[1-\exp(\Delta G_{rxn}/RT)]\}$ and is the growth rate normalized by the magnitude of the free energy of reaction. Data sources: diamonds, Liu et al. (1998), with reinterpretation based on delayed nucleation (Fig. 2, see also text); squares, in situ experiments of Kubo (1999); triangle, Brearley et al. (1992). Also shown (circles) are the in situ "wet" results of Kubo et al. (1998a), obtained using boron–epoxy pressure medium, showing a clear discrepancy from the in situ experiments conducted under relatively dry conditions. The line shows the fit to data (excluding the "wet" in situ experiments) corresponding to kinetic model 2 in Table 2; fits corresponding to the other kinetic models are non-linear. Compositions: filled symbols, $Mg_{1.8}Fe_{0.2}SiO_4$; open symbols, Mg_2SiO_4.

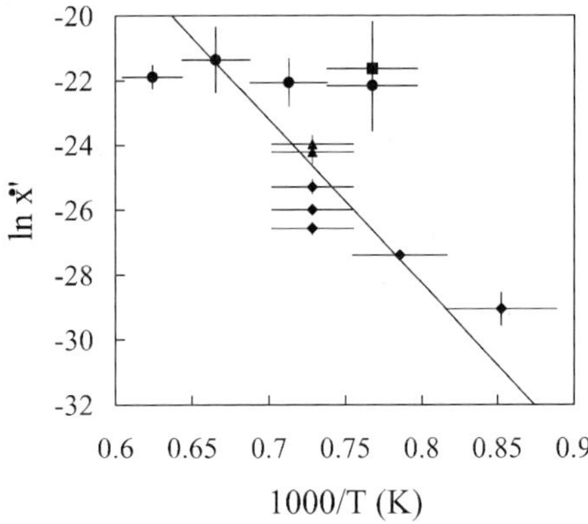

Fig. 4. Growth rates obtained from reaction rims around olivine single crystals, compared to the fit to the polycrystalline data. The line is the same fit shown in Fig. 3 and is not a fit to the data shown here. Circles, "nominally dry" experiments from Kubo et al. (1998b); square, wet experiment of Kubo et al. (1998c); triangles, sphere experiments from Mosenfelder et al. (2000); diamonds, single crystal data from Liu et al. (1998).

discussed above. The upper limit determined by Kubo et al. (1998a) is consistent with the model of Rubie and Ross (1994).

Because of the uncertainties discussed above, we have used two models for V^* in this study: (1) the empirical model of Rubie and Ross (1994), and (2) an assumption of $V^* = 0$ at all pressures. The fitted parameters in Eq. (2) are thus k_0 and ΔH_a. We have also fitted two sets of growth rates, with and without the alternate interpretation of the data of Liu et al. (1998) regarding delayed nucleation, thus resulting in four distinct models. Fitting was conducted using non-linear least-squares regression via the refinement algorithm of Deming (1964). Note that the in situ "wet" experiments of Kubo et al. (1998a), shown as open circles in Fig. 3, were not included in the fits. The fit based on the reinterpreted data of Liu et al. (1998) and an activation volume of zero is shown in Fig. 3. Refined values for the parameters for all four fits are listed in Table 2. A similar analysis was also performed by Kubo (1999) but without estimating uncertainties.

Growth rates derived from the single crystal data are shown in Fig. 4 together with the fit of Fig. 3. The single crystal data from Kubo et al. (1998b,c) deviate significantly from the regression line for the polycrystalline data, defining a much lower activation energy (Fig. 4). This is most likely a result of the

Table 2
Refined values for parameters in growth rate equation (Eq. (2))[a]

Model	$\ln k_0$	ΔH_a (kJ mol^{-1})	Assumptions
1	13.421 (9.346)	369 (96)	Delayed nucleation, V^* from Rubie and Ross (1994)[b]
2	10.725 (8.145)	404 (84)	Delayed nucleation, $V^* = 0$[b]
3	11.462 (8.882)	348 (91)	No nucleation delay, V^* from Rubie and Ross (1994)[b]
4	9.609 (6.638)	392 (68)	No nucleation delay[b], $V^* = 0$

[a] Uncertainties listed in parentheses.
[b] The nucleation models refer to different interpretations of the results of Liu et al. (1998), as discussed in the text.

effect of hydrogen on growth rates. In contrast, the single crystal data of Mosenfelder et al. (2000) and Liu et al. (1998) lie closer to the regression line for the polycrystalline data. Deviations from the regression line for the latter two data sets may be related to the transformation stress effect discussed above, the presence of hydrogen, and/or the effect of activation volume on growth rates.

3.2. Kinetics of intracrystalline transformation

The mechanisms and kinetics of intracrystalline transformation of $Mg_{1.8}Fe_{0.2}SiO_4$ olivine to wadsleyite and ringwoodite have been studied as a function of time at 1173–1373 K and 18–20 GPa (Kerschhofer et al., 2000). In particular, the rates of intracrystalline nucleation and growth of semi-coherent ringwoodite platelets which formed on $(1\,0\,0)_\alpha$ stacking faults in olivine were determined. The rates of semicoherent growth were found to be one to three-orders of magnitude slower than the rates of incoherent growth (see also Dupas-Bruzek et al., 1998), even though the activation energies for these processes are similar. Extrapolation of these results implies that semi-coherent growth during intracrystalline transformation is unlikely to be an important contributor to overall transformation rates in subduction zones. However, by creating new nucleation sites in olivine for incoherent nucleation and growth of the high-pressure phases, intracrystalline transformation may have a large indirect effect on transformation kinetics. The increase in nucleation sites dramatically decreases the growth distance required for transformation to proceed to completion. This effect is considered further in Section 4.3.

4. Transformation rates in subducting slabs

In extrapolating the experimental kinetic data to predict the extent of olivine metastability in subducting slabs, we have followed several previous studies with regard to calculating the thermal structure of the subducting lithosphere and assumptions about the thermodynamics and kinetics of the transformation. Two-dimensional thermal structures were computed using a finite difference algorithm, with initial lithospheric temperatures from the GDH1 plate model (Stein and Stein, 1992; see Marton et al., 1999, for details). Latent heat feedback was incorporated into the thermal model as well as the kinetics, in contrast to the study of Kirby et al. (1996) in which latent heat was only incorporated into the kinetics. Latent heat was calculated assuming 60% modal olivine, with transformations in pyroxenes not being taken into account. As in previous studies, the thermodynamics of the transformation were simplified by fixing the mantle to a composition of $Mg_{1.8}Fe_{0.2}SiO_4$, i.e. transformation in two-phase stability fields was not considered (Rubie and Ross, 1994; Kirby et al., 1996).

4.1. Grain boundary nucleation and interface-controlled growth

As a first approximation, we consider only grain boundary nucleation and interface-controlled growth and ignore the effects of intracrystalline transformation and transformation stress, which are considered in Sections 4.2 and 4.3. This approach enables our results to be compared directly to those of previous studies (e.g. Kirby et al., 1996; Devaux et al., 1997). An initial olivine grain size of 5 mm was used for most calculations (cf. Rubie and Ross, 1994); the effect of varying the grain size is considered in Section 4.2. We assumed that nucleation kinetics are fast under all *P–T* conditions of interest, such that reaction is controlled by growth kinetics (Rubie and Ross, 1994; Kirby et al., 1996). Extent of transformation is thus determined by modifying Eq. (1) to evaluate non-isobaric, non-isothermal transformation (Kirby et al., 1996; Marton et al., 1999):

$$\xi = 1 - \exp\left[-2S\int_0^t \dot{x}(\theta)\mathrm{d}\theta\right] \tag{3}$$

where $\dot{x}(\theta)$ is the growth rate calculated at each time step θ.

Calculations were performed for model slabs by varying either convergence rate (v) or lithospheric age such that the thermal parameter φ (vertical descent rate × age of lithosphere at the trench; see Kirby et al., 1996) varied from 520 to 16,974 km. As shown by previous studies, the depth of maximum olivine metastability generally increases progressively as φ increases (Fig. 5). Two of these models were calculated using parameters corresponding to slabs C and D of Kirby et al. (1996) coupled to kinetic models 1–4 (Tables 2 and 3). Slab C has $v = 8$ cm per year,

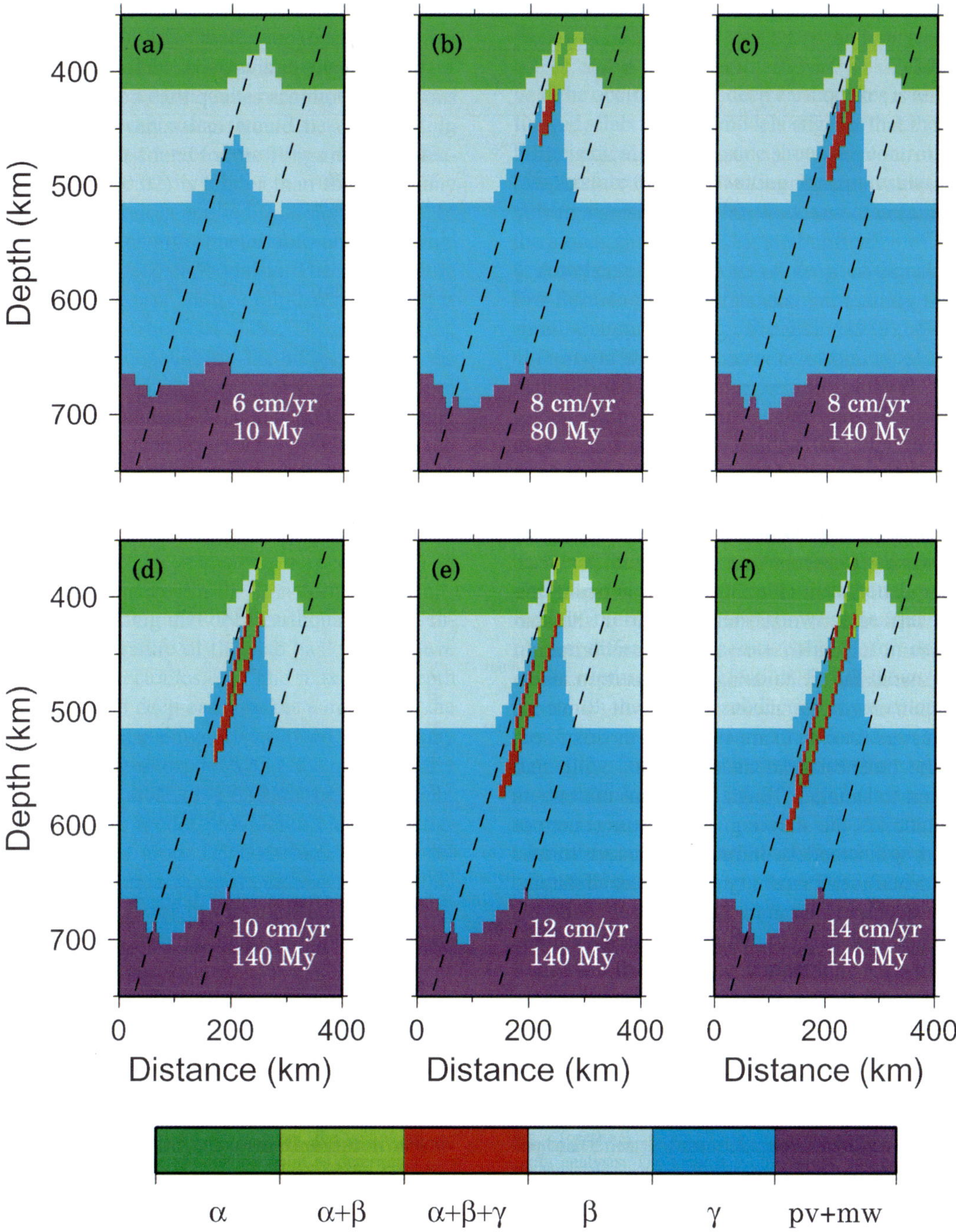

Fig. 5. Thermo-kinetic model results for slabs with varying convergence rates and ages, as shown on the figure. Colors represent phase fields for olivine (α), wadsleyite (β), ringwoodite (γ), and (Mg, Fe)SiO_3 perovskite + magnesiowüstite (pv + mw). All of these models were calculated using the growth rate parameters of kinetic model 1 (see Table 2), (a) Young, warm slab ($\varphi = 520$ km) in which growth rates are fast enough for transformation to occur close to the equilibrium phase boundary; (b–f) successively colder slabs ($\varphi = 5543$, 9699, 12124, 14549, and 16974 km, respectively) showing increasingly greater depth of olivine metastability. Parts (c) and (f) show slabs with parameters corresponding to slabs C and D, respectively, of Kirby et al. (1996), as discussed in the text.

Table 3
Thermo-kinetic model results for slabs C and D[a]

Kinetic model (see Table 2)	Depth to 1% ξ (km)	Depth to 99% ξ (km)	Area of wedge (km^2)
Slab C			
1	464 (412–520)	500 (449–540)	4204 (2496–5783)
2	447 (412–500)	485 (440–520)	3847 (2349–5167)
3	446 (412–509)	489 (439–530)	3765 (2208–5381)
4	438 (412–482)	480 (440–510)	3608 (2414–4758)
Slab D			
1	585 (496–640)	600 (540–650)	7555 (5075–9914)
2	552 (478–600)	574 (528–614)	6612 (4800–8571)
3	564 (474–620)	586 (529–630)	6805 (4550–9074)
4	541 (481–590)	568 (529–600)	6350 (4866–7895)

[a] Depths are maximum depth to 1 and 99% transformation. Uncertainties are determined by propagating the uncertainties in growth rate parameters (see Table 2). Ranges of uncertainties are given by the upper and lower limiting values shown in parentheses.

age = 140 million years, and a dip angle of 60°, corresponding to φ = 9699 km (Fig. 5c). Slab D is identical to slab C except that v = 14 cm per year, yielding φ = 16974 km (Fig. 5f). Uncertainties in the cross-sectional area of metastable olivine in the slab and maximum depths to 1 and 99% transformation were calculated by propagating the uncertainties in the activation energy parameters for growth (Table 3). The parameters for slab C are similar to those of sections of the Indonesian subduction zone, while slab D is chosen to simulate Tonga. Note that the effects of deformation of the slab (e.g. slab bending) are not considered in our models. In order to explore further the dependence of metastability on the thermal parameter of the slab, we also ran models for groups of slabs with dip angles of 60° in which either lithospheric age or convergence rate was systematically varied. Fig. 6 shows the depth to the tip of the metastable wedge (99% transformation contour) as a function of thermal parameter for three different groups of slabs: Group 1, with v = 8 cm per year, age = 50–170 million years; Group 2, with v = 6–20 cm per year, age = 70 million years; and Group 3, with v = 6–20 cm per year, age = 100 million years.

Our results are consistent with previous models (Kirby et al., 1996; Däßler et al., 1996; Devaux et al., 1997) in predicting that a significant metastable wedge can form in slabs with thermal parameters greater than about 5000 km. However, our models predict significantly smaller wedges of metastable olivine than Kirby et al. (1996), because (1) our new growth rate parameters result in growth rates that are faster by about a factor of 5 compared with those of the previous model (Rubie and Ross, 1994) and (2) we have taken latent heat feedback into the thermal model into account. The new results show that, regardless of the kinetic model used, a metastable

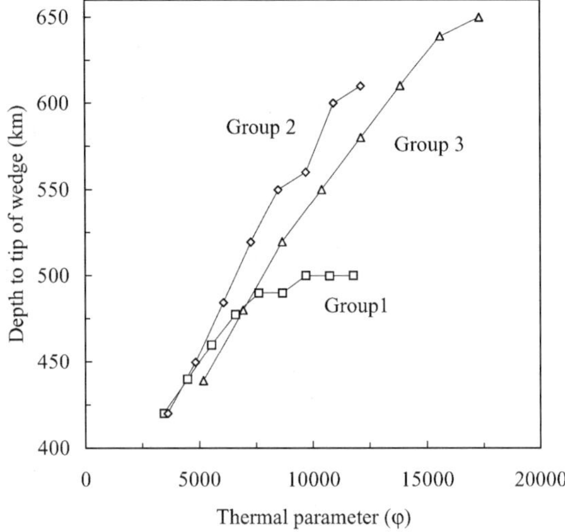

Fig. 6. Maximum depth of olivine metastability (depth to 99% transformation) as a function of thermal parameter for three groups of slabs. All models were calculated using the model 1 kinetic parameters (see Table 2). Squares (Group 1): convergence rate (v): 8 cm per year, lithospheric age: 50–170 million years; diamonds (Group 2): v: 6–20 cm per year, age: 70 million years; triangles (Group 3): v: 6–20 cm year, age: 100 million years.

olivine wedge should persist in slabs with high thermal parameters, but the maximum depth of metastability is reduced substantially compared with previous estimates, consistent with the conclusions of Devaux et al. (1997). The maximum depth of olivine metastability is further reduced by 20–30 km if an activation volume of zero is used, as in our kinetic models 2 and 4. Finally, the depth to the tip of the metastable olivine wedge reaches a maximum of about 500 km for group 1 slabs, which have constant velocity but different lithospheric ages (Fig. 6). This reflects the fact that the initial thermal structure reaches a steady state (when using a plate model) for old slabs (cf. Marton et al., 1999).

4.2. Intracrystalline transformation

Although uncertainties in the kinetics of intracrystalline transformation are large, the effect on overall transformation rates in subducting slabs can be modeled semi-quantitatively using some simple assumptions. The most critical factor concerns the nucleation kinetics of the ringwoodite platelets upon which further incoherent nucleation and growth occurs. On experimental timescales, a pressure of at least 18 GPa is required at 1173 K for intracrystalline ringwoodite platelets to develop (Kerschhofer et al., 2000). This corresponds to a ΔG_{rxn} of about −14 kJ/mol. Because nucleation kinetics are much more sensitive to ΔG_{rxn} than temperature (e.g. Rubie and Thompson, 1985), we can use this value to estimate the approximate pressure at which nucleation kinetics become significant at lower temperatures. At 873 K, for instance, the same ΔG_{rxn} is achieved at a pressure of about 17 GPa (~500 km depth). This is roughly consistent with the theoretical model of Liu and Yund (1995), which predicts that rates of coherent homogeneous nucleation become high at a depth of ~450 km in a cold slab. Furthermore, the extrapolation of Kerschhofer et al. (2000) gives a nucleation rate of $5 \times 10^8\,m^{-3}\,s^{-1}$ at this temperature, which is close to the value of $\sim 1 \times 10^9\,m^{-3}\,s^{-1}$ predicted by Liu and Yund (1995).

The above calculations suggest that the nucleation rate of coherent ringwoodite platelets, the first step in the intracrystalline transformation mechanism, reaches a high value (e.g. $10^9\,m^{-3}\,s^{-1}$) at a depth of 450–500 km. The effect on overall transformation rates can now be estimated by calculating

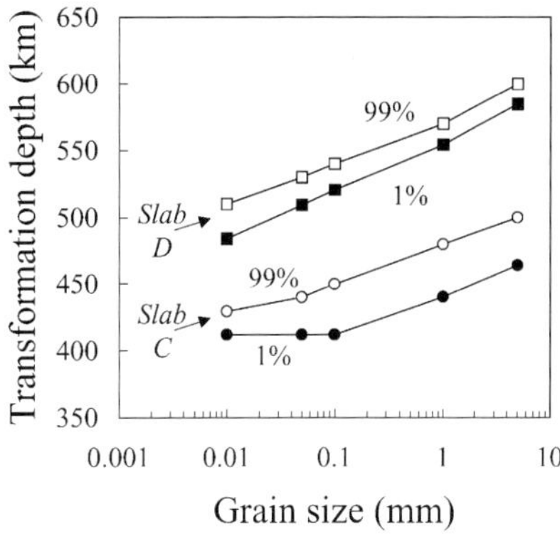

Fig. 7. Effect of olivine grain size on the calculated depth of olivine metastability. All models were calculated using kinetic model 1 (Table 2). Circles: slab C (φ = 9699 km); squares: slab D (φ = 16974 km). Closed symbols: depth to 1% transformation; open symbols: depth to 99% transformation. See text for more explanation.

the additional grain boundary area (provided by the ringwoodite platelets) that becomes available for incoherent nucleation, converting this value to an effective grain size d (Eq. (3)), and incorporating it into the thermo-kinetic model (Fig. 7). The extrapolation of Kerschhofer et al. (2000) indicates that a grain boundary area of $67{,}000\,m^{-1}$ can develop in 27 years (corresponding to a subduction distance of only 2–4 m). This is equivalent to a grain size of ~50 μm. The effect on overall transformation is demonstrated in Fig. 7, which shows the depth of transformation for slabs C and D for a range of grain sizes. For a grain size of 50 μm the depth to 99% transformation for slab D is ~530 km. Therefore, as a preliminary estimate, we conclude that intracrystalline transformation can cause transformation to reach completion at depths of 500–550 km even in the interior of a cold slab such as Tonga.

4.3. Effect of transformation stress

Although inhibition of growth by transformation stress in experiments has been quantitatively modeled (Liu et al., 1998; Mosenfelder et al., 2000), extrapolation to lower temperatures of subduction zones requires rheological flow laws for the high-pressure phases that are currently not available. Mosenfelder et al. (2000) presented a preliminary extrapolation of their model for ringwoodite rims growing around olivine with an initial grain size of 5 mm, assuming different values for the "yield strength" of ringwoodite. Note that this model is formulated assuming perfect plasticity, i.e. strength is independent of strain rate and temperature. This is a reasonable approximation in the case of low-temperature plasticity, the likely deformation mechanism under these conditions (e.g. Karato, 1997). At 873 K and 15 GPa, the extrapolation (Fig. 8 in Mosenfelder et al., 2000) indicates that the transformation can effectively cease after the rim has reached a thickness of between ∼1 mm (for a yield strength of 5 GPa) and 2 mm (yield strength of 3 GPa), corresponding to ∼78 and 99% transformation, respectively. This suggests that inhibition of growth by transformation stress can broaden the depth interval of transformation and possibly allow a some amount of metastable olivine to persist throughout the transition zone. However, if intracrystalline transformation occurs, as discussed above, this phenomenon is unlikely to be important because the growth rate of isolated grains or domains does not decrease with time (Liu et al., 1998); this conclusion is also supported by experimental data (Kerschhofer et al., 1998, Fig. 5b).

5. Discussion

Our analysis of recent experimental growth rate data for wadsleyite and ringwoodite with compositions applicable to the mantle yields activation enthalpies for growth in the range of 350–400 kJ mol^{-1}. Although there are still uncertainties in the growth kinetics, modeling using the newly derived parameters confirms the results of previous studies in predicting the occurrence of a metastable olivine wedge in cold subducting slabs ($\varphi \gtrsim 5000$ km). However, an important difference between our results and those of some previous models is that the depth of transformation is significantly reduced using the new kinetic parameters and when latent heat is fed back into both the thermal model and kinetics.

Our new results have important implications for the viability of the transformational faulting hypothesis for deep-focus earthquakes (Kirby, 1987; Green and Burnley, 1989; Kirby et al., 1996). Our results suggest that a wedge of metastable olivine can only persist to the maximum observed depths of deep-focus earthquakes in slabs with unusually high thermal parameters. The 1994 Bolivian earthquake, which occurred at a depth of about 630 km in the subducting Nazca plate (with a thermal parameter of 5000 km or less; see Kirby et al., 1996), is difficult to explain using the transformational faulting hypothesis because the maximum depth of olivine metastability according to our models is ∼450 km (Fig. 6). Likewise, the occurrence of earthquakes at depths of 680 km in the Indonesian and Marianas subduction zones, which are warmer than Tonga, is also difficult to explain by transformational faulting. However, our models do not take complicated slab geometries into account, which would result in more complicated thermal structures. In our thermal models, we have assumed that thermal diffusivity is constant (8.04×10^{-7} $m^2 s^{-1}$) and does not vary with mineralogy, temperature or pressure. Recent results (Hofmeister, 1999; Xu, pers. commun.) show that wadsleyite and ringwoodite have higher thermal diffusivities than olivine. This could further reduce the extent of olivine metastability as a result of slab warming by thermal conduction (Hauck et al., 1999).

The recent studies of reaction mechanisms reviewed in this paper reveal a number of complexities in the transformation behavior of olivine that have not previously been included in thermo-kinetic models. Most importantly, intracrystalline transformation can significantly reduce the maximum depth of metastability and affect the depth interval over which transformation occurs. If this mechanism operates, the effect of transformation stress in inhibiting transformation is likely to be negligible. However, more data are needed to confirm the relative importance of these mechanisms. In particular, the dependence of the rates of nucleation of ringwoodite platelets on P, T and differential stress, and the rheology of the high-pressure phases of olivine need to be better understood. Another factor that may be important is the presence of fluids, either incorporated in olivine or derived by dehydration of hydrous phases in hydrothermally altered mantle lithosphere (see, e.g. Angel et al., 2001). If hydrous phases are inhomogeneously distributed in the slab, dehydration may result in rapid transformation in localized regions.

Acknowledgements

We thank S. Karato and an anonymous referee for helpful reviews that helped to greatly improve the manuscript, E. Ohtani for discussion and T. Kubo for generously providing a copy of his Ph.D. thesis.

References

Akaogi, M., Ito, E., Navrotsky, A., 1989. Olivine-modfied spinel-spinel transitions in the system Mg_2SiO_4–Fe_2SiO_4: calorimetric measurements, thermochemical calculation, and geophysical application. J. Geophys. Res. 94, 15671–15685.

Angel, R.J., Frost, D.J., Ross, N.L., Hemley, R., 2001. Stabilities and EoS of dense hydrous magnesium silicates. Phys. Earth Planet. Inter. 127, 179–194.

Bina, C.R., 1996. Phase transition buoyancy contributions to stresses in subducting lithosphere. Geophys. Res. Lett. 23, 3563–3566.

Bina, C.R., 1997. Patterns of deep seismicity reflect buoyancy stresses due to phase transitions. Geophys. Res. Lett. 24, 3301–3304.

Bina, C.R., Stein, S., Marton, F.C., Van Ark, E.M., 2001. Implications of slab mineralogy for subduction dynamics. Phys. Earth Planet. Inter. 127, 51–66.

Bowen, N.L., Schairer, J.F., 1935. The system MgO–FeO–SiO_2. Am. J. Sci. 29, 151–217.

Brearley, A.J., Rubie, D.C., 1994. Transformation mechanisms of San Carlos olivine to (Mg, $Fe)_2SiO_4$ β-phase under subduction zone conditions. Phys. Earth Planet. Inter. 86, 45–67.

Brearley, A.J., Rubie, D.C., Ito, E., 1992. Mechanisms of the transformations between the alpha, beta and gamma polymorphs of Mg_2SiO_4 at 15 GPa. Phys. Chem. Miner. 18, 343–358.

Burnley, P.C., 1995. The fate of olivine in subducting slabs: a reconnaissance study. Am. Mineral. 80, 1293–1301.

Cahn, J.W., 1956. The kinetics of grain boundary nucleated reactions. Acta Metall. 4, 449–459.

Chen, J., Inoue, T., Weidner, D.J., Wu, Y., Vaughan, M., 1998. Strength and water weakening of mantle minerals, olivine, wadsleyite, and ringwoodite. Geophys. Res. Lett. 25, 575–578.

Collier, J.D., Hellfrich, G.R., 1997. Topography of the 410 and 660 km seismic discontinuities in the Izu-Bonin subduction zone. Geophys. Res. Lett. 24, 1535–1538.

Däßler, R., Yuen, D.A., Karato, S., Riedel, M.R., 1996. Two-dimensional thermo-kinetic model for the olivine-spinel phase transition in subducting slabs. Phys. Earth Planet. Inter. 94, 217–239.

Deming, W.E., 1964. Statistical adjustment of data, Dover, New York, p. 261.

Devaux, J.P., Schubert, G., Anderson, C., 1997. Formation of a metastable olivine wedge in a subducting slab. J. Geophys. Res. 102, 24627–24637.

Dupas-Bruzek, C., Sharp, T.G., Rubie, D.C., Durham, W.B., 1998. Mechanisms of transformation and deformation in $Mg_{1.8}Fe_{0.2}SiO_4$ olivine and wadsleyite under non-hydrostatic stress. Phys. Earth Planet. Inter. 108, 33–48.

Goto, K., Suzuki, Z., Hamaguchi, H., 1987. Stress distribution due to olivine-spinel phase transition in descending plate and deep-focus earthquakes. J. Geophys. Res. 92, 13811–13820.

Green, H.W., Burnley, P.C., 1989. A new self-organizing mechanism for deep-focus earthquakes. Nature 341, 733–737.

Green, H.W., Young, T.E., Walker, D., Scholz, C.H., 1992. The effect of nonhydrostatic stress on the $\alpha \rightarrow \beta$ and $\alpha \rightarrow \gamma$ olivine phase transformations. In: Syono, Y., Manghnani, M.H. (Eds.), High Pressure Research: Application to Earth and Planetary Sciences, American Geophysical Union, Washington, pp. 229–235.

Hauck II, S.A., Philips, R.J., Hofmeister, A.M., 1999. Variable conductivity: effects on the thermal structure of subducting slabs. Geophys. Res. Lett. 26, 3257–3260.

Hofmeister, A.M., 1999. Mantle values of thermal conductivity and the geotherm from phonon lifetimes. Science 283, 1699–1706.

Hornstra, J., 1960. Dislocations, stacking faults and twins in the spinel structure. J. Phys. Chem. Sol. 15, 311–323.

Iidaka, T., Suetsugu, D., 1992. Seismological evidence for metastable olivine inside a subducting slab. Nature 356, 593–595.

Karato, S., 1997. Phase transformations and rheological properties of mantle minerals. In: Crossley, D.J. (Ed.), Earth's Deep Interior, Vol. 7, Gordon and Breach, Amsterdam, pp. 223–272.

Karato, S., Riedel, M.R., Yuen, D.A., 2001. Rheological structure and deformation of subducted slabs in the mantle transition zone: implications for mantle circulation and deep earthquakes 127, 83–108.

Kerschhofer, L., Sharp, T.G., Rubie, D.C., 1996. Intracrystalline transformation of olivine to wadsleyite and ringwoodite under subduction zone conditions. Science 274, 79–81.

Kerschhofer, L., Dupas, C., Liu, M., Sharp, T.G., Durham, W.B., Rubie, D.C., 1998. Polymorphic transformations between olivine, wadsleyite and ringwoodite: mechanisms of intracrystalline nucleation and the role of elastic strain. Mineral. Mag. 62, 617–638.

Kerschhofer, L., Rubie, D.C., Sharp, T.G., McConnell, J.D.C., Dupas-Bruzek, C., 2000. Kinetics of intracrystalline olivine-ringwoodite transformation. Phys. Earth Planet. Inter. 121, 59–76.

Kirby, S.H., 1987. Localized polymorphic phase transformations in high-pressure faults and applications to the physical mechanism of deep earthquakes. J. Geophys. Res. 92, 13789–13800.

Kirby, S.H., Stein, S., Okal, E.A., Rubie, D.C., 1996. Metastable mantle phase transformations and deep earthquakes in subducting lithosphere. Rev. Geophys. 34, 261–306.

Kohlstedt, D.L., Keppler, H., Rubie, D.C., 1996. Solubility of water in the α-, β-, and γ-phases of (Mg, $Fe)_2SiO_4$. Contrib. Mineral. Petrol. 123, 345–357.

Koper, K.D., Wiens, D.A., 2000. The waveguide effect of metastable olivine in slabs. Geophys. Res. Lett. 27, 573–576.

Koper, K.D., Wiens, D.A., Dorman, L.M., Hildebrand, J.A., Webb, S.C., 1998. Modeling the Tonga slab: can travel time data resolve a metastable olivine wedge? J. Geophys. Res. 103, 30079–30100.

Kubo, T., 1999. Transformation kinetics in the earth's interior. Rev. High Press. Sci. Technol. 9, 26–33.

Kubo, T., Ohtani, E., Kato, T., Morishima, H., Yamazaki, D., Suzuki, A., Mibe, K., Kikegawa, T., Shimomura, O., 1998a. An in situ X-ray diffraction study of the α–β transformation kinetics of Mg_2SiO_4. Geophys. Res. Lett. 25, 695–698.

Kubo, T., Ohtani, E., Kato, T., Shinmei, T., Fujino, K., 1998b. Experimental investigation of the α–β transformation of San Carlos olivine single crystal. Phys. Chem. Mineral. 26, 1–6.

Kubo, T., Ohtani, E., Fujino, K., 1998c. Effects of water on the α–β transformation kinetics in San Carlos olivine. Science 281, 85–87.

Liu, M., Yund, R.A., 1995. The elastic strain energy associated with the olivine-spinel transformation and its implications. Phys. Earth Planet. Inter. 89, 177–197.

Liu, M., Kerschhofer, L., Mosenfelder, J., Rubie, D.C., 1998. The effect of strain energy on growth rates during the olivine-spinel transformation and implications for olivine metastability in subducting slabs. J. Geophys. Res. 103, 23897–23909.

Martinez, I., Wang, Y., Guyot, F., Liebermann, R.C., 1997. Microstructures and iron partitioning in (Mg, Fe)SiO_3 perovskite and (Mg, Fe)O assemblages: an analytical transmission electron microscope study. J. Geophys. Res. 102, 5265–5280.

Marton, F.C., Bina, C.R., Stein, S., Rubie, D.C., 1999. Effects of slab mineralogy on subduction rates. Geophys. Res. Lett. 26, 119–122.

Morishima, H., Kato, T., Suto, M., Ohtani, E., Urukawa, S., Utsumi, W., Shimomura, O., 1994. The phase boundary between α- and β-Mg_2SiO_4 determined by in situ X-ray diffraction. Science 265, 1202–1203.

Morris, S., 1992. Stress relief during solid-state transformations in minerals. Proc. R. Soc. Lond. Ser. A. 436, 203–216.

Mosenfelder, J.L., Connolly, J.A.D., Rubie, D.C., Liu, M., 2000. Strength of $(Mg, Fe)_2SiO_4$ wadsleyite determined by relaxation of transformation stress. Phys. Earth Planet. Inter. 120, 63–78.

O'Connell, R.J., 1977. On the scale of mantle convection. Tectonophysics 38, 119–136.

Ohtani, E., Kumazawa, M., Kato, T., Irifune, T., 1982. Melting of various silicates at elevated pressures. In: Akimoto, S., Manghnani, M.H. (Eds.), High Pressure Research in Geophysics. D. Reidel, Dordrecht, pp. 259–270.

Poirier, J.P., 1981. On the kinetics of olivine-spinel transition. Phys. Earth Planet. Inter. 26, 179–187.

Riedel, M.R., Karato, S., 1997. Grain-size evolution in subducted oceanic lithosphere associated with the olivine-spinel transformation and its effects on rheology. Earth Planet. Sci. Lett. 148, 27–43.

Rubie, D.C., 1984. The olivine-spinel transformation and the rheology of subducting lithosphere. Nature 308, 505–508.

Rubie, D.C., 1986. The catalysis of mineral reactions by water and restrictions on the presence of aqueous fluid during metamorphism. Mineral. Mag. 50, 399–415.

Rubie, D.C., 1993. Mechanisms and kinetics of reconstructive phase transformations in the earth's mantle. In: Luth, R.W. (Ed.), Short Course on Experiments at High Pressure and Applications to the Earth's Mantle, Vol. 21, Mineralogical Association of Canada, Edmonton, pp. 247–303.

Rubie, D.C., Thompson, A.B., 1985. Kinetics of metamorphic reactions at elevated temperatures and pressures: an appraisal of experimental data. In: Thompson, A.B., Rubie, D.C. (Eds.), Metamorphic Reactions: kinetics, textures, and deformation, Springer, New York.

Rubie, D.C., Champness, P.E., 1987. The evolution of microstructure during the transformation of Mg_2GeO_4 olivine to spinel. Bull. Mineral. 110, 471–480.

Rubie, D.C., Ross II, C.R., 1994. Kinetics of the olivine-spinel transformation in subducting lithosphere: experimental constraints and implications for deep slab processes. Phys. Earth Planet. Inter. 86, 223–241.

Schmeling, H., Monz, R., Rubie, D.C., 1999. The influence of olivine metastability on the dynamics of subduction. Earth Plant. Sci. Lett. 165, 55–66.

Sharp, T.G., Rubie, D.C., 1995. Catalysis of the olivine to spinel transformation by high clinoenstatite. Science 269, 1095–1098.

Sharp, T.G., Bussod, G.Y.A., Katsura, T., 1994. Microstructures in β-$Mg_{1.8}Fe_{0.2}SiO_4$ experimentally deformed at transition-zone conditions. Phys. Earth Planet. Inter. 86, 69–83.

Solomatov, V.S., Stevenson, D.J., 1994. Can sharp seismic discontinuities be caused by non-equilibrium phase transformations? Earth Planet. Sci. Lett. 125, 267–279.

Stein, C.A., Stein, S., 1992. A model for the global variation in oceanic depth and heat flow with lithospheric age. Nature 359, 123–129.

Stein, S.A., Rubie, D.C., 1999. Deep earthquakes in real slabs. Science 286, 909–910.

Sung, C., Burns, R.G., 1976a. Kinetics of high-pressure phase transformations: implications to the evolution of the olivine $\rightarrow$ spinel transition in the downgoing lithosphere and its consequences on the dynamics of the mantle. Tectonophysics 31, 1–32.

Sung, C., Burns, R.G., 1976b. Kinetics of the olivine $\rightarrow$ spinel transition: implications to deep-focus earthquake genesis. Earth Planet. Sci. Lett. 32, 165–170.

Turnbull, D., 1956. Phase changes. Solid State Phys. 3, 225–306.

Walter, M.J., Thibault, Y., Wei, K., Luth, R.W., 1995. Characterizing experimental pressure and temperature conditions in multi-anvil apparatus. Can. J. Phys. 73, 273–286.

Yoshioka, S., Daessler, R., Yuen, D.A., 1997. Stress fields associated with metastable phase transitions in descending slabs and deep-focus earthquakes. Phys. Earth Planet. Inter. 104, 345–361.

Young, T.E., Green II, H.W., Hofmeister, A.M., Walker, D., 1993. Infrared spectroscopic investigation of hydroxyl in β-$(Mg, Fe)_2SiO_4$ and coexisting olivine: implications for mantle evolution and dynamics. Phys. Chem. Mineral. 19, 409–422.

ELSEVIER

Physics of the Earth and Planetary Interiors 127 (2001) 181–196

PHYSICS OF THE EARTH AND PLANETARY INTERIORS

www.elsevier.com/locate/pepi

Stabilities and equations of state of dense hydrous magnesium silicates

R.J. Angel [a,*], D.J. Frost [a], N.L. Ross [b,1], R. Hemley [c]

[a] *Bayerisches Geoinstitut, Universität Bayreuth, D-95440 Bayreuth, Germany*

[b] *Department of Geological Sciences, University College London, Gower St., London WC1E 6BT, England, UK*

[c] *Geophysical Laboratory, Carnegie Institution of Washington, 5251 Broad Branch Rd. NW, Washington, DC 20015-1305, USA*

Received 7 February 2000; received in revised form 4 August 2000; accepted 12 August 2000

Abstract

The thermodynamic stability of high-pressure phases in the ternary system MgO–SiO_2–H_2O is reviewed with special emphasis on hydrated phases including the "alphabet phases". On the basis of recent experimental data, a stability diagram for these phases along a geotherm appropriate for a subducting slab is constructed. This suggests that at realistic water contents, the breakdown of antigorite will produce olivine plus phase A in the upper mantle, but that the ability of wadsleyite to accommodate wt.% levels of water means that phases E and D are unlikely to occur in the upper portion of the transition zone. For the lower part of the transition zone in which ringwoodite becomes stable, the coexisting hydrous phase would be superhydrous phase B, provided ringwoodite is not able to accommodate significant amounts of water. At the top of the lower mantle, phase D is stable but only at temperatures below 1300°C. As the maximum solubility of H_2O in magnesium silicate perovskite is also quite low slabs may dehydrate somewhat as they enter the lower mantle.

The available thermodynamic data for these phases is also reviewed. From the most recent measurements of equations of state, it is concluded that the elasticity of the phases in the MgO–SiO_2–H_2O ternary system is primarily dependent upon the density and not upon the water content. It is, therefore, concluded that the presence of even significant amounts of hydrated phases in subducting slabs could not be unequivocally identified from seismological observations.

Keywords: Hydrous phases; Equations of state; Thermodynamic properties phase equilibria; Subduction

1. Introduction

Water may be present in earth's mantle in a number of different forms. It may exist as a free fluid phase, it may dissolve in a silicate melt or a nominally anhydrous mineral such as olivine, or it could be incorporated in a hydrous silicate mineral, such as phlogopite mica. In addition, the oxygen fugacity in some regions of the earth's interior may be such that hydrogen exists not as H_2O but as the reduced species H_2 or CH_4.

The bulk of earth's mantle is believed to be peridotitic in composition. Estimates of the water content of MORB source-type peridotite range from 100 to 500 ppm (Javoy and Pineau, 1991; Jambon, 1994; Sobolev and Chaussidon, 1996). At pressures compatible with the top of the upper mantle, this water

* Corresponding author. Present address: Crystallography Laboratory, Department of Geological Sciences, Virginia Polytechnic and State University, Blacksburg, VA 24061-0420, USA.
Tel.: +1-540-231-7974; fax: +1-540-231-3386.
E-mail address: rangel@vt.edu (R.J. Angel).

[1] Crystallography Laboratory, Department of Geological Sciences, Virginia Polytechnic and State University, Blacksburg, VA 24061-0420, USA.

0031-9201/01/$ – see front matter
PII: S0031-9201(01)00227-8

content may reside as free fluid or as partial melt, but at higher pressures it might be entirely hosted by nominally anhydrous minerals like olivine, pyroxene and garnet (Bell and Rossman, 1992; Kohlstedt et al., 1996). Subduction zones, in comparison to the bulk mantle, present a more complex and intriguing prospect for water storage. The subducting oceanic lithosphere has been extensively altered by interaction with sea water and will contain significant amounts of H_2O. The lithosphere is also chemically heterogeneous with an upper basaltic and gabbroic crust and a more extensive lower depleted peridotite mantle region. Evidence from ophiolite deposits suggests that hydration of both crustal and mantle components can take place. Furthermore, thermal models of subduction zones predict much lower temperatures in the interiors of subducting slabs than in the bulk of the mantle. Some models (e.g. Furukawa, 1993) predict temperatures of only 600°C in slabs at pressures of 10 GPa, whereas the bulk mantle temperature at this pressure is probably closer to 1300°C. At such low temperature conditions H_2O will reside to some extent in hydrous silicate minerals which may therefore transport H_2O into the mantle.

A knowledge of the depth to which hydrous silicates could transport H_2O in a subduction zone is important for determining whether there is a positive or negative flux of H_2O between the surface and the mantle. If water is entirely released by dehydration reactions within the slab beneath the island arc, then the water released will ultimately return to the earth's surface and will have a relatively short mantle residence time (on the order of 10 million years). If, on the other hand, hydrous minerals remain stable in the slab then water could be transported to much greater depths and may be subsequently released into the transition zone or the lower mantle. It has been proposed that dehydration within a subducting slab may act as a trigger mechanism for deep focus earthquakes (Silver et al., 1995) and that the fluids released may be important metasomatic agents.

A significant number of experimental studies have been performed in order to identify hydrous phases that may form in subducting lithosphere and to determine their stability fields. The high pressure stability of hydrous minerals that are known to exist in hydrated peridotite and basaltic compositions, such as lawsonite, amphibole, antigorite, phengite, and phlogopite (Pawley, 1994; Schmidt, 1995; Ulmer and Trommsdorff, 1995; Wunder et al., 1995; Domanik and Holloway, 1996; Sato et al., 1997) have been extensively studied. In addition, many exploratory high-pressure studies have been aimed at identifying new hydrous minerals that may be stable at higher pressures. Such studies are frequently based in simple chemical systems such as $MgO–SiO_2–H_2O$ and the total H_2O contents are generally high. A host of so-called "dense hydrous magnesium silicate" (DHMS) phases have been identified at high pressure, including the alphabet phases, A, B, superhydrous B, C, D, E, F and G (Ringwood and Major, 1967; Liu, 1987; Kanzaki, 1991; Gasparik, 1993; Ohtani et al., 1997), 10 Å phase and 3.65 Å phase (Sclar et al., 1965; Rice et al., 1989) and, in more complicated systems, phase X (Tronnes, 1990) and phase Egg (Eggleton et al., 1978; Schmidt et al., 1998). In parallel with these developments, measurements have also been made on the maximum H_2O content of nominally hydrous minerals in the mantle (Kohlstedt et al., 1996; Lu and Keppler, 1997). For example, wadsleyite, a major mineral of the transition zone, is capable of containing up to 3.3 wt.% H_2O (Smyth, 1987; Inoue et al., 1995) and must also be viewed as a potentially hydrous mineral.

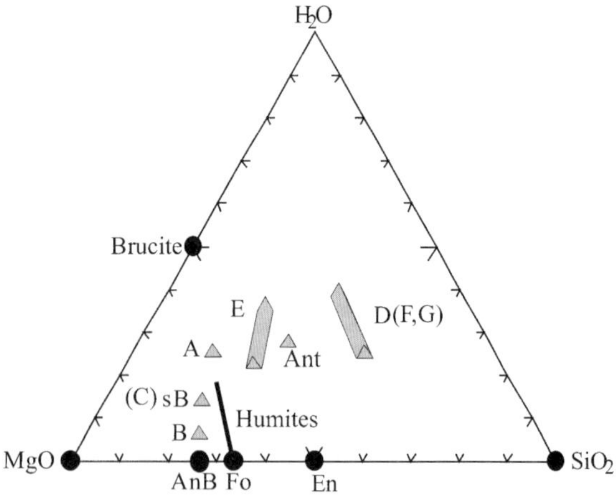

Fig. 1. The compositions of the dense hydrous magnesium silicate phases shown on a molar ternary diagram for $MgO–SiO_2–H_2O$. Phases F and G are identical to phase D and phase C is identical to superhydrous phase B (sB). Phases D and E are variable in composition. Anhydrous phase B (AnB), antigorite (Ant), Mg_2SiO_4 and $MgSiO_3$ (En) are also shown.

In this contribution, we provide a brief critical review of the experimental constraints on the phase equilibria of the DHMS phases (Fig. 1). From the available data we construct for the first time a tentative phase diagram that applies to the moderate levels of hydration that are likely to be appropriate for the cold and wet portions of subducted slabs. In principle, such phase equilibria could also be calculated if the physical and thermodynamic properties of both the solid and fluid phases were known. We review the available data and show that the compressional moduli of the pure DHMS phases are simple linear functions of density, a result that makes their presence in subducted slabs difficult to detect by seismological observations.

2. Experimental phase equilibria

Recently, Schmidt and Poli (1999) presented a calculation of water budgets in a subducting slab to depths of 300 km. They suggested that water is released continually from the slab over this depth range by a series of dehydration reactions in the sedimentary, mafic and ultramafic horizons of the slab. The mafic component of the slab could store water mainly in lawsonite to depths of about 300 km, with maximum water contents at these depths estimated to be on the order of 1000 ppm. There is an equal if not greater potential for high pressure water storage in the peridotite portion of the slab where the serpentine mineral antigorite forms. Antigorite is stable to approximately 6 GPa (Ulmer and Trommsdorff, 1995) but for water to be passed from antigorite to the next most likely stable dense hydrous magnesium silicate phase, phase A, temperatures would have to be below 700°C at this pressure. The pressure region, around 6 GPa, where hydrous phases seem to be at a minimum in thermal stability has been previously referred to as the 'choke point' (Kawamoto et al., 1995). If, in the inner portions of a cold slab, temperatures are low enough for the reaction of antigorite to phase A + enstatite to occur, up to 4 wt.% H_2O could be retained in the hydrous phase, that could then be subducted to higher pressures. A pervasively serpentinised slab at these depths is unlikely due to buoyancy problems. A more realistic scenario may be regions of serpentinisation, perhaps concentrated around faults that originated at the surface. If this is the case then H_2O released from the antigorite breakdown reaction could also produce phase A or other hydrous phases in unhydrated parts of the slab, thus, redistributing H_2O. Some thermal models do predict subduction environments capable of stabilising phase A (e.g. Furukawa, 1993). After phase A has been produced it becomes stable to higher temperatures with increasing pressure and could potentially transport water into the transition zone. Furthermore, the pressure interval between 5 and 10 GPa has been quite poorly explored experimentally and in the future other hydrous minerals capable of bridging this 'choke point' and transporting H_2O deep into earth's interior may be identified.

It is then necessary to determine what further reactions may take place for a hydrous peridotite composition that could be subducted into the transition zone or lower mantle. It is pertinent, therefore, to examine phase diagrams that are applicable to hydrated ultramafic compositions at depths greater than 180 km or about 6 GPa. Perhaps the simplest and most extensively studied starting composition is that of Mg_2SiO_4 + 20 wt.% H_2O for which multi-anvil experimental results from Kanzaki (1991), Inoue (1994), Luth (1995), Ohtani et al. (1995) and Frost and Fei (1998) are summarised in Fig. 2. There are still many areas of the diagram with poor data coverage, and the indicated boundaries are tentative. These experiments were in general performed in order to synthesise and characterise the hydrous phases, but this diagram serves as a first approximation of the possible stability fields of hydrous minerals in a hydrated ultramafic slab. A further complicating element in the study of H_2O-rich systems at high pressure is the definition and recognition of fluid and melt at these conditions. A super-critical hydrous fluid at high pressures will contain a significant amount of dissolved species, so that in quenched samples it becomes difficult, if not impossible, to differentiate between a hydrous fluid and a hydrous silicate melt. It may be that a second critical end-point is reached in the Mg_2SiO_4–H_2O system at high pressure as observed for many hydrous–silicate systems at lower pressures (Shen and Keppler, 1997; Bureau and Keppler, 1999).

Phase A is stable to approximately 15 GPa in this H_2O-rich composition (Fig. 2) and displays maximum temperature stability of 1000°C at 11 GPa. Clinohumite is observed in experiments at temperatures higher than the stability of phase A but is replaced

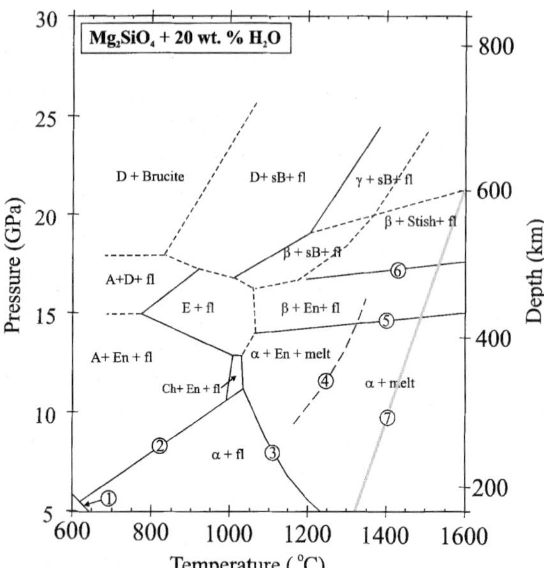

Fig. 2. Synthesis conditions of dense hydrous magnesium silicates for a Mg_2SiO_4 + 20 wt.% H_2O starting composition. Data are from Kanzaki (1991), Ohtani et al. (1995) and Frost and Fei (1998). The more tentative boundaries are shown as broken lines. The line labelled 1 is the breakdown curve of antigorite (Ulmer and Trommsdorff, 1995), line 2 is for the reaction A + En = α + fl (Luth, 1995), line 3 is the wet solidus and line 4 is the enstatite out curve (Inoue, 1994), line 5 is the α–β-Mg_2SiO_4 transition and line 6 the En = β + stish reaction (Fei et al., 1990). Line 7 is an average mantle adiabat from Jeanloz and Morris (1986). Fluid is denoted by "fl".

by an assemblage containing forsterite, enstatite and liquid above 1100°C. Between 13 and 17 GPa and to temperatures around 1100°C phase E is observed with a coexisting fluid phase. Phase E is an unusually water-rich mineral with a disordered structure (Kudoh et al., 1993). The composition of phase E shows little obvious dependence on either pressure or temperature and can be compositionally zoned within a single experimental charge (Frost, 1999). One possibility is that the composition of phase E is modified during quenching of the temperature and it remains to be tested as to whether phase E is a stable mineral at high temperature and is not a product entirely of the quenching process.

At pressures above 17 GPa assemblages containing phase D and superhydrous phase B occur. Phase D, a Si-rich hydrous phase with Si solely in octahedral coordination, also shows compositional variation. However, in contrast to phase E, the Mg/Si ratio and water content of phase D show clear variations with pressure and temperature (Ohtani et al., 1997; Frost, 1999). In situ measurements have been made on phase D at high pressure and temperature that confirm that it is not a quench product (Irifune et al., 1996; Frost and Fei, 1998).

Some confusion surrounds the identification of Si-rich hydrous phases at high pressures. Phases D, F and G (Liu, 1987; Kanzaki, 1991; Ohtani et al., 1997) have all been named from experiments above 17 GPa and all seem to possess a similar range in composition and structure. It is very likely that these three phases are identical (Frost, 1999) although a study by Liu (1986, 1987) raises the possibility that phase D may develop a superlattice at pressures above 22 GPa. This has not, however, been confirmed. Phase D, as we shall refer to D, F and G henceforth, is stable to temperatures of approximately 1400°C at 25 GPa (Fig. 2), and the laser heated diamond anvil cell experiments of Shieh et al. (1998) demonstrate that phase D is in fact stable to 44 GPa.

Superhydrous phase B is a magnesium-rich hydrous phase (Mg/Si = 3.2) which is stable to temperatures slightly higher than phase D. It has been proposed (Kanzaki, 1993) that superhydrous phase B, first identified by Gasparik (1993), may be identical to phase C reported by Ringwood and Major (1967). The X-ray diffraction pattern originally reported for phase C was from a mixed experimental charge and some of the lines may have been incorrectly indexed.

The possibility is raised, therefore, that the dense hydrous magnesium silicates A, E, and superhydrous B could be stable in slabs and transport water into the lower mantle, albeit only if subduction geotherms are cool. The experiments summarised in Fig. 2, however, are for a bulk composition more water-rich than can be applicable to the subducting slab. If the subducting peridotite mantle were initially entirely serpentinised approximately 12 wt.% H_2O could be stored in antigorite to 6 GPa. The reaction of antigorite to produce phase A and enstatite, however, would release water from the slab and only a maximum of 4 wt.% H_2O could be stored in the solid assemblage and subducted further. For this reason high-pressure phase equilibria determinations that employ serpentine as a starting material (e.g. Irifune et al., 1998; Shieh et al., 1998) over-estimate plausible water contents above

6 GPa and are probably not directly applicable to a subducting slab. As stated previously complete serpentinisation of the slab peridotite at these depths also seems unlikely. Even H_2O concentrated in small regions of intense serpentinisation might have a propensity to diffuse into the surrounding anhydrous slab as the H_2O solubility of the nominally anhydrous minerals increases with pressure (Kohlstedt et al., 1996). If we consider that 10% of the peridotite may be serpentinised this would result in a more reasonable H_2O content of ~4000 ppm in the solid assemblage after antigorite breakdown.

A simple assessment of hydrous phase stability for lower water concentrations can be made by interpolating between the results of experiments shown in Fig. 2 and measurements made of the maximum solubility of water in nominally anhydrous phases such as olivine, wadsleyite, ringwoodite and perovskite. Fig. 3 shows a pseudobinary Mg_2SiO_4–H_2O phase diagram along a hypothetical geotherm that passes through 600°C at 5 GPa and 1400°C at 30 GPa. The diagram incorporates some data from experiments performed at lower water concentrations by Gasparik (1993) and Ohtani et al. (1997) but is in the main estimated from observed coexistences in the H_2O-rich experiments. Water contents of olivine and ringwoodite are from Kohlstedt et al. (1996) but the theoretical maximum water content of wadsleyite (3.3 wt.%) is assumed. There are a significant number of uncertainties in the diagram. The measurements of Kohlstedt et al. (1996), for example, were made at higher temperatures. The degree to which the phase boundaries of wadsleyite and ringwoodite are affected by the presence of H_2O are speculative in the absence of any data and there are no data on the solubility of H_2O in ringwoodite and wadsleyite as a function of pressure. The shaded regions result from uncertainty in the compositions of phases E and D for a given pressure and temperature and these phases may also transform through divariant loops. Using Fig. 3, however, a number of predictions can be made as to the likely phases stable for more realistic water concentrations (<1 wt.% or 2.5 mol%). After the breakdown of antigorite, phase A will be produced as the solubility of H_2O in olivine is low. But in the transition zone, due to the high solubility of H_2O in wadsleyite, it is unlikely that phase E will form. Only a slab or a region of the slab that was pervasively serpentinised (>80%) could result in the formation of phase E. The solubility of H_2O in ringwoodite is not as well understood as it is for wadsleyite. The infrared spectroscopy measurements of Kohlstedt et al. (1996) indicate that ringwoodite may

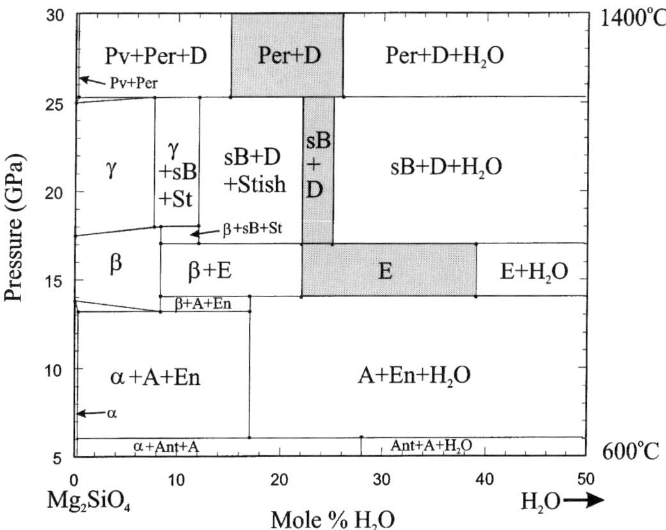

Fig. 3. Phase relations along a Mg_2SiO_4–H_2O binary between 5 and 30 GPa and temperatures corresponding to a cold subduction geotherm. Shaded regions arise from uncertainties in the composition of phases D and E at given conditions. Plausible slab water contents are most likely below 2.5 mol% (1 wt.%) at these conditions.

dissolve up to 2.7 wt.% H_2O, as shown in Fig. 3 but the site of hydration has not been clearly identified and the solubility of H_2O does not appear to be associated with a change in the Mg/Si ratio. If this is correct then it is unlikely that superhydrous phase B would form in the slab. If, on the other hand, the measurements have over-estimated the water content of ringwoodite, then superhydrous phase B could still be important. Fig. 3 indicates that phase D could not form within slabs in the transition zone for realistic water contents. Current measurements of the water content of silicate perovskite (Lu et al., 1994; Meade et al., 1994) seem to indicate a low maximum solubility (200 ppm) although Bolfan-Casanova et al. (2000) were unable to detect any water in pure $MgSiO_3$ perovskite (i.e. <1 ppm) and suggested that the previous measurements might have been performed on un-equilibrated samples. Clearly, further work is required in order to explore water solubility over the full range of perovskite chemistry. If the solubility is indeed low, then phase D would be the dominant water host in the slab at lower mantle conditions and may be stable to depths of around 1200 km (Shieh et al., 1998).

Fig. 3 is based mainly on water saturated experiments in the $MgO–SiO_2–H_2O$ system. It is possible that for lower water concentrations the maximum thermal stabilities of the hydrous phases are affected. The mantle also contains significant FeO, Al_2O_3 and CaO, which could similarly affect the phase stability fields. As a preliminary test of hydrous phase stability in a realistic mantle composition, Frost (1999) equilibrated a natural harzburgite composition with 4 wt.% H_2O at conditions applicable to slabs in the transition zone and lower mantle. The maximum stabilities of hydrous phases determined in this study are shown in Fig. 4. The inferences made from Fig. 3 are confirmed, to the most part, by the results of Frost (1999), although it was found that phase D was only stable at temperatures below 1300°C at lower mantle pressures, as opposed to temperatures up to 1400°C in the simpler system shown in Fig. 2. The reduced water contents or the solubility of FeO and Al_2O_3 in phase D, into which they are partitioned in preference to coexisting perovskite, may be responsible for this reduced thermal stability. The important point, however, is that while in the transition zone there are a number of phases that can store large quantities of water up to high temperatures, slabs entering the lower mantle encounter a decrease in hydrous phase stability. Combining this with the probable low water content of magnesium silicate perovskite raises the possibility that slabs may dehydrate to some degree as they enter the lower mantle.

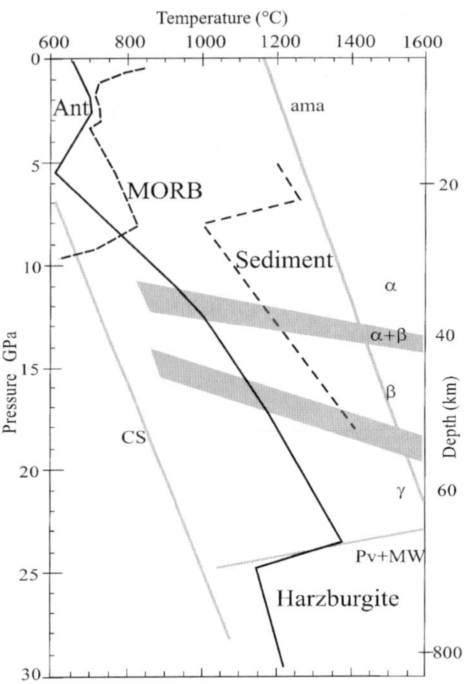

Fig. 4. The maximum thermal stability of hydrous phases in sediment (Ono, 1998), MORB basalt (Schmidt and Poli, 1999), and ultramafic harzburgite (Frost, 1999) bulk compositions. Positions of the α–β-, β–γ- and γ-Pv + MW transformations for $(Mg_{1.8}Fe_{0.2})SiO_4$ are shown in grey (Ito and Takahashi, 1989; Katsura and Ito, 1989). An average mantle adiabat (ama), an estimate of the minimum temperature in a subducting slab (cs), and the antigorite breakdown curve (ant) are also shown (Jeanloz and Morris, 1986; Ulmer and Trommsdorff, 1995; Irifune et al., 1996; Stein and Stein, 1996).

3. Thermodynamic properties

The direct experimental determination of the equilibrium phase boundaries involving hydrous phases is complicated by a number of factors. Not least is the necessity to undertake the experiments with the correct total water content, which requires the time-consuming prior synthesis of all of the phases to be studied. Furthermore, while in principle, the fluid phase from dehydration reactions is water, at the high pressures and temperatures involved it dissolves significant other components, not least silica and alumina; for complete determination of reaction boundaries, the activities of these components need to be determined, a task beyond current experimental capabilities. Lastly, for application to the earth, a large number of additional chemical components outside of the $MgO–SiO_2–H_2O$ system must be considered, of which the most important are CaO and iron oxides, the latter bringing the added requirement of control of oxygen fugacity in experiments.

A complementary approach to experimental determinations, that can be used to confirm the experimental results and to extend them, is the calculation of the phase stabilities and the positions of reaction boundaries in $P–T$ space from measured or calculated thermodynamic properties of the mineral phases. The most serious hurdle to be overcome in such an approach is the synthesis of pure phases; many synthesis attempts result in mixtures of co-existing phases (Burnley and Navrotsky, 1996) for reasons of either thermal gradients, intrinsic metastability or pressure buffering in multi-anvil presses. Metastability is enhanced when the differences in thermodynamic properties between two phases (which provides the driving force for reactions to proceed) is small, as is likely to be the case for many DHMS phases whose structures and bonding are very similar, as we discuss in more detail below. Conversely, the presence of phases of similar composition that have large differences in thermodynamic properties, especially density, may enhance pressure-buffering, the phenomenon of the pressure being held at the pressure of the equilibrium phase boundary in a closed system of constant total volume (e.g. Rubie, 1999).

Once pure samples have been obtained, enthalpy contents can be measured by high-temperature solution calorimetry. Techniques have been specially developed for the typical small samples obtained from multi-anvil syntheses, while such measurements become more routine and more accurate as developments in multi-anvil techniques result in larger sample volumes. At the time of writing, of the DHMS phases only the enthalpy of formation of phase A has been determined, although talc and antigorite were measured in the same study (Bose and Navrotsky, 1998). The sizes of samples recovered from the multi-anvil press generally preclude the use of conventional DSC instruments to measure heat capacities, from which estimates of the entropy contribution to the free energy of the phase can be derived. Instead, the mode frequencies obtained from infra-red and Raman absorption spectra (which can be measured on milligram quantities of sample) can be used in vibrational models to estimate the total vibrational density of states of the material. If performed with care this procedure in turn yields the entropy and heat capacity of the phase to better than 5% (e.g. Kieffer, 1979, 1980). Cynn et al. (1996) provide a detailed example of this procedure applied to phases A, B, superhydrous phase B and chondrodite. The principal potential problem with these measurements at room conditions is whether the material is in the same structural state as it has at the high P, T conditions of its stability field. But a check is possible, in that structural changes on quenching will result in measured thermodynamic properties at room conditions that cannot be reconciled with phase equilibrium boundaries determined in situ at high pressures and temperatures.

3.1. Equations of state

A critical element in the calculation of high-pressure, high-temperature phase stabilities in the DHMS system is the volume contribution to the free energy. The EoS of most of the DHMS phases, together with their anhydrous counterparts, have been measured by single-crystal high-pressure X-ray diffraction at room temperature, which provides the most precise and most accurate measurements of volume variation with pressure. High-pressure powder diffraction measurements can provide complementary data to higher pressures, although of lower precision. These isothermal compression experiments yield values for the room-pressure, room-temperature bulk modulus K_{0T} and its pressure derivative K'_{0T}. Because these compression experiments are performed under hydrostatic conditions, the specimen is under a state of uniform stress, and the values of the elastic moduli obtained therefore correspond to the Reuss bound on the bulk modulus of a polycrystalline aggregate of the same material. Ultrasonic interferometry can be performed on polycrystalline aggregates held at high temperatures and pressures in multi-anvil presses in order to obtain direct measurements of elastic wave speeds of mantle materials in situ at pressure and/or temperature conditions approaching those of the mantle (e.g. Knoche et al., 1997; Liebermann and Li, 1999). From the measured wave speeds, the adiabatic bulk and shear moduli can be obtained. The values of room-pressure bulk moduli are generally in good agreement with the results obtained from static compression measurements, after the conversion of adiabatic to isothermal quantities is made, but usually lie to slightly higher values. This is a result of the experimental polycrystalline samples having aggregate properties lying between the Reuss bound of uniform stress and the higher Voigt bound of uniform strain. Elastic properties would be at the latter limit if the inter-granular contacts remain fully-welded (slip-free) at high pressures. The Reuss and Voigt bounds on the bulk modulus typically differ by 1–2% for denser phases such as wadsleyite, but can change with pressure (e.g. Zha et al., 1997) and temperature. No ultrasonic measurements of the hydrated phases of the DHMS system are known to the authors except for brucite, $Mg(OH)_2$ (Xia et al., 1998), a material not relevant to subduction zones. But extensive measurements of the anhydrous phase olivine and its high-pressure polymorphs have been made because the velocity jumps at the 410 and 520 km discontinuities in the bulk mantle provide the best constraints on its chemical and mineralogical composition. There are also a number of Brillouin scattering measurements on single-crystal specimens at room pressure and temperature which provide not only the room-pressure adiabatic bulk modulus K_{0S}, but also the components of the complete elastic constant tensor. From these data not only the Reuss and the Voigt bounds on the adiabatic bulk modulus of a polycrystalline aggregate can be calculated, but the shear modulus can also be obtained; elastic wave velocities follow from combining these moduli with the density.

While there is extensive thermal expansion data for the anhydrous $MgO–SiO_2$ phases (e.g. olivine, Matsui and Manghnani, 1985; enstatite, Hugh-Jones, 1997; wadsleyite, Suzuki et al., 1980; ringwoodite, Suzuki et al., 1979), direct measurements of the thermal expansion of the DHMS phases are restricted to talc and phase A (Pawley et al., 1995). The temperature dependencies of the bulk moduli of olivines have been measured by resonance ultrasound spectroscopy (Isaak et al., 1989; Isaak, 1992; Sumino, 1979), but no such data have been measured for hydrous phases. Future high-temperature measurements of these properties at ambient pressure will be hampered by the metastability of the high-pressure phases, and even measurements well below the temperature at which they decompose back to their low-pressure polymorphs may well be biased by anharmonic effects related to such decomposition. One way to overcome these difficulties is to measure the low-temperature thermal expansion coefficients as was done for $MgSiO_3$ perovskite (Ross and Hazen, 1989), but has yet to be performed for the DHMS phases. An alternative is to measure the volume variation of the phase with pressure and temperature by in situ powder diffraction and to extract thermal expansion coefficients simultaneously with compressional moduli by fitting an appropriate thermal equation of state. In the past, this approach has been hampered by the low resolution and low precision of energy-dispersive powder diffraction techniques that precluded the reliable determination of EoS parameters. Significant improvements in the precision of volume determinations have now been obtained by the recent introduction of angle-dispersive powder diffraction techniques for high-P, T studies. But significant uncertainties in the refined EoS parameters still remain, and arise from two sources. First, the various EoS parameters (e.g. K_{0T}, K'_{0T}, dK_0/dT) are highly correlated and are, therefore, difficult to determine independently. Second, current estimates of the inaccuracy of the NaCl pressure scale commonly used to determine pressure in high-P, T experiments are of the order of 3% at 10 GPa and room-temperature, and only slightly less at simultaneous high pressures and high temperatures (Brown, 1999). Such uncertainties in pressure scales (in all experiments to measure EoS) propagate directly into additional uncertainties in the value of K_0, although they are often not considered in experimental reports.

In summary, room-temperature EoS data are available for all of the DHMS phases, but measurements of high-temperature properties either remain to be made, or are of relatively low precision. Therefore, in discussing the elasticity of phases in this system, and some of the structural principles that control the elasticity, we will restrict ourselves to ambient condition densities and room-temperature equations of state. For the sake of consistency, most of the data will be drawn from single-crystal diffraction experiments, as these form the most precise and most extensive dataset.

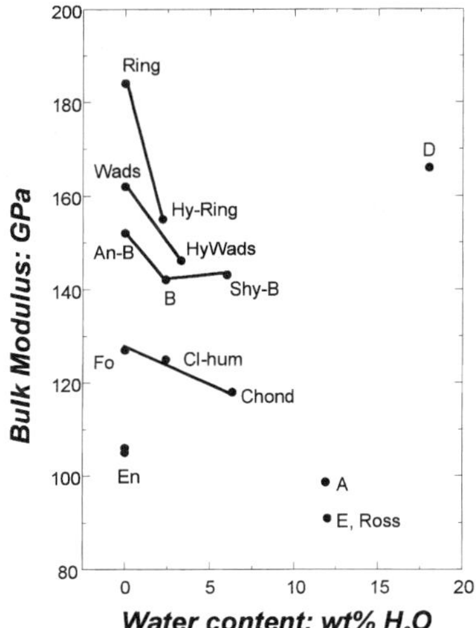

Fig. 5. The variation with water content of the isothermal bulk moduli, K_{0T}, of series of related compounds in the MgO–SiO_2–H_2O ternary (see Fig. 1). Data sources: ringwoodite (Weidner et al., 1984); hydrous–ringwoodite (Inoue et al., 1998); wadsleyite, hydrous–wadsleyite (Ross et al., 1998); anhydrous B, B and superhydrous B (Crichton and Ross, 2000); forsterite (Isaak et al., 1989); OH-clinohumite and OH-chondrodite (Ross and Crichton, 2001); enstatite (Angel and Hugh-Jones, 1994; Jackson et al., 1999); A and E (Ross and Crichton, 2000).

Except where explicitly noted, the measurements by other techniques are in good agreement with the data presented here.

A question that can be addressed by the available data is to what extent are the elastic properties of the DHMS phases controlled by their water content. Fig. 5 summarises the available data for the pure system. The low-pressure phases enstatite pyroxene ($MgSiO_3$) and forsterite olivine only accommodate H_2O up to the ~1000 ppm level, and this probably does not have a measurable effect on the elasticity. By contrast, the remainder of the minerals shown in Fig. 5 exhibit significant softening of their compressional moduli when H_2O is added at wt.% levels. But the magnitude of the effect is much stronger in wadsleyite and ringwoodite than in either the B-series of structures or the olivine–humite series. As pointed out by Sinogeiken and Bass (1999), this is a reflection of the two very different ways in which H_2O is accommodated within the various structural series.

In the humite series of structures (humite, clinohumite, chondrodite), the water is accommodated by the incorporation of a new type of structural layer containing the OH groups in an ordered manner between olivine-structured layers. The result is a series of compounds of specific stoichiometry (lying on the olivine–brucite join, Fig. 1) in which the water content is determined by the number of each kind of layer in the regular stacking sequence. Complete structural integrity is maintained in each phase, and there is relatively little elastic softening (<10%) with increasing water content. Substitution of F can occur for OH in these phases, but the effect on the bulk moduli is not known. Complete replacement of OH in brucite to produce MgF_2 results in a large increase in bulk modulus from 40 (Xia et al., 1998) to 102 GPa (Jones, 1977). But F (and Cl) substitution for OH in the apatite structure has no significant effect on the bulk modulus, and only affects the anisotropy of compression (Brunet et al., 1999). The bulk moduli of natural chondrodites with different OH/F contents range from 136 GPa for a specimen with $OH/(OH + F) = 0.27$ (Faust and Knittle, 1994; Kuribayashi et al., 1998) to 118 GPa for a sample with $OH/(OH + F) = 0.68$ (Sinogeiken and Bass, 1999). Ross and Crichton (2001) reported a bulk modulus of 115.7(8) GPa for a synthetic specimen of pure OH-chondrodite. Thus, the natural chondrodites tend to have higher bulk moduli than those obtained from the pure Mg + OH end-members (Fig. 5), suggesting that the incorporation of F and/or Fe is stiffening the structure. Nonetheless, a similar degree of softening in both bulk modulus and shear modulus (Sinogeiken and Bass, 1999) occurs with increase in water content.

The composition of phase A lies on an extension of the olivine–chondrodite line (Fig. 1), and possesses a related structure, but not one that can be described exactly as a further stacking sequence of olivine and humite layers (Horiuchi et al., 1979). The first measurement of the compression of phase A was performed by energy-dispersive powder diffraction (Pawley et al., 1995), and the data suggested that phase A is significantly stiffer than the humite minerals, even though it has a higher water content. This apparent discrepancy has been resolved by more recent single-crystal X-ray diffraction measurements (Kuribayashi et al., 1999; Ross and Crichton, 2000) which show that phase A is only slightly softer than a value obtained by extrapolation of the forsterite–chondrodite line in Fig. 5. Similarly, recent compression data

(Crichton and Ross, 2000a; Shieh et al., 2000) for the enigmatic disordered phase E suggest that it is softer than earlier Brillouin measurements (Bass et al., 1991), although this difference may well be explained by different H_2O contents between samples.

The three "B" phases anhydrous B, B, and superhydrous B are also closely related structures, based on the alternate stacking of "TO" layers and "O" layers. The "O" layers consist of sheets of edge-shared MgO_6 and SiO_6 octahedra (Finger et al., 1991; Pacalo and Parise, 1992). The "TO" layer of anhydrous B is a sheet of olivine structure containing MgO_6 octahedra and SiO_4 tetrahedra. In the structures of phase B and superhydrous B, this sheet is sheared on a periodic basis which results in the removal of $[MgSiO_2]^{2+}$ groups from the layer and charge balance is maintained by the introduction of H^+ pairs. The detailed arrangement of the octahedral layer is different in each phase, as the vacancies in the "O" layer have to lie above the remaining SiO_4 tetrahedra in the adjacent "TO" layers. But, as in the olivine–humite series of minerals, each structure has a specific stoichiometry and full structural integrity so that the softening induced by higher water contents (Fig. 5), although significant, is only 4.5% for superhydrous B with a water content of >5 wt.% compared to the anhydrous B end-member (Crichton and Ross, 2000). Crichton and Ross (2000) found that phase B has $K'_{0T} = 7.0(5)$ which is significantly higher than either anhydrous B ($K'_{0T} = 5.5(3)$) or superhydrous B ($K'_{0T} = 5.1(2)$). Crichton et al. (1999) and Crichton and Ross (2000) found that the cell parameter compressibility data yielded valuable insights into the structural controls on the response of the structures to compression. The axial compressibilities parallel to the layers are controlled by the stiffness of the edge-sharing octahedra, and show no significant change with increasing water content. The compressibility normal to the layers clearly depends on the individual compressibilities of both types of layers, and the softening of this direction with increasing water content makes the biggest contribution to the softening of the bulk modulus. Most interestingly, these effects result in a reduction by half of the compressional anisotropy on addition of water to anhydrous B to form superhydrous B.

The high-pressure polymorphs of Mg_2SiO_4, wadsleyite (frequently mis-named as "beta-phase" or "modified-spinel") and ringwoodite, the spinel-structured phase, are also structurally related to one another. The wadsleyite structure consists of spinel-structured layers but, while in spinel every layer is inverted with respect to its two neighbours, the layers in wadsleyite are inverted in pairs. This pairing of layers gives rise to Si_2O_7 groups formed of two SiO_4 tetrahedra sharing a common apex, whereas in spinel the tetrahedra are completely isolated from one another. The important consequence for hydration is that in the structure of wadsleyite there is one oxygen atom that is not coordinated to a silicon atom, and is therefore under-bonded. Hydration of wadsleyite therefore occurs by the protonation of this oxygen with charge balance being maintained by the removal of one Mg to form an octahedral vacancy (Kudoh et al., 1996; Smyth et al., 1997; Kudoh and Inoue, 1999). Up to 3.3 wt.% of water can be substituted into the structure in this way.

Much experimental effort has been expended to measure the elastic properties of wadsleyite, but there are still significant discrepancies between the values of the room-pressure elastic moduli measured by various techniques. Some discrepancies may be related to differences in composition, whose effect on the elastic properties were not initially appreciated. For polycrystalline ultrasonic measurements there is also a small contribution from the uncertainty in the stress–strain states of samples. Some discrepancies also arise from the difficulty in resolving values of K_0 and K'_0 independently from lower precision data, and the resulting truncation of the EoS to $K'_0 = 4$. Such truncations always result in an overestimate of K_0 because more recent studies reveal that like many of the DHMS phases, K'_0 is ~5, and K'_0 and K_0 are always negatively correlated. A self-consistent and internally precise dataset of the bulk moduli of various wadsleyites is provided by the single-crystal diffraction studies of Ross et al. (1998). These data show that hydrous wadsleyites with approx. 2.5 wt.% H_2O synthesised from a hydrated harzburgitic composition (Frost, 1999) have bulk moduli that are about 10% less than the "dry" Mg-end-member while K'_{0T} is not significantly changed from 5. The softening is not isotropic; the *b*-axis which is the stacking direction of the spinel-like layers (and also the axis of the Si_2O_7 groups) softens by 13%, whereas the other axes within the layers soften by ~7%.

At higher levels of H_2O substitution in wadsleyite, it appears that some form of ordering of vacancies and protonation sites can occur, leading to a reduction in symmetry of the structure from orthorhombic to monoclinic (Smyth et al., 1997; Kudoh and Inoue, 1999). Similarly, ordering of defects can also result in an orthorhombic structure whose *b*-unit-cell parameter is 2.5 times that of the pure Mg_2SiO_4 phase (Smyth and Kawamato, 1997). How much of these small symmetry distortions are dependent upon composition (for instance upon Fe content) and whether they are purely the result of structural changes during quenching following synthesis, is not yet clear. The only EoS measurement so far made on one of these ordered structures was performed on a monoclinic polymorph of pure Mg-wadsleyite with approximately the same water content, 2.5 wt.%, as the orthorhombic sample discussed above, and yielded the same K_{0T} (Ross et al., 1998). It can, therefore, be tentatively concluded that these small structural changes make little significant difference to the elasticity of wadsleyites.

No such under-bonded oxygen position is available in the densely-packed spinel structure, and the details of the hydration mechanism are not known. But recent reports suggest levels of hydration of at least 2 wt.% are possible (Kohlstedt et al., 1996, Inoue et al., 1998), although some of the reported IR spectra for H_2O in ringwoodite show broad peaks for OH which may indicate that much of this water is not stored in the crystal structure at all (see discussion in Bolfan-Casanova et al., 2000, who synthesised pure Mg ringwoodite with 7500 ppm H_2O). Preliminary results of a single-crystal X-ray study suggest that H_2O is incorporated in Fe-containing ringwoodite via a complex mechanism involving Fe oxidation, vacancies on both tetrahedral and octahedral sites, and partial occupancy of normally vacant tetrahedral interstices within the close-packed oxygen array (Smyth et al., 1999). One possible interpretation would be that the hydration mechanism involves the production of stacking faults within the spinel. Because of the polytypic relationship between the structures, these stacking faults would locally have the wadsleyite structure, and hydration may then take place on these faults in the same manner as in bulk wadsleyite samples. However, the bulk modulus of a hydrated magnesium silicate ringwoodite (Inoue et al., 1998) is 16% less than that of the anhydrous end-member, proportionately greater than that observed for similar levels of substitution in wadsleyite, which is suggestive of another substitution mechanism.

Whatever the details of the substitution mechanism in ringwoodite, both wadsleyite and ringwoodite appear to be softened by hydration proportionately more than the other DHMS phases. This has been suggested to be the result of the differences in hydration mechanisms at the atomic level. Whereas the "alphabet phases", as we have described, generally accommodate water by changing the structure or proportions of stacking modules and so maintain structural integrity, ringwoodite and wadsleyite must accommodate protons by creation of vacancies on cation sites, thereby breaking structural integrity and weakening the structures. The same is true for hydration of garnets. The mineral hibschite, with 40% of the tetrahedral Si substituted by 4H (equivalent to ~11 wt.% of H_2O) has a bulk modulus approximately 60% of that of its anhydrous analogue grossular (O'Neill et al., 1993).

Prediction on the basis of Fig. 5 of the effects of hydration on the elastic moduli of magnesium silicate phases, therefore, requires prior knowledge of the mechanism of hydration in the mineral phase of interest. Fortunately, based upon the most recent measurements available, there appears to be a strong correlation between the density of the DHMS phases and their bulk modulus (Fig. 6), of the sort previously postulated for a variety of structures (e.g. Birch's law, Birch, 1947) and subsequently shown by more precise measurements not to hold in general. The reason why a general linear trend appears to hold for the DHMS phases, irrespective of the level of hydration, is related to the common elements in their structures. All contain arrays of edge-sharing octahedra, some with Si in addition to Mg. Such layers of octahedra contain no degrees of internal freedom and so cannot compress by relative tilting of the octahedra as, for example, occurs in perovskites which have corner-linked octahedra. Similarly, the topology of the cross-linking of the octahedra by SiO_4 tetrahedra, where present, is always such that these cannot tilt either. All of the DHMS phases, therefore, possess no rigid-unit modes (e.g. Dove et al., 1995) that would allow compression to be accommodated by tilting. Instead, they all have to compress by the direct shortening of cation–oxygen bonds and/or deformation of the polyhedra. The resistance to such compression is increased with the degree

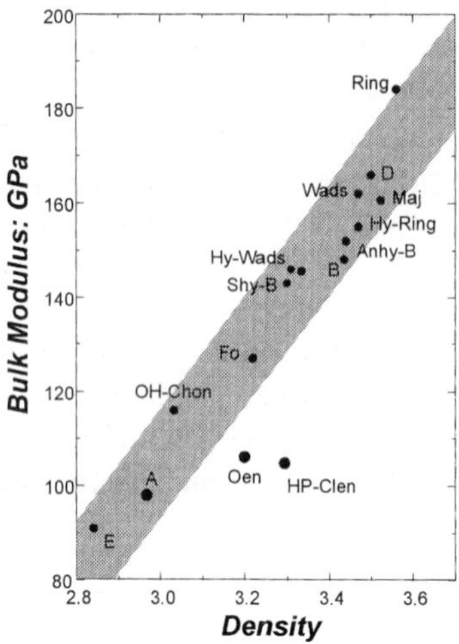

Fig. 6. The variation with density of the isothermal bulk moduli, K_{0T}, of series of related compounds in the MgO–SiO_2–H_2O ternary (see Fig. 1). Data sources the same as Fig. 5, except $MgSiO_3$ garnet (Maj) from Hazen et al. (1994) and the high-pressure clinoenstatite (HP-Clen) from Shinmei et al. (1999).

of edge-sharing between the polyhedra because of the reduction in cation–cation distances and corresponding increase in cation–cation repulsions. An increase in shared edges also leads to an increase in density, so the relationship shown in Fig. 6 follows directly. It should be noted that $MgSiO_3$ garnet, a structure that is a framework but also possesses no rigid-unit modes (Hammonds et al., 1998), also lies on the trend shown in Fig. 6. By contrast, the two pyroxene structures which can accommodate compression, at least in part, by polyhedral rotation are softer than would be expected from their density alone.

Shared edges between polyhedra result in higher coordination number for the oxygen anions by cations. Thus, polyhedral edge-sharing usually results in fully-bonded or slightly over-bonded oxygen anions, which are not suitable as sites for protonation. The structural changes accompanying hydration in most of the DHMS phases, therefore, result in a reduction of shared polyhedral edges to provide oxygen sites suitable for protonation, and this results in increased cation–cation distances and hence both the observed elastic softening and the reduction in density. The common structural features of the DHMS phases also presumably account for the narrow range in values found for the pressure derivative of the bulk modulus.

For all of the phases that have been measured at room temperature with sufficiently high precision, the K'_0 values range from between 5 and 6 for the humite minerals, wadsleyites, superhydrous B and phase A to $K'_0 \geq 7$ for phases B and E. Note that these values are all significantly greater than the commonly assumed value of 4 for dense-packed structures, a number derived from a truncation of a power series in the derivation of equations of state based upon Eulerian strain. But this range falls significantly below that found for the more open structures stable at lower pressures; for example, $MgSiO_3$ *ortho*-pyroxene has $K'_0 \approx 9$ (Ito et al., 1979; Angel and Hugh-Jones, 1994; Flesch et al., 1998).

4. Conclusions

In summary, dense hydrous magnesium silicate phases could form along cool subduction geotherms where the stability of antigorite overlaps with that of phase A. Under such conditions, if a slab were fully serpentinised, it could transport 4 wt.% H_2O into the transition zone in an assemblage containing phase A. Serpentinisation could not possibly exceed 10%, however, as the slab would become buoyant. Therefore, water contents of 4000 ppm maximum are more reasonable. Wadsleyite, and possibly also ringwoodite, will dominate water storage in slabs in the transition zone. However, superhydrous phase B may also form if current measurements of the H_2O content of ringwoodite are over estimated. Due to the low temperature stability of phase D in the lower mantle and the possible low water content of magnesium silicate perovskite, slabs may have a propensity to release water as they enter the lower mantle.

Whether the transport of water into the mantle by dense hydrous magnesium silicates is plausible or not remains uncertain due to factors that cannot be determined by high-pressure experiments. The main uncertainty is whether antigorite is stable to high enough pressures for phase A to form and this will depend on the depth of hydration of the lithosphere as it resides under the ocean, and the geothermal gradient followed during subduction. If hydration occurs to significant depth into the slab (>10 km) then it may be plausible that such regions could remain cool enough for the antigorite + phase A 'choke point' to be passed. It is unlikely that pervasive serpentinisation could take place to such depth in the slab, but if deep faults become hydrated, such as those that occur at the outer rise as the slab bends to subduct, then a narrow zone of hydrated slab could be created and preserved into the deep mantle. One possibility is that such a zone of weaker hydrous material could act as a focal plane for deep earthquakes. The previously discussed propensity for slabs to dehydrate as they enter the lower mantle could then explain the absence of deep focus earthquakes at depths greater than 660 km.

Could volumes of hydrated material be detected seismologically? As we have alluded to earlier, there are very few data on the high-temperature EoS of the DHMS phases apart from olivine and its high-pressure polymorphs wadsleyite and ringwoodite. Similarly, there are few shear moduli data for the DHMS phases, and what is available (e.g. Sinogeiken and Bass, 1999) is data for either natural material with complex chemistry or synthetic Fe-bearing material. However, from this limited dataset, it appears that the trends of shear modulus with hydration and density parallel those that have been demonstrated for bulk modulus. Similarly, there are few thermal expansion data, and the only data on the temperature dependence of bulk moduli come from high P, T diffraction studies in which the values of the derived EoS parameters are highly correlated with one another. The absence of such data makes it impossible to calculate seismic wave velocities for these DHMS phases at the temperatures and pressures at which they might occur within subduction zones. However, if we make the reasonable assumption that the structural controls on the bulk moduli of the DHMS phases extends to their temperature dependence, then one can estimate that the values of dK/dT and α_V for all of these phases are similar to those determined for forsterite olivine. In this case, the scaling of the bulk modulus of these phases with density (Fig. 6) means that their seismic velocities will vary approximately in proportion to $\rho^{-1/2}$. Thus, they will be seismically undetectable because any anomaly in wave velocities in the subducting slab may be equally attributed to hydration, higher temperature, or lower pressure; all three will lead to a reduction in density and an increase in velocities.

These issues of prediction are, in principle, further complicated by the influence of chemical substitutions into the pure DHMS phases. In olivine, for example, complete substitution of Mg by Fe leads to a ~5% increase in the room-pressure bulk modulus (Isaak et al., 1993). By contrast, no effect of Fe substitution, at least up to expected mantle levels of 10–15 mol% is found for wadsleyite. Minor amounts of other chemical substituents may well have disproportionate effects on elastic properties of at least this order of magnitude, and the distribution coefficients between the phases are not yet known. A further uncertainty of 1–3% (depending upon the elastic anisotropy of the mineral) is introduced as we do not know the distribution of stress and strain within the rocks within the subducting slab. Under dry, low-temperature conditions, the grains may well be welded together, in which case the seismic wave velocities correspond to the Voigt bound on the elastic properties of the aggregate. As temperature is increased and fluids are released, one might expect the elasticity of the rock to evolve towards that of the Reuss bound, at which all grains are free to deform under a state of uniform stress. Furthermore, grain boundary sliding at high temperatures has been recently experimentally demonstrated to cause reduction almost to zero of shear moduli for low frequencies in the seismic regime (Webb et al., 1999).

Acknowledgements

We would like to thank I. Stretton and S. Mackwell for discussions, and B.J. Wood and J. Smyth for reviews. Synthesis of single-crystal specimens, and this collaboration, were supported by the EU "TMR — Large Scale Facilities" programme in Bayreuth (Contract no. ERBFMGECT980111 to D.C. Rubie). Single-crystal high-pressure X-ray diffraction studies at University College London were supported by NERC Grant GR3/09782 to N.L. Ross.

References

Angel, R.J., Hugh-Jones, D.A., 1994. Equations of state and thermodynamic properties of enstatite pyroxenes. J. Geophys. Res. 99B, 19777–19783.

Bass, J.D., Kanzaki, M., Howell, D.A., 1991. Sound velocities and elastic properties of phase E: a high pressure hydrous silicate. EOS, 72, 499 (abstract).

Bell, D.R., Rossman, G.R., 1992. Water in earth's mantle: the role of nominally anhydrous minerals. Science 255, 1391–1396.

Birch, F., 1947. Finite elastic strain of cubic crystals. Phys. Rev. 71, 809–824.

Bolfan-Casanova, N., Keppler, H., Rubie, D.C., 2000. Water partitioning between nominally anhydrous minerals in the $MgO–SiO_2–H_2O$ system up to 24(GPa: implications for the distribution of water in the earth's mantle. Earth Planet. Sci. Lett. 182, 209–221.

Bose, K., Navrotsky, A., 1998. Thermochemistry and phase equilibria of hydrous phases in the system $MgO–SiO_2–H_2O$: implications for volatile transport to the mantle. J. Geophys. Res. 103B, 9713–9719.

Brown, J.M., 1999. The NaCl pressure standard. J. Appl. Phys. 86, 5801–5808.

Brunet, F., Allan, D.R., Redfern, S.A.T., Angel, R.J., Miletich, R., Reichmann, H., Sergent, J., Hanfland, M., 1999. Compressibility and thermal expansivity of synthetic apatites, $Ca_5(PO_4)_3X$ with X = OH, F and Cl. Euro. J. Mineral. 11, 1023–1036.

Bureau, H., Keppler, H., 1999. Complete miscibility between silicate melts and hydrous fluids in the upper mantle: experimental evidence and geochemical implications. Earth Planet. Sci. Lett. 165, 187–196.

Burnley, P.C., Navrotsky, A., 1996. Synthesis of high pressure hydrous magnesium silicates: observations and analysis. Am. Mineral. 81, 317–326.

Crichton, W.A., Ross, N.L., 2000a. Equation of state of phase E. Miner. Mag. 424, 561–567.

Crichton, W.A., Ross, N.L., 2000. Single crystal equation of state measurements on Mg-End-members of the B-group minerals. In: Manghnani, M.H., Nellis, W.J., Nicol, M.F. (Eds.), AIRAPT-17 Proceeding-Science and Technology of High Pressure, Volume II, Universities Press, Hydrabad, India, pp. 587–590.

Crichton, W.A., Ross, N.L., Gasparik, T., 1999. Equations of state of magnesium silicates anhydrous B and superhydrous B. Phys. Chem. Minerals 26, 570–575.

Cynn, H., Hofmeister, A.M., Burnley, P.C., Navrotsky, A., 1996. Thermodynamic properties and hydrogen speciation from vibrational spectra of dense hydrous magnesium silicates. Phys. Chem. Minerals 23, 361–376.

Domanik, K.J., Holloway, J.R., 1996. The stability and composition of phengitic muscovite and associated phases from 5.5 to 11 GPa: implications for deeply subducted sediments. Geochim. Cosmochim. Acta 60, 4133–4150.

Dove, M.T., Heine, V., Hammonds, K.D., 1995. Rigid unit modes in framework silicates. Miner. Mag. 59, 629–639.

Eggleton, R.A., Boland, J.N., Ringwood, A.E., 1978. High pressure synthesis of a new aluminium silicate: $Al_5Si_5O_{17}(OH)$. Geochem. J. 12, 191–194.

Faust, J., Knittle, E., 1994. Static compression on chondrodite: implications for water in the upper mantle. Geophys. Res. Lett. 21, 1935–1938.

Fei, Y., Saxena, S.K., Navrotsky, A., 1990. Internally consistent thermodynamic data and equilibrium phase relations for compounds in the system $MgO–SiO_2$ at high pressure and high temperature. J. Geophys. Res. 95B, 6915–6928.

Finger, L.W., Hazen, R.M., Prewitt, C.T., 1991. Crystal structures of $Mg_{12}Si_4O_{19}(OH)_2$ (phase B) and $Mg_{14}Si_5O_{24}$ (phase AnB). Am. Mineral. 76, 1–7.

Flesch, L.M., Li, B., Liebermann, R.C., 1998. Sound velocities of polycrystalline $MgSiO_3$-*ortho*-pyroxene to 10 GPa at room temperature. Am. Mineral. 83, 444–450.

Frost, D.J., 1999. The stability of dense hydrous magnesium silicates in earth's transition zone and lower mantle. In: Fei, Y., Bertka, C.M., Mysen, B.O. (Eds.), Mantle Petrology: Field observations and High-pressure Experimentation. Geochemical Society Special Publication no. 6, pp. 283–296.

Frost, D.J., Fei, Y., 1998. Stability of phase D at high pressure and high temperature. J. Geophys. Res. 103B, 7463–7474.

Furukawa, Y., 1993. Depth of the decoupling plate interface and thermal structure under arcs. J. Geophys. Res. 98B, 20005–20013.

Gasparik, T., 1993. The role of volatiles in the transition zone. J. Geophys. Res. 98B, 4287–4299.

Hammonds, K.D., Bosenick, A., Dove, M.T., Heine, V., 1998. Rigid unit modes in crystal structures with octahedrally coordinated atoms. Am. Mineral. 83, 476–479.

Hazen, R.M., Downs, R.T., Conrad, P.G., Finger, L.W., Gasparik, T., 1994. Comparative compressibilities of majorite-type garnets. Phys. Chem. Minerals 21, 344–349.

Horiuchi, H., Morimoto, N., Yamamoto, K., Akimoto, S.-I., 1979. Crystal structure of $2Mg_2SiO_4{\cdot}3Mg(OH)_2$, a new high-pressure structure type. Am. Mineral. 64, 593–598.

Hugh-Jones, D.A., 1997. Thermal expansion of $MgSiO_3$ and $FeSiO_3$ *ortho*- and clino-pyroxenes. Am. Mineral. 82, 689–696.

Inoue, T., 1994. Effect of water on melting phase relations and melt composition in the system $Mg_2SiO_4–MgSiO_3–H_2O$ up to 15 GPa. Phys. Earth Planet. Inter. 85, 237–263.

Inoue, T., Yurimoto, H., Kudoh, Y., 1995. $Mg_{1.75}SiH_{0.5}O_4$: a new water reservoir in the mantle transition region. Geophys. Res. Lett. 22, 117–120.

Inoue, T., Weidner, D.J., Northrup, P.A., Parise, J.B., 1998. Elastic properties of hydrous ringwoodite (γ-phase) in Mg_2SiO_4. Earth Planet. Sci. Lett. 160, 107–113.

Irifune, T., Kuroda, K., Funamori, N., Uchida, T., Yagi, T., Inoue, T., Miyajima, N., 1996. Amorphization of serpentine at high pressure and high temperature. Science 272, 1468–1470.

Irifune, T., Kubo, T., Isshiki, M., Yamasaki, Y., 1998. Phase transformations in serpentine and transportation of water into the lower mantle. Geophys. Res. Lett. 22, 117–120.

Isaak, D.G., 1992. High temperature elasticity of iron-bearing olivines. J. Geophys. Res. 97B, 1871–1885.

Isaak, D.G., Anderson, O.L., Goto, T., 1989. Elasticity of single-crystal forsterite measured to 1700 K. J. Geophys. Res. 94B, 5895–5906.

Isaak, D.G., Graham, E.K., Bass, J.D., Wang, H., 1993. The elastic properties of single-crystal fayalite as determined by dynamical measurement techniques. Pure App. Geophys. 141, 393–414.

Ito, E., Takahashi, E., 1989. Postspinel transformations in the system $Mg_2SiO_4–Fe_2SiO_4$ and some geophysical implications. J. Geophys. Res. 94B, 10637–10646.

Ito, H., Mizutani, H., Ichinose, K., Akimoto, S., 1979. Ultrasonic wave velocity measurements in solids under high pressure using solid pressure media. In: Manghnani, M.H., Akimoto, S.-I. (Eds.), High-pressure Research: Applications in Geophysics. Academic Press, New York, NY, pp. 603–620.

Jackson, J.M., Sinogeikin, S.V., Bass, J.D., 1999. Elasticity of $MgSiO_3$ *ortho*-enstatite. Am. Mineral. 84, 677–680.

Jambon, A., 1994. Earth degassing and large scale geochemical cycling of volatile elements. In: Carroll, M.R., Holloway, J.R. (Eds.), Volatiles in Magmas, MSA Rev. Min. 30, 479–509.

Javoy, M., Pineau, F., 1991. The volatiles record of a popping rock from the Mid-Atlantic Ridge at 14°N: chemical and isotopic composition of gas trapped in the vesicles. Earth Planet. Sci. Lett. 107, 598–611.

Jeanloz, R., Morris, S., 1986. Temperature distribution in the crust and mantle. Ann. Rev. Earth Planet. Sci. 14, 377–415.

Jones, L.E.A., 1977. High-temperature elasticity of rutile-structure MgF_2. Phys. Chem. Minerals 1, 179–197.

Kanzaki, M., 1991. Stability of hydrous magnesium silicates in the mantle transition zone. Phys. Earth Planet. Int. 66, 307–312.

Kanzaki, M., 1993. Calculated powder X-ray patterns of phase B, anhydrous B and superhydrous B: re-assessment of previous studies. Mineral. J. 16, 278–285.

Katsura, T., Ito, E., 1989. The system Mg_2SiO_4–Fe_2SiO_4 at high pressures and temperatures: precise determination of stabilities of olivine, modified spinel and spinel. J. Geophys. Res. 94, 15663–15670.

Kawamoto, T., Leinenweber, K., Hervig, R.L., Holloway, J.R., 1995. Stability of hydrous minerals in H_2O-saturated KLB-1 peridotite up to 15 GPa. In Farley, K.A. (Ed.), Volatiles in the Earth and Solar System. American Institute of Physics, Woodbury, New York, p. 229.

Kieffer, S.W., 1979. Thermodynamics and lattice vibrations of minerals: (3) lattice dynamics and an approximation for minerals with application to simple substances and framework silicates. Rev. Geophys. Space Phys. 17, 35–39.

Kieffer, S.W., 1980. Thermodynamics and lattice vibrations of minerals: (4) application to chain and sheet silicates and *ortho*-silicates. Rev. Geophys. Space Phys. 18, 862–886.

Knoche, R., Webb, S.L., Rubie, D.C., 1997. Experimental determination of acoustic wave velocities at earth-mantle conditions using a multi-anvil press. Phys. Chem. Earth 22, 125–130.

Kohlstedt, D.L., Keppler, H., Rubie, D.C., 1996. Solubility of water in the α-, β- and γ-phases of $(Mg, Fe)_2SiO_4$. Contrib. Mineral. Petrol. 123, 345–357.

Kudoh, Y., Inoue, T., 1999. Mg-vacant structural modules and dilution of the symmetry of hydrous wadsleyite, β-$Mg_{2-x}SiH_{2x}O_4$ with $0.00 < x < 0.25$. Phys. Chem. Minerals 26, 382–388.

Kudoh, Y., Finger, L.W., Hazen, R.M., Prewitt, C.T., Kanzaki, M., Veblen, D.R., 1993. Phase E: a high pressure hydrous silicate with unique crystal chemistry. Phys. Chem. Minerals 19, 357–360.

Kudoh, Y., Inoue, T., Arashi, H., 1996. Structure and crystal chemistry of hydrous wadsleyite, $Mg_{1.75}SiH_{0.5}O_4$: possible hydrous magnesium silicate in the mantle transition zone. Phys. Chem. Minerals 23, 461–469.

Kuribayashi, T., Kudoh, Y., Akizuki, M., 1998. The anisotropy of compression of chondrodite, $Mg_5Si_2O_8(OH, F)_2$ in relation to its crystal structure. EOS, 79, F867 (abstract).

Kuribayashi, T., Kudoh, Y., Kagi, H., 1999. High pressure X-ray diffraction study on phase A, $Mg_7Si_2H_6O_{14}$, up to 11.2 GPa. EOS Trans. AGU Fall Meeting, 80, F938.

Liebermann, R.C., Li, B., 1999. Elasticity at high pressures and temperatures. In: Hemley, R.J. (Ed), Ultrahigh-pressure Mineralogy. MSA Rev. Mineral. Vol. 37.

Liu, L.-G., 1986. Phase transformations in serpentine at high pressures and temperature. Phys. Earth Planet. Inter. 42, 255–262.

Liu, L.-G., 1987. Effects of H_2O on the phase behaviour of the forsterite–enstatite system at high pressures and temperatures and implications for the earth. Phys. Earth Planet. Inter. 49, 142–167.

Lu, R., Keppler, H., 1997. Water solubility in pyrope to 100 kbar. Contrib. Mineral. Petrol. 129, 35–42.

Lu, R., Hofmeister, A.M., Wang, Y.B., 1994. Thermodynamic properties of ferromagnesium silicate perovskites from vibrational spectroscopy. J. Geophys. Res. 99B, 11795–11804.

Luth, R.W., 1995. Is phase A relevant to the earth's mantle? Geochim. Cosmochim. Acta 59, 679–682.

Matsui, T., Manghnani, M.H., 1985. Thermal expansion of single-crystal forsterite to 1023 K by Fizeau interferometry. Phys. Chem. Minerals 12, 201–210.

Meade, C., Reffner, J.A., Ito, E., 1994. Synchrotron infrared absorbance measurements of hydrogen in $MgSiO_3$ perovskite. Science 264, 1558–1560.

O'Neill, B., Bass, J.D., Rossman, G.R., 1993. Elastic properties of hydrogrossular garnet and implications for water in the upper mantle. J. Geophys. Res. 98, 20031–20037.

Ohtani, E., Shibata, T., Kubo, T., Kato, T., 1995. Stability of hydrous phases in the transition zone and the upper most part of the lower mantle. Geophys. Res. Lett. 22, 2553–2556.

Ohtani, E., Mitzobata, H., Kudoh, Y., Nagase, T., 1997. A new hydrous silicate, a water reservoir, in the upper part of the lower mantle. Geophys. Res. Lett. 24, 1047–1050.

Ono, S., 1998. Stability limits of hydrous minerals in sediment and mid-ocean ridge basalt compositions: implications for water transport in subduction zones. J. Geophys. Res. 103B, 18253–18267.

Pacalo, R.E.G., Parise, J.B., 1992. Crystal structure of superhydrous B, a hydrous magnesium silicate synthesised at 1400°C and 20 GPa. Am. Mineral. 77, 681–684.

Pawley, A.R., 1994. The pressure and temperature stability limits of lawsonite: implications for H_2O recycling in subduction zones. Contrib. Mineral. Petrol. 118, 99–108.

Pawley, A.R., Redfern, S.A.T., Wood, B.J., 1995. Thermal expansivities and compressibilities of hydrous phases in the system MgO–SiO_2–H_2O: talc, phase A and 10 Å phase. Contrib. Mineral. Petrol. 22, 301–307.

Rice, S.B., Benimoff, A.I., Sclar, C.B., 1989. 3.65 phase in system MgO–SiO_2–H_2O at pressures greater than 90 kbars: crystallochemical implications for mantle phases. Proceedings of the International Geological Congress, Meeting 2 (abstract), Washington, DC, p. 694.

Ringwood, A.E., Major, A., 1967. High-pressure reconnaissance investigations in the system Mg_2SiO_4–MgO–H_2O. Earth Planet. Sci. Lett. 2, 130–133.

Ross, N.L., Hazen, R.M., 1989. Single crystal X-ray diffraction study of $MgSiO_3$ perovskite from 77 to 400 K. Phys. Chem. Minerals 16, 415–420.

Ross, N.L., Crichton, W.A., 2001. Compression of hydroxy-clinohumite ($Mg_9Si_4O_{16}(OH)_2$) and hydroxy-chondrodite ($Mg_5Si_2O_8(OH)_2$). Am. Mineral., in press.

Ross, N.L., Crichton, W.A., 2000. Equation of state of phase A and elasticity systematics along the forsterite–brucite join. EOS, 81, S50 (abstract).

Ross, N.L., Crichton, W.A., Frost, D., Kung, J., 1998. Equations of state of hydrous wadsleyites determined by high-pressure single-crystal X-ray diffraction. EOS, 80, F861 (abstract).

Rubie, D.C., 1999. Characterising the sample environment in multi-anvil high-pressure experiments. Phase Transitions 68, 431–451.

Sato, K., Katsura, T., Ito, E., 1997. Phase relations of natural phlogopite with and without enstatite up to 8 GPa: implication for mantle metasomatism. Earth Planet. Sci. Lett. 146, 511–526.

Schmidt, M.W., 1995. Lawsonite: upper pressure stability and formation of higher density hydrous phases. Am. Mineral. 80, 1286–1292.

Schmidt, M.W., Poli, S., 1999. Experimentally-based water budgets for dehydrating slabs and consequences for arc magma generation. Earth Planet. Sci. Lett. 163, 361–379.

Schmidt, M.W., Finger, L.W., Angel, R.J., Dinnebier, R.E., 1998. Synthesis, crystal structure, and phase relations of $AlSiO_3OH$, a high-pressure hydrous phase. Am. Mineral. 83, 881–888.

Sclar, C.B., Carrison, L.C., Schwartz, C.M., 1965. High pressure synthesis and stability of a new hydronium bearing layer silicate in the system $MgO–SiO_2–H_2O$. EOS, 46, 184 (abstract).

Shen, A.H., Keppler, H., 1997. Direct observation of complete miscibility in the albite–H_2O system. Nature 285, 710–712.

Shieh, S.R., Mao, H.-K., Hemley, R.J., Ming, L.C., 1998. The decomposition of phase D in the lower mantle and the fate of dense hydrous silicates in subducting slabs. Earth Planet. Sci. Lett. 159, 12–23.

Shieh, S.R., Mao, H.-K., Konzett, J., Hemley, R.J., 2000. In situ high pressure X-ray diffraction of Phase E to 15 GPa. Am. Mineral. 85, 765–769.

Shinmei, T., Tomioka, N., Fujino, K., Kuroda, K., Irifune, T., 1999. In situ X-ray diffraction study of enstatite up to 12 GPa and 1473 K and equations of state. Am. Mineral. 84, 1588–1594.

Silver, P.G., Beck, S.L., Wallace, T.C., Meade, C., Myers, S.C., James, D.E., Kuhnel, R., 1995. Rupture characteristics of the deep Bolivian earthquake of 9 June 1994 and the mechanism of deep-focus earthquakes. Science 268, 69–73.

Sinogeiken, S.V., Bass, J.D., 1999. Single-crystal elsatic properties of chondrodite: implications for wayer in the upper mantle. Phys. Chem. Minerals 26, 297–303.

Smyth, J.R., 1987. β-Mg_2SiO_4: a potential host for water in the mantle? Am. Mineral. 72, 1051–1055.

Smyth, J.R., Kawamato, T., 1997. Wadsleyite II: a new high-pressure hydrous phase in the peridotite–H_2O system. Earth Planet. Sci. Lett. 146, E9–E16.

Smyth, J.R., Kawamoto, T., Jacobsen, S.D., Swope, R.J., Hervig, R., Holloway, J.R., 1997. Crystal structure of monoclinic hydrous wadsleyite [β-$(Mg, Fe)_2SiO_4$]. Am. Mineral. 82, 270–275.

Smyth, J.R., Frost, D.J., McCammon, C., 1999. Hydration mechanism in ringwoodite, γ-$Mg_{1.63}Fe_{0.22}H_{0.4}Si_{0.95}O_4$. EOS, 80, F927 (abstract).

Sobolev, A.V., Chaussidon, M., 1996. H_2O concentrations in primary melts from supra-subduction zones and mid-ocean ridges: implications for H_2O storage and recycling in the mantle. Earth Planet. Sci. Lett. 137, 45–55.

Stein, S., Stein, C.A., 1996. Thermo-mechanical evolution of oceanic lithosphere: implications for the subduction process and deep earthquakes. In: Bebout, G.E., Scholl, D.W., Kirby, S.H., Platt, J.P. (Eds.), Subduction: Top to Bottom, Geophys. Monogr. Ser., Vol. 96, AGU, Washington, DC, pp. 1–17.

Sumino, Y., 1979. The elastic constants of Mn_2SiO_4, Fe_2SiO_4 and Co_2SiO_4 and the elastic properties of olivine group minerals at high temperatures. J. Phys. Earth 27, 209–238.

Suzuki, I., Ohtani, E., Kumazawa, M., 1979. Thermal expansion of γ-Mg_2SiO_4. J. Phys. Earth 27, 53–61.

Suzuki, I., Ohtani, E., Kumazawa, M., 1980. Thermal expansion of modified spinel, γ-Mg_2SiO_4. J. Phys. Earth 28, 273–280.

Tronnes, R.G., 1990. Low-Al, high-K amphiboles in subducted lithosphere from 200 to 400 km depth: experimental evidence. EOS, 71, 1587 (abstract).

Ulmer, P., Trommsdorff, V., 1995. Serpentine stability to mantle depths and subduction-related magnetism. Science 268, 858–861.

Webb, S.L., Jackson, I., Fitz Gerald, J., 1999. Viscoelasticity of the titanite perovskites $CaTiO_3$ and $SrTiO_3$ at high temperature. Phys. Earth Planet. Int. 115, 259–291.

Weidner, D.J., Sawamoto, H., Sasaki, S., Kumazawa, M., 1984. Elasticity of Mg_2SiO_4 spinel. J. Geophys. Res. 89, 7852–7860.

Wunder, B., Medenbach, O., Daniels, P., Schreyer, W., 1995. First synthesis of the hydroxyl-end member of humite, $Mg_7Si_3O_{12}(OH)_2$. Am. Mineral. 80, 638–640.

Xia, X., Weidner, D.J., Zhao, H., 1998. Equation of state of brucite: single-crystal Brillouin spectroscopy study and polycrystalline pressure–volume–temperature measurement. Am. Mineral. 83, 68–74.

Zha, C.S., Duffy, T.S., Mao, H.K., Downs, R.T., Hemley, R.J., Weidner, D.J., 1997. Single-crystal elasticity of β-Mg_2SiO_4 to the pressure of the 410 km seismic discontinuity in the earth's mantle. Earth Planet. Sci. Lett. 147, E9–15.

ELSEVIER

Physics of the Earth and Planetary Interiors 127 (2001) 197–214

PHYSICS
OF THE EARTH
AND PLANETARY
INTERIORS

www.elsevier.com/locate/pepi

Seismological structure of subduction zones and its implications for arc magmatism and dynamics

Dapeng Zhao*

Geodynamics Research Center, Ehime University, Matsuyama 790-8577, Japan

Received 12 January 2000; accepted 27 July 2000

Abstract

In this paper I review recent seismological findings on the structure, magmatism, and dynamics of subduction zones. High-resolution seismic tomography has revealed prominent low-velocity (low-V) and high-attenuation (low-Q) zones that exist in the crust and uppermost mantle just beneath active arc volcanoes and extend to 400 km depth in the mantle wedge. The low-V/low-Q zones are located in the central portion of the mantle wedge and lie 30–50 km above the subducted oceanic slab. The mantle wedge low-V/low-Q zones also exhibit strong seismic anisotropy. These results suggest that arc magmatic systems are not limited to the near-surface areas, but are related to the deep processes, such as the convective circulation in the mantle wedge and dehydration reactions in the subducted slab. The low-V/low-Q zones beneath the arc and back-arc are separated at shallow levels but merge at depths >100 km, indicating that the slab components of the arc and back-arc magmas occur through mixing at these depths. These low-V/low-Q bodies form the deep roots and sources of the arc magmatism and volcanism. Large crustal earthquakes in Japan are found to occur around low-V zones that may represent weak sections of the seismogenic crust. The crustal weakening is thought to be closely related to the subduction of the Pacific and Philippine Sea plates in this region. Along the volcanic front and in back-arc areas, the crustal weakening may be caused by the active volcanism and the presence of magma chambers. In the forearc areas, fluids were detected in the earthquake source areas, which may have contributed to the rupture nucleation and may be related to the dehydration of the subducted slab. These results suggest that large crustal earthquakes may not strike anywhere, but only anomalous areas which may be detected with geophysical methods. Suggestions are also provided for future directions of seismological research of subduction zones.

Keywords: Subduction zones; Seismic tomography; Arc magmatism; Dynamics; Seismic velocity; Attenuation; Anisotropy

1. Introduction

According to plate tectonics, new oceanic plates are formed at mid-ocean ridges by the upwelling of hot mantle materials, producing mid-ocean ridge magmatism. The same amount of materials returns back to the Earth's interior at subduction zones where the heavier oceanic plates are descending beneath the more buoyant continental plates, causing the most active seismicity and volcanism on the Earth. Subduction zones are convergent plate boundaries characterized geomorphologically by deep ocean trenches and island arcs or continental margins, seismically by landward dipping Wadati–Benioff zones of deep earthquakes, tectonically by regional-scale crustal faulting and terrane movements, and magmatically by arcuate and linear belts of eruptive centers, the so-called volcanic front. Subduction zones have long been recognized as key

* Tel.: +81-89-927-9652;
fax: +81-89-927-9640; URL: http://www.ehime-u.ac.jp/.
E-mail address: zhao@sci.ehime-u.ac.jp (D. Zhao).

PII: S0031-9201(01)00228-X

elements in plate tectonics. Subduction and arc magmatism are fundamental processes in the evolution of the Earth. They play critical roles in the present day differentiation of Earth's material and are believed to be the major sites of generation of continental crust. Subduction is also significant in the water and carbon cycles.

In contrast to magmatism at mid-ocean ridges where it is generally agreed that the responsible process is adiabatic upwelling, arc magmatism at subduction zones has led to a range of proposals for its source region and process. The fundamental paradox is why the high heat flows and the abundance of melt generation take place right above the relatively cold subducting slab. With the advent of plate tectonics, some researchers proposed that shear and frictional heating led to melting of the oceanic crust (Oxburgh and Turcotte, 1970; Turcotte and Schubert, 1973). Hsui et al. (1983) suggested that the oceanic crust melts as a result of high temperatures arising from the hot induced mantle flow impinging on it. Some researchers proposed that the melting is the result of adiabatic upwelling of either ecologitic crust (Brophy and Marsh, 1986) or mantle wedge (Tatsumi et al., 1983; Plank and Langmuir, 1988), while others argued that the melting results from fluxing of the peridotitic mantle wedge by water from the subducting oceanic crust (Gill, 1981; Tatsumi, 1989; Davies and Stevensen, 1992; Iwamori, 1998).

One of the striking observations of arc volcanism is that the active volcanoes nearest the trench form a line parallel to the trench: the volcanic front (Sugimura, 1960). In addition, these volcanoes lie approximately 100 km above the Wadati–Benioff deep seismic zone within the subducting slab (Hasegawa et al., 1978; Tatsumi, 1989; Zhao et al., 1997a,b). These constraints are quite robust, and any model hoping to explain subduction zone volcanism must be able to account for them.

Once an earthquake occurs, it emits seismic waves that propagate through the Earth's interior and can be recorded by seismographs installed at seismic stations when they finally arrive at the Earth's surface. Analyses of the observed seismic waves provide information on the rupture process of the earthquake as well as physical properties of the materials along the wave trajectories in the Earth. The classic studies of seismic waves have led to the discoveries of the layered structure of the Earth. Since the advent of seismic tomography two decades ago (Aki and Lee, 1976; Aki et al., 1977), seismologists have been mapping the heterogeneous structure of the Earth on a broad range of scales. These three-dimensional (3-D) modelings of the Earth's structure promise to answer some basic questions of geodynamics and signify a revolution in Earth science (Dziewonski and Anderson, 1984). Seismic tomography is still developing at an accelerated pace and is expected to be an active field of seismological research for several decades to come (Iyer and Hirahara, 1993). Seismic tomography has also been the most powerful tool for the imaging of subduction zones.

During the last decade, thanks to the advances of seismic imaging techniques like tomography, and high seismic activity and dense seismographic networks installed at some island arcs such as Japan, Alaska, and Tonga, seismologists have made significant progresses in understanding the structure, magmatism and dynamics of subduction zones. Here, I address the recent seismological findings on this topic. Numerous such studies of subduction zones have been published in literature in the last decade, and it would be impossible to make a thorough review of all of them in this limited space. Therefore, I will mainly focus on my own studies and those I am familiar with.

2. Seismic velocity structure

There are basically three seismological parameters to describe the Earth structure: seismic velocity, attenuation, and anisotropy. Seismic velocity has been the best estimated physical parameter of the Earth's interior because it is determined mainly from seismic wave travel times that can be measured with a high precision and a great quantity. Hence, seismic tomography methods have been mainly applied to travel time data to estimate the 3-D velocity structure. Before describing the major findings on the seismic velocity structure of subduction zones, let me first make a brief review of the recent advances in the methodology of seismic tomography at local and regional scales.

2.1. Advances in seismic tomography methodology

Most of the early tomographic studies used the method of Aki and Lee (1976) and Aki et al. (1977), who divided the medium under study into many cubic blocks and take seismic velocities in every blocks as unknown parameters. Unavoidably, the first tomography method has several drawbacks, e.g. artificial velocity discontinuities are introduced into the model between blocks, seismic rays are assumed to be straight lines within each block, and only one iteration of inversion is conducted. Later, the block method was modified by a number of researchers (see Iyer and Hirahara, 1993, for a detailed review). Thurber (1983) adopted grid nodes, instead of blocks, to model the Earth structure. Velocities at the nodes are taken as unknown parameters; the velocity at any point in the model is calculated by interpolating the velocities at the eight nodes surrounding that point. Thus, the velocity is continuous everywhere in the model; no velocity discontinuity is allowed to exist in the model even when discontinuities are actually detected in the study area. Thurber (1983) adopted the Cartesian coordinates, hence his method has been applied mainly to local-scale studies.

Japan provides a useful illustration of some of the weaknesses of these earlier methods. Japan is located in a complex region where four plates are interacting with each other (Ishida, 1992). The upper boundary of the subducting Pacific slab is found to be a sharp seismic discontinuity with a complex geometry by detailed studies with reflected and converted waves (Fukao et al., 1978; Hasegawa et al., 1978; Nakanishi et al., 1981; Matsuzawa et al., 1986, 1990; Zhao et al., 1997a; Ohmi and Hori, 2000). The Moho and the Conrad discontinuities also have large lateral depth variations (Horiuchi et al., 1982a,b; Zhao et al., 1990, 1992b). These discontinuities generate clear converted and reflected waves observable in seismograms and should be taken into account when studying the 3-D Earth structure. In principle, when the block or grid size is sufficiently small, the effects of the discontinuity undulations can be evaluated. The size in actual studies, however, is not small enough because of the sparse distribution of seismic stations. Hence, the tomographic images obtained not only represent velocity variations themselves, but also contain the effects of discontinuity undulations. Without introducing discontinuities into the model, the observed later (converted and reflected) phase data cannot be used. Later phase data contain important information on the Earth structure. The use of them can improve the ray path coverage and the accuracy of earthquake locations. However, neither Aki nor Thurber's methods can handle the complex discontinuities and the later phase data.

To tackle these problems, Zhao et al. (1992a) developed an improved tomography method. They divided the medium under study into a number of layers bounded by the discontinuities that are known to exist. 3-D grid nets are set individually in every layers. They also developed a fast 3-D ray tracing technique to compute travel times and ray pathes in such a model by combining the pseudo-bending technique (Um and Thurber, 1987) and Snell's law. First P-and S-waves and later phase data can all be used in the tomographic inversion with the LSQR algorithm (Paige and Saunders, 1982). Later, Zhao et al. (1994, 1997b) improved this method to make it use data from local, regional and teleseismic events simultaneously to determine the 3-D velocity structure under a region. Recently, Zhao (1999) extended this method to the global scale to estimate the whole mantle 3-D structure with the data set compiled by Engdahl et al. (1998). He took into account the depth variations (up to 36 km) of the 410 and 660 km discontinuities (Flanagan and Shearer, 1998) in the inversion and found that the discontinuity topography has a significant influence on the tomography of the mantle transition zone.

In the early 1980s, the introduction of iterative matrix solvers (e.g. SIRT, LSQR) by Clayton and Comer (1983) and Nolet (1985) proved to be an important step toward the refining of the Earth structure. It allowed for the implementation of larger numbers of both seismic data and model parameters in tomographic inversions. Recently, a few researchers have attempted to estimate resolution and covariance matrices using the iterative algorithms (like LSQR) without explicitly performing the operations involving the large and complex matrices (Zhang and McMechan, 1995; Nolet et al., 1999; Yao et al., 1990). Although there are still some debates on the technical details, this approach seems very promising and will become a useful tool in future tomographic studies if a consensus can be achieved.

Great advances have been made in the last two decades on the development of fast 3-D ray tracing algorithms (Thurber and Ellsworth, 1980; Thurber, 1983; Um and Thurber, 1987; Vidale, 1990; Moser, 1991; Zhao et al., 1992a; Koketsu and Sekine, 1998; Sadeghi et al., 1999), which make it possible to conduct nonlinear and iterative tomographic inversions. Some researchers adopted irregular parameterizations to account for the inhomogeneous sampling of the Earth by seismic rays (Abers and Roecker, 1991; Vesnaver, 1996; Bijwaard et al., 1998). Hu et al. (1994) and Steck (1995) combined teleseismic polarization data with arrival times to conduct tomographic inversions. Hirahara (1988) and Wu and Lees (1999) proposed to detect 3-D velocity anisotropy in the study area using travel time data, and formulated the problem of travel-time inversion for both velocity heterogeneity and anisotropy in each block. Other examples include Lees and van Decar (1991), who used gravity data to constrain travel-time inversion, and Neele et al. (1993), who used teleseismic P-wave amplitudes together with travel times to determine the upper mantle velocity structure.

2.2. *Subducting slab and arc magmatism*

Zhao et al. (1992a, 1994) applied their tomography method to over 50,000 arrival times from local, regional and teleseismic events to determine a detailed 3-D P-wave velocity structure of the Japan subduction zone. In addition to first P-and S-wave data, they also used converted waves at the upper slab boundary and the Moho discontinuity. They picked or double checked all the arrival times directly from the seismograms, which ensured the high accuracy of their data. Fig. 1 shows the P-wave tomography in northeast Japan they obtained. The spatial resolution of the tomographic images is 25–30 km in horizontal direction and 10–25 km in depth. The subducting Pacific plate is clearly imaged as a cold slab that is 90 km thick and has a velocity 4–6% higher than the normal mantle. Intermediate-depth earthquakes occur in the subducting slab and form two distinct seismic planes, the so-called double seismic zone (Hasegawa et al., 1978). The upper-plane earthquakes occur at the top of the slab; the lower-plane earthquakes occur in the central portion of the slab. Low velocity (low-V) zones are visible just beneath the active volcanoes, and extend to 150 km depth in the mantle wedge. Anomalous low-frequency microearthquakes occur around the low-V zones beneath active volcanoes in the depth range of 25–47 km, which are caused by the magmatic and volcanic activity (Hasegawa and Zhao, 1994). Distinct shear-wave reflectors are also found near the low-V zones in the lower crust, representing the upper surface of hot magma chambers with a thickness of about 100 m (Hasegawa and Zhao, 1994; Matsumoto and Hasegawa, 1996). Based on these observations, the low-V zones in the crust and mantle wedge are considered to be associated with the magma chambers which form the roots of the arc volcanoes.

Mineral physics studies of phase changes of some mantle minerals, such as olivine, wadsleyite and ringwoodite at high-temperature and high-pressure conditions found that the dehydration of the subducted oceanic crust plays an important role in the generation of melts and magma in the mantle wedge of subduction zones (Ringwood, 1982; Irifune, 1993; Iwamori, 1998).

Fig. 1 shows some heterogeneities within the subducting Pacific slab. It is found from map views that earthquakes within the slab (the lower-plane events in the double seismic zone) occurred in the relatively higher velocity areas of the slab (Fig. 18 in Zhao et al., 1992a). There may be two origins for the heterogeneity in the slab. One is the heterogeneity in the oceanic plate before subduction, for instance, seamounts, ridges, and fracture zones, as can be seen from maps of sea floors. The other may be the phase changes of the rocks within the slab due to the increasing temperature and pressure during the subduction of the oceanic plate (e.g. Kao and Liu, 1995).

Analyses of converted waves at the upper boundary of the subducting Pacific slab revealed a thin low-V layer on the top of the slab, which is interpreted to be the subducted oceanic crust (Nakanishi et al., 1981; Matsuzawa et al., 1986; Abers, 2000). The low-V layer is thinner than 10 km and about 6% slower than the normal mantle. The depth extent of the low-V layer is unclear, but is estimated to be deeper than 100 km, and earthquakes in the upper plane of the double seismic zone occur within it (Matsuzawa et al., 1986). The tomographic images in Fig. 1 do not have enough spatial resolution to detect such a thin low-V layer on the top of the slab.

Iidaka and Suetsugu (1992) suggested the existence of a low-V metastable olivine wedge within the subducting Pacific slab in the Izu–Bonin region. In the Tonga slab, however, Koper et al. (1998) did not find such an olivine wedge though they have used a much better data set and it is more likely that such an olivine wedge would exist in the Tonga slab than in

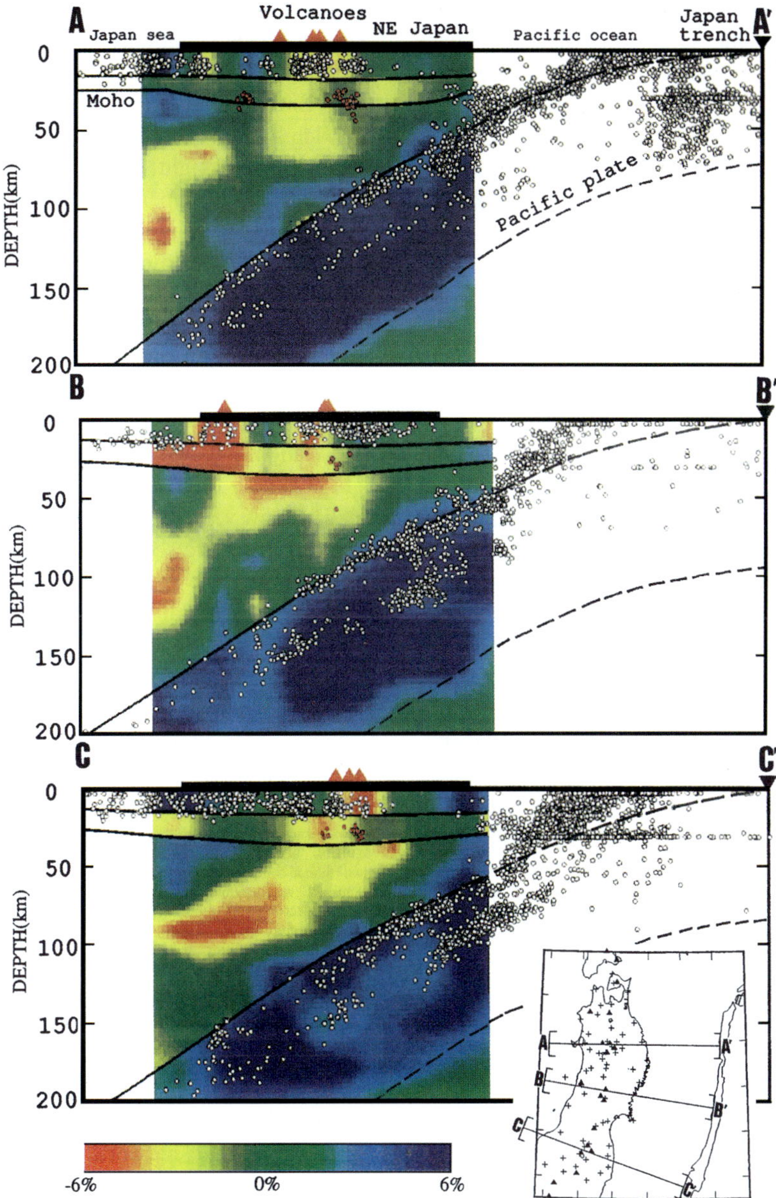

Fig. 1. Vertical cross sections of P-wave velocity structure beneath northeast Japan from the Earth's surface to 200 km depth along the profiles shown in the insert map. Red and blue colors denote slow and fast velocities, respectively. The velocity perturbation scale is shown at the bottom. Open circles denote earthquakes that occurred within a 20-km width along each profile. Red circles show low-frequency microearthquakes that occurred around the Moho discontinuity, due to the magmatic and volcanic activity. Red triangles, active volcanoes; reverse triangle, the location of the Japan Trench. The horizontal bar at the top of each cross section shows the land area where seismic stations exist. The three thick curved lines denote the Conrad and Moho discontinuities and the upper boundary of the subducting Pacific slab. The dashed line shows the lower boundary of the slab (after Zhao et al., 1992a).

the Izu–Bonin slab because of the much faster convergence rate. Although the feasibility of a metastable olivine wedge within the subducting slab has been established by mineral physicists (Green and Burnley, 1989; Kirby, 1991), it is still a controversial issue to seismologists. This is an important subject for future seismological research since it has important implications for the mechanism of deep earthquakes and dynamics of subduction zones.

2.3. Deep dehydration and back-arc spreading

Most of the seismic stations in Japan are located on the narrow land area of approximately 200 km in width, hence high-resolution tomographic images are determined only down to about 200 km depth (Fig. 1). The depth extent of the low-*V* zones in the mantle wedge is unclear in the back-arc region. Recently, Zhao et al. (1997b) used the data recorded by land seismic stations and ocean bottom seismographs (OBS) to determine a detailed 3-D structure down to 700 km depth beneath the Tonga arc and the Lau back-arc (Fig. 2). The subducting Tonga slab is imaged as a 100-km thick zone with a P-wave velocity 4–6% higher than the surrounding mantle. Beneath the Tonga arc and the Lau back-arc, low-*V* anomalies of up to 6% are visible. The low-*V* anomaly beneath the Tonga arc represents a dipping zone about 30–50 km above the slab, extending from the surface to about 140 km depth. This feature is similar to the low-*V* zones found beneath the Japan and Alaska volcanic fronts (Zhao et al., 1992a, 1995). Beneath 100 km depth, the amplitude of the back-arc anomalies is reduced, but a moderately slow anomaly (−2 to −4%) exists down to a depth of at least 400 km. This deep extent of the mantle wedge slow anomalies has been confirmed by detailed resolution analyses (Zhao et al., 1997b), seismic attenuation tomography (Roth et al., 1999, 2000) and waveform modeling studies (Xu and Wiens, 1997). These results indicate that geodynamic systems associated with the back-arc spreading are not limited to the near-surface areas, but are related to deep processes.

The slow velocity anomalies at depths of 300–400 km in the mantle wedge (Fig. 2) could be caused by upwelling flow patterns in the mantle wedge or by volatiles resulting from the deep dehydration reactions occurring in the subducting slab (Nolet, 1995). Volatiles would have the effect of lowering the melting temperature and the seismic velocity, and may produce small amounts of partial melt (Collier and Sinha, 1992). Temperatures in fast subducting slabs like Tonga are low enough for water to reach the stability depths of dense hydrous magnesian silicate phases (Parson and Wright, 1996; Taylor et al., 1996), which may allow water penetration down to depths of 660 km. The phase diagrams of important hydrous phases, the associated reaction kinetics, and the relevant mantle conditions (slab temperature and composition) are not known sufficiently well to predict the depth at which dehydration would occur. Partial melting of the mantle wedge by volatiles from the deep slab may be important in localizing low seismic velocities; the slow anomalies we observe at depths of 300–400 km (Fig. 2) may represent this process.

The slow velocity regions beneath the Tonga arc and the Lau back-arc seem to be separated at the shallow levels, but merge at depths >100 km (Fig. 2). This suggests that although the arc and back-arc magma systems are separated at shallow levels, where most of the magma is generated, there may be some interchange between the magma systems at depths >100 km. Interchange with slab-derived volatiles at depths >100 km may help to explain some of the unique features in the petrology of back-arc magmas relative to typical mid-ocean ridge basalts, including excess volatiles and large ion lithophile enrichment (Faul et al., 1994).

The spatial resolution of a tomographic experiment depends on the density and homogeneity of crisscrossing seismic rays available in the inversion. The reliability (quality) of a tomographic solution depends on the data accuracy as well as the density of rays used. Rays used in the Japan tomography are those from the local events in the crust and the subducting slab as well as teleseismic rays which were recorded by a dense seismic network located only on the island area of about 200 km in width. In contrast, rays in the Tonga tomography are from local and teleseismic events recorded by a seismic array that is less denser but has a much wider aperture than that in Japan. In addition, the abundant deep earthquakes in Tonga provided many rays sampling the deep parts of the mantle wedge. The data in both experiments have high quality. Therefore, the resolution of the Japan tomography is very high (10–30 km) down to 200 km depth right beneath the volcanic arc (Fig. 1), but very low under

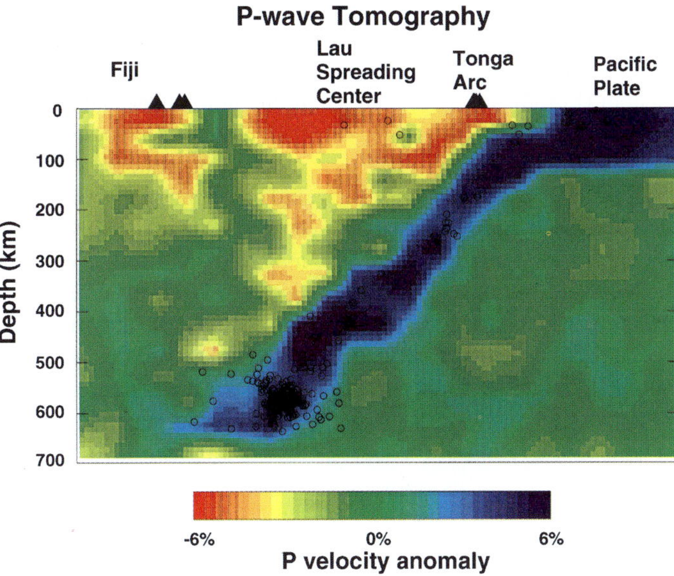

Fig. 2. East–west vertical cross section of P-wave velocity image from 0 to 700 km depth beneath the Tonga arc and Lau back-arc region. Red and blue colors denote slow and fast velocities, respectively. Solid triangles denote active volcanoes. Earthquakes within a 40-km width from the cross section are shown in open circles. The velocity perturbation scale is shown at the bottom (after Zhao et al., 1997b).

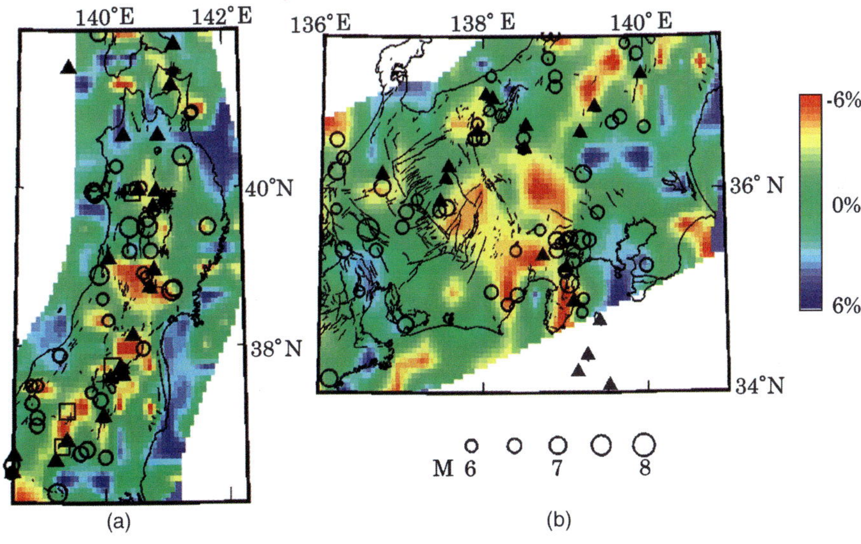

Fig. 3. P-wave velocity image at a depth of 40 km beneath (a) northeast and (b) central Japan. Red and blue colors denote low and high velocities, respectively. Circles denote earthquakes (M 5.7–8.0, depths 0–20 km) that occurred during a period of 115 years from 1885 to 1999. Solid triangles denote active volcanoes. Active faults are shown by thick lines. The velocity perturbation scale and the earthquake magnitude scale are shown on the right and at the bottom, respectively. Crosses and open squares in (a) show low-frequency microearthquakes and S-wave reflectors in the mid-crust, respectively.

the forearc, back-arc and deeper areas. In contrast, the Tonga tomography has a resolution of 40–50 km for a much wider area under the Tonga arc and the Lau back-arc down to 700 km depth (Fig. 2).

2.4. Large earthquakes in arc and back-arc: influence of magma

Countries located in subduction zone regions, such as Japan, have suffered heavily and frequently from earthquake hazards. Interplate earthquakes occur in mega-thrust zones along oceanic trenches. Intraplate earthquakes within the continental plate take place in the crust beneath the island arc or continental margin. Although large intraplate earthquakes do not occur so frequently as the interplate ones, they generally inflict greater damages because they are shallow and near the densely populated areas. A recent example of an inland crustal earthquake is the 1995 Kobe earthquake (M 7.2) in southwest Japan, which caused over 6400 fatalities and tremendous property losses. A good understanding of where large crustal earthquakes occur is important for the clarification of physics of earthquake generation and for the mitigation of seismic hazards.

Zhao et al. (2000a) investigated the relationship between the 3-D crustal structure and the distribution of large crustal earthquakes in Japan in the recent history. They found that most of the 160 large crustal earthquakes (M 5.7–8.0, depth 0–20 km) during a period of 115 years from 1885 to 1999 in Japan occurred around zones of low seismic velocity revealed by seismic tomography (Fig. 3). The low-V zones may represent weak sections of the seismogenic crust. Along the volcanic front and in back-arc areas, the crustal weakening may be caused by active volcanism and the presence of magma chambers. Low-V zones in the uppermost mantle may be the manifestation of mantle diapirs associated with the ascending flow of subduction-induced convection in the mantle wedge and dehydration reactions in the subducting slab (Zhao et al., 1992a) (Fig. 1). Magmas further rising from the mantle diapirs to the crust may cause low-frequency microearthquakes at levels of the lower crust and uppermost mantle, and make their appearance as S-wave reflectors at midcrustal levels. Their upward intrusion raises the temperature and reduces the seismic velocity of crustal materials around them, causing the brittle seismogenic layer above them to become locally thinner and weaker. This can be seen clearly from the distribution of crustal seismicity in volcanic areas (Figs. 4 and 5). The cut-off depth of the crustal earthquakes is elevated toward the volcanoes, indicating a higher temperature and a thinner brittle seismogenic layer beneath the active volcanoes (Matsuo, 1985).

Subject to horizontally compressional stresses in the plate convergence direction, contractive deformations will take place mainly in the low-V, low-Q areas because of the thinner brittle seismogenic layer and the weaker crust and uppermost mantle there due to the higher temperature. The deformation proceeds partially in small earthquakes but mainly in plastic deformation, causing the crustal shortening, upheaval and mountain building there, as evidenced by the geodetic measurements and GPS studies (Hasegawa et al., 2000). Large crustal earthquakes cannot occur within the weak low-V zones but in their edge portions where

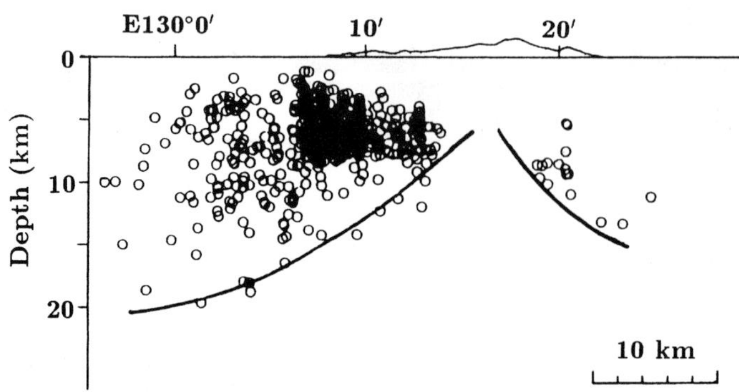

Fig. 4. East–west vertical cross section of microearthquakes and their cutoff depth (solid lines) beneath Unzen Volcano in Kyushu (after Matsuo, 1985).

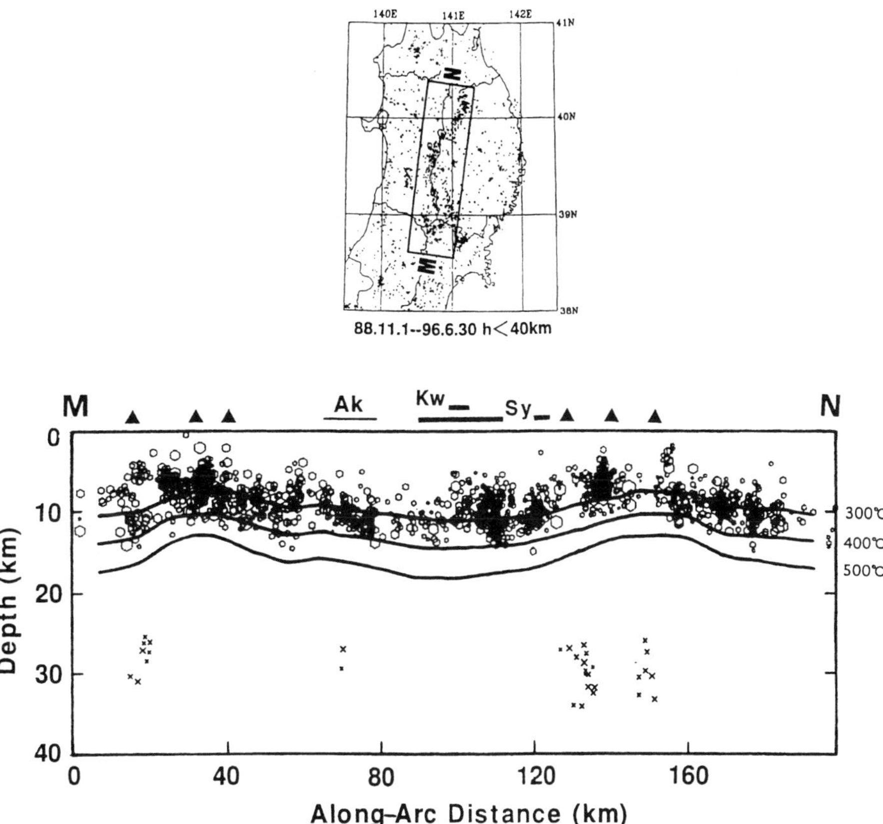

Fig. 5. Along-arc vertical cross section of microearthquakes that occurred during 1 November 1988 to 30 June 1996 and the estimated crustal temperature in the central part of northeast Japan. Events located in the rectangular area MN in the insert map are plotted in the cross section. Solid triangles and crosses show active volcanoes and deep, low-frequency microearthquakes, respectively. On the top the thin line (Ak) denotes the aftershock area of the 17 October 1970, Southeast-Akita earthquake (M 6.2); the short bar (Kw) and two bars (Sy) denote Kawafune Fault and Senya Fault which were ruptured by the 31 August 1896, Rikuu earthquake (M 7.2). The temperature of the crust is shown in isotherms of 300, 400 and 500°C, which was estimated from seismic velocities determined by seismic tomography (after Hasegawa et al., 2000; Zhao et al., 2000a).

the mechanical strength of materials is stronger than those of the low-V zones but still weaker than the normal sections of the seismogenic layer. Thus, the edge portion of the low-V areas becomes the ideal locations to generate large earthquakes (Fig. 6).

2.5. Large earthquakes in forearc: effects of slab dehydration

Zhao et al. (2000a) found that large crustal earthquakes in the forearc regions in Japan are also located in or around low-V zones. Those low-V zones, however, may not be caused by high temperature because no volcano exists there (Yokoyama et al., 1987) and the surface heat flow is low (Okubo et al., 1989). The 1995 Kobe earthquake (M 7.2) is such an example. Kobe is located in the forearc region where the Philippine Sea plate is subducting beneath the Eurasian plate (Fig. 7). Zhao et al. (1996) and Zhao and Negishi (1998) determined 3-D P- and S-velocity and Poisson's ratio structures in the Kobe source area with a spatial resolution of 4–5 km (Fig. 8). They found that the Kobe mainshock is located in a distinctive zone characterized by low velocity and high Poisson's ratio, which is interpreted to be a fluid-filled, fractured rock matrix that contributed to the initiation of the Kobe earthquake.

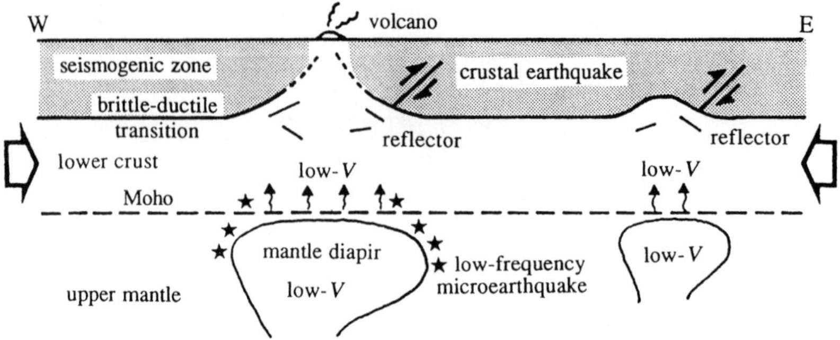

Fig. 6. Schematic illustration of across-arc vertical cross section of the crust and uppermost mantle in volcanic areas of Japan, showing the cause of large crustal earthquakes and its relation to low-V zones and magma chambers in the uppermost mantle (after Hasegawa and Zhao, 1994).

There may be two origins of fluids in the Kobe fault zone: one has a shallow origin, such as fluids trapped in the pore space, crustal mineral dehydration, and the permeation of the meteoric and sea waters down to the deep crust through the active faults that would have been ruptured during many earthquake cycles (Zhao and Mizuno, 1999); the other has a deep origin, such as the dehydration of the subducting Philippine Sea slab.

Zhao et al. (2000a) determined the detailed 3-D crust and upper mantle structure under Shikoku and Chugoku (Fig. 9). The subducting Philippine Sea slab is imaged clearly. It has a thickness of 30–35 km and a P-velocity 3–5% higher than that of the normal

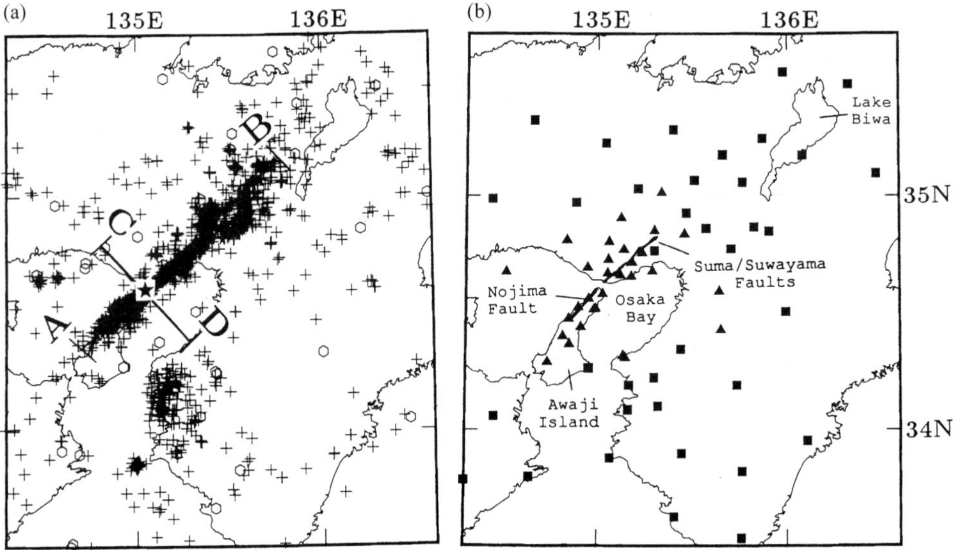

Fig. 7. (a) Epicentral distribution of 3634 events used in the tomographic imaging of the source area of the 1995 Kobe, Japan, earthquake (star). Crosses denote the events that occurred after 17 January 1995; most of them were aftershocks of the Kobe earthquake along the fault zone (parallel to cross section line A–B). Circles denote microearthquakes that occurred from January 1990 to December 1994. (b) Distribution of seismic stations that recorded the earthquakes in (a). Solid triangles denote portable stations that were set up following the Kobe mainshock. Solid squares denote permanent stations. Solid lines represent the surface traces of the Nojima, Suma, and Suwayama faults (after Zhao et al., 1996).

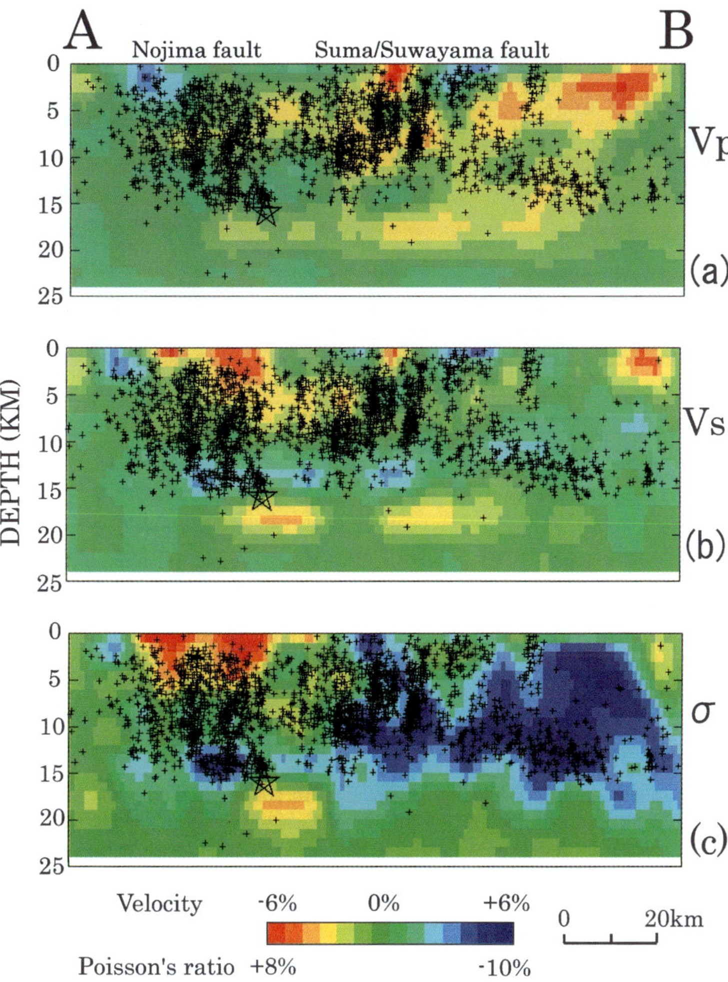

Fig. 8. Vertical cross sections of (a) P-wave velocity (V_P), (b) S-wave velocity (V_S), and (c) Poisson's ratio along the line A–B in Fig. 7. Slow velocity and high Poisson's ratio are shown in red; fast velocity and low Poisson's ratio are shown in blue. V_P and V_S perturbations range from −6 to 6% from the 1-D velocity model. Poisson's ratio ranges from 0.225 to 0.27 (−10 to 8% from the average value). Small crosses denote the Kobe aftershocks within a 5-km width from the line A–B. Star denotes the hypocenter of the Kobe mainshock. The vertical exaggeration is 2:1 (after Zhao et al., 1996).

mantle. A prominent low-V zone exists in the lower crust and right above the Philippine Sea slab. This low-V zone has properties as the anomaly at the Kobe hypocenter that Zhao et al. (1996) detected in their high-resolution imaging which shows low V_P, low V_S and high Poisson's ratio (Fig. 8). These results suggest that the fluids that contributed to the initiation of the 1995 Kobe earthquake may be related to the dehydration process of the subducted Philippine Sea slab, though the fluids may also have the shallow origins as mentioned above. It is generally considered that fluids widely exist in the crust and uppermost mantle in the forearc regions of subduction zones (Tatsumi, 1989; Iwamori, 1998). The Philippine Sea slab is descending at a very small dip angle in Shikoku and eastern Kii Peninsula, and the slab is located right under the crust (Ishida, 1992), thus the fluids from the slab dehydration may easily migrate up to the crust. When the fluids enter the active faults (such as the Nojima Fault which generated the 1995 Kobe earthquake) pore

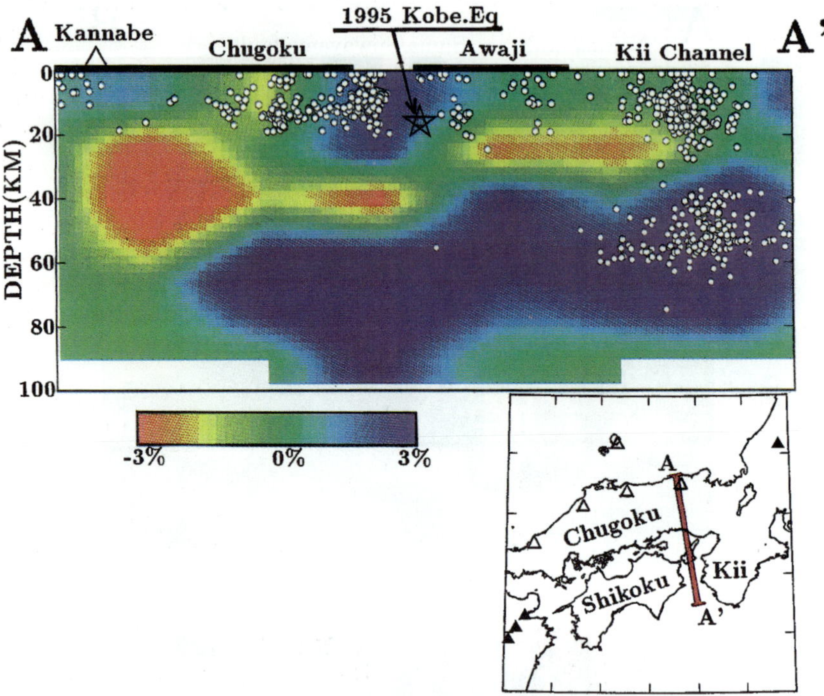

Fig. 9. Vertical cross section of P-wave velocity structure down to a depth of 100 km along the line AA′ in the insert map. Blue and red colors denote fast and slow velocities, respectively. The velocity perturbation scale is shown at the bottom. The star symbol shows the hypocenter of the 1995 Kobe mainshock (M 7.2). White dots show the microearthquakes within a 20-km width from the line AA′, which occurred during 1985 to 1993. The thick lines on the top show the land areas, the Chugoku District and Awaji Island. The open triangle denotes the Kannabe Quaternary volcano in Chugoku.

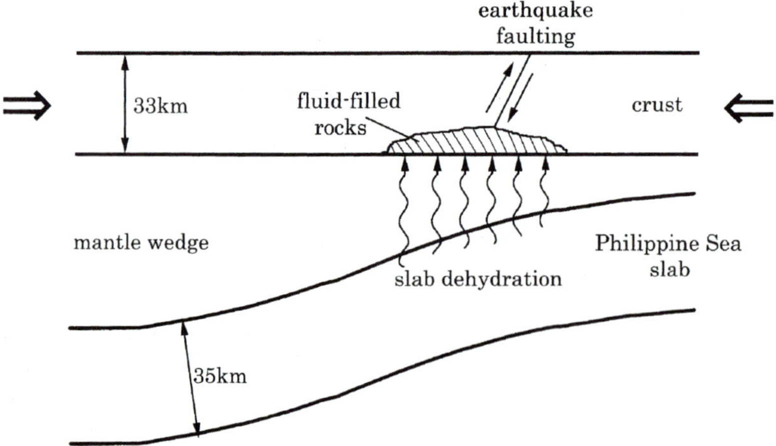

Fig. 10. Schematic illustration on the effects of slab dehydration on the generation of large crustal earthquakes in the forearc region of the Nankai subduction zone. According to mineral physics studies, dehydration reactions would take place in the subducting oceanic lithosphere when it descends into the mantle due to the increasing temperature and pressure. The Philippine Sea slab is descending at a small dip angle beneath southwest Japan, and it is located right under the crust, thus the fluids from the slab dehydration may easily migrate up to the crust. When the fluids enter the active faults in the crust, pore pressures will increase and fault zone frictions will decrease. Thus, active faults may be triggered to move to generate large crustal earthquakes.

pressures will increase and fault zone friction will decrease. Thus, active faults can be triggered to move to generate large crustal earthquakes (Fig. 10).

These results suggest that the generation of a large crustal earthquake is closely related to the surrounding tectonic environment such as plate subduction and physical/chemical properties of crustal materials, such as magmas, fluids, etc. The rupture nucleation zone should have a 3-D spatial extent, not just limited to the 2-D surface of a fault, as suggested earlier by Tsuboi (1956) in the concept of "earthquake volume". Complex physical and chemical reactions may take place in the source zone of a future earthquake, causing heterogeneities in the material property and stress field. The source zone of a M 6–8 earthquake extends from about 10 to over 100 km (Kanamori and Anderson, 1975). The resolution of our tomographic imaging is close to that scale of the earthquake sources, which may have enabled us to image the earthquake-related heterogeneities in the crust and uppermost mantle in Japan.

3. Seismic attenuation structure

Seismic attenuation and its variations in the Earth are useful for determining the type and state of the rocks and minerals composing the Earth. The attenuation of seismic waves is due to three effects: geometric spreading, intrinsic attenuation, and scattering attenuation. Geometric spreading is simply the energy density decrease that occurs as an elastic wavefront expands. Intrinsic attenuation is energy lost to heat and internal friction during the passage of an elastic wave. Scattering attenuation is not true energy loss in this sense. Elastic energy is not converted into heat but is redistributed into angular directions away from the receiver or converted into wave types arriving in different time windows at the receiver. Scattering takes place by reflection, refraction, and conversion of elastic energy by the medium heterogeneities that are discontinuous or rapid variations in the velocity and/or density of the medium.

Intrinsic attenuation structure of the Earth can be estimated from observed data of seismic wave attenuation with a method similar to velocity tomography. The data used are amplitudes or amplitude spectra of seismic waves. Attenuating bodies reduce the amplitude or change the spectral content of seismic waves that pass through them, just as low-V bodies increase the travel times of seismic waves. The parameter used to quantify seismic attenuation is seismic wave quality factor, Q, value. High-Q and low-Q values indicate weak and strong seismic attenuations, respectively. Compared with velocity tomography, attenuation (Q) tomography has, in general, a lower spatial resolution because the amplitude or amplitude spectrum data are usually measured by investigators manually or semi-automatically and thus their measurements are much fewer than arrival time measures. The Q images are usually less accurate than velocity images because the amplitude (spectra) measurements are noisier. Moreover, these inaccuracies are somewhat compensated by the much larger variations in Q than velocity anomalies. The velocity may vary 5–10% between the slab and the mantle wedge, but Q varies by 1000% (factors of 10). There are also much fewer Q studies than velocity studies.

During the recent years, a few 3-D attenuation models have been determined for subduction zone regions, mainly in the Japanese Islands (e.g. Umino and Hasegawa, 1984; Furumura and Moriya, 1990; Sekiguchi, 1991; Tsumura et al., 2000). These studies revealed that the subducting Pacific slab shows very small attenuation and has Q values of up to 1000 to 1500. In contrast, the mantle wedge and the crust beneath active volcanoes show very high attenuation with Q values of 100 or smaller. Recently, Tsumura et al. (2000) obtained an updated 3-D attenuation structure in northeast Japan. Their model has a spatial resolution of about 40 km, and shows clearly the high-Q Pacific slab and the low-Q anomalies in the crust and mantle wedge beneath active volcanoes. The general pattern of the Q variations is quite similar to that of velocity variations in this region (Fig. 1) determined by Zhao et al. (1992a). Roth et al. (1999) estimated the 3-D Q structure of the Tonga–Fiji region and found that low-Q anomalies extend to 400 km depth in the mantle wedge above the high-Q Tonga slab, in good agreement with the velocity images by Zhao et al. (1997b).

Attenuation structure of subduction zones has also been estimated from the seismic intensity data (e.g. Hashida, 1989). Seismic intensity is usually measured on the basis of human perception and movement of objects observed by experts without instruments. The estimated attenuation structure shows similarity to the Q models determined with the precise seismological data such as seismic wave amplitudes and amplitude spectra. Apparently, however, the intensity data cannot be expected to determine high-resolution and accurate Q structure because of its subjective nature.

As a whole, the images of the low-Q zones in the crust and mantle wedge beneath active volcanoes and the high-Q zones corresponding to the subducting slab are similar to those of the low-V and high-V zones revealed by travel time tomography (Tsumura et al., 2000; Roth et al., 2000). These 3-D seismic attenuation models have provided additional information on the physical properties of subduction zones, and show the mantle wedge anomalous bodies associated with the magmatism of the island arc and back-arc.

4. Seismic anisotropy

Seismic anisotropy is the direction-dependent nature of the propagation velocity of seismic waves. Natural minerals usually have some crystallographic structure. Under the physical conditions of high temperature and high pressure in the Earth's deep interior, rocks undergo slow plastic deformation. Through the deformation process, crystallographic axes of minerals are realigned to a particular direction due to uniaxial tectonic stress. Rocks may thus exhibit petrofabic structure, which in turn will produce seismic anisotropy on a macroscopic scale. Anisotropy can also form due to the presence of aligned cracks. It is found that seismic anisotropy seems to exist in almost every portions of the Earth, from the crust, mantle, to inner core.

Shear wave splitting has been the most useful tool to detect seismic anisotropy, and numerous researchers have made studies using this approach for subduction zones (e.g. Ando et al., 1980; Kaneshima and Silver, 1992; Okada et al., 1995; Yang et al., 1995; Hiramatsu et al., 1997; Fischer et al., 1998). Shear wave splitting studies show the existence of anisotropy in the crust, mantle wedge and the subducted slab, among them the mantle wedge seems to be the most anisotropic portion of subduction zones.

Okada et al. (1995) made a detailed investigation of seismic anisotropy in northeast Japan using shear wave splitting data from local shallow and deep earthquakes. They found that shear waves from the earthquakes in the subducting slab show significant splittings (strong anisotropy) of up to 1 s when the waves pass through the low-V and low-Q zones in the mantle wedge as imaged by the velocity and attenuation tomography (Zhao et al., 1992a; Tsumura et al., 2000). The shear waves show little splitting when they propagate the normal areas in the crust and mantle wedge. They suggested that the anisotropy was caused by the planar alignment of melts within the low-V/low-Q zones in the crust and mantle wedge beneath the active volcanoes. The fraction of melts is estimated to be about 2% from the degree of anisotropy and attenuation, and from the velocity reduction.

The local seismic rays from the shallow and deep events used by Okada et al. (1995) show that the anisotropy of the crust and the subducting slab is insignificant, while Hiramatsu et al. (1997) suggested that the subducting slab shows strong anisotropy resulting from phase changes in the slab. Hiramatsu et al. (1997) used ScS waves which travel from the events in the slab down to the deep mantle and are bounced back from the core-mantle boundary. ScS waves propagate a long distance in the mantle and may be affected by the anisotropy in the deep mantle and/or the D″ layer above the core-mantle boundary (Iidaka and Niu, 1998; Vinnik et al., 1998). It is also possible that the local rays of Okada et al. (1995) did not sample the deep and inner portion of the slab where phase changes take place and seismic anisotropy exists. Future studies are needed to clarify whether the subducting slab is anisotropic or not.

5. Concluding remarks

Although great advances have been made in the seismic imaging of subduction zones in the last decade, a great deal remains to be done in the future in both the theoretical and observational aspects of seismology. The tomographic methods that have been the most powerful tool to imaging subduction zones need further improvement. The existing schemes for model parameterization, ray tracing and inversion need to be thoroughly compared and examined (e.g. Spakman and Nolet, 1988; Boschi and Dziewonski, 1999); new and more efficient techniques should be explored continuously. New theory and technologies are needed to use amplitudes, polarizations and waveforms together with travel times in the tomographic inversion for seismic velocity, attenuation and anisotropy structures.

Much more work remains to be done on the application of tomography and other seismic methods to the imaging of subduction zones, which largely depends on the installation of new seismic networks or the expansion of the existing networks. Most of the existing seismic stations are located on island arcs or continental margins, thus only the structure in or around the volcanic front can be well imaged. To study the forearc and back-arc regions, installing of OBS stations is crucial. Precise images of the forearc regions of subduction zones are important for clarifying the initiation of subduction, seismic and mechanic coupling of the subducting oceanic slab and the overlying continental plate, and the frequent occurrence of destructive thrust-type earthquakes (Zhao et al., 1997a; Ito et al., 2000). Detailed structure of the back-arc regions of subduction zones is needed to understand the formation of back-arc spreading centers and its relationship to the subduction process (Zhao et al., 1997b; Roth et al., 1999, 2000).

Other seismological puzzles and research topics of subduction zones include: the formation of the double seismic zone and its relationship to the stress regime and structural heterogeneities in the shallow portion of the subducting slab (Kao and Liu, 1995); the existence or absence of a metastable olivine wedge within the deep portion of the slab and its relationship to the generation of deep earthquakes (Iidaka and Suetsugu, 1992; Koper et al., 1998); the geometry of the mantle discontinuities near the subducting slab (Collier and Helffrich, 1997); the detailed morphology and structure of the subducting slab around the 670 km discontinuity (Zhao and Clayton, 1990; van der Hilst et al., 1991; van der Hilst, 1995); high-resolution imaging of the mantle wedge to clarify the relationship between the arc magmatism and slab dehydration (Iwamori and Zhao, 2000; Zhao et al., 2000b); the detailed structure of the mantle below the subducting slab and the nature of the lower boundary of the slab (Hasegawa et al., 1994; Zhao et al., 1994), among others.

So far, most of the seismological studies on large earthquake sources have concentrated on the coseismic rupture process on the fault plane through analyzing the seismic waves generated during the faulting, which provides little information on the preparatory process of the earthquake generation. To tackle this problem, it is useful to conduct high-resolution tomographic imagings of the earthquake source areas and active fault zones. The effects of inelastic structures and processes such as magmas and fluids in the crust and lithosphere need to be paid special attention on the generation of earthquakes.

The challenge is great, but with ingenuity, striving, and collaboration, we can anticipate exciting new advances in understanding the structure and dynamics of subduction zones as well as the entire Earth's interior in the 21st century.

Acknowledgements

This work was partially supported by a Grant-in-Aid for Scientific Research from the Japanese Ministry of Education, Science and Culture (Monbusho) (No. B-11440134). The author appreciates the helpful discussion and collaborations in recent years with A. Hasegawa, S. Horiuchi, H. Kanamori, I.S. Sacks, D. Christensen, D. Wiens, T. Matsuzawa, E. Humphreys, H. Iwamori, H. Negishi, T. Mizuno, K. Asamori, and F. Ochi. R. van der Hilst, G. Abers and H. Zhou provided thoughful reviews, which improved the manuscript.

References

Abers, G., 2000. Hydrated subducted crust at 100–250 km depth. Earth Planet. Sci. Lett. 176, 323–330.

Abers, G., Roecker, S., 1991. Deep structure of an arc-continent collision: earthquake relocation and inversion for upper mantle P-and S-wave velocities beneath Papua New Guinea. J. Geophys. Res. 96, 6379–6401.

Aki, K., Lee, W., 1976. Determination of three-dimensional velocity anomalies under a seismic array using first P arrival times from local earthquakes, 1. A homogeneous initial model. J. Geophys. Res. 81, 4381–4399.

Aki, K., Christoffersson, A., Husebye, E., 1977. Determination of the three-dimensional seismic structure of the lithosphere. J. Geophys. Res. 82, 277–296.

Ando, M., Ishikawa, Y., Wada, H., 1980. S-wave anisotropy in the upper mantle under a volcanic area in Japan. Nature 268, 43–46.

Bijwaard, H., Spakman, W., Engdahl, E., 1998. Closing the gap between regional and global travel time tomography. J. Geophys. Res. 103, 30055–30078.

Boschi, L., Dziewonski, A., 1999. High and low-resolution images of the Earth's mantle: implications of different approaches to tomographic modeling. J. Geophys. Res. 104, 25567–25594.

Brophy, J., Marsh, B., 1986. On the origin of high-alumina arc basalts and the mechanics of melt extraction. J. Petrol. 27, 763–789.

Clayton, R., Comer, R., 1983. A tomographic analysis of mantle heterogeneities from body wave travel time data. EOS Trans. AGU 64, 776.

Collier, J., Helffrich, G., 1997. Topography of the 410 and 660 km seismic discontinuities in the Izu–Bonin subduction zone. Geophys. Res. Lett. 24, 1535–1538.

Collier, J., Sinha, M., 1992. Seismic mapping of a magma chamber beneath the Valu Fa Ridge, Lau Basin. J. Geophys. Res. 97, 14031–14053.

Davies, J., Stevensen, D., 1992. Physical model of source region of subduction zone volcanics. J. Geophys. Res. 97, 2037–2070.

Dziewonski, A., Anderson, D., 1984. Seismic tomography of the Earth's interior. Am. Sci. 72, 483–494.

Engdahl, E., van der Hilst, R., Buland, R., 1998. Global teleseismic earthquake relocation with improved travel times and procedures for depth determination. Bull. Seismol. Soc. Am. 88, 722–743.

Faul, U., Toomey, D., Waff, H., 1994. Intergranular basaltic melt is distributed in thin, elongated inclusions. Geophys. Res. Lett. 21, 29–32.

Fischer, K., Fouch, M., Wiens, D., 1998. Anisotropy and flow in subduction zone back-arcs. Pure Appl. Geophys. 151, 463–475.

Flanagan, M., Shearer, P., 1998. Global mapping of topography on transition zone velocity discontinuities by stacking SS precursors. J. Geophys. Res. 103, 2673–2692.

Fukao, Y., Kanjo, K., Nakamura, I., 1978. Deep seismic zone as an upper mantle reflector of body waves. Nature 272, 606–608.

Furumura, T., Moriya, T., 1990. Three-dimensional Q structure in and around the Hidaka mountains, Hokkaido, Japan. J. Seismol. Soc. Japan 43, 121–132.

Gill, J., 1981. Orogenic Andesites and Plate Tectonics. Springer, New York.

Green, H., Burnley, P., 1989. A new self-organizing mechanism for deep-focus earthquakes. Nature 341, 733–737.

Hasegawa, A., Zhao, D., 1994. Deep structure of island arc magmatic regions as inferred from seismic observations. In: Ryan, M.P. (Ed.), Magmatic Systems. Academic Press, San Diego, CA.

Hasegawa, A., Umino, N., Takagi, A., 1978. Double-planed deep seismic zone and upper-mantle structure in the northeastern Japan arc. Geophys. J. R. Astron. Soc. 54, 281–296.

Hasegawa, A., Horiuchi, S., Umino, N., 1994. Seismic structure of the northeastern Japan convergent margin a synthesis. J. Geophys. Res. 99, 22295–22311.

Hasegawa, A., Yamamoto, A., Umino, N., Miura, S., Horiuchi, S., Zhao, D., Sato, H., 2000. Seismic activity and deformation process of the overriding plate in the northeastern Japan subduction zone. Tectonophysics 319, 225–239.

Hashida, T., 1989. Three-dimensional seismic attenuation structure beneath the Japanese Islands and its tectonic and thermal implications. Tectonophysics 159, 163–180.

Hirahara, K., 1988. Detection of three-dimensional velocity anisotropy. Phys. Earth Planet. Inter. 51, 71–85.

Hiramatsu, Y., Ando, M., Ishikawa, Y., 1997. ScS wave splitting of deep earthquakes around Japan. Geophys. J. Int. 128, 409–424.

Horiuchi, S., Ishii, H., Takagi, A., 1982a. Two-dimensional depth structure of the crust beneath the Tohoku district, the northeastern Japan arc, I. Method and Conrad discontinuity. J. Phys. Earth 30, 47–69.

Horiuchi, S., Yamamoto, A., Ueki, S., 1982b. Two-dimensional depth structure of the crust beneath the Tohoku district, the northeastern Japan arc, II. Moho discontinuity and P-wave velocity. J. Phys. Earth 30, 71–86.

Hsui, A., Marsh, B., Toksoz, M., 1983. On the melting of the subducted oceanic crust: effects of subduction induced mantle flow. Tectonophysics 99, 207–220.

Hu, G., Menke, W., Powell, C., 1994. Polarization tomography in southern California. J. Geophys. Res. 99, 15245–15256.

Iidaka, T., Niu, F., 1998. Evidence for an anisotropic lower mantle beneath eastern Asia: comparison of shear-wave splitting data of SKS and P660s. Geophys. Res. Lett. 25, 675–678.

Iidaka, T., Suetsugu, D., 1992. Seismological evidence for metastable olivine inside a subducting slab. Nature 356, 593–595.

Irifune, T., 1993. Phase transformations in the earth's mantle and subducting slabs: implications for their compositions seismic velocity and density structures and dynamics. The Island Arc 2, 55–71.

Ishida, M., 1992. Geometry and relative motion of the Philippine Sea plate and Pacific plate beneath the Kanto-Tokai district Japan. J. Geophys. Res. 97, 489–513.

Ito, S., Hino, R., Matsumoto, S., 2000. Deep seismic structure of the seismogenic plate boundary in the off-Sanriku region northeastern Japan. Tectonophysics 319, 261–274.

Iwamori, H., 1998. Transportation of H_2O and melting in subduction zones. Earth Planet. Sci. Lett. 160, 65–80.

Iwamori, H., Zhao, D., 2000. Melting and seismic structure beneath the northeast Japan arc. Geophys. Res. Lett. 27, 425–428.

Iyer, H., Hirahara, K., 1993. Seismic Tomography: Theory and Practice. Chapman and Hall, London.

Kanamori, H., Anderson, D., 1975. Theoretical basis of some empirical relations in seismology. Bull. Seismol. Soc. Am. 65, 1073–1095.

Kaneshima, S., Silver, P., 1992. A search for source side mantle anisotropy. Geophys. Res. Lett. 19, 1049–1052.

Kao, H., Liu, L., 1995. A hypothesis for the seismogenesis of double seismic zone. Geophys. J. Int. 123, 71–84.

Kirby, S., 1991. Mantle phase changes and deep-earthquake faulting in subducting lithosphere. Science 252, 216–224.

Koketsu, K., Sekine, S., 1998. Pseudo-bending method for three-dimensional seismic ray tracing in a spherical Earth with discontinuities. Geophys. J. Int. 132, 339–346.

Koper, K., Wiens, D., Dorman, L., Hildebrand, J., Webb, S., 1998. Modeling the Tonga slab: can travel time data resolve a metastable olivine wedge? J. Geophys. Res. 103, 30079–30100.

Lees, J., van Decar, J., 1991. Seismic tomography constrained by Bouguer gravity anomalies: applications in western Washington. Pure Appl. Geophys. 135, 31–52.

Matsumoto, S., Hasegawa, A., 1996. Distinct S-wave reflector in the mid-crust beneath Nikko–Shirane volcano in the northeastern Japan arc. J. Geophys. Res. 101, 3067–3083.

Matsuo, N., 1985. Seismicity in and near the Shimabara Peninsula. Sci. Rep. Shimabara Earthquake Volcanol. Obs. Kyushu Univ. 13, 51–62.

Matsuzawa, T., Umino, N., Hasegawa, A., Takagi, A., 1986. Upper mantle velocity structure estimated from PS-converted wave beneath the northeastern Japan arc. Geophys. J. R. Astron. Soc. 86, 767–787.

Matsuzawa, T., Kono, T., Hasegawa, A., Takagi, A., 1990. Subducting plate boundary beneath northeastern Japan arc estimated from SP converted waves. Tectonophysics 181, 123–133.

Moser, T., 1991. Shortest path calculation of seismic rays. Geophysics 56, 59–67.

Nakanishi, I., Suyehiro, K., Yokota, T., 1981. Regional variations of amplitudes of ScSp phases observed in the Japanese Islands. Geophys. J. R. Astron. Soc. 67, 615–634.

Neele, F., van Decar, J., Snieder, R., 1993. The use of P-wave amplitude data in a joint inversion with travel times for upper mantle velocity structure. J. Geophys. Res. 98, 12033–12054.

Nolet, G., 1985. Solving or resolving inadequate and noisy tomographic systems. J. Comput. Phys. 61, 463–482.

Nolet, G., 1995. Deep dehydration of the subducting lithosphere. In: Farley, K. (Ed.), Processes of Deep Earth and Planetary Volatiles. American Institute of Physics, New York, pp. 22–32.

Nolet, G., Montelli, R., Virieux, J., 1999. Explicit, approximate expressions for the resolution and a posteriori covariance of massive tomographic systems. Geophys. J. Int. 138, 36–44.

Ohmi, S., Hori, S., 2000. Seismic wave conversion near the upper boundary of the Pacific plate beneath the Kanto district, Japan. Geophys. J. Int. 141, 136–148.

Okada, T., Matsuzawa, T., Hasegawa, A., 1995. Shear-wave polarization anisotropy beneath the northeastern part of Honshu, Japan. Geophys. J. Int. 123, 781–797.

Okubo, Y., Tsu, H., Ogawa, K., 1989. Estimation of Curie point and geothermal structure of island arcs of Japan. Tectonophysics 159, 279–290.

Oxburgh, E., Turcotte, D., 1970. Thermal structure of island arcs. Geol. Soc. Am. Bull. 81, 1665–1688.

Paige, C., Saunders, M., 1982. LSQR: an algorithm for sparse linear equations and sparse least squares. ACM Trans. Math. Soft. 8, 43–71.

Parson, L., Wright, I., 1996. The Lau-Havre-Taupo backarc basin A southward-propagating, multi-stage evolution from rifting to spreading. Tectonophysics 263, 1–22.

Plank, T., Langmuir, C., 1988. An evaluation of the global variations in the major element chemistry of arc basalts. Earth Planet. Sci. Lett. 90, 349–370.

Ringwood, A., 1982. Phase transformations and differentiation in subducted lithosphere: implications for mantle dynamics, basalt petrogenesis, and crustal evolution. J. Geol. 90, 611–643.

Roth, E., Wiens, D., Dorman, L., Hildebrand, J., Webb, S., 1999. Seismic attenuation tomography of the Tonga–Fiji region using phase pair methods. J. Geophys. Res. 104, 4795–4810.

Roth, E., Wiens, D., Zhao, D., 2000. An empirical relationship between seismic attenuation and velocity anomalies in the upper mantle. Geophys. Res. Lett. 27, 601–604.

Sadeghi, H., Suzuki, S., Takenaka, H., 1999. A two-point three-dimensional seismic ray tracing using genetic algorithms. Phys. Earth Planet. Inter. 113, 355–365.

Sekiguchi, S., 1991. Three-dimensional Q structure beneath the Kanto-Tokai district, Japan. Tectonophysics 195, 83–104.

Spakman, W., Nolet, G., 1988. Imaging algorithms, accuracy and resolution in delay time tomography. In: Vlaar, N., Nolet, G., Wortel, M., Cloetingh, S. (Eds.), Mathematical Geophysics. Reidel, Norwell, pp. 155–187.

Steck, L., 1995. Simulated annealing inversion of teleseismic P-wave slowness and azimuth for crustal velocity structure at Long Valley caldera. Geophys. Res. Lett. 22, 497–500.

Sugimura, A., 1960. Zonal arrangement of some geophysical and petrological features in Japan and its environs. J. Fac. Sci. Univ. Tokyo 12, 133–153.

Tatsumi, Y., 1989. Migration of fluid phases and genesis of basalt magmas in subduction zones. J. Geophys. Res. 94, 4697–4707.

Tatsumi, Y., Sakuyama, M., Fukuyama, H., Kushiro, I., 1983. Generation of arc basalt magmas and thermal structure of the mantle wedge in subduction zones. J. Geophys. Res. 88, 5815–5825.

Taylor, B., Zellmer, K., Martinez, F., Goodliffe, A., 1996. Sea-floor spreading in the Lau back-arc basin. Earth Planet. Sci. Lett. 144, 35–40.

Thurber, C., 1983. Earthquake locations and three-dimensional crustal structure in the Coyote Lake area, central California. J. Geophys. Res. 88, 8226–8236.

Thurber, C., Ellsworth, W., 1980. Rapid solution of ray tracing problems in heterogeneous media. Bull. Seismol. Soc. Am. 70, 1137–1148.

Tsuboi, C., 1956. Earthquake energy earthquake volume aftershock area and strength of the earth's crust. J. Phys. Earth 4, 63–66.

Tsumura, N., Matsumoto, S., Horiuchi, S., Hasegawa, A., 2000. Three-dimensional attenuation structure beneath the northeastern Japan arc estimated from spectra of small earthquakes. Tectonophysics 319, 241–260.

Turcotte, D., Schubert, G., 1973. Frictional heating of the descending lithosphere. J. Geophys. Res. 78, 5786–5876.

Um, J., Thurber, C., 1987. A fast algorithm for two-point seismic ray tracing. Bull. Seismol. Soc. Am. 77, 972–986.

Umino, N., Hasegawa, A., 1984. Three-dimensional Qs structure in the northeastern Japan arc. J. Seismol. Soc. Japan 37, 217–228.

van der Hilst, R., 1995. Complex morphology of subducted lithosphere in the mantle beneath the Tonga trench. Nature 374, 154–157.

van der Hilst, R., Engdahl, E., Spakman, W., Nolet, G., 1991. Tomographic imaging of subducted lithosphere below northwest Pacific island arcs. Nature 353, 37–43.

Vesnaver, A., 1996. Irregular grids in seismic tomography and minimum-time ray tracing. Geophys. J. Int. 126, 147–165.

Vidale, J., 1990. Finite-difference calculation of traveltime in three dimensions. Geophysics 55, 521–526.

Vinnik, L., Breger, L., Romanowicz, B., 1998. On the inversion of Sd particle motion for seismic anisotropy in D''. Geophys. Res. Lett. 25, 679–682.

Wu, H., Lees, J., 1999. Cartesian parameterization of anisotropic traveltime tomography. Geophys. J. Int. 137, 64–80.

Xu, Y., Wiens, D., 1997. Upper mantle structure of the southwest Pacific from regional waveform inversion. J. Geophys. Res. 102, 27439–27451.

Yang, X., Fischer, K., Abers, G., 1995. Seismic anisotropy beneath the Shumagin Islands segment of the Aleutian–Alaska subduction zone. J. Geophys. Res. 100, 18165–18177.

Yao, Z., Roberts, R., Tryggvason, A., 1990. Calculating resolution and covariance matrices for seismic tomography with the LSQR method. Geophys. J. Int. 138, 886–894.

Yokoyama, I., Aramaki, S., Nakamura, K., 1987. Volcanoes. Iwanami Press, Tokyo, 294 pp.

Zhang, J., McMechan, G., 1995. Estimation of resolution and covariance for large matrix inversions. Geophys. J. Int. 121, 409–426.

Zhao, D., 1999. Whole mantle tomography with grid parameterization, 3-D ray tracing and topography of mantle discontinuities. Eos Trans. AGU 80, 716.

Zhao, D., Mizuno, T., 1999. Crack density and saturation rate in the 1995 Kobe earthquake region. Geophys. Res. Lett. 26, 3213–3216.

Zhao, D., Negishi, H., 1998. The 1995 Kobe earthquake: seismic image of the source zone and its implications for the rupture nucleation. J. Geophys. Res. 103, 9967–9986.

Zhao, D., Horiuchi, S., Hasegawa, A., 1990. 3-D seismic velocity structure of the crust and uppermost mantle in the northeastern Japan arc. Tectonophysics 181, 135–149.

Zhao, D., Hasegawa, A., Horiuchi, S., 1992a. Tomographic imaging of P-and S-wave velocity structure beneath northeastern Japan. J. Geophys. Res. 97, 19909–19928.

Zhao, D., Horiuchi, S., Hasegawa, A., 1992b. Seismic velocity structure of the crust beneath the Japan Islands. Tectonophysics 212, 289–301.

Zhao, D., Hasegawa, A., Kanamori, H., 1994. Deep structure of Japan subduction zone as derived from local, regional, and teleseismic events. J. Geophys. Res. 99, 22313–22329.

Zhao, D., Christensen, D., Pulpan, H., 1995. Tomographic imaging of the Alaska subduction zone. J. Geophys. Res. 100, 6487–6504.

Zhao, D., Kanamori, H., Negishi, H., Wiens, D., 1996. Tomography of the source area of the 1995 Kobe earthquake: evidence for fluids at the hypocenter? Science 274, 1891–1894.

Zhao, D., Matsuzawa, T., Hasegawa, A., 1997a. Morphology of the subducting slab boundary in the northeastern Japan arc. Phys. Earth Planet. Int. 102, 89–104.

Zhao, D., Xu, Y., Wiens, D., Dorman, L., Hildebrand, J., Webb, S., 1997b. Depth extent of the Lau back-arc spreading center and its relation to subduction processes. Science 278, 254–257 .

Zhao, D., Ochi, F., Hasegawa, A., Yamamoto, A., 2000a. Evidence for the location and cause of large crustal earthquakes in Japan. J. Geophys. Res. 105, 13579–13594.

Zhao, D., Asamori, K., Iwamori, H., 2000b. Seismic structure and magmatism of the young Kyushu subduction zone. Geophys. Res. Lett. 27, 2057–2060.

Zhou, H., Clayton, R., 1990. P and S travel time inversions for subducting slab under the island arcs of the northwest Pacific. J. Geophys. Res. 95, 6829–6851.

ELSEVIER

Physics of the Earth and Planetary Interiors 127 (2001) 215–232

PHYSICS OF THE EARTH AND PLANETARY INTERIORS

www.elsevier.com/locate/pepi

Partial melting in the mantle wedge — the role of H_2O in the genesis of mantle-derived 'arc-related' magmas

Peter Ulmer*

Department of Earth Sciences, ETH-Zentrum, CH-8092 Zurich, Switzerland

Received 4 April 2000; received in revised form 15 September 2000; accepted 30 September 2000

Abstract

The fundamental role of H_2O in the generation of supra-subduction arc-related mantle derived magmas has long been recognized (e.g. Yoder and Tilley, 1962). This review of available experimental data attempts to highlight some principal factors controlling magma generation and magma chemistry that can be attributed to the presence of a hydrous component during partial melting in the mantle wedge. Arc-related igneous rocks display a typical trace element abundance spectrum, the so-called 'arc-signature', characterized by the enrichment of highly mobile large ion lithophile elements (LILE) relative to high field strength elements (HFSE). This signature can be explained by a two-component mantle source consisting of variably depleted asthenospheric mantle and a hydrous fluid (or H_2O-saturated low percentage melt) that originates from the breakdown of hydrous phases transported in the cold part of the partly hydrated oceanic lithosphere. The principal effect of H_2O on the partial melting is a reduction of the melting temperature by 100–150°C for moderate to substantial amounts of partial melting (10–25 wt.%) compared to anhydrous melting of a lherzolite source. Such a reduction of the melting temperature requires the presence of 0.1–0.5 wt.% H_2O in the source and results in 1–7 wt.% H_2O in the basaltic to picritic primary liquid. The majority of primitive arc magmas are basaltic and do not represent near-solidus H_2O-saturated liquids that are highly alkalic and mostly nepheline-normative. The decrease of the melting temperature on the order of 100–150°C compared to anhydrous peridotite melting requires high melting temperatures of 1250–1300°C at 1.5 GPa (45 km depth) and 1350–1400°C at 2.5 GPa (70 km depth) that are close to the average current mantle adiabat (ACMA). Therefore, partial melting in the mantle wedge is confined to the hottest region where temperatures approach undisturbed asthenospheric conditions. At a given pressure (depth) hydrous melts are less magnesian and more silica-rich than melts produced under anhydrous conditions. Arc magmas (and their mantle sources) are more oxidized than MORB or OIB melts (and their respective mantle sources). It is still controversial whether or not the oxidation state and the more silica-rich nature of hydrous arc-melts are indeed the principal factors controlling their predominantly calc-alkaline nature as opposed to the predominantly tholeiitic character of MORB and OIB basalts.

Keywords: Hydrous magmas; Subduction zone; Partial melting; Experimental petrology

1. Introduction

'Arc' magmatism, defined as the magmatism generated in convergent plate margins during active subduction of oceanic lithosphere, has been the subject of ongoing debate since the 1960s when the concept of plate tectonics was formulated in the earth sciences. The observation that subduction of cold oceanic lithosphere is accompanied by the generation of basic to intermediate to acidic (basalt/tholeiite–andesite–dacite/rhyolite) plutonism

* Fax: +41-16321294.
E-mail address: peter.ulmer@erdw.ethz.ch (P. Ulmer).

0031-9201/01/$ – see front matter
PII: S0031-9201(01)00229-1

and volcanism has attracted a lot of attention. Partial melting by (near-) adiabatic decompression of asthenospheric mantle is the currently accepted model for the generation of MOR-basalts and/or plume-related magmatism associated with the generation of ocean island basalts (OIBs) and 'flood' basalts of large igneous provinces (LIPs). This model cannot account for the generation of arc-related magmas. The magmas intruded or erupted in arc settings have a few particular and even unique characteristics in common:

1. They are mostly hydrous, expressed by the presence of hydrous phases, amphibole and mica, even in little differentiated basic rocks. The explosive eruption style of most differentiated 'arc' volcanics is attributed to their high H_2O-contents that is also indicated by the predominance of H_2O in the volcanic gases and the hydrothermal systems associated with 'arc' plutonics.
2. Arc magmas are predominantly calc-alkaline in character. Tholeiitic and alkaline varieties occur but are subordinate. Calc-alkaline rocks, younger than Palaeozoic occur nearly exclusively in 'arc' settings. One of the rare exceptions is the andesitic to rhyolitic volcanism of Iceland that occurs in a MOR-setting.[1] Calc-alkaline arc-related igneous rocks are not only characterized by silica enrichment, in contrast to Fe-enrichment in the tholeiitic series, but they are also distinctly more oxidized than MORBs or OIBs (e.g. Carmichael, 1991). These relations do not only apply to the magmas, but also hold for their mantle sources as testified by the oxidation state of rare mantle xenoliths in arc environments (e.g. Ballhaus, 1993; Brandon and Draper, 1996; Mysen et al., 1998). The oxidized nature of arc magmas and their mantle sources could be the result of the presence of H_2O in their source region during partial melting (e.g. Brandon and Draper, 1996) but is still a matter of debate.
3. Arc magmas exhibit characteristic trace element patterns when normalized to chondrite or MORB concentrations (Fig. 1; e.g. Pearce, 1982; McCulloch and Gamble, 1991). Relative to MORB they are *enriched* in large ion lithophile elements (LILE, e.g. Rb, K, Cs) and light REE (La, Ce, Nd) and *depleted* in high field strength elements (HFSE, i.e. Nb, Ta, Zr, Hf) and the heaviest REE (Yb, Lu). Fig. 1 illustrates the generally accepted explanation for this typical trace element abundance pattern. A two-component source for the generation of primary mantle-derived basaltic magmas is proposed: An asthenospheric mantle source that has previously been depleted by a small to moderate degree of melt extraction (asthenospheric component in Fig. 1) and a slab-derived hydrous fluid (e.g. Ringwood, 1974; Pearce, 1982; McCulloch and Gamble, 1991) that transports the mobile LILE into the peridotitic mantle source region above the descending dehydrating oceanic lithosphere. Identical geochemical characteristics have not only been found for most arc related igneous rocks, but also for the rare mantle xenoliths that occur in primitive arc basalts (e.g. Maury et al., 1992). There

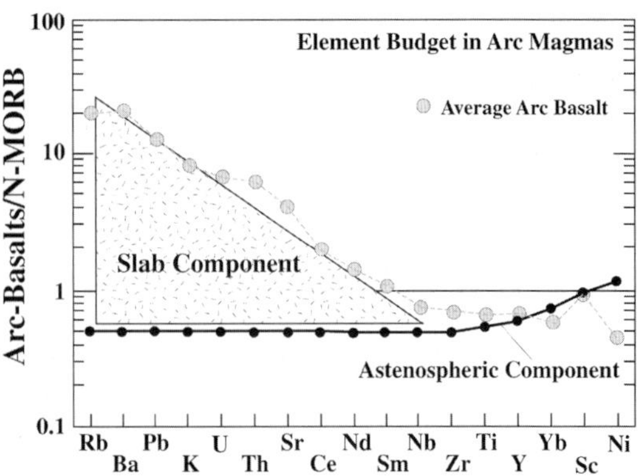

Fig. 1. Schematic 'spidergram' modified after McCulloch and Gamble (1991). The relative contributions for slab and asthenospheric components of 'arc' basalts are shown schematically. The grey dots represent the MORB-normalized trace element abundances of average arc basalts.

[1] A discussion of the fundamental differences between Precambrian and less frequent Palaeozoic TTG (tonalite–trondhjemite–granite) series igneous rocks as opposed to present day arc-related calc-alkaline gabbro–tonalite–granodiorite–granite suites is beyond the scope of this contribution. In general it is accepted that the TTG-series are the result of H_2O-undersaturated anatexis of basaltic rocks in deep-crustal processes (e.g. Rudnick and Taylor, 1986; Martin, 1987; Rapp et al., 1991; van der Laan and Wyllie, 1992) or during hot subduction (e.g. Barker and Arth, 1976; Drummond and Defant, 1990; Martin, 1999). In contrast, the post-Palaeozoic island-arc calc-alkaline suites are mostly regarded as the products of differentiation (and assimilation/mixing) of hydrous mantle-derived basaltic liquids. Only in exceptional geotectonic settings, such as ridge-subduction, slab-derived hydrous intermediate to acidic magmas are formed (e.g. Defant and Drummond, 1990).

are, however, several different explanations for how these hydrous fluids (or hydrous low-percentage melts of basaltic oceanic crust) acquire their trace element characteristics:

3.1. The trace element characteristics are produced directly by the partitioning behavior of trace elements between the major constituents of peridotitic (olivine, pyroxene) or basaltic (omphacite, garnet) rocks of the subducted oceanic lithosphere and the fluid during dehydration (breakdown) reactions of hydrosilicates (e.g. Tatsumi et al., 1986; Kogiso et al., 1997).
3.2. The HFSE depletion of the fluid is the result of residual titanium phases (rutile or ilmenite) present in the eclogitic rocks that constitute the oceanic basaltic crust of the subducted slab. HFSE strongly partition into the titanium phases, but LILE do not. Consequently, fluids in equilibrium with an ilmenite or rutile bearing solid assemblage display arc-like trace element abundances and interaction with depleted asthenospheric mantle can produce the observed trace element characteristics of arc magmas. Experimental evidence for the role of titanium phases has been provided for example by Green (1981), Brenan et al. (1995) and Stalder et al. (1998).
3.3. Ayers and co-workers (Ayers and Watson, 1991; Ayers, 1998) propose on the basis of their experimental results that decreasing rutile solubility with decreasing pressure in hydrous fluids leads to crystallization of rutile from an ascending fluid in the mantle wedge and results in a decrease (depletion) of the HFSE elements relative to the LILE in the fluid. It has, however, to be stressed that rutile saturation (crystallization) in a peridotitic mantle assemblage requires TiO_2-contents of several thousand ppm of Ti because otherwise Ti is accommodated in pyroxenes and garnet/spinel. A similar model based on the presence of Ti-pargasite amphibole in the mantle wedge (in veins) and retention of the HFSE by amphibole has been proposed by Ionov and Hofmann (1995).
3.4. Keppler (1996) proposed that chlorine-rich fluids are responsible for the HFSE-depleted, LILE-enriched character of arc magmas. Based on low-pressure partition experiments he observed increased solubilities of LILE elements in an alkali-chlorine rich fluid relative to the HFSE elements that were not affected by the presence or absence of chlorine in the hydrous fluid. It should, however, be noted that the results and interpretation of Keppler are in contradiction with high-pressure (2–6 GPa) experimental results by Brenan et al. (1995), Adam et al. (1997) and Stalder et al. (1998) who found no significant effect of chlorine (even at high concentrations) on the relative partitioning of HFSE and LILE between fluid and solid residue.
3.5. Kelemen et al. (1990a) proposed that the relative depletion of the HFSE is a result of extensive mantle–magma interaction during the passage of the hydrous primary arc-magma through the overlying lithosphere.

Most of the specific characteristics of arc magmas outlined above (e.g. the early crystallization of hydrous minerals and the relative depletion of the HFS elements) require the presence of H_2O either as a fluid phase or dissolved in the magma. The next section provides a short summary of the H_2O-repositories and transport capacities in subduction zones.

2. The source of H_2O in subduction zone environments and arc-related magmas

The hydrous fluid is most probably derived from dehydration of altered oceanic lithosphere composed of serpentinized ultramafic rocks, hydrated basic igneous rocks (gabbros and basalts) and variable amounts of mostly pelitic metasediments. The degree of hydrous alteration strongly varies in the oceanic lithosphere. Hydrothermal alteration in the vicinity of mid-ocean ridges, hydration in fracture zones and exposure of mafic and ultramafic rock on the sea floor are three possible processes that produce strong alteration of mafic and ultramafic rocks in the oceanic lithosphere. Field and experimental studies over more than 30 years have provided a framework of potential hydrous minerals that are stable to considerable depths at low to moderate temperatures typical for subduction zone metamorphism (e.g. Peacock, 1991; Davies and Stevenson, 1992; Deal et al., 1999). Until recently, most models on H_2O-transport and release in subduction zones were focussed on the occurrence and stability of amphibole in the basaltic oceanic crust and amphibole and phlogopite in the hydrated part of the ultramafic oceanic lithosphere (e.g. Davies and Bickle, 1991; Davies and Stevenson, 1992; Tatsumi and Eggins, 1995). The restriction to amphibole necessitated rather complicated models (e.g. Davies and Bickle, 1991) of H_2O transport because amphibole is only stable to $\leq$2.5 GPa in the basaltic (eclogitic) crustal rocks and to ~3.0 GPa in the overlying (and underlying) mantle at temperatures up to 1100°C. The volcanic front, defined by the alignment of arc volcanoes approximately parallel to the plate boundary, is located 166 ± 60 km away from the plate boundary (trench) and 124 ± 38 km above the seismic zone (Wadati–Benioff zone) (Gill, 1981) reflecting the surface of the descending slab. If amphibole constitutes the main H_2O-source, this would require a horizontal transport of H_2O from the descending slab to the magma source region because the maximum pressure stability of amphibole is exceeded at the slab–mantle interface below the volcanic front. Magmas that are produced at even greater depth therefore are either anhydrous or phlogopite alone provides the required H_2O according to the models discussed by the above authors, Tatsumi and Eggins (1995) and Tatsumi and Kogiso (1997).

Pawley et al. (1993), Poli and Schmidt (1995), Domanik and Holloway (1996), Schmidt (1996), Ono (1998) and Schmidt and Poli (1998) provided experimental data on the stability of hydrous phases in basaltic and pelitic (oceanic) crustal compositions and demonstrated that other minerals such as lawsonite, zoisite, chloritoid and phengite are stable to considerable depths even exceeding 300 km (lawsonite and phengite). Numerous studies in simplified synthetic (MgO–SiO_2–H_2O) and natural hydrous mantle compositions revealed that a large number of hydrous silicates could be stable in the entire upper mantle, the transition zone and into the lower mantle (e.g. Yamamoto and Akimoto, 1977; Kawamoto et al., 1995; Luth, 1995; Pawley and Wood, 1995; Ulmer and Trommsdorff, 1995; Wunder and Schreyer, 1997; Wunder, 1998, see also reviews by Frost (1999) and Ulmer and Trommsdorff, 1999). The large number of potentially stable hydrous minerals in subducted oceanic lithosphere led Schmidt and Poli (1998) and Iwamori (1998) to conclude that H_2O is supplied continuously from the subducted lithosphere to the overlying mantle wedge (in a decreasing amount with increasing depth). The location of the magma source region in the mantle wedge and hence the location of the volcanic arc is therefore controlled by the temperature distribution in the wedge and not by discrete H_2O-pulses as proposed by Davies and Bickle (1991) and Tatsumi and Kogiso (1997). A continuous dehydration scenario with temperature control of the partial melting process has been proposed previously by Ringwood (1974) and Kushiro (1987).

3. The effect of H_2O on the phase relations and compositions of primary partial melts from a peridotitic source

The potential role of H_2O as a flux in melting processes has long been recognized (Bowen, 1928; Yoder and Tilley, 1962) and its role in the generation of basaltic magmas in the mantle wedge overlying subducted oceanic lithosphere has been the subject of numerous studies since the late 1960s and 1970s (e.g. Kushiro et al., 1968; Kushiro, 1974; Green and Ringwood, 1968; Nicholls and Ringwood, 1973a,b; Nicholls, 1974; Green, 1973; Mysen and Boettcher, 1975a,b). Most studies concluded that near-solidus melts in H_2O-saturated peridotite have a broadly andesitic character and hence andesites, the most abundant composition erupted in arc volcanoes, could represent primary mantle-derived liquids. Temperatures as low as 900°C (e.g. Mysen and Boettcher, 1975a,b) or between 1050 and 1100°C (e.g. Green, 1973, 1976; Kushiro, 1974) have been derived for the generation of primary andesite-like magmas under H_2O-saturated conditions from a peridotitic source at approximately 1.5 GPa (50 km depth). This is 250–400°C lower than the anhydrous (dry) solidus at the same pressure conditions (Fig. 2; e.g. Takahashi and Kushiro, 1983). Green (1973) challenged the conclusion that H_2O-saturated near solidus liquids are qtz-normative (andesitic) and instead inferred that they are Ne-normative. Until very recently, fluid-undersaturated partial melting experiments were nearly lacking (with the exception of a small number of experiments performed by Mysen and Boettcher (1975a,b)). The experimental difficulties of analyzing near-solidus melts and the control of the H_2O-content in the charges at low abundances made these experiments difficult to perform. Several recent experimental studies (Kushiro, 1990; Hirose and Kawamoto, 1995; Gaetani and Grove, 1998; Falloon and Danyushevsky, 2000) provide first data that can be used to evaluate the potential role of H_2O in the generation of arc-magmas. All the above mentioned studies were partial melting, sandwich or forced saturation[2] experiments where the compositions of partial melts in equilibrium with a residual harzburgite or lherzolite residue was determined at a given pressure, temperature and H_2O-content of a mantle (-like) composition.

A different approach to determine the composition and formation conditions of primary mantle magmas is the inverse technique of multiple saturation experiments: A liquid composition that is potentially in equilibrium with mantle olivine (fo_{88}–fo_{92})[3] as indicated by the Mg-number (Mg# = MgO/(MgO + FeO)) and Ni-content is subjected to a series of melting/crystallization experiments to determine the pressure–temperature conditions where the liquidus is multiply saturated with mantle minerals. Any composition that truly represents a primary (undifferentiated) mantle melt should saturate with the potential residual phases ol and opx and, depending on the degree of partial melting, additionally with cpx and an aluminous phase (plag, sp, gar) over a narrow pressure range on its liquidus. Such studies have been performed on likely hydrous primary andesitic, basaltic and picritic compositions by Yoder and Tilley (1962), Nicholls and Ringwood (1973a), Stern et al. (1975), Tatsumi (1982), Tatsumi et al. (1983, 1994), Ulmer (1989), Foden and Green (1992),

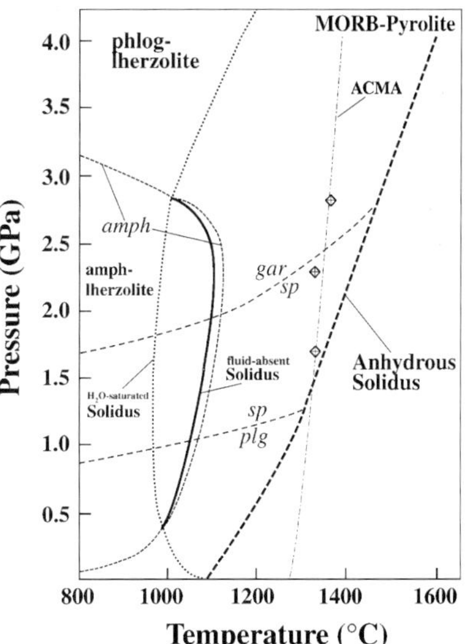

Fig. 2. Pressure–temperature diagram for H_2O-saturated, H_2O-undersaturated (0.2–0.5 wt.% H_2O) and anhydrous partial melting of MORB-pyrolite mantle composition. Data are from Kushiro et al. (1968), Green (1973), Olafsson and Eggler (1983), Takahashi and Kushiro (1983), Robinson and Wood (1998), Niida and Green (1999) and Klemme and O'Neill (2000). The solid line indicates the H_2O-undersaturated (fluid-absent) solidus; the dotted line represents the H_2O-saturated solidus; the thick broken line corresponds to the anhydrous solidus, the other broken lines indicate the stability limits of plagioclase (plag), spinel (sp), garnet (gar), and amphibole (amph) in the MORB-pyrolite mantle composition. The dotted-broken line corresponds to the average current-mantle adiabat with a potential temperature of 1280°C (Thompson, 1992). Crossed-diamonds are multiply saturated liquidus of hydrous primary magmas from island-arc settings (Tatsumi et al., 1983; Ulmer, 1989) that plot closely to the ACMA.

Baker et al. (1994) and Kawamoto (1996). The results of these studies directly constrain the conditions of last equilibration with a mantle residue that most probably reflects the depth of extraction of the partial melt from their solid harzburgite or lherzolite residue. These studies, however, do not provide crucial data on the melt fraction produced under a given set of P–T–$f(H_2O)$–$f(O_2)$ conditions. One important result of the multiple saturation experiments is that for H_2O contents of 2–6 wt.%, regarded as the range of H_2O-contents of primary arc-related calc-alkaline to tholeiitic basalts and picrites (e.g. Sisson and Layne, 1993; Baker et al., 1994), the temperature of multiple saturation exceeds 1200°C. Temperatures cluster around 1250–1300°C at 1.5 GPa and

[2] Forced saturation experiments (Gaetani and Grove, 1998; Falloon and Danyushevsky, 2000) are similar to sandwich experiments: instead of using a layer with near primary liquid composition (basalt or boninite) that is 'sandwiched' between layers of peridotite, the near primary liquid is mixed with lherzolite or harzburgite to force the liquid into saturation with ol + opx ± cpx ± sp/garnet.

[3] Mineral abbreviations used throughout this paper: ol: olivine, opx: low-Ca (ortho)pyroxene, cpx: high-Ca (clino)pyroxene, plag: plagioclase, sp: spinel, gar: garnet, amph: amphibole, liq: liquid/melt, fo: forsterite, en: enstatite.

1350–1400°C at pressures in excess of 2.2 GPa. This implies that olivine-enstatite normative basaltic melts that constitute by far the largest volume of erupted primitive basaltic compositions in arcs are not produced at conditions close to the H_2O-undersaturated mantle solidus (Fig. 2). The H_2O-undersaturated (amphibole-dehydration) melting starts just below the amphibole-out curve around 1050°C (e.g. Niida and Green, 1999). The temperatures obtained from H_2O-underaturated multiple saturation experiments are 200–300°C higher than the H_2O-undersaturated solidus and 100–150°C less than the anhydrous (dry) solidus. Therefore, voluminous melt production in the mantle wedge takes place at high temperatures close to the ACMA (average current-mantle adiabat, with a potential temperature of 1280°C). Further implications are that the generation of basaltic melts in the mantle wedge is predominantly temperature controlled and the major effect of H_2O is the reduction of the melting temperature of the anhydrous mantle solidus by 100–150°C for moderate to substantial melt fractions (10–25 wt.% melting).

The experimental data available for the anhydrous systems are presented first, followed by data on the hydrous system and a comparison and discussion of the effect of H_2O on partial melts. Fig. 3a shows the results of a partial melting study by Hirose and Kushiro (1993) on the KLB-1 lherzolite performed with the diamond-aggregate technique in the pressure range 1–3 GPa: The temperature of partial melting for a given melt fraction increases with pressure and $\partial F/\partial T$ (sometimes called 'melting efficiency') increases with increasing pressure. The temperature increase for a given melt fraction with increasing pressure reflects the positive slope of the anhdyrous mantle solidus (Fig. 2). Increasing 'melting efficiency' is caused by the increasing amount of olivine and opx components that contribute to the partial melts. This leads to a shift of the composition of the primary liquid closer to the bulk peridotite composition, hence the system becomes more 'eutectoid' and larger amounts of melt can be produced prior to exhaustion of a phase. It should however be noted that (anhydrous) mantle melting does not become eutectic over the entire upper mantle and transition zone (e.g. Takahashi, 1986; Herzberg and Zhang, 1996). Fig. 3b shows the bulk composition effect at a pressure of 1.5 GPa. Decreasing source fertility, expressed as the amount of 'basaltic component' represented by cpx and the aluminous phase(s) plag, sp and/or gar present in the bulk peridotite, causes a temperature increase for a given melt fraction. The temperatures for 15% melting at 1.5 GPa increase from 1310°C for the fertile MORB-pyrolite to 1330°C for the KLB-1 (Kilborne Hole Crater lherzolite xenolith) to 1380°C for the depleted Tinaquillo lherzolite. A summary of melt-fraction versus H_2O-content of the resulting basaltic liquid is shown in Fig. 3c based on the experimental study of Hirose and Kawamoto (1995) at 1.0 GPa pressure. With increasing H_2O-content of the basaltic liquid the temperature required to produce a given melt fraction decreases considerably. For H_2O-contents of 2–5 wt.% (thick solid line in Fig. 3c) the temperature decrease is on the order of 140°C, similar to the value deduced from the multiple saturation experiments. Falloon and Danyushevsky (2000) have proposed an exponential equation to describe the effect of H_2O on the liquidus temperature of primary boninitic liquid. Their equation yields a liquidus depression for the primary liquid of 110–130°C for 3–5 wt.% H_2O in the liquid, in the same range as deduced from multiple-saturation and lherzolite melting experiments. The effect of H_2O can also be expressed by the increase in melt-fraction per amount H_2O added to the mantle source. Based on data by Hirose and Kawamoto (1995) at 1.0 GPa and by Gaetani and Grove (1998) at 1.2–2.0 GPa the increase in melt fraction at constant temperature per 0.1 wt.% H_2O added to the mantle source is approximately 3–4 wt.%. This is comparable to the estimate by Stolper and Newman (1994) of 2.7 wt.% increase in melt fraction per 0.1 wt.% H_2O added, based on an inversion of geochemical data on primitive arc lavas from the Mariana trough.

The temperature–pressure-composition effects of H_2O on the generation of primary magmas is expressed as MgO content (wt.%) of the liquid as a function of temperature (Fig. 4) and pressure (Fig. 5). Figs. 4 and 5 are based on a reduced dataset that is limited to near cpx-out olivine–opx–cpx ± aluminous phase assemblages. This dataset has been chosen to eliminate strong bulk composition control on the melt compositions at low percentage melting. At substantial melt fractions (10–25 wt.% depending on fertility), the effect of bulk composition is

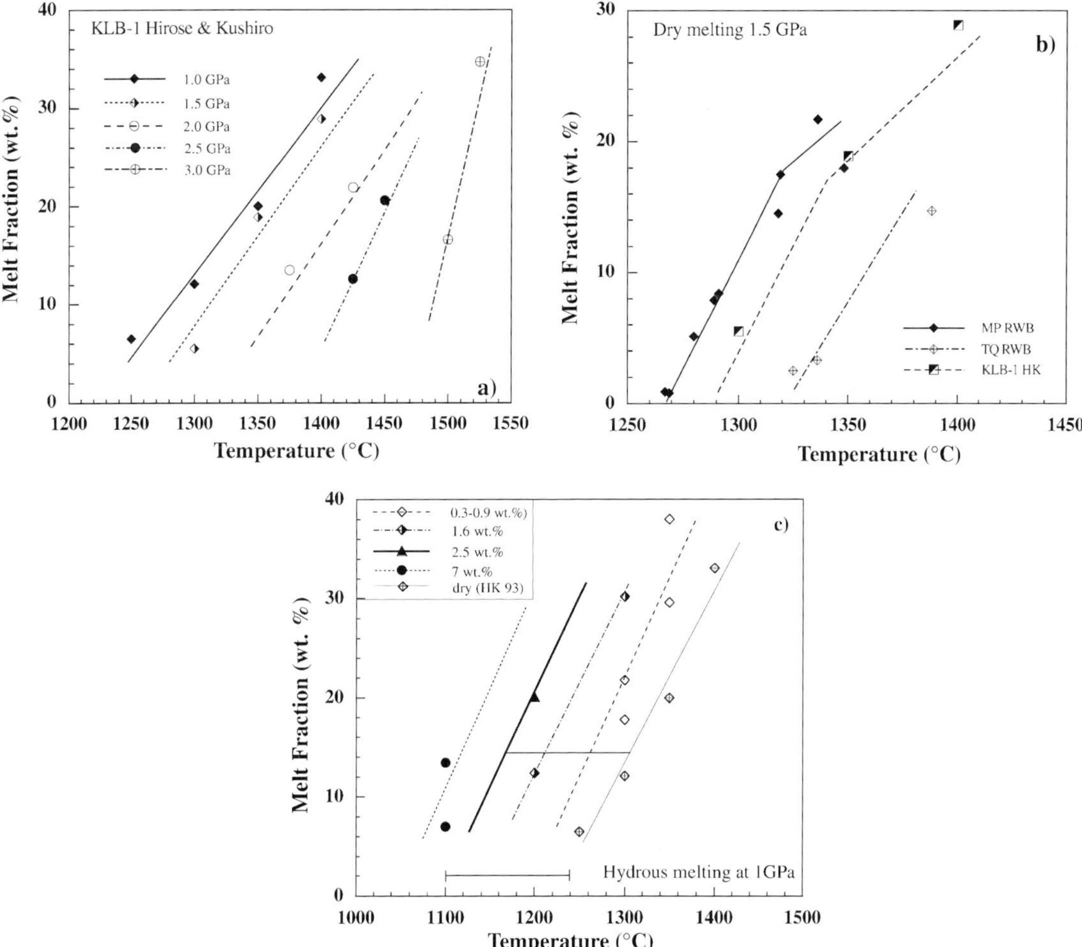

Fig. 3. Melt fraction versus temperature relationships. (a) Anhydrous experiments on KLB-1 composition (data from Hirose and Kushiro, 1993) at variable pressure from 1 to 3 GPa. (b) Anhydrous melting at 1.5 GPa and variable 'fertility' of the bulk composition. MP RWB–MORB Pyrolite (fertile) and TQ RWB–Tinaquillo Lherzolite (depleted) from Robinson et al. (1998); KLB-1 HK–Kilborne Hole Crater lherzolite xenolith (intermediate fertility) from Hirose and Kushiro (1993). Temperature required for a given melt fraction increases with decreasing fertility (= amount basaltic component present in the peridotite starting material). The change in slope for MP RWB and KLB-1 HK indicates the cpx-out point; at higher temperature olivine + opx alone form the solid residue. (c) Hydrous experiments (Hirose and Kawamoto, 1995) at 1 GPa and variable amount of H_2O in the generated basaltic liquids. Crossed diamonds correspond to the results of anhydrous partial melting experiments performed on the same starting material at 1 GPa by Hirose and Kushiro (1993). The horizontal bar close to the temperature axis indicates the temperature difference of ~140°C between hydrous (2–5 wt.% H_2O in the liquid) and anhydrous partial melting for an equal amount of partial melt.

minimized but phase equilibria exert a relatively good control on melt compositions that are 'buffered' by 3–4 residual phases in the peridotite. In other words; in a multi-component projection, the selected data plot on the olivine–opx–cpx cotectic close to the cpx-out point. In addition, near cpx-out compositions are directly comparable with multiple-saturation experiments that saturate with olivine–opx–cpx.

Fig. 4a reveals a linear relationship between temperature and the MgO-content of the primary liquid for *anhydrous* compositions, including both forward lherzolite melting, forced saturation and sandwich

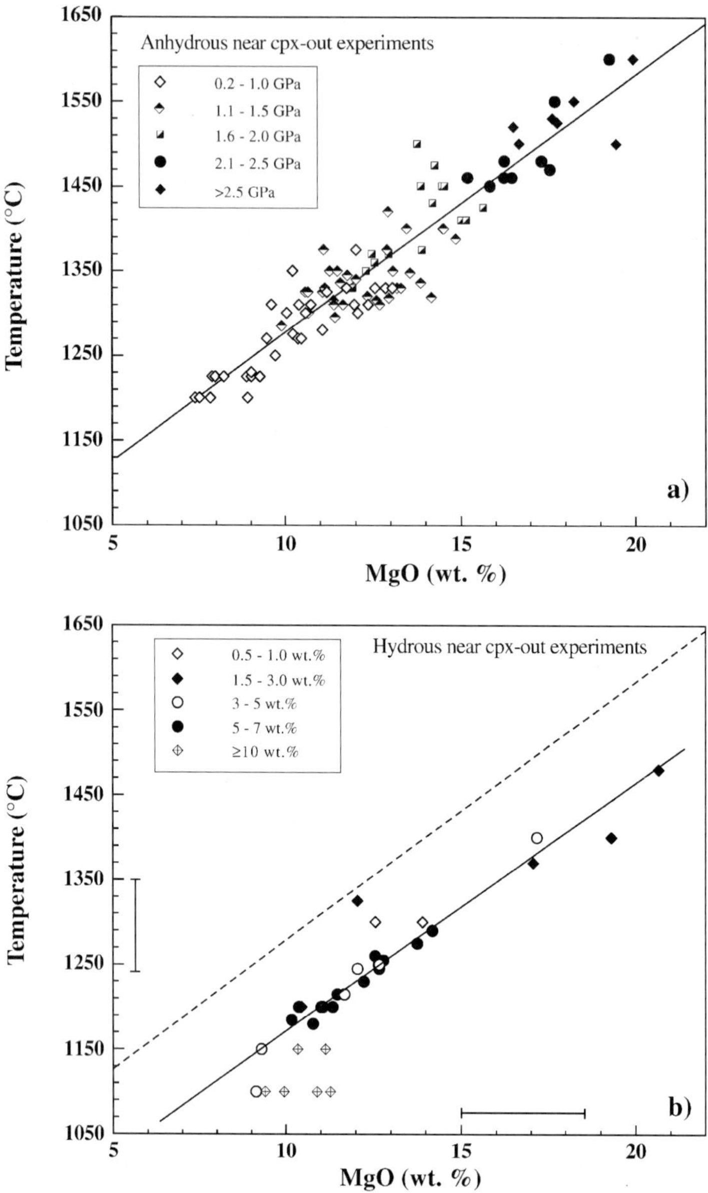

Fig. 4. Temperature–MgO relationships of near cpx-out melt compositions produced in partial melting, sandwich, 'forced saturation' and multiple saturation experiments. All melts are olivine–opx–cpx saturated and close to the *P–T* conditions of cpx-out. Pressure ranges from 0.2 to 3.0 GPa for anhydrous experiments and 1.0–2.8 GPa for hydrous experiments: (a) anhydrous experiments. Data sources: Green and Ringwood (1967), Kushiro et al. (1972), Mysen and Kushiro (1977), Bender et al. (1978), Tatsumi (1982), Takahashi and Kushiro (1983), Tatsumi et al. (1983, 1994), Elthon and Scarfe (1984), Gust and Perfit (1987), Kushiro (1987), Bartels et al. (1991), Draper and Johnston (1992), Kinzler and Grove (1992a,b), Falloon and Green (1988), Hirose and Kushiro (1993), Baker and Stolper (1994), Falloon et al. (1988, 1997, 1999a,b), Gaetani and Grove (1998), Robinson et al. (1998) and Falloon and Danyushevsky (2000). (b) Hydrous experiments. Data sources: Additionally to some of the above references — Ulmer (1989), Hirose and Kawamoto (1995) and Baker et al. (1994). Solid line indicates approximate T-MgO relationship for hydrous compositions containing 1–7 wt.% H_2O. Broken line corresponds to anhydrous melt relationship from (a). The difference between anhydrous and hydrous (undersaturated) is about 120°C at equal MgO-content or about 3.5 wt.% MgO at constant temperature.

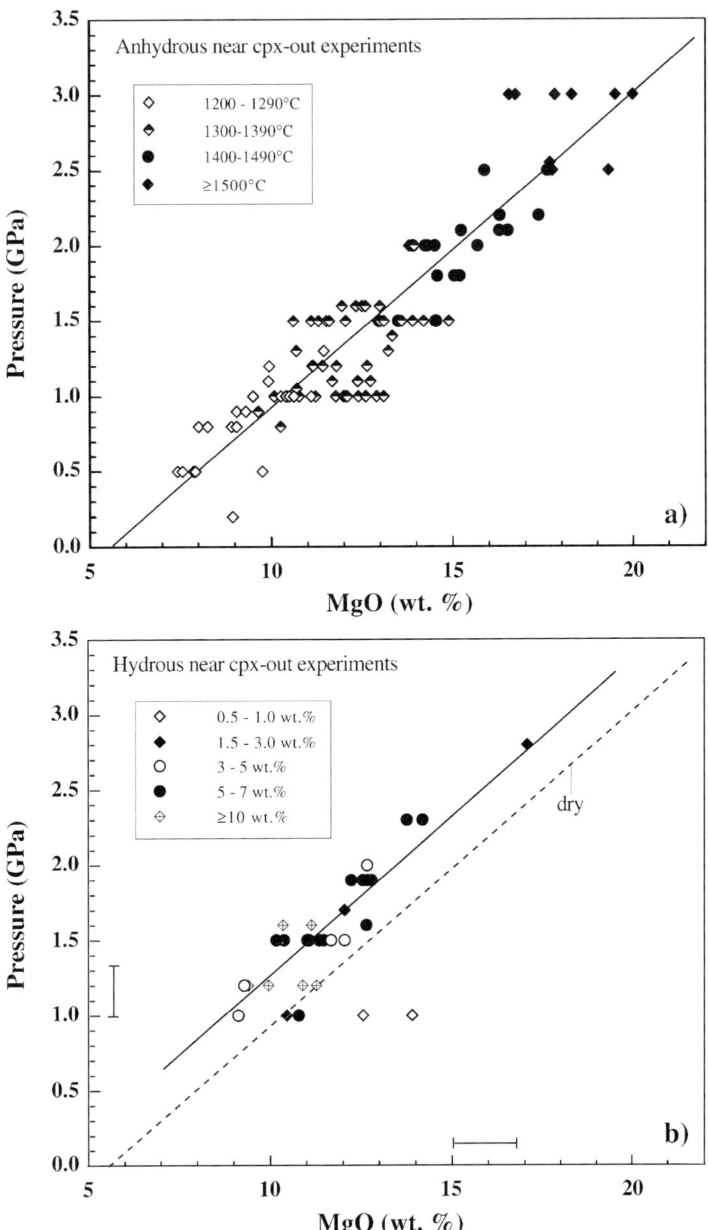

Fig. 5. Pressure–MgO relationships of near cpx-out compositions produced in partial melting, sandwich, 'forced saturation' and multiple saturation experiments. All compositions are olivine–opx–cpx saturated and close to the *P–T* conditions of cpx-out. Temperatures range from 1200 to 1600°C for anhydrous experiments and from 1100 to 1480°C for hydrous experiments: (a) anhydrous experiments. Solid line represents a linear fit of the experimental dataset: (b) hydrous experiments. Data sources: see Fig. 4. Solid line indicates approximate P–MgO relationship for hydrous compositions containing 1–7 wt.% H_2O. Broken line corresponds to anhydrous melt relationship from (a). The difference between anhydrous and hydrous (undersaturated) is about 0.3 GPa at equal MgO-content or about 2 wt.% MgO at constant pressure.

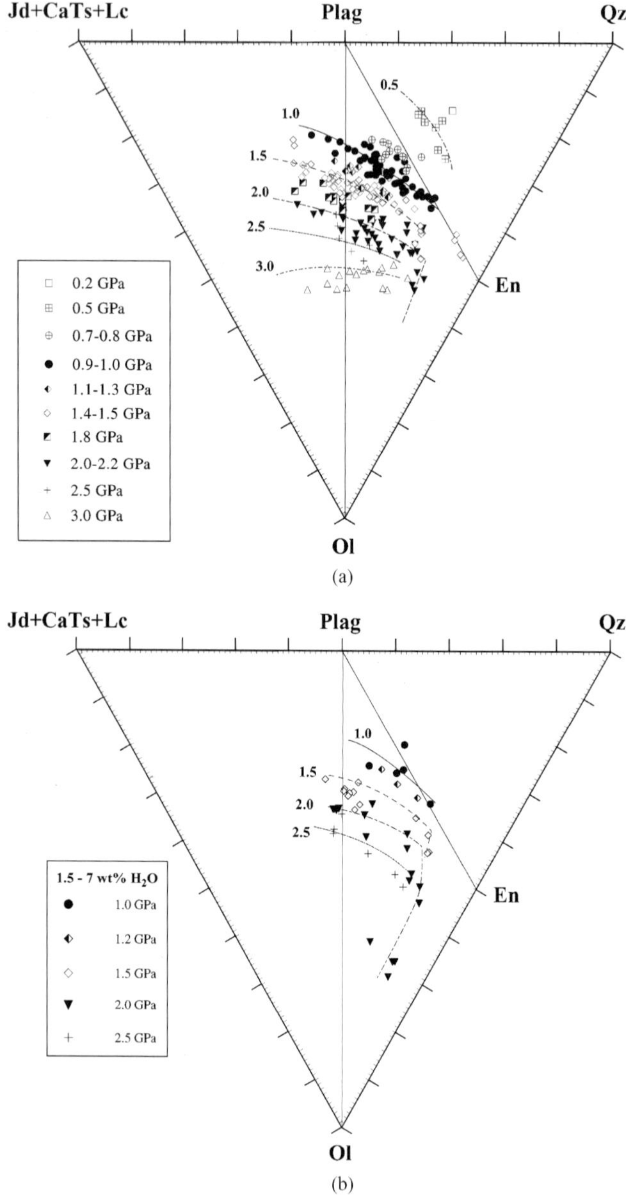

Fig. 6. Molecular normative projection from olivine onto the base [Jd+CaTs+Lc]–Qz–Ol of the basalt tetrahedron. Normalization according to Falloon and Green (1988) except that anhydrous CIPW-norm was calculated with $Fe_2O_3/FeO = 0.10$: (a) anhydrous experiments. Data sources are the same as Fig. 4a plus additional experiments by Cohen et al. (1967), Green (1969), Thompson (1975) and Eggins (1992). (b) Hydrous fluid-undersaturated experiments. All liquid compositions plotted in (b) contain between 1.5 and 7 wt.% H_2O and are undersaturated at the experimental pressures. Data sources are the same as in Fig. 4b plus additional experiments by Yoder and Tilley (1962), Nicholls and Ringwood (1973a), Stern et al. (1975), Green (1976), Foden and Green (1992) and Kawamoto (1996). Lines pointing towards the olivine apex for 1.5 and 2.0 GPa connect compositions that are olivine saturated only. (c) Comparison of the locations of the (cpx)–opx–ol cotectics in anhydrous and hydrous fluid-undersaturated systems. Curves are taken from (a) and (b). The shift of the hydrous cotectics towards the qz-apex, reflecting an expansion of the olivine primary phase volume, corresponds to a pressure shift of $\approx$0.3 GPa. Hydrous incipient melts are less Jd + CaTs + Lc normative than corresponding anhydrous melts, but still silica undersaturated at pressures exceeding 1 GPa. Silica-normative melts can only be produced at pressures of less than 0.7 GPa in anhydrous and less than 1.0 GPa in hydrous systems from a peridotitic source.

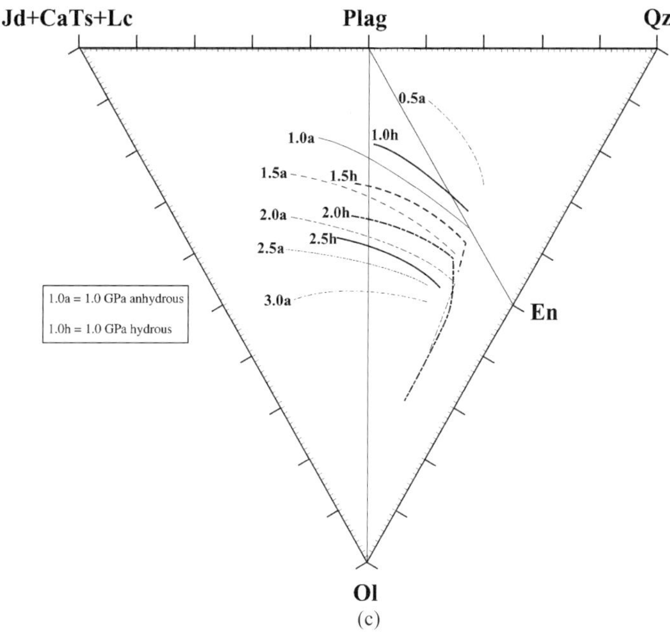

(c)

Fig. 6. (*Continued*).

experiments as well as multiple saturation experiments on primary liquid compositions. The increase in MgO content with temperature reflects predominantly the reduction of the olivine primary phase volume with increasing pressure (compare also Fig. 6a: shift of the ol–opx cotectic towards the ol-apex with increasing pressure) and the positive slope of the lherzolite solidus in the anhydrous system. Fig. 4b presents the results from hydrous melting and multiple-saturation experiments close to the cpx-out point. The temperatures for the ol–opx–cpx cotectic compositions are lowered by the addition of H_2O. In this diagram there is no systematic decrease of temperature with increasing H_2O-content in the liquid from 1.5 to 7 wt.%, but the data points cluster around the solid line drawn in Fig. 4b. Only at very high and possibly unrealistic H_2O contents of 10 or more wt.% H_2O in the liquid a further decrease of the temperature of the cotectic is observed. The decrease of the liquidus temperature of olivine–opx–cpx saturated primary basaltic liquids is again on the order of 120°C (difference between broken line = anhydrous and solid line = H_2O-undersaturated). At present the lack of a clear correlation of H_2O-content and temperature cannot be evaluated. If this lack of correlation is not an artifact, it has some implications for the generation of the voluminous arc-related basalts. The addition of relatively small amounts of H_2O, on the order 0.2 wt.%, to the peridotitic source, as for example proposed by Kushiro (1987) is sufficient to decrease the liquidus temperature of the olivine–opx–cpx cotectic by approximately 100–150°C. The addition of a larger amount of H_2O, up to approximately 0.5 wt.%, would not affect the bulk composition (at least the MgO-content) of the generated liquid but most probably would lead to an increased amount of partial melting (if the source is fertile enough to maintain residual cpx to higher melt percentages). Hence, the amount of H_2O present in the mantle source mainly controls the extent of partial melting and therefore the concentrations of minor and trace elements such as the alkalis, Ti and P, but not the major elements. Similar conclusions have been reached by Stolper and Newman (1994) and Singer et al. (1992) on the basis of geochemical data of primitive basalts in the Mariana trough and the Aleutian arc. They clearly observed that the degree of melting increased with additional H_2O in the source but that inferred liquidus temperatures and major element compositions of the erupted basaltic lavas only varied in a narrow range.

In Fig. 5a and b the pressure effect on the primary melt compositions expressed by the MgO-content is displayed. Anhydrous melting experiments show a positive correlation between pressure and MgO content for the ol–opx–cpx cotectic compositions. This is the result of the reduction of the olivine primary phase volume with pressure leading to more MgO-rich, SiO_2-poorer liquid compositions with increasing pressure. The addition of H_2O (Fig. 5b) leads to a reduction of the MgO content at a given pressure. This is the well-known effect of the expansion of the olivine primary phase volume, shifting the ol–opx $\pm$ cpx cotectic away from the ol-apex in triangular projections (e.g. Fig. 6c) with the addition of H_2O (Kushiro, 1972; Green, 1973; Mysen and Boettcher, 1975a). There is considerable spread in the available dataset and no attempt has been made to further quantify the effect of H_2O. The difference between the broken (= anhydrous) and solid (= H_2O-undersaturated) lines is approximately 3–4 kbar for the addition of 0.1–0.5 wt.% H_2O (1–7 wt.% in the liquid). The addition of H_2O produces liquids that are richer in SiO_2 and poorer in MgO at a given pressure; i.e. the effect of adding H_2O is opposite to the effect of increasing the pressure.

The compositional changes as a function of pressure and H_2O-content of primary liquids generated from a peridotitic source are presented in more detail in Fig. 6a–c. The complete datasets covering primary melt compositions from incipient melting to extensive partial melting with ol–opx or even ol-only residues are displayed on the ol–qz–jd + CaTs + lc ternary diagram. This diagram represents the base of the basalt tetrahedron (Yoder and Tilley, 1962) projected from diopside according to the scheme proposed by Falloon and Green (1988). Fig. 6a shows the locations of the cotectic curves for anhydrous compositions from a large number of experimental studies referred to in the caption of Fig. 6. Despite the large variation of investigated bulk compositions a high degree of consistency in the positions of the cotectic curves is observed. Important observations are: (i) with increasing pressure cotectic curves consistently shift to higher normative olivine contents (reduction of the olivine primary phase volume); (ii) near-solidus melts that are located at the Jd + CaTs + Lc rich end of the curves are all silica-poor and mostly nepheline-normative. SiO_2-rich liquids that are qtz-normative (plotting to the right of the En-Plag tie-line) can only be produced at pressures of less than 0.7 GPa. Melting of asthenospheric mantle sources in an arc environment is generally restricted to depths in excess of 30 km (1.0 GPa) and therefore anhydrous melting does not produce any qtz-normative primary liquids in arc environments.

Experimental data on hydrous melting of peridotites and multiple saturation experiments on hydrous primary liquids are very limited compared to the large dataset available on anhydrous melting/crystallization. Fig. 6b displays the available data for an H_2O-content ranging from 1.5 to 7 wt.%. A similar trend of increasing normative olivine component with increasing pressure is observed. The lowest pressure data at 1.0 GPa just straddle the field of qtz-normative compositions. Near-solidus compositions remain Jd + CaTS + Lc rich but to a lesser extend than the incipient melts of anhydrous melting experiments. It is important to note that near-solidus melts, which are H_2O-undersaturated are not silica-normative. In the 1970s controversial results and opinions were discussed concerning whether silica-saturated andesite-like liquids can be produced at moderate depths (1.0–1.5 GPa) from H_2O-rich peridotitic sources (Kushiro et al., 1968; Kushiro, 1974; Green, 1973, 1976; Mysen and Boettcher, 1975a,b). The results of recent H_2O-undersaturated melting experiments on lherzolite compositions (Hirose and Kawamoto, 1995; Gaetani and Grove, 1998; Falloon and Danyushevsky, 2000) indicate near-solidus melts are Jd + CaTs + Lc rich, or even Ne-normative (Gaetani and Grove, 1998). In contrast, very refractory harzburgitic sources can produce nearly silica-saturated liquids by hydrous partial melting, but these compositions are quite distinct from calc-alkaline andesites. These experimental liquids are MgO-rich boninites (often referred to as 'high-Mg andesites') that can occur in island-arcs at the initial stage of inner-oceanic subduction. They also require extremely high temperatures in excess of 1450°C at a depth of 45 km (1.5 GPa) (Falloon and Danyushevsky, 2000), conditions generally not attained in subduction zone environments. However, there are other scenarios based on field, geochemical and experimental data to explain silica-rich high-Mg andesites that occur at the initial stage of subduction. Tatsumi (1982) and Baker et al. (1994) have proposed that high degrees of melting from an H_2O-rich mantle source (6 wt.%

H_2O in the high-Mg andesite at 20–30% partial melting of a peridotitic mantle source) at shallow depths (1.0 GPa) and temperatures close to 1200°C can yield high-silica high-MgO liquids comparable to high-Mg andesites. Kelemen et al. (1990a,b) have proposed that wall-rock interaction between either slab-derived andesitic liquids or mantle-derived basaltic liquids and peridotite at shallow depths can also produce calc-alkaline high-Mg andesitic compositions.

Fig. 6c compares the location of hydrous and anhydrous cotectic curves. The cotectic curves shift away from olivine by the addition of H_2O at a given pressure. This diagram clearly shows the expansion of the olivine primary phase volume relative to enstatite with the addition of H_2O. Similar relationships have been observed in synthetic systems (e.g. Kushiro, 1975; Taylor and Green, 1987; Foley, 1992; Edgar and Vukadinovic, 1992). The shift in the cotectic curves between anhydrous and H_2O-undersaturated experiments corresponds to approximately 3 kbar. The shift towards the quartz apex implies that peritectic melting relationships between ol–opx–liq persist to higher pressures in hydrous systems than in the anhydrous system. This is supported by partial melting experiments: Gaetani and Grove (1998) proposed a melting reaction cpx + opx + gar = ol + liq at 2.3 GPa, 1275–1290°C in the hydrous system as opposed to cpx + ol + gar = opx + liq at 2.3 GPa, 1430–1440°C as derived by Kinzler (1997) in the anhydrous system. Ulmer (1989) proposed a similar reaction (gar + opx + cpx = ol + liq) at 2.8 GPa, 1370°C based on multiple saturation experiments on a hydrous primary picrobasalt composition.

4. The influence of volatiles other than H_2O on primary partial melts of peridotite

The previous discussion is confined exclusively to the role of H_2O as volatile phase. Other volatile species, such as CO_2 and CH_4, could exert important additional controls on partial melting. It is beyond the scope of this contribution to discuss in detail the influence of additional volatile components that might be involved in the generation of arc magmas. It should, however, be noted that fluid inclusions in arc magmas as well as the composition of gases emanating from arc volcanoes are predominantly hydrous with Cl as the second most important fluid constituent. CO_2 and CH_4 only occur in minor quantities unlike in alkaline and tholeiitic systems where CO_2 might even represent the most important volatile species. For overviews of the principal effects of CO_2 and CH_4 on the phase relations involving olivine and opx the reader is referred to Wyllie (1978, 1979), Wendtland and Mysen (1980), Olafsson and Eggler (1983), Taylor and Green (1987, 1988), Edgar and Vukadinovic (1992) or Foley (1992). CH_4 behaves similarly to H_2O; i.e. it expands the olivine primary phase volume. CO_2 has the opposite effect: it increases the opx primary phase volume thereby producing silica-poor mostly silica-undersaturated liquids. With increasing pressure near-solidus melts become carbonatitic (at pressures $\geq$2.5 GPa) in the presence of CO_2. Peridotite–CO_2–H_2O systems at pressures of $\leq$2.5 GPa produce highly alkalic near-solidus liquids that become tholeiitic with increasing degree of partial melting. At pressures exceeding 2.5 GPa near-solidus melts are also carbonatitic and become more silica-rich but still undersaturated with increasing degree of melting. CH_4 is not an abundant volatile species in the oxidized mantle wedge that generates arc magmas and CO_2 produces silica-undersaturated alkaline magmas that are not very common in arc-related systems.

5. Summary and outlook

Water is a fundamental requirement to explain the generation and the geochemical characteristics of arc-related mantle derived basaltic magmas. Small to moderate amounts of H_2O in the mantle source (0.1–0.5 wt.%) lead to a decrease of 100–150°C of the melting temperature for moderate to substantial amounts of partial melting (5–20 wt.%). A considerably larger decrease of the melting temperature in the order of 300–400°C is observed for H_2O-saturated magmas. However, such magmas are alkalic, nepheline-normative and do not correspond to the large volume enstatite-normative calc-alkaline to tholeiitic compositions that produce the igneous rock suites typical of arc settings. Melt fractions in the order of 5–25 wt.% of a lherzolitic source (e.g. KLB-1) are required to produce the observed compositions. Temperatures on the order of 1250–1300°C at

1.5 GPa and 1350–1400°C at 2.5 GPa are required to produce the composition of magmas observed in convergent plate margin magmatism. Such temperature estimates are close to the average current mantle adiabat (ACMA); hence the production of large volumes of igneous rocks in the mantle wedge is restricted to the hottest area where temperatures approach those of undisturbed asthenospheric mantle. Melt compositions produced by partial melting in the absence and presence of H_2O are similar, but cotectic compositions are shifted to more silica-rich, MgO-poor compositions at a given pressure. This corresponds to the effect of 3–4 kbar; i.e. hydrous melts produced at 1.5 GPa are similar to anhydrous melts produced at 1.2 GPa.

As outlined in the introduction, the unique trace element characteristics observed for convergent plate magmas, the enrichment of LILE relative to the HFSE is most probably the result of fluids emanating from the subducted oceanic lithosphere.

The currently available dataset favors a melting model similar to that proposed by Ringwood (1974), Kushiro (1987, 1990), Schmidt and Poli (1998) or Iwamori (1998). Models differ in detail, however, and one significant aspect that has not been discussed in this contribution is the question of whether melting occurs as a dynamic process driven by diapiric decompression (Ringwood, 1974; Kushiro, 1987, 1990) or as a static process exclusively controlled by the advection of H_2O and the temperature structure of the wedge (e.g. Iwamori, 1998). Inverted melt compositions or multiple saturation experiments on primary liquid compositions provide the pressure (depth) of last equilibration with a peridotitic mantle residue and hence the depth where the partial melt was separated from its residue. Pressures inferred from compositions and multiple saturation experiments, therefore, cannot provide the means to distinguish between dynamic melting over a large depth interval and static melting confined to a narrow depth interval.

To date the majority of experimental studies on hydrous partial melting of peridotites are restricted to pressures of 1.0–1.6 GPa, corresponding to depths of 30–50 km. This is appropriate for intra-oceanic island-arcs and some continental arcs with relatively thin continental crust in the order of 30 km. However, continental margins and collision belts such as the Andes, the Alps and the Himalayas require extraction depths in excess of 70 km as the continental crust already reaches 50–70 km in some areas where subduction-related magmatism occurs. The few examples available, such as the Adamello Intrusion, Southern Alps (Ulmer, 1989) indicate indeed that hydrous primary liquids in such environments have separated from their mantle residue at pressures in excess of 2.5 GPa, within the stability field of garnet-lherzolite. Currently, only a very limited set of experimental data on hydrous partial melting of peridotitic mantle at pressures in excess of 2 GPa is available (e.g. Mysen and Boettcher, 1975a,b) that provides crucial information on temperature, residual mineralogy, melt fraction and melt compositions for small to moderate amounts of H_2O. In order to understand the generation of voluminous calc-alkaline igneous products in regions with thick continental crust additional experiments are required to evaluate quantitatively the role of H_2O on the partial melting of garnet and phlogopite/K-richterite bearing mantle sources at pressures of 2.5 GPa and higher. The increase of the K_2O-content of primary magmas with increasing horizontal distance from the trench (Dickenson, 1968), still awaits experimental testing of the various models proposed. These are: (i) breakdown of phlogopite in the subducted slab and production of K-rich hydrous fluid that generates partial melting in the mantle wedge (Tatsumi and Eggins, 1995; Tatsumi and Kogiso, 1997). (ii) Breakdown of phengitic mica in the subducted slab at 250–300 km (Schmidt, 1996; Domanik and Holloway, 1996; Ono, 1998) and production of K-rich hydrous fluids. (iii) Reduction of the amount of partial melting due to high pressure and small H_2O-concentration in the source region as the result of decreasing H_2O-transport capacity in the slab (e.g. Luhr, 1992).

Progress in experimental petrology, geochemistry, thermal modeling and seismic tomography has clearly provided deeper and more quantitative insights into the generation of supra-subduction magmas in the mantle wedge. However, the restricted experimental dataset on hydrous compositions, crucial to understand the role of hydrous versus anhydrous melting, still makes it difficult to quantify the role of H_2O for the magma production and the compositions that can be produced by partial melting in the mantle wedge. Additional experimental studies that provide complete information regarding compositions and melt fractions would considerably contribute to our current understanding of melt generation in arcs.

Acknowledgements

I thank Dave Rubie for inviting me to participate in the Symposium and in this volume, for his tolerance with respect to deadlines, and his editorial comments. I would like to acknowledge Othmar Müntener, Bjorn Mysen, Ikuo Kushiro, Peter Kelemen, Roland Stalder, Ralf Kägi, Alan Thompson, Max Schmidt and Stefano Poli for their many (controversial) discussions on the subject of arc magma generation that have influenced the content of this contribution. I am grateful to Othmar Müntener for his comments prior to submission and to Max Schmidt and Dan Frost for their critical and helpful reviews.

References

Adam, J., Green, T.H., Sie, S.H., Ryan, C.G., 1997. Trace element partitioning between aqueous fluids, silicate melts and minerals. Eur. J. Mineral. 9, 569–584.

Ayers, J.C., 1998. Trace element modeling of aqueous fluid–peridotite interaction in the wedge of subduction zones. Contrib. Mineral. Petrol. 132, 390–404.

Ayers, J.C., Watson, E.B., 1991. Solubility of apatite, monazite, zircon, and rutile in supercritical fluids with implications for subduction zone geochemistry. Phil. Trans. R. Soc. London A 335, 365–375.

Baker, M.B., Stolper, E.M., 1994. Determining the composition of high-pressure mantle melts using diamond aggregates. Geochim. Cosmochim. Acta 58, 2811–2827.

Baker, M.B., Grove, T.L., Price, R., 1994. Primitive basalts and andesites from the Mt. Shasta region, N. California: products of varying melt fraction and water content. Contrib. Mineral. Petrol. 118, 111–129.

Ballhaus, C., 1993. Redox state of lithospheric and asthenospheric upper mantle. Contrib. Mineral. Petrol. 114, 331–348.

Barker, F., Arth, J.G., 1976. Generation of trondhjemitic-tonalitic liquids and Archean bimodal trondhjemite-basalt suites. Geology 4, 596–600.

Bartels, K.S., Kinzler, R.J., Grove, T.L., 1991. High pressure phase relations of primitive high-alumina basalts from Medicine Lake volcano, northern California. Contrib. Mineral. Petrol. 108, 253–270.

Bender, A., Hodges, F.N., Bence, A.E., 1978. Petrogenesis of basalts from the Project FAMOUS area. Earth Planet. Sci. Lett. 41, 277–302.

Bowen, N.L., 1928. The Evolution of the Igneous Rocks. Dover Publications, New York, 332 pp.

Brandon, A.D., Draper, D.S., 1996. Constraints on the origin of the oxidation state of mantle overlying subduction zones: an example from Simcoe, Washington, USA. Geochim. Cosmochim. Acta 60, 1739–1749.

Brenan, J.M., Shaw, H.F., Ryerson, F.J., Phinney, D.L., 1995. Mineral-aqueous fluid partitioning of trace elements at 900°C and 2.0 GPa: Constraints on the trace element chemistry of mantle and deep crustal fluids. Geochim. Cosmochim. Acta 59, 3331–3350.

Carmichael, I.S.E., 1991. The redox states of basic and silicic magmas: a reflection of their source regions? Contrib. Mineral. Petrol. 106, 129–141.

Cohen, L.H., Ito, K., Kennedy, G.C., 1967. Melting and phase relations in an anhydrous basalt to 40 kbar. Am. J. Sci. 265, 475–518.

Davies, J.H., Bickle, M.J., 1991. A physical model for the volume and composition of melt produced by hydrous fluxing above subduction zones. Phil. Trans. R. Soc. London A 335, 355–364.

Davies, J.H., Stevenson, D.J., 1992. Physical model of source region of subduction zone volcanics. J. Geophys. Res. 97, 2037–2070.

Deal, M.M., Nolet, G., van der Hilst, R.D., 1999. Slab temperature and thickness from seismic tomography: 1. Method and application to Tonga. J. Geophys. Res. 104, 28789–28802.

Defant, M.J., Drummond, M.S., 1990. Derivation of some modern arc magmas by melting of young subducted lithosphere. Nature 347, 662–665.

Dickenson, W.R., 1968. Circum-Pacific andesite types. J. Geophys. Res. 73, 2261–2269.

Domanik, K.J., Holloway, J.R., 1996. The stability and composition of phengitic muscovite and associated phases from 5.5 to 11 GPa: implications for deeply subducted sediments. Geochim. Cosmochim. Acta 60, 4133–4150.

Draper, D.S., Johnston, A.D., 1992. Anhydrous PT phase relations of an Aleutian high-MgO basalt: an investigation of the role of olivine-liquid reaction in the generation of arc high-alumina basalts. Contrib. Mineral. Petrol. 112, 501–519.

Drummond, M.S., Defant, M.J., 1990. A model for trondhjemite-tonalite-dacite genesis and crustal growth via slab melting: Archean to modern comparison. J. Geophys. Res. 95, 21503–21521.

Edgar, A.D., Vukadinovic, D., 1992. Implications of experimental petrology to the evolution of ultrapotassic rocks. Lithos 28, 205–220.

Eggins, S.M., 1992. Petrogenesis of Hawaiian tholeiites: 1, phase equilibria constraints. Contrib. Mineral. Petrol. 110, 387–397.

Elthon, D., Scarfe, C.M., 1984. High pressure phase equilibria of a high-magnesia basalt and the genesis of primary oceanic basalts. Am. Mineral. 69, 1–15.

Falloon, T.J., Danyushevsky, L.V., 2000. Melting of refractory mantle at 1.5, 2 and 2.5 GPa under anhydrous and H_2O-undersaturated conditions: Implications for the petrogenesis of high-Ca boninites and influence of subduction components on mantle melting. J. Petrol. 41 (2), 257–283.

Falloon, T.J., Green, D.H., 1988. Anhydrous partial melting of peridotite from 8 to 35 kbar and the petrogenesis of MORB. J. Petrol. Special Lithosphere Issue 29, 379–414.

Falloon, T.J., Green, D.H., Hatton, C.J., Harris, K.L., 1988. Anhydrous partial melting of a fertile and depleted peridotite from 2 to 30 kbar and application to basalt petrogenesis. J. Petrol. 29, 1257–1282.

Falloon, T.J., Green, D.H., O'Neill, H.S.C., Hibberson, W.O., 1997. Experimental tests of low degree peridotite partial melt compositions: implications for the nature of anhydrous near-solidus peridotite melts at 1 GPa. Earth Planet. Sci. Lett. 152, 149–162.

Falloon, T.J., Green, D.H., Danyushevsky, L.V., Faul, U.H., 1999a. Peridotite melting at 1.0 and 1.5 GPa: an experimental evaluation of techniques using diamond aggregates and mineral mixes for determination of near-solidus melts. J. Petrol. 40, 1343–1375.

Falloon, T.J., Green, D.H., Jaques, A.L., Hawkins, J.W., 1999b. Refractory magmas in back-arc basin settings — experimental constraints on the petrogenesis of a Lau Basin example. J. Petrol. 40, 255–277.

Foden, J.D., Green, D.H., 1992. Possible role of amphibole in the origin of andesites: some experimental and natural evidence. Contrib. Mineral. Petrol. 109, 479–493.

Foley, S., 1992. Petrological characterization of the source components of potassic magmas: geochemical and experimental constraints. Lithos 28, 187–204.

Frost, D.J., 1999. The stability of dense hydrous magnesium silicates in earth's transition zone and lower mantle. In: Fei, Y., Bertka, C.M. Mysen, B.O. (Eds.), Mantle Petrology: Field Observations and High Pressure Experiments. The Geochemical Society, Special Publication 6, pp. 283–296.

Gaetani, G.A., Grove, T.L., 1998. The influence of water on melting of peridotite. Contrib. Mineral. Petrol. 131, 323–346.

Gill, J.B., 1981. Orogenic Andesites and Plate Tectonics. Springer, Berlin, 1981.

Green, D.H., 1969. The origin of basaltic and nephelinitic magmas in the earth's mantle. Tectonophysics 7, 409–422.

Green, D.H., 1973. Experimental melting studies on a model upper mantle composition at high pressure under water-saturated and water undersaturated conditions. Earth Planet. Sci. Lett. 19, 37–53.

Green, D.H., 1976. Experimental testing of equilibrium partial melting of peridotite under water saturated, high pressure conditions. Can. Mineral. 14, 255–268.

Green, T.H., 1981. Experimental evidence for the role of accessory phases in magma genesis. J. Volcanol. Geotherm. Res. 10, 405–422.

Green, D.H., Ringwood, A.E., 1967. The genesis of basaltic magmas. Contrib. Mineral. Petrol. 15, 103–190.

Green, T.H., Ringwood, A.E., 1968. Genesis of the calc-alkaline igneous rock suite. Contrib. Mineral. Petrol. 18, 105–162.

Gust, D.A., Perfit, M.R., 1987. Phase relations of a high-Mg basalt from the Aleutian Island Arc: implications for primary island arc basalts and high-Al basalts. Contrib. Mineral. Petrol. 97, 7–18.

Herzberg, C., Zhang, J., 1996. Melting experiments on anhydrous peridotite KLB-1: compositions of magmas in the upper mantle and transition zone. J. Geophys. Res. 101 (B4), 8271–8295.

Hirose, K., Kawamoto, T., 1995. Hydrous partial melting of lherzolite at 1 GPa: The effect of H_2O on the genesis of basaltic magmas. Earth Planet. Sci. Lett. 133, 463–473.

Hirose, K., Kushiro, I., 1993. Partial melting of dry peridotites at high pressures: determination of composition of melts segregated from peridotite using aggregates of diamonds. Earth Planet. Sci. Lett. 114, 477–489.

Ionov, D.A., Hofmann, A.W., 1995. Nb-Ta-rich mantle amphiboles and micas implications for subduction-related metasomatic trace element fractionations. Earth Planet. Sci. Lett. 131, 341–356.

Iwamori, H., 1998. Transportation of H_2O and melting in subduction zones. Earth Planet. Sci. Lett. 160, 65–80.

Kawamoto, T., 1996. Experimental constraints on differentiation and H_2O abundance of calc-alkaline magmas. Earth Planet. Sci. Lett. 144, 577–589.

Kawamoto, T., Leinenweber, K., Hervig, R.L., Holloway, J.R., 1995. Stability of hydrous minerals in H_2O-saturated KLB-1 peridotite up to 15 GPa. In: Farley, K.A. (Ed.), Volatiles in the Earth and Solar System. Am. Inst. Phys., pp. 229–239.

Kelemen, P.B., Johnson, K.T.M., Kinzler, R.J., Irving, A.J., 1990a. High-field-strength element depletions in arc basalts due to mantle-magma interaction. Nature 345, 521–524.

Kelemen, P.B., Joyce, D.B., Webster, J.D., Holloway, J.R., 1990b. Reaction between ultramafic rock and fractionating basaltic magma. II. Experimental investigation of reaction between olivine tholeiite and harzburgite at 1150–1050°C and 5 kb. J. Petrol. 31, 99–134.

Keppler, H., 1996. Constraints from partitioning experiments on the composition of subduction-zone fluids. Nature 380, 237–240.

Kinzler, R.J., 1997. Melting of mantle peridotite at pressures approaching the spinel to garnet transition: Application to mid-ocean ridge basalt petrogenesis. J. Geophys. Res. 102 (B1), 853–874.

Kinzler, R.J., Grove, T.L., 1992a. Primary magmas of mid-ocean ridge basalts 1. Experiments and methods. J. Geophys. Res. 97, 6885–6906.

Kinzler, R.J., Grove, T.L., 1992b. Primary magmas of mid-ocean ridge basalts 2. Applications. J. Geophys. Res. 97, 6907–6926.

Klemme, S., O'Neill, H.S.C., 2000. The near-solidus transition from garnet lherzolite to spinel lherzolite. Contrib. Mineral. Petrol. 138, 237–248.

Kogiso, T., Tatsumi, Y., Nakano, S., 1997. Trace element transport during dehydration processes in the subducted oceanic crust: 1. Experiments and implications for the origin of ocean island basalts. Earth Planet. Sci. Lett. 148, 193–205.

Kushiro, I., 1972. Effects of H_2O on the composition of magmas formed at high pressures. J. Petrol. 13, 311–344.

Kushiro, I., 1974. Melting of hydrous upper mantle and possible generation of andesite magma. An approach from synthetic systems. Earth Planet. Sci. Lett. 22, 294–299.

Kushiro, I., 1975. On the nature of silicate melts and its significance in magma genesis: regularities in the shift of liquidus boundaries involving ol–opx and silica minerals. Am. J. Sci. 275, 411–431.

Kushiro, I., 1987. A petrologic model of the mantle wedge and lower crust in the Japanese island arc. In: Mysen, B.O. (Ed.), Magmatic Processes: Physiochemical Principles. The Geochemical Society, Special Publication 1, pp. 165–181.

Kushiro, I., 1990. Partial melting of mantle wedge and evolution of island arc crust. J. Geophys. Res. 95, 15929–15939.

Kushiro, I., Syono, Y., Akimoto, S., 1968. Melting of a peridotite nodule at high pressures and high water pressures. J. Geophys. Res. 73, 6023–6029.

Kushiro, I., Shimizu, N., Nakamura, Y., Akimoto, S., 1972. Composition of coexisting liquid and solid phases formed upon melting of natural garnet and spinel lherzolite at high pressure. A preliminary report. Earth Planet. Sci. Lett. 14, 19–25.

Luhr, J.F., 1992. Slab-derived fluids and partial melting in subduction zones: insights from two contrasting Mexican volcanoes (Colima and Ceboruco). J. Volcanol. Geotherm. Res. 54, 1–18.

Luth, R.W., 1995. Is phase A relevant to the Earth's mantle? Geochim. Cosmochim. Acta 59, 679–682.

Martin, H., 1987. Petrogenesis of Archean trondhjemites, tonalites, granodiorites from Eastern Finland: Major and trace element geochemistry. J. Petrol. 28, 921–953.

Martin, H., 1999. Adakitic magmas: modern analogues of Archean granitoids. Lithos 46, 411–429.

Maury, R.C., Defant, M.J., Joron, J.-L., 1992. Metasomatism of the sub-arc mantle inferred from trace elements in Philippine xenoliths. Nature 360, 661–663.

McCulloch, M.T., Gamble, J.A., 1991. Geochemical and geodynamical constraints on subduction zone magmatism. Earth Planet. Sci. Lett. 102, 358–374.

Mysen, B.O., Boettcher, A.L., 1975a. Melting of a hydrous mantle. II. Geochemistry of crystals and liquids formed by anatexis of mantle peridotite at high pressure and high temperature as a function of controlled activities of water, hydrogen and carbon dioxide. J. Petrol. 16, 549–593.

Mysen, B.O., Boettcher, A.L., 1975b. Melting of hydrous mantle. I. Phase relations of natural peridotite at high p and T and with controlled addition of water, carbon dioxide and hydrogen. J. Petrol. 16, 520–548.

Mysen, B.O., Kushiro, I., 1977. Compositional variations of coexisting phases with degree of melting of peridotite in the upper mantle. Am. Mineral. 62, 843–865.

Mysen, B.O., Ulmer, P., Konzett, J., Schmidt, M.W., 1998. The upper mantle near convergent plate boundaries. In: Hemley, R.J. (Ed.), Ultrahigh-Pressure Mineralogy. Reviews in Mineralogy. Mineralogical Society of America, Washington, DC, pp. 97–138.

Nicholls, I.A., 1974. Liquids in equilibrium with peridotitic mineral assemblages at high water pressure. Contrib. Mineral. Petrol. 45, 289–316.

Nicholls, I.A., Ringwood, A.E., 1973a. Effect of water on olivine stability in tholeiites and production of silica-saturated magmas in the island arc environment. J. Geol. 81, 285–306.

Nicholls, I.A., Ringwood, A.E., 1973b. Production of silica-saturated tholeiitic magmas in island arcs. Earth Planet. Sci. Lett. 17, 243–246.

Niida, K., Green, D.H., 1999. Stability and chemical composition of pargasitic amphibole in MORB pyrolite under upper mantle conditions. Contrib. Mineral. Petrol. 135, 18–40.

Olafsson, M., Eggler, D.H., 1983. Phase relations of amphibole, amphibole–carbonate, and phlogopite–carbonate peridotite: petrologic constraints on the asthenosphere. Earth Planet. Sci. Lett. 64, 305–315.

Ono, S., 1998. Stability limits of hydrous minerals in sediment and mid-ocean ridge basalt compositions: implications for water transport in subduction zones. J. Geophys. Res. 103 (B8), 18253–18267.

Pawley, A.R., Wood, B.J., 1995. The high-pressure stability of talc and 10 Å phase: potential storage sites for H_2O in subduction zones. Am. Mineral. 80, 998–1003.

Pawley, A.R., Holloway, J.R., McMillan, P.F., 1993. The effect of oxygen fugacity on the solubility of carbon–oxygen fluids in basaltic melts. Earth Planet. Sci. Lett. 110, 213–225.

Peacock, S.M., 1991. Numerical simulation of subduction zone pressure–temperature–time paths: constraints on fluid production and arc magmatism. Phil Trans. R. Soc. London A 335, 341–353.

Pearce, J.A., 1982. Trace element characteristics of lavas from destructive plate boundaries. In: Thorpe, R.S. (Ed.), Andesites. Wiley, Chichester, pp. 525–548.

Poli, S., Schmidt, M.W., 1995. H_2O transport and release in subduction zones: experimental constraints on basaltic and andesitic systems. J. Geophys. Res. 100, 22299–22314.

Rapp, R.P., Watson, E.B., Miller, C.F., 1991. Partial melting of amphibolite/eclogite and the origin of Archean trondhjemites and tonalites. Precambrian Res. 51, 1–25.

Ringwood, A.E., 1974. The petrological evolution of island arc systems. J. Geol. Soc. London 130, 183–204.

Robinson, J.A., Wood, B.J., 1998. The depth of the spinel to garnet transition at the peridotite solidus. Earth Planet. Sci. Lett. 164, 277–284.

Robinson, J.A.C., Wood, B.J., Blundy, J.D., 1998. The beginning of melting of fertile and depleted peridotite at 1.5 GPa. Earth Planet. Sci. Lett. 155, 97–111.

Rudnick, R.L., Taylor, S.R., 1986. Geochemical constraints on the origin of Archean tonalitic-trondhjemitic rocks and implications for lower crustal composition. In: Dawson, J.B., Carswell, D.A., Hall, J., Wedepohl, K.H. (Eds.), The nature of the lower continental crust. Geological Society Special Publication. Geological Society of London, pp. 179–191.

Schmidt, M.W., 1996. Experimental constraints on recycling of potassium from subducted oceanic crust. Science 272, 1927–1930.

Schmidt, M.W., Poli, S., 1998. Experimentally based water budgets for dehydrating slabs and consequences for arc magma generation. Earth Planet. Sci. Lett. 163, 361–379.

Singer, B.S., Myers, J.D., Frost, C.D., 1992. Mid-Pleistocene basalt from the Seguam volcanic center, central Aleutian Arc, Alaska: Local lithospheric structures and source variability in the Aleutian arc. J. Geophys. Res. 97, 4561–4578.

Sisson, T.W., Layne, G.D., 1993. H_2O in basalt and basaltic andesite glass inclusions from four subduction-related volcanoes, Earth Planet. Sci. Lett. 619–635.

Stalder, R., Foley, S.F., Brey, G.P., Horn, I., 1998. Mineral-aqueous fluid partitioning of trace elements at 900–1200°C and 3.0–5.7 GPa: new experimental data for garnet, clinopyroxene, and rutile, and implications for mantle metasomatism. Geochim. Cosmochim. Acta 62, 1781–1801.

Stern, C.R., Huang, W.-L., Wyllie, P.J., 1975. Basalt–andesite–rhyolite–H_2O: crystallization intervals with excess H_2O and H_2O-undersaturated liquidus surfaces to 35 kbars, with implications for magma genesis. Earth Planet. Sci. Lett. 28, 189–196.

Stolper, E., Newman, S., 1994. The role of water in the petrogenesis of Mariana trough magmas. Earth Planet. Sci. Lett. 121, 293–325.

Takahashi, E., 1986. Melting of dry peridotite KLB-1 up to 14 GPa. Implications on the origin of peridotitic upper mantle. J. Geophys. Res. 91, 9367–9382.

Takahashi, E., Kushiro, I., 1983. Melting of a dry peridotite at high pressures and basalt magma genesis. Am. Mineral. 68, 859–879.

Tatsumi, Y., 1982. Origin of high-magnesian andesites in the Setouchi volcanic belt, SW Japan. II. Melting phase relations at high pressure. Earth Planet. Sci. Lett. 60, 305–315.

Tatsumi, Y., Eggins, S., 1995. Subduction Zone Magmatism. Blackwell, Cambridge, 211 pp.

Tatsumi, Y., Hamilton, D.L., Nesbitt, R.W., 1986. Chemical characteristics of fluid phase released from a subducted lithosphere and origin of arc magmas: evidence from high pressure experiments and natural rocks. J. Volcanol. Geotherm. Res. 29, 293–310.

Tatsumi, Y., Kogiso, T., 1997. Trace element transport during dehydration processes in the subducted oceanic crust: 2. Origin of chemical and physical characteristics in arc magmatism. Earth Planet. Sci. Lett. 148, 207–221.

Tatsumi, Y., Sakuyama, M., Fukuyama, H., Kushiro, I., 1983. Generation of Arc basalt magmas and thermal structure of the mantle wedge in subduction zones. J. Geophys. Res. 88, 5815–5825.

Tatsumi, Y., Furukawa, Y., Yamashita, S., 1994. Thermal and geochemical evolution of the mantle wedge in the northeast Japan arc. 1. Contribution from experimental petrology. J. Geophys. Res. 99 (B11), 22275–22283.

Taylor, W.R., Green, D.H., 1987. The petrogenetic role of methane: effect on liquidus phase relations and the solubility of reduced C–H volatiles. In: Mysen, B.O. (Ed.), Magmatic Processes: Physiochemical Principles, Vol. 1. The Geochemical Society, Special Publication 1, pp. 121–138.

Taylor, W.R., Green, D.H., 1988. Measurement of reduced peridotite–C–O–H solidus and implications for redox melting of the mantle. Nature 332, 349–352.

Thompson, R.N., 1975. Primary basalts and magma genesis. II. Snake River Plain, Idaho, USA. Contrib. Mineral. Petrol. 52, 213–232.

Thompson, A.B., 1992. Water in the earth's upper mantle. Nature 358, 295–302.

Ulmer, P., 1989. High pressure phase equilibria of a calc-alkaline picro-basalt: implications for the genesis of calc-alkaline magmas. Carnegie Institution of Washington Yb.. Annual Report of the Director of the Geophysical Laboratory 88, 28–35.

Ulmer, P., Trommsdorff, V., 1995. Serpentine stability to mantle depths and subduction-related magmatism. Science 268, 858–861.

Ulmer, P., Trommsdorff, V., 1999. Phase relations of hydrous mantle subducting to 300 km. In: Fei, Y., Bertka C.M., Mysen, B.O. (Eds.), Mantle Petrology: Field Observations and High Pressure Experiments. The Geochemical Society, Special Publication 6, pp. 259–281.

van der Laan, S.R., Wyllie, P.J., 1992. Constraints on Archean trondhjemite genesis from hydrous crystallization experiments on Nuk Gneiss at 10–17 kbar. J. Geol. 100, 57–68.

Wendtland, R.F., Mysen, B.O., 1980. Melting phase relations of natural peridotite $+ CO_2$ as a function of degree of partial melting at 15 and 30 kbar. Am. Mineral. 65, 37–44.

Wunder, B., 1998. Equilibrium experiments in the system $MgO–SiO_2–H_2O$ (MSH): stability of clinohumite-OH [$Mg_9Si_4O_{16}(OH)_2$], chondrodite-OH [$Mg_5Si_2O_8(OH)_2$] and phase A [$Mg_7Si_2O_8(OH)_6$]. Contrib. Mineral. Petrol. 132, 111–120.

Wunder, B., Schreyer, W., 1997. Antigorite: high-pressure stability in the system $MgO–SiO_2–H_2O$ (MSH). Lithos 41, 213–227.

Wyllie, P.J., 1978. Mantle fluid compositions buffered in peridotite–CO_2–H_2O by carbonates, amphiboles and phlogopite. J. Geol. 86, 687–713.

Wyllie, P.J., 1979. Magmas and volatile components. Am. Mineral. 64, 469–500.

Yamamoto, K., Akimoto, S.-I., 1977. The system $MgO–SiO_2–H_2O$ at high pressures and temperatures — stability field for hydroxyl-chondrodite, hydroxyl-clinohumite and 10 Å-phase. Am. J. Sci. 277, 288–312.

Yoder, H.S., Tilley, C.E., 1962. Origin of basalt magmas: an experimental study of natural and synthetic rock systems. J. Petrol. 3, 342–532.

ELSEVIER

Physics of the Earth and Planetary Interiors 127 (2001) 233–252

PHYSICS
OF THE EARTH
AND PLANETARY
INTERIORS

www.elsevier.com/locate/pepi

Boron isotope geochemistry of metasedimentary rocks and tourmalines in a subduction zone metamorphic suite

Toshio Nakano, Eizo Nakamura*

The Pheasant Memorial Laboratory (PML) for Geochemistry and Cosmochemistry, Institute for Study of the Earth's Interior, Okayama University at Misasa, Tottori 682-0193, Japan

Received 24 January 2000; accepted 11 October 2000

Abstract

In order to understand the behavior of boron (B) and its isotope fractionation during subduction zone metamorphism, B contents and isotopic compositions together with major element compositions were determined for metasedimentary rocks and tourmalines from the Sambagawa Metamorphic Belt, central Shikoku, Japan. No systematic changes in whole-rock B content and isotope composition of the metasediments were observed among the different metamorphic grades, indicating the lack of a bulk fluid-rock B isotope fractionation as a result of devolatilization.

Both modal abundance and grain size of tourmaline increase with increasing metamorphic grade. In contrast, B contents in muscovite and chlorite decrease with increasing metamorphic grade. These observations combined with mass balance calculations of B suggest the formation of tourmaline during progressive metamorphism from metamorphic fluids containing B mainly derived from muscovite and subordinately from chlorite without allowing significant net removal of B from the metasedimentary rocks. Tourmalines in the higher-grade metasedimentary rocks have zonal structure of B isotope and major element composition with decreasing δ^{11}B and increasing Mg/(Mg+Fe) from the inner rim (core) to the outer rim. The change of Mg/(Mg + Fe) in the tourmalines with increasing grade is paralleled by similar variation in chlorite. These observations suggest that the growing tourmalines record the progressive evolution of the B isotopic composition of the metamorphic fluid, in the outermost rims preserving the isotope signature of peak metamorphic *P*–*T*-fluid conditions.

Based on the above observations, the δ^{11}B of the tourmaline is thought to have been nearly identical to that of the metamorphic fluid resulting in the "apparent" B isotopic fractionation factor between metamorphic fluid and whole-rock ($\alpha = (^{11}B/^{10}B)_{fluid}/(^{11}B/^{10}B)_{whole\text{-}rock}$) which decreases from 1.007 ± 0.003 to 1.001 ± 0.003 from chlorite to biotite zone metamorphism. Such results together with the formation of tourmaline from (and sequestering of) B in metamorphic fluids may lead to less B isotopic fractionation as a result of subduction zone devolatilization than noted in suites containing less tourmaline. This, therefore, makes it possible to transport B isotopic signatures, which ultimately reflect Earth's surface materials, to the deep mantle, perhaps resulting in mantle B isotope anomalies near convergent margins. © 2001 Elsevier Science B.V. All rights reserved.

Keywords: Boron isotope; TIMS; SIMS; Metasedimentary rocks; Tourmaline; Metamorphic fluid; Subduction zone

1. Introduction

Fluids probably carry components from the slab (oceanic crust and sediment) to the mantle wedge in deep-seated portions of subduction zones and induce

* Corresponding author. Tel.: +81-858-43-1215;
fax: +81-858-43-2184.
E-mail address: eizonak@misasa.okayama-u.ac.jp (E. Nakamura).

PII: S0031-9201(01)00230-8

partial melting of the mantle wedge to form island arc magmas (e.g. Saundaers et al., 1980; Perfit et al., 1980; Nakamura et al., 1985; Sakuyama and Nesbitt, 1986). Based on multi-isotope systematics including boron (B), island arc volcanics are believed to originate through addition of oceanic slab components to island arc basalt (IAB) sources by aqueous fluids released from the subducting slab (e.g. Shibata and Nakamura, 1997; Moriguti and Nakamura, 1998).

Boron is considered to be an excellent tracer of the recycling of subducted oceanic sediment and/or altered oceanic crust at convergent margins (Morris et al., 1990; Palmer, 1991; Ryan and Langmuir, 1993; Ishikawa and Nakamura, 1994). This is because of its extremely high mobility during both magmatic and fluid-related processes (e.g. Seits, 1972; Mottl and Holland, 1978; Malinko et al., 1979; Seyfried et al., 1984; Ryan and Langmuir, 1993) and its extreme enrichment in the subducted oceanic sediments and altered oceanic crust compared with unaltered mid-ocean ridge basalts (MORB) and upper mantle materials (Spivack and Edmond, 1987; Ishikawa and Nakamura, 1992, 1993). Ishikawa and Nakamura (1994) investigated B and Pb isotope systematics and trace element geochemistry for Izu arc volcanic rocks, which show clear across-arc variation of $\delta^{11}B$ and B/Nb values. They suggested that hydrous fluid from the subducting slab, enriched in H_2O-soluble elements such as B and Pb, extensively metasomatized the originally depleted mantle wedge with the degree of metasomatism decreasing with increasing depths of the down-going slab beneath the Izu arc.

Boron isotopic compositions and the concentrations of B in oceanic sediments and oceanic crustal materials are reasonably well characterized (e.g. Spivack and Edmond, 1987; Ishikawa and Nakamura, 1992, 1993). Recently, Peacock and Hervig (1999) reported B isotopic compositions of some minerals in subduction-related metamorphic rocks. Less is known, however, about the effects of slab metamorphism on B isotopic composition and the concentrations of B in subducted rocks. Boron may be efficiently transferred from subducted slab to arc magma source regions by aqueous fluid through metamorphic processes (e.g. Moran et al., 1992; Bebout et al., 1993). Hence, it is essential to clarify the behavior of B and B isotope fractionation in materials such as oceanic sediments and altered oceanic crust as the result of prograde metamorphism and related devolatilization. Subduction zone metamorphic processes may also affect the trace element compositions and isotope compositions of rocks subducted to greater depths in the mantle beyond arcs (e.g. Hofmann and White, 1982; Hart and Reid, 1991). It is, therefore, important to characterize the metamorphic fluid liberated in the subduction processes as a function of the metamorphic grade.

Tourmaline, which is extremely resistant mechanically and chemically, is the most common B mineral found in many rock types and is stable over a wide range of metamorphic conditions, from diagenesis through the granulite facies. Once formed, tourmaline does not readily readjust its composition by volume diffusion (Henry and Dutrow, 1996). Tourmaline requires large amounts of B (ca. 3 wt.% of B) to form. However, B is a trace constituent in the altered oceanic crust and sedimentary rocks, which contain up to 70 ppm (Spivack and Edmond, 1987; Ishikawa and Nakamura, 1992) and up to 130 ppm of B (Ishikawa and Nakamura, 1993), respectively. Therefore, B must be concentrated by geochemical processes, such as percolation of metamorphic fluid along grain boundaries of minerals during prograde metamorphism, in order to form tourmaline. Tourmaline compositions are likely very responsive to the physico-chemical environment of formation, in particular to the chemical and B isotopic composition of coexisting minerals and fluids, activities of H_2O and dissolved species, pressure and temperature (Henry and Guidotti, 1985; Morgan and London, 1989; Palmer et al., 1992). Tourmaline may thus effectively record the physico-chemical history of the rocks and fluids during metamorphism. However, B isotopic data for tourmaline in metamorphic rocks are limited (Swihart and Moore, 1989; Chaussidon and Appel, 1997), making it difficult to ascertain the behavior and isotope fractionation of B in subduction processes.

In this article, we present B isotope data for metasedimentary whole-rocks and their contained tourmalines from the Sambagawa Metamorphic Belt, central Shikoku, Japan and discuss the B isotopic fractionation between fluid and host rock during progressive metamorphism. In addition, we evaluate the utility of B isotopes as a geochemical tracer of material recycling between the mantle and the crust through subduction zones.

2. Geological background

The Sambagawa (Sanbagawa) Metamorphic Belt is a regionally coherent high-P–T metamorphic belt that stretches for some 800 km throughout the southwestern part of Japan and represents processes of underplating at depths of 15–30 km in an Early Cretaceous accretionary complex (Isozaki and Itaya, 1990). The Sambagawa Metamorphic Belt is divided into four zones based on the appearance of index minerals in graphite-bearing metapelites (Miyashiro and Banno, 1958; Kurata and Banno, 1974; Higashino, 1975, 1990) consisting of the chlorite, garnet, albite–biotite and oligoclase–biotite zones in increasing order of metamorphic grade (Fig. 1a). These zones are equivalent to a range of metamorphic facies from blueschist to epidote–amphibolite facies. Peak metamorphic temperature and pressure have been estimated at 250–620°C and 0.5–1.1 GPa, respectively (Banno and Sakai, 1989; Enami et al., 1994). In the Sambagawa Metamorphic Belt, the Asemigawa (Asemi River) region, located in central Shikoku, has been most intensively studied (e.g. Kurata and Banno, 1974; Higashino, 1975, 1990; Higashino et al., 1982; Enami et al., 1994; Banno, 1964; Ernst et al., 1970; Nakajima et al., 1977; Itaya and Banno, 1980; Itaya, 1981; Banno et al., 1986). Moreover, this area contains continuous exposure of low to high-grade metamorphic rocks consisting of abundant metasedimentary rocks and subordinate amounts of metabasite. The metasedimentary rocks are dominantly pelitic schists with small amounts of associated psammitic and siliceous schists and rare calcareous schist.

The Shimanto Belt is a Cretaceous to Tertiary accretionary prism, which is distributed in the south to the Sambagawa Belt along the Pacific Ocean, and is composed of broken flysch units and tectonic melange (Taira et al., 1988).

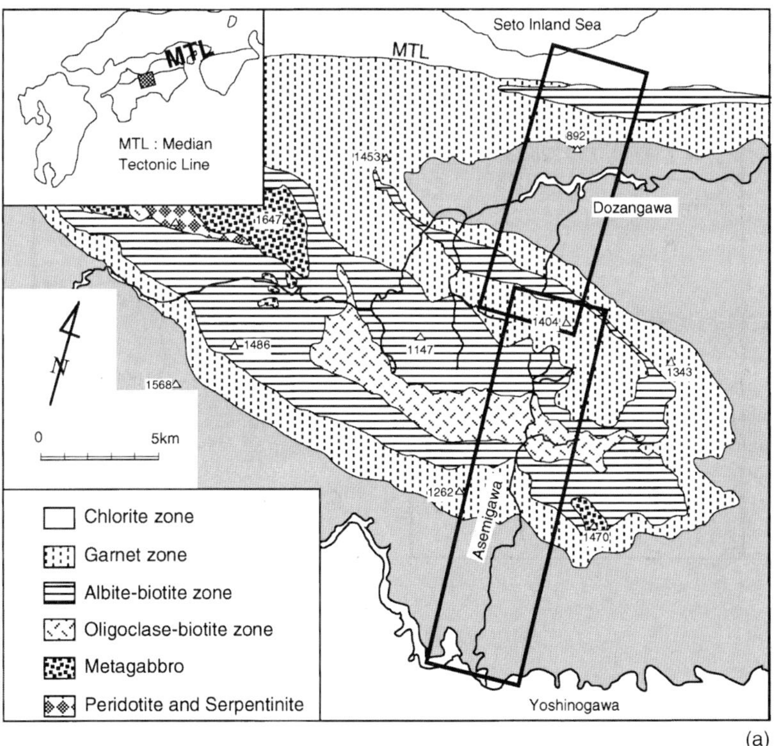

Fig. 1. (a) Metamorphic zonal map of the Sambagawa Metamorphic Belt in central Shikoku (after Higashino, 1990). The study area is shown by the rectangular outlines. (b) Sample locality maps, the localities of which are shown in boxes on (a).

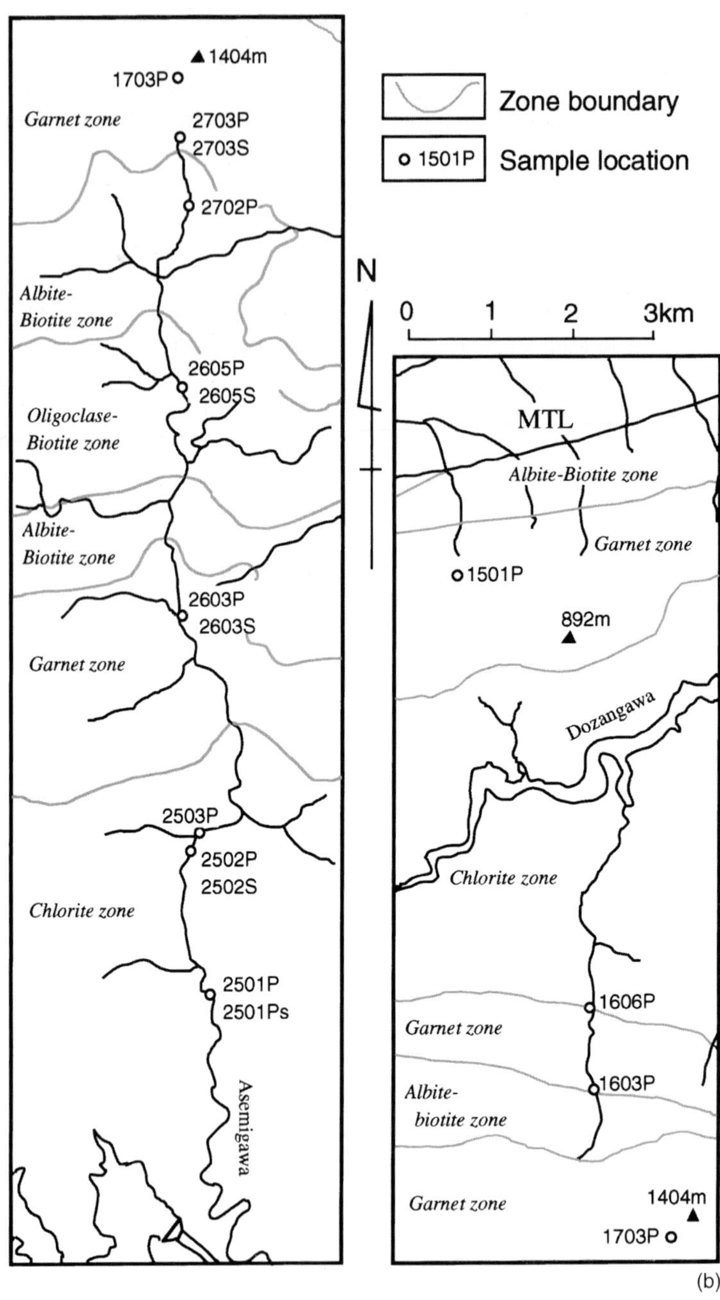

Fig. 1. (*Continued*).

Table 1
Whole-rock boron concentrations, boron isotope compositions, and major element compositions, and modal abundances of metamorphic minerals in metasedimentary rocks of the Sambagawa Metamorphic Belt

Sample name	Metamorphic grade[a]	B (ppm)	$\delta^{11}B$ (per mil)	Tourmaline mode (%)	Points counted		SiO_2	TiO_2	Al_2O_3	Fe_2O_3	MnO	MgO	CaO	Na_2O	K_2O	P_2O_5	LOI	Total	Modal abundance (%)			
					Tourmaline	Total													Muscovite	Biotite	Garnet	Chlorite
2501P	Chl	26.6	−11.9	0.01	1	10000	71.36	0.48	13.34	3.80	0.05	1.41	0.90	2.58	2.64	0.09	3.2	99.81	31	–	–	0.4
2501Ps	Chl	20.6	−11.9	0.01	1	10000	71.03	0.38	13.09	2.93	0.05	1.06	2.26	4.04	1.79	0.07	3.2	99.88	20	–	–	0.7
2502P	Chl	57.4	−11.7	<0.01	0	10000	67.22	0.54	15.77	4.39	0.15	1.64	0.59	2.37	3.86	0.09	3.3	99.90	45	–	–	0.9
2502S	Chl	3.4	−11.6	–[b]	–	–	72.16	0.62	9.56	6.60	0.29	2.47	3.58	2.14	0.41	0.18	2.2	100.19	ND[c]	ND	ND	ND
2503P	Chl	38.2	−9.0	<0.01	0	10000	74.58	0.37	11.83	4.49	0.12	1.50	0.22	2.78	1.87	0.06	2.3	100.17	ND	ND	ND	ND
1606P	Chl	42.9	−11.2	<0.01	0	10000	65.99	0.52	15.18	4.50	0.07	1.28	0.37	3.39	2.87	0.08	5.5	99.76	21	–	–	0.5
2603P	Gar	56.9	−10.2	0.04	4	10000	71.06	0.45	14.12	4.04	0.15	1.49	0.75	2.76	2.75	0.09	2.4	100.05	36	–	0.9	1.1
2603S	Gar	11.1	−8.8	–	–	–	92.73	0.15	2.49	1.70	0.06	0.61	0.35	0.39	0.63	0.08	0.6	99.80	ND	ND	ND	ND
2703P	Gar	19.6	−14.2	0.01	1	9139	61.57	1.06	15.24	6.22	0.31	3.25	2.10	3.26	2.26	0.24	4.2	99.66	20	–	1.2	11.4
2703S	Gar	3.4	−13.4	–	–	–	85.84	0.26	5.23	2.91	0.32	1.71	0.77	1.13	0.53	0.14	1.5	100.31	6	–	1.5	7.3
1703P	Gar	38.2	−11.8	0.12	18	15000	68.37	0.58	14.77	4.46	0.12	1.71	0.70	2.24	3.06	0.09	3.6	99.74	28	–	+	5.9
1501P	Gar	19.6	−12.7	0.01	1	10000	76.36	0.32	11.59	2.39	0.04	0.71	0.89	3.00	2.08	0.04	2.2	99.63	22	–	+	2.2
2702P	Ab–Bt	26.7	−10.8	0.03	4	15000	65.57	0.57	15.01	4.97	0.11	1.65	2.10	3.38	2.56	0.09	3.8	99.82	20	–	1.7	6.9
1603P	Ab–Bt	58.9	−10.4	0.19	37	20000	66.26	0.61	15.27	5.03	0.11	1.75	1.63	3.10	2.70	0.10	3.6	100.16	37	+[d]	6.5	1.8
2605P	Ol–Bt	52.1	−8.9	0.10	10	10000	64.93	0.62	16.36	5.92	0.12	1.88	1.45	2.62	3.00	0.12	2.7	99.74	31	0.9	4.9	1.4
2605S	Ol–Bt	19.1	−10.3	–	–	–	92.59	0.14	3.30	1.75	0.14	0.58	0.22	0.40	0.79	0.09	0.8	100.75	14	–	0.4	0.5

[a] Chl: chlorite zone; Gar: garnet zone; Ab–Bt: albite–biotite zone; Ol–Bt: oligoclase–biotite zone.
[b] Absent.
[c] Not determined.
[d] Present (<0.1%).

3. Samples

3.1. Whole-rock samples

Sixteen metasedimentary rocks from the Asemi-gawa region were used in this study. Pelitic schists contain quartz+sodic plagioclase (albite, oligoclase)+ muscovite + epidote + carbonaceous material as well as combinations of chlorite, garnet, biotite, sphene, rutile and ilmenite. The psammitic and siliceous schists have mineral assemblages similar to those of the pelitic schists, but with different mineral modal abundances (Table 1). Both the pelitic and psammitic schists contain tourmaline at all metamorphic grades. No systematic difference is observed in the major element compositions of metasedimentary rocks as a function of metamorphic grade (Table 1).

3.2. Tourmalines

3.2.1. Tourmalines in metasedimentary rocks from the Sambagawa

The modal abundance of tourmaline was determined by point counting with 9100–20 000 points examined petrographically for each sample (Table 1). Tourmaline in rocks from the chlorite zone is rare (<0.01 vol.%), fine-grained (<100 μm diameter perpendicular to the long dimension) and variably

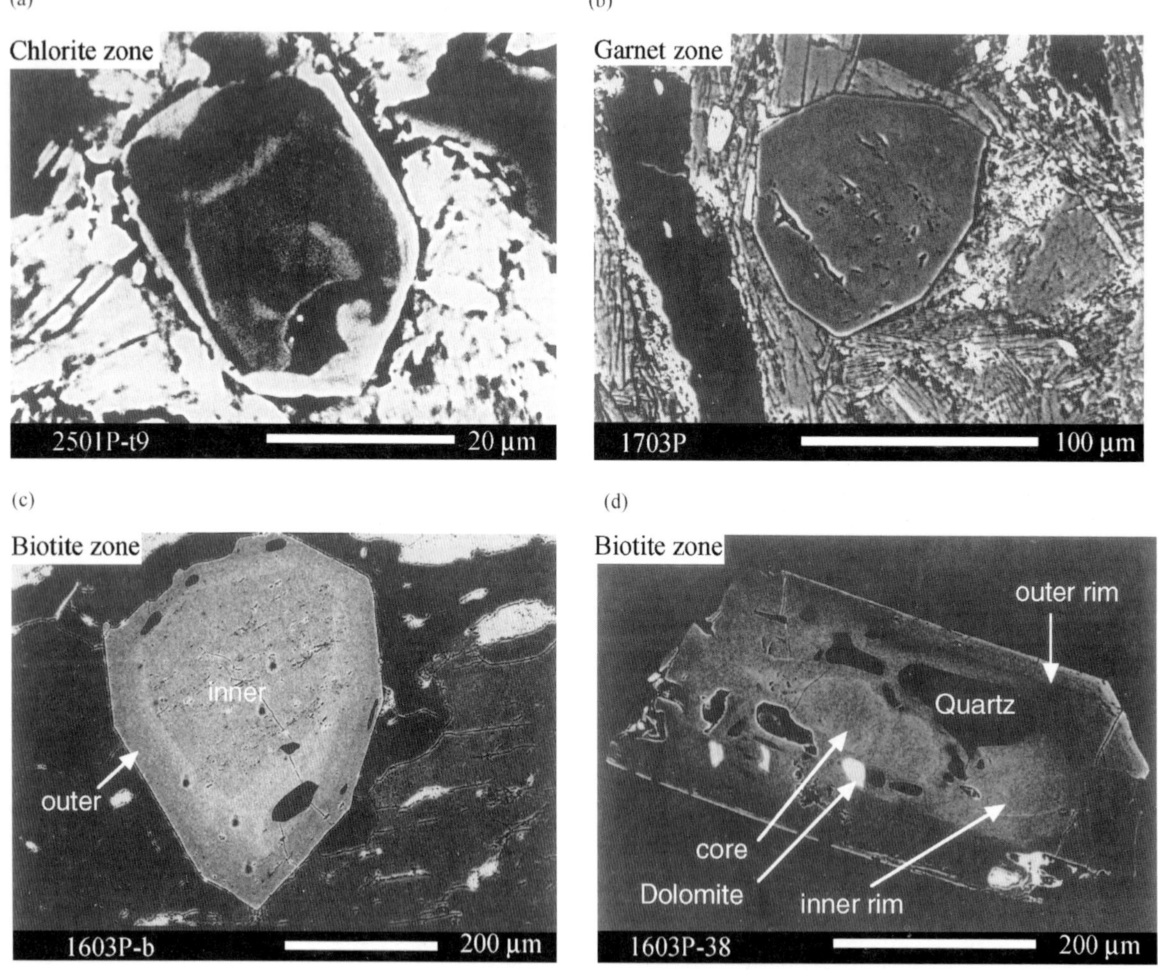

Fig. 2. Back-scattered electron images of tourmalines in metasedimentary rocks from the Sambagawa Metamorphic Belt. The tourmalines presented are from (a) chlorite zone (2501P), (b) garnet zone (1703P), and (c) and (d) biotite zone (1603P).

euhedral. In rocks from the chlorite zone, overgrowths of metamorphic tourmaline on detrital cores are very narrow (Fig. 2a), making it difficult to determine the B isotope composition by SIMS. In the garnet zone, tourmaline is euhedral, more abundant (up to 0.12 vol.%) and larger (up to 150 μm diameter perpendicular to the long dimension) than in the chlorite zone. Tourmaline in the biotite zone is euhedral, more abundant (up to 0.2%) and coarser-grained (up to 1 mm diameter) than in the garnet zone. Tourmaline grains in the garnet and biotite zones are typically zoned with yellowish brown rims and brown to bluish green cores. The color changes from core to rim occur both as gradational and sharp optical discontinuities. Some tourmaline grains with sharp optical discontinuities have irregular cores overgrown by euhedral rims (Fig. 2d), whereas for others both cores and rims are euhedral (Fig. 2c and d). The most common mineral inclusion in the tourmaline is quartz, and others are sphene, rutile and dolomite. At all grades, inclusions occur only in the rims. In this article, the cores having irregular boundaries with the overgrown rims are defined as an "irregular cores", whereas the other cores showing euhedral zoning are defined as "inner rims" (see Fig. 2c and d).

3.2.2. Tourmaline in sedimentary rocks from the Shimanto Belt

The sedimentary rock samples were collected from Miocene horizons of the Shimanto Belt in Shikoku Island (Taira et al., 1988), and are free from metamorphic minerals under the optical microscope. Tourmaline is rare in the Shimanto sedimentary rocks (nine grains in eight thin sections). The Shimanto tourmalines are irregular in shape and extremely fine-grained (up to 70 μm diameter). The colors of the tourmalines are variable (greenish blue, yellowish brown and brown) presumably reflecting their different sedimentary provenance (e.g. Henry and Dutrow, 1996).

4. Experiments

4.1. Sample preparation

4.1.1. Whole-rock samples

The rock samples from the Asemigawa region were crushed to less than 5 mm in diameter with a hammer, and then carefully rinsed with B-free distilled deionized water (Nakamura et al., 1992). After drying, the samples were finally ground to powder (<200 mesh) with an alumina ceramic mill or silicon nitride mortar.

To assess possible grain boundary enrichment of B content and any effect of this possible enrichment on whole-rock B isotopic compositions, an acid leaching test was carried out prior to the sample dissolution. A powdered rock sample (1603P) was leached with 1 ml of 0.1 M HCl in an ultrasonic bath for 20 min. Then, the sample was centrifuged and the B content of the supernatant was measured by isotope dilution mass spectrometry (IDMS) by TIMS (Nakamura et al., 1992). The leached B from the sample was 0.9% of the unleached whole-rock B, and the difference in B isotopic composition between the leached (−10.7‰) and unleached (−10.4‰) samples is not significant when the analytical errors for the natural rock samples are considered (see discussion below). This result indicates that the grain boundary enrichment of B is negligible in the Sambagawa metamorphic rocks. Therefore, such acid leaching was not performed for the whole-rock samples for which data are presented in this article.

4.1.2. Tourmaline samples

Tourmaline samples were prepared by polishing on thin section or in resin, then rinsed with B-free distilled deionized water in an ultrasonic bath. After drying, the samples were coated with gold (Au) and carbon (C) for SIMS and EDX–SEM analysis, respectively.

4.2. Analytical methods

4.2.1. Boron content and isotope ratios of whole-rock samples by TIMS

The procedures for the whole-rock B isotope analyses applied in this study are the same as those described in Nakamura et al. (1992). Prior to the isotopic ratio measurement, the B content was determined for 10–200 mg of sample by IDMS. The analytical error in the IDMS analyses was lower than 1% (R.S.D.). A sample containing 1–3 μg B was then used for the determination of B isotopic composition. Boron in the sample was chemically isolated using cation- and anion-exchange chromatography in the H^+ and F^- form, respectively. The total procedural blank for the analyses of metasedimentary rocks was 4.3–7.9 ng in the course of this study. The amount of B used in individual isotope analysis was usually larger than 1 μg, so the possible blank contamination was never larger than 1%, and thus negligible. The B isotopic ratios were determined in a thermal ionization mass spectrometer (TIMS: Finnigan MAT® 261) using the $Cs_2BO_2^+$ method with graphite (Xiao et al., 1988; Nakamura et al., 1992). The B isotopic composition is expressed as a per mil deviation ($\delta^{11}B$) relative to NIST SRM 951 boric acid:

$$\delta^{11}B = \left[\frac{(^{11}B/^{10}B)_{sample}}{(^{11}B/^{10}B)_{SRM\,951}} - 1 \right] \times 1000.$$

A mean $^{11}B/^{10}B$ ratio of 4.0530 ± 0.0002 ($2\sigma_{mean}$) was obtained for 62 individual analyses of SRM 951 during this study. Analytical reproducibility of the $\delta^{11}B$ values in natural samples was ±0.2‰ ($2\sigma_{mean}$) based on replicate analyses.

4.2.2. *Major element compositions of whole-rock samples and tourmalines*

Major element compositions of the whole-rock samples were analyzed by X-ray fluorescence (XRF: PHILIPS PW2400), following the method established by Takei and Nakamura (paper in preparation). Major element compositions of tourmalines were determined by an energy dispersive X-ray method (EDX) with a HORIBA EMAX-7000 energy-dispersive spectrometer on a HITACHI S-3100H SEM, based on the method established by Sakaguchi and Nakamura (paper in preparation). This latter instrument was operated at an accelerating voltage of 20 kV and a current of 0.3 nA. The relative uncertainty in the precision of the Mg/Fe ratios of olivine is ±1% (1σ).

4.2.3. *Boron isotope analysis of tourmaline with SIMS*

A single euhedral tourmaline initially of >15 mm length and 7 mm × 4 mm in size perpendicular to the long dimension was sliced into two pieces with a thickness of 1–1.5 mm perpendicular to the *c*-axis. One of the tourmaline slices was mounted in epoxy resin, then polished for EDX–SEM and SIMS analyses with Baikalox® alumina polishing suspension with grain size down to 0.3 μm. Another tourmaline slice was cut into several 1 mm cubes. Individual cubes of tourmaline were crushed in ethanol using a silicon nitride mortar, then transferred to Teflon® containers. After drying, the tourmaline was leached with 3 M HF for 20 min in an ultrasonic bath. Subsequently the HF-leached tourmaline was washed twice with 6 M HCl for 10 min and then rinsed three times with B-free distilled deionized water. After drying, the B isotopic composition of the tourmaline was determined by TIMS following chemical separation of B (Table 2).

The $\delta^{11}B$ values of the standard tourmalines were measured using a Cameca ion microprobe, ims 5f. Tourmaline samples were sputtered with an O^- primary beam of 8–10 nA intensity, resulting in a beam size of about 5 μm in diameter. Positive secondary ions were collected by ion counting using an energy offset of −45 V from 4500 V acceleration with an energy bandpass of ±10 V in order to separate the ^{10}B and ^{11}B peaks from possible $^{30}Si^{3+}$ and $^{30}BH^+$. A routine measurement consisted of five blocks of 10 cycles per block, with 5 and 3 s counting times on the mass peaks of ^{10}B and ^{11}B, respectively. The entire measurement lasted approximately 13 min, including pre-sputtering

Table 2
Mass discrimination factors for various tourmalines[a]

	TIMS		SIMS	Discrimination factor	Major elements	
	$^{11}B/^{10}B \pm 2\sigma_{mean}$	$\delta^{11}B$	$^{11}B/^{10}B \pm 2\sigma_{mean}$		Al/(Mg + Fe + Al)	Mg/(Mg + Fe + Al)
NIST SRM 951	4.0536 ± 0.0003					
TO-1	3.9983 ± 0.0008	−13.6 ± 0.2	3.763 ± 0.005	1.063 ± 0.002	0.838	0.044
TO-2	3.9968 ± 0.0005	−14.0 ± 0.1	3.773 ± 0.007	1.059 ± 0.002	0.996	0.003
TO-3	3.9976 ± 0.0005	−13.8 ± 0.1	3.768 ± 0.003	1.061 ± 0.001	0.972	0.015
TO-4	3.9972 ± 0.0005	−13.9 ± 0.1	3.763 ± 0.006	1.062 ± 0.002	0.975	0.002
TO-5	3.9947 ± 0.0004	−14.5 ± 0.1	3.761 ± 0.006	1.062 ± 0.002	0.900	0.015

[a] Instrumental mass discrimination factor: $(F_{inst}) = (^{11}B/^{10}B)_{TIMS}/(^{11}B/^{10}B)_{SIMS}$.

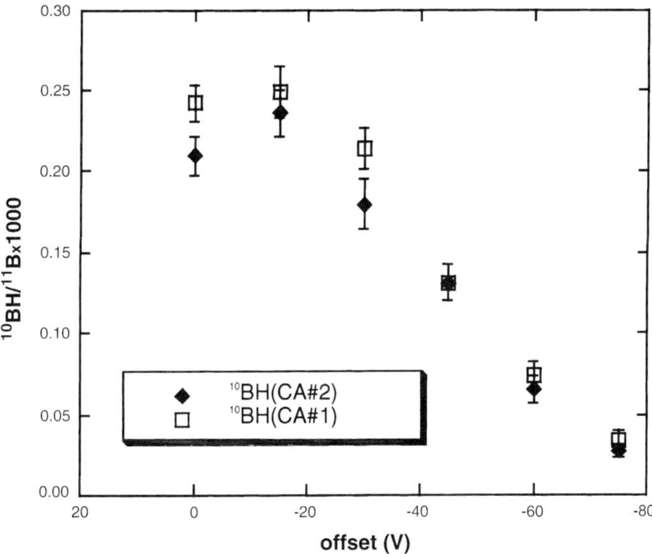

Fig. 3. Energy offset (−V) vs. $^{10}BH/^{11}B$ ratio. The plotted symbols show the $^{10}BH/^{11}B$ ratio at contrast aperture no. 1 (400 μm, CA#1; open square) and no. 2 (150 μm, CA#2; diamond). The vertical error bars show standard deviations.

time (180 s) and waiting times between measurements of the different masses (2 s). The typical intensity on the ^{11}B peak is 2×10^4 counts per second (cps). Measured $^{11}B/^{10}B$ ratios are corrected for instrumental mass discrimination as described below, and are given in $\delta^{11}B$ notation relative to the NIST SRM 951.

The effect of energy offset on the measured $^{11}B/^{10}B$ ratio was examined using a tourmaline standard. Since, in the analysis of tourmaline, the intensity of $^{30}Si^{3+}$ is extremely small and negligible compared to that for ^{10}B, only an interference of ^{10}BH on ^{11}B is considered in this study. Fig. 3 shows the relationship between energy offset (V) and $^{10}B/^{11}B$ ratio at mass resolution of ~2000 capable of separating ^{10}BH and ^{11}B. The $^{10}B/^{11}B$ ratio decreases with decreasing energy offset. This indicates that most of the ^{10}BH peak was removed from the ^{11}B peak by an energy offset smaller than −45 V even when using different contrast aperture sizes. The extent of ^{10}BH against ^{11}B at −45 V offset is smaller than 0.15‰, significantly less than the analytical precision of 0.8–1.2‰ ($2\sigma_{mean}$). The analytical reproducibility for repeat analyses ($n = 5$) for tourmalines was 0.4–0.8‰ ($2\sigma_{mean}$).

The $^{11}B/^{10}B$ ratios of five different tourmaline standards with major element compositions ranging from Schorl to Elbaite were measured to examine the matrix effect on B isotope measurements and the instrumental mass discrimination (Table 2). No significant difference in the instrumental mass discrimination factor ($F_{inst} = (^{11}B/^{10}B)_{TIMS}/(^{11}B/^{10}B)_{SIMS}$) was observed for the standard tourmalines during a single day, when the tuning of the instrument was fixed. This indicates that there is no matrix effect on the analyses of B isotope ratios of tourmaline. However, the instrumental mass discrimination factors were different day-to-day because the measured $^{11}B/^{10}B$ ratio is extremely sensitive to the instrumental conditions. Before measurement of unknown samples, therefore, the instrumental mass discrimination factor (F_{inst}) was precisely obtained by measuring $^{11}B/^{10}B$ ratios of standard tourmalines.

4.2.4. Quantitative analyses of B concentrations in minerals by SIMS

Boron contents of minerals in metasedimentary rock were also determined by employing a Cameca ims 5f. Analytical techniques are the same as those described in Nakamura et al. (1998) and Nakamura and Kushiro (1998).

5. Results and discussion

The EDX and SIMS analyses for all data points in Figs. 5–10 are available from the author (Nakamura) as an Excel file upon request.

5.1. *Whole-rock compositions*

The B contents and isotopic compositions, and major element compositions, of whole-rock samples are given in Table 1, and B data are plotted in Fig. 4 against metamorphic grade. Also plotted are B data for modern and ancient marine sediments (from Ishikawa and Nakamura, 1993) that are considered to be the protoliths for the metasedimentary rocks. The B contents of metasedimentary rocks range from 3.4 to 58.9 ppm, reflecting their lithological characteristics; i.e. siliceous metasedimentary rocks contain 3.4–19.1 ppm, whereas pelitic and psammitic metasedimentary rocks contain 19.6–58.9 ppm. Moran et al. (1992) and Bebout et al. (1993, 1999) observed successive depletion of B, with increasing temperature and extents of metamorphic devolatilization, that correlate with depletions of other trace elements (e.g. Cs, As and Sb). However, such a systematic decrease of B was not observed in the metasedimentary rocks from the Sambagawa.

The overall B contents of metasedimentary rocks (3.4–58.9 ppm) are lower than those of modern (70.4–132 ppm) and ancient (44.9–123 ppm) marine sediments (Fig. 4). Boron in modern and ancient marine sediments is largely distributed in illite (100–2000 ppm; Harder, 1970) and/or smectite (>100 ppm; Ishikawa and Nakamura, 1993). These clay minerals may react to form white-mica + chlorite + H_2O-rich fluid during diagenesis and subsequent low-grade metamorphism. The sedimentary rocks further dehydrate through consumption of chlorite to form garnet and other phases by the various reactions, the simplest of which is

$$\text{chlorite} + \text{quartz} \rightarrow \text{garnet} + H_2O\ (\text{fluid})$$

Thus, the modal abundance ratio of chlorite/garnet decreases with increasing metamorphic grade in the Sambagawa, as suggested by Higashino (1990). In the Sambagawa metasediments, the white-mica and chlorite contain 27–113 ppm and 1–27 ppm of B with modal abundances of 6–45 and 0.4–11.4%, respectively (Tables 1 and 3). In contrast, the B contents in plagioclase and garnet from the Sambagawa are less than 1 ppm (Table 3). The H_2O-rich fluid containing some B, which is liberated during prograde metamorphism, may be mobilized toward the surface, as was suggested by Bebout (1995).

Both modal abundance and grain size of tourmaline increase with increasing metamorphic grade (Table 4). This results in an increase in the proportion of whole-rock B residing in tourmaline with increasing metamorphic grade. When we assume that the density of tourmaline and the host metasediment are 3.0 and 2.5 g/cm^3, respectively, and the B content of tourmaline is 3%, and the averaged B contents of

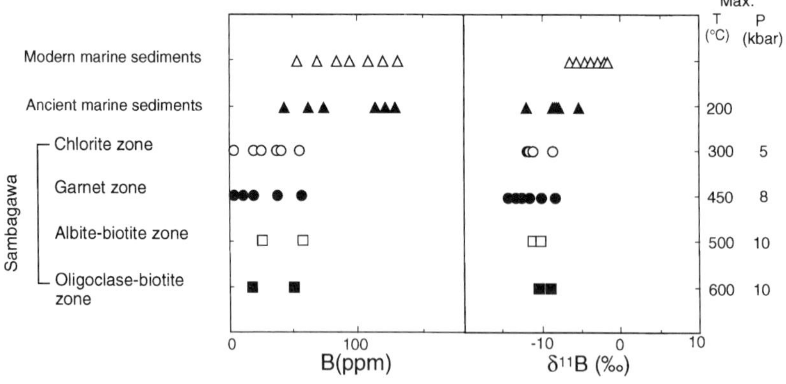

Fig. 4. Boron contents and δ^{11}B of Sambagawa metasedimentary rocks, as a function of metamorphic grade, and comparison with the boron contents and δ^{11}B of ancient and modern marine sediments (data of modern and ancient marine sediments from Ishikawa and Nakamura (1993)).

Table 3
B concentrations of minerals in metasedimentary rocks in the Sambagawa Metamorphic Belt, as a function of metamorphic grade

Sample	Metamorphic grade (ppm)	Whole-rock (ppm)	Chlorite					Muscovite					Garnet			Biotite			Plagioclase		
			Average	S.D.	Minimum	Maximum	*n*	Average	S.D.	Minimum	Maximum	*n*	Average	S.D.	*n*	Average	S.D.	*n*	Average	S.D.	*n*
2501P	Chlorite zone	26.6	16	7	3	27	10	101	9	91	113	7	–[a]	–	–	–	–	–	0.6	0.1	2
1703P	Garnet zone	38.2	4	3	1	10	15	69	10	56	90	14	0.3	0.2	2	–	–	–	0.3	0.0	1
1603P	Albite–biotite zone	58.9	3	1	2	5	5	68	18	50	91	5	0.2	0.1	4	–	–	–	0.5	0.3	2
2605P	Oligoclase–biotite zone	52.1	6	2	4	10	13	47	16	27	89	17	0.5	0.5	6	3	1	2	0.6	0.0	1

[a] Absent.

Table 4
Size and modal abundance of tourmalines, and estimated proportion of whole-rock boron incorporated in tourmaline for metasedimentary samples as a function of metamorphic grade

Metamorphic grade	Size (mm)	Average modal abundance (%)[a]	Average B contents of whole-rock (ppm)[a]	Proportion of B in tourmaline for whole-rock (%)
Shimanto sediments	Rare[b]	100[c]	–	–
Chlorite zone	<100	0.01	37	6
Garnet zone	<150	0.05	34	48
Biotite zone	<1000	0.11	46	84

[a] Excluding siliceous schists.
[b] Nine grains in eight thin sections.
[c] Ishikawa and Nakamura (1993). Densities of tourmaline and metasedimentary rocks used in the mass-balance calculation are 3.0 and 2.5 g/cm^3, respectively.

whole-rocks from each grade are used, the proportion of whole-rock B in tourmaline increases from 6 to 84% from the chlorite zone to the biotite zone through the garnet zone. Boron contents of muscovite and chlorite in the same rocks tend to decrease with increasing metamorphic grade (Fig. 5). Assuming that the densities of white-mica and metasediments are 2.8 and 2.5 g/cm^3, respectively, the proportion of whole-rock B residing in muscovite decreases from 90 to 32% from the chlorite zone to the oligoclase–biotite zone (Table 5). The phengite component in muscovite similarly decreases with increasing metamorphic grade (Radvanec et al., 1994). These observations suggest that the white-mica in the Sambagawa may have been re-equilibrated, with respect to B contents, with coexisting tourmalines (via metamorphic fluids), without net breakdown of white-mica. In such a way, little or no whole-rock B was lost to fluids leaving the system. Therefore, the metamorphic tourmalines in metamorphic rocks could conceivably provide a record of the fluid history of the rocks during metamorphism.

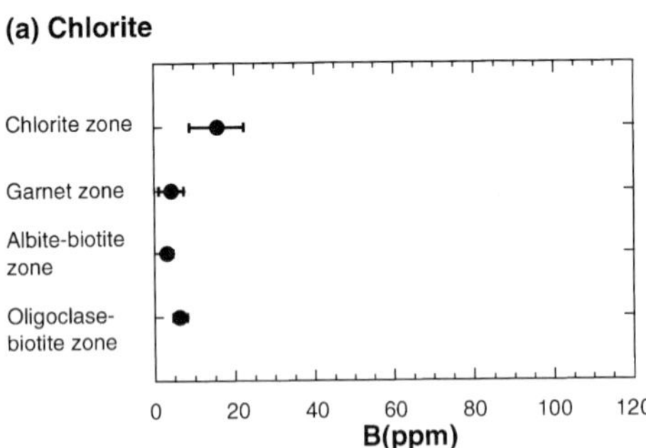

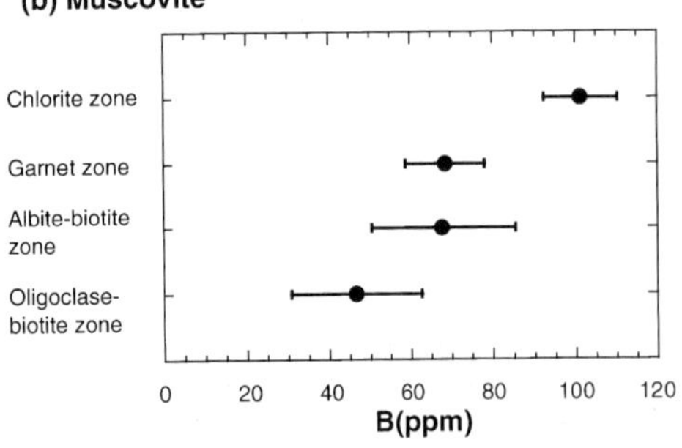

Fig. 5. Average B contents of chlorite (a) and muscovite (b) in each metamorphic grade. Error bars show the standard deviations.

The above discussions lead to the conclusion that the lack of systematic change in B contents with increasing grade in the Sambagawa metasedimentary rocks has been accomplished by the sequestering of B in newly formed tourmaline with the necessary intergranular transport of B occurring via metamorphic fluids. This process was evidently efficient enough to prevent significant removal of B from the metasedimentary rocks during progressive metamorphism. This conclusion is consistent with the observation that there is no significant difference in the whole-rock $\delta^{11}B$ of the metasedimentary rocks as a function of metamorphic grade in the Sambagawa (Fig. 4).

5.2. *Tourmalines in metasedimentary and sedimentary rocks*

5.2.1. *Major element composition*

Most of the tourmalines in metasedimentary rocks from the Sambagawa have compositions falling within the schorl–dravite solid solution:

$$Na(Fe^{2+}, Mg)_3Al_6(BO_3)Si_6O_{18}(OH)_{14}$$

Table 5
Estimated proportions of whole-rock boron incorporated in white-mica as a function of metamorphic grade

Metamorphic grade	Average B content of whole-rock (ppm)[a]	Average modal abundance of muscovite (%)[a]	Average B content of muscovite (ppm)	Proportion of B in white-mica for whole-rock (%)
Chlorite zone	37	29	101	90
Garnet zone	34	26	69	60
Albite–biotite zone	43	29	68	51
Oligoclase–biotite zone	52	32	47	32

[a] Excluding siliceous schists. Densities of muscovite and metasedimentary rocks used in the mass-balance calculation are 2.8 and 2.5 g/cm^3, respectively.

Fig. 6 shows Al–Fe–Mg ternary diagrams (after Henry and Guidotti, 1985) for the tourmalines in sedimentary rocks and metasedimentary rocks of the Shimanto Belt and the Sambagawa, respectively.

Most tourmalines in the chlorite zone metasedimentary rocks and sedimentary rocks from the Shimanto Belt are fine-grained and anhedral. The major element compositions of tourmalines from the Shimanto and the chlorite zone in the Sambagawa show extreme variability even within individual grains, indicating that these tourmalines were formed under variable physico-chemical conditions and derived from different detrital sedimentary sources (Fig. 6). The irregularly-shaped cores surrounded by metamorphic tourmaline overgrowths in the garnet and biotite zones may also be detrital tourmaline (see discussion

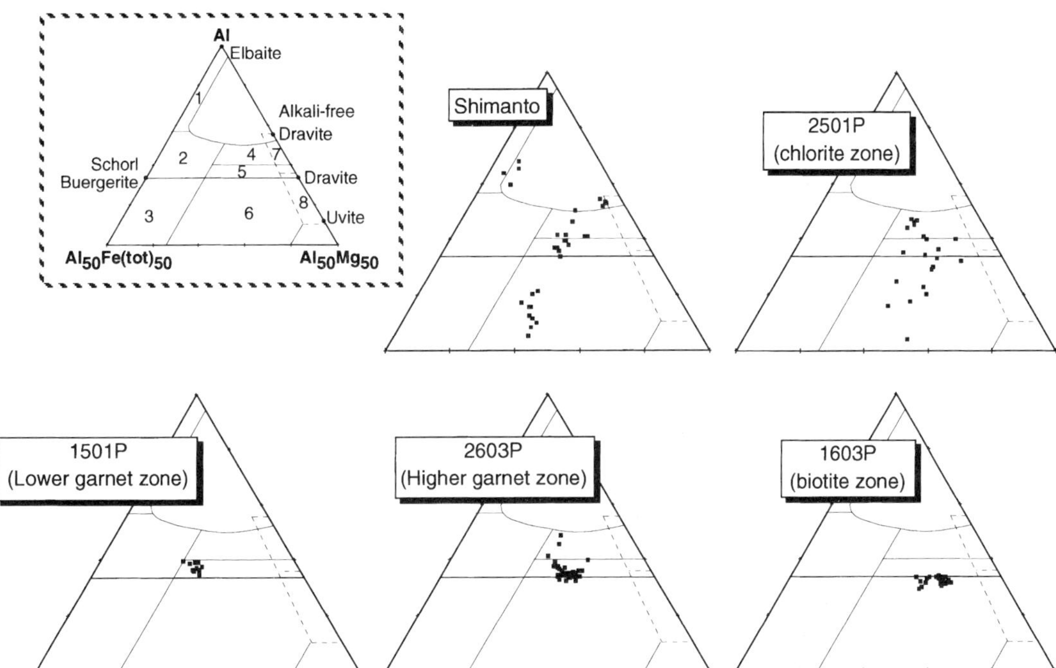

Fig. 6. Al–Fe–Mg ternary diagram for tourmaline in sedimentary and metasedimentary rocks from the Shimanto and Sambagawa Belts, respectively. Fields (defined by Henry and Guidotti (1985)) represent the compositions of tourmaline for the following rock types: (1) Li-rich granitoid pegmatites and aplites; (2) Li-poor granitoids and their associated pegmatites and aplites; (3) Fe^{3+}-rich quartz–tourmaline rocks (hydrothermally-altered granites); (4) metapelites and metapsammites coexisting with an Al-saturating phase; (5) metapelites and metapsammites not coexisting with an Al-saturating phase; (6) Fe^{3+}-rich quartz–tourmaline rocks, calc-silicate rocks, and metapelites; (7) low-Ca metaultramafics and Cr, V-rich metasediments; (8) metacarbonates and metapyroxenites.

by Henry and Guidotti (1985)). Metamorphic tourmaline overgrowths in the chlorite zone are very narrow (Fig. 2a). The compositional variations of tourmalines from samples of the lower garnet zone (1501P), higher garnet zone (2603P) and biotite zone (1603P) are relatively small compared with that of the chlorite zone samples (2501P; Fig. 6). Most grains in the garnet and biotite zones show remarkable chemical zoning. The dravite-component [Mg/(Mg + Fe) ratio] of the tourmalines in the metasedimentary rocks from the biotite zone shows an increase from inner to outer rim of the single grains (Fig. 7a). The dravite component in the outermost rims clearly shows an increase as a function of increasing metamorphic grade (Fig. 7b). The trend of Mg/(Mg + Fe) ratios in the outermost rims of tourmalines is similar to that in the coexisting chlorites in the metasedimentary rocks (Fig. 8). It is therefore suggested that the outermost rims of the tourmalines are equilibrated with coexisting Fe–Mg minerals as has been inferred by Henry and Guidotti (1985) for other suites. Higashino (1975) previously documented that the Mg/(Mg + Fe) ratio of chlorite in metasedimentary rocks from the Sambagawa increases with increasing metamorphic grade throughout the exposures. Chlorite appears in the metasedimentary rocks of all metamorphic grades and, at higher grades, reacted to form garnet-bearing assemblages by continuous metamorphic reactions of the type described above. Its breakdown at higher grades provided an important source of aqueous fluid to the reacting rock system, and tourmaline may be formed by reactions of the following general type:

$$\text{chlorite} + \text{albite} + B_2O_3 \rightarrow \text{tourmaline} + \text{quartz} + \text{Fe oxides} + H_2O.$$

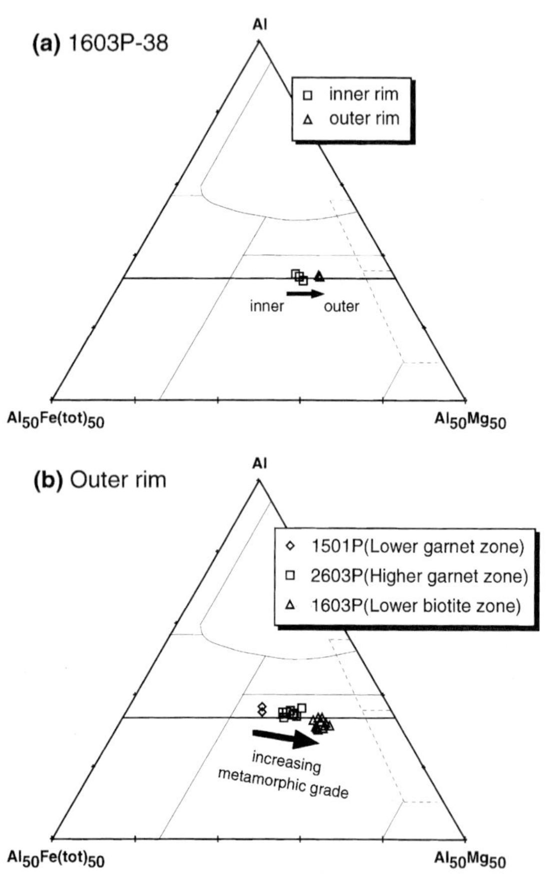

Fig. 7. Al–Fe–Mg ternary diagram (a) for inner and outer rim of a single tourmaline grain (1603P-38) and (b) for outer rims of tourmaline from each metamorphic grade.

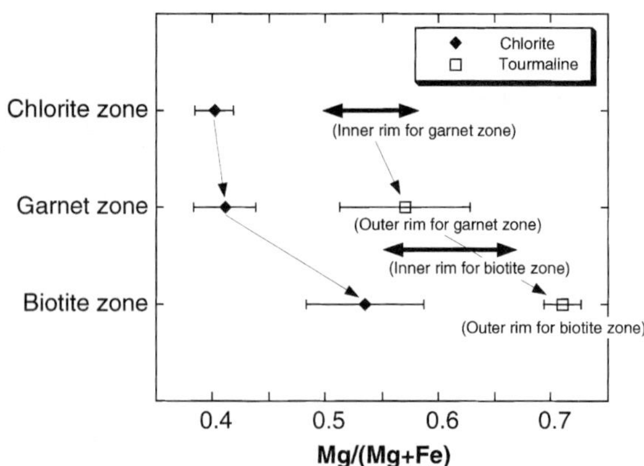

Fig. 8. Mg/(Mg + Fe) ratio of chlorite and tourmaline for each metamorphic grade. Symbols show average values of Mg/(Mg+Fe) ratios with error bars as standard deviations. The ranges of Mg/(Mg + Fe) ratios, shown by thick arrow, are deduced from those of inner rims in tourmalines from the higher metamorphic grades. Because metamorphic overgrowths of tourmaline in the chlorite zone are too narrow to analyze, and inner rims preserve compositions resulting from lower grade metamorphism. Data for chlorites are from Higashino et al. (1982).

5.2.2. *B isotopic composition*

The B isotopic compositions ($\delta^{11}B$) of tourmalines, determined by SIMS, in the metasedimentary rocks in the Sambagawa show a range from −14.3

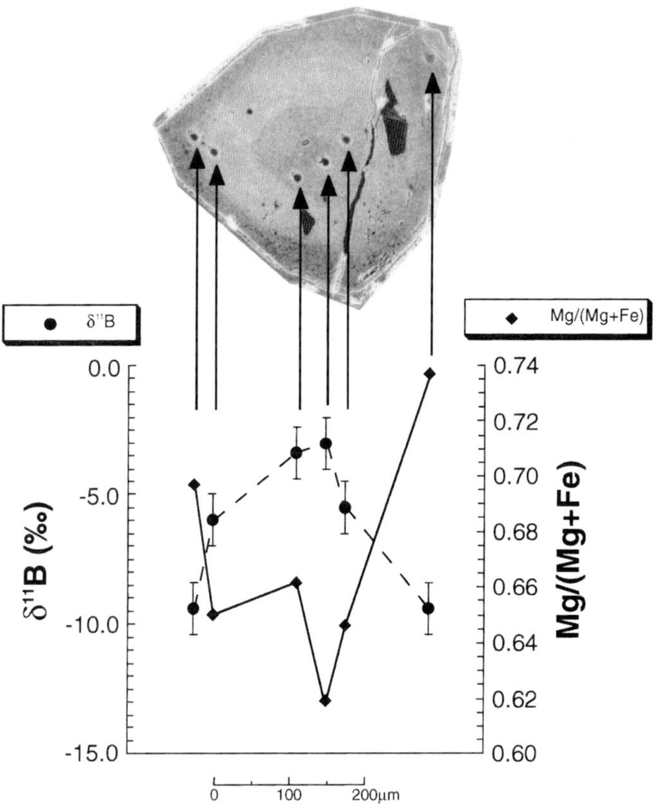

Fig. 9. Zoning in Mg/(Mg + Fe) and δ^{11}B in tourmaline in the biotite zone sample 1603P, as a function of distance along a rim-to-rim traverse.

to −0.3‰ from the garnet to the biotite zone and the sedimentary rocks from the Shimanto Belt have values of −11.1 to −3.4‰.

Some grains in the biotite zone show zoning in both Mg/(Mg + Fe) and δ^{11}B (Fig. 9). The δ^{11}B values decrease with increasing Mg/(Mg + Fe) from inner to outer rim in a single grain (Fig. 9). There is a broad negative correlation between δ^{11}B and Mg/(Mg + Fe) throughout the Sambagawa across metamorphic grade (Fig. 10). Because the systematic change of Mg/(Mg+ Fe) in tourmaline is indicative of metamorphic grade as shown in Fig. 8, δ^{11}B values of the tourmaline are also regarded as being related to the metamorphic conditions, showing an overall decrease with increasing metamorphic grade. This implies that the δ^{11}B of the tourmaline is highly sensitive to its crystallization environment. Previous studies (e.g. Ishikawa and Nakamura, 1994; Ishikawa and Tera, 1998) assumed that B isotope fractionation does not occur at temperatures greater than 200°C in the subducting oceanic crust. As discussed in the previous section based on the mass balance calculations, most of the B released from the white-mica and chlorite into the metamorphic fluid appears to have been retained in tourmaline in the Sambagawa metasedimentary rocks (Tables 4 and 5). Thus, the growing tourmaline may record the progressive evolution of the B isotopic composition of the metamorphic fluid with the outermost rim preserving the isotopic signature of peak metamorphic conditions.

5.3. Boron isotope composition of metamorphic fluid and isotopic fractionation between metamorphic fluids and rocks

The B isotopic fractionation between aqueous vapor/fluid and tourmaline ($\alpha_{f-t} = (^{11}B/^{10}B)_{fluid}/$

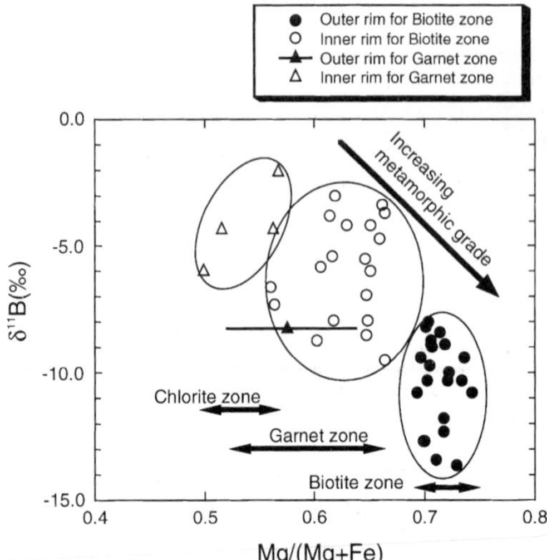

Fig. 10. Correlation of tourmaline Mg/(Mg + Fe) ratio and δ^{11}B for outer and inner rims. The horizontal line with a filled triangle indicates the variation of Mg/(Mg + Fe) ratio in the outer rims of the garnet zone tourmalines. Thick bars with arrows indicate the variations of Mg/(Mg + Fe) in tourmalines from metasedimentary rocks at each of the metamorphic grades.

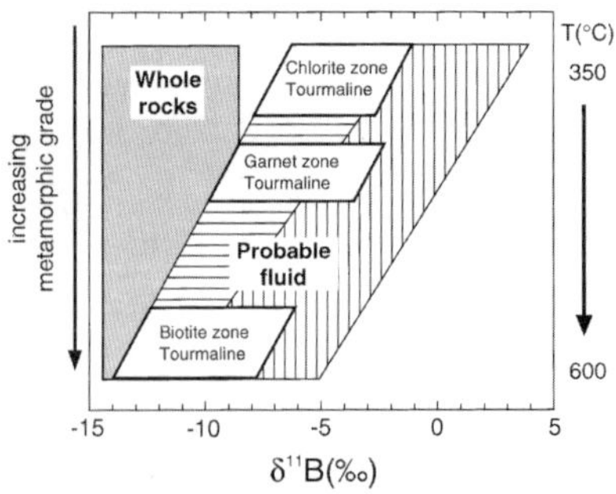

Fig. 11. Schematic diagram of δ^{11}B of tourmaline, metasedimentary whole-rocks and probable metamorphic fluid as a function of metamorphic grade. Horizontally and vertically shaded regions show the δ^{11}B of tourmalines and the δ^{11}B estimated for the fluids based on the fractionation factors of Palmer et al. (1992), respectively. The ranges in δ^{11}B for tourmaline in chlorite, garnet, and biotite zone metasedimentary rocks correspond to inner rims in garnet zone rocks (data for chlorite zone), both outer rims in garnet zone rocks and inner rims for biotite zone rocks (data for garnet zone), and outer tourmaline rims for biotite zone rocks (data for biotite zone).

$(^{11}B/^{10}B)_{tourmaline}$) has been investigated experimentally as a function of temperature and pressure up to 750°C and 0.2 GPa, respectively (Palmer et al., 1992). At 0.2 GPa, the isotopic fractionation factor decreases from 1.006 to 1.003 with increasing temperature from 350 to 750°C (Palmer et al., 1992). When the regression line on the relation between the temperature and the $\alpha_{f\text{–}t}$ at 0.2 GPa from Palmer et al. (1992) is applied to the tourmalines from the Sambagawa metasediments, the δ^{11}B of probable fluids equilibrated with the tourmalines are obtained as given in Fig. 11. The δ^{11}B of metamorphic fluid decreases with increasing metamorphic grade towards the whole-rock δ^{11}B in the biotite zone. According to Palmer et al. (1992), the $\alpha_{f\text{–}t}$ becomes smaller with increasing pressure, which may be important because the metamorphic conditions estimated for the studied metasediments are 0.6–1 GPa and 330–610°C (Enami et al., 1994). If this is the case, the δ^{11}B of metamorphic fluid becomes closer to those of the tourmalines. Since most of the B liberated from mica and chlorite minerals to metamorphic fluid is thought to have partitioned into tourmaline based on the mass balance calculation as discussed above, the δ^{11}B of tourmaline must be nearly identical to that of the metamorphic fluid as also shown in Fig. 11. Therefore, it may be adequate to assume that δ^{11}B of metamorphic fluids are equal to those of the coexisting tourmaline. Consequently, the δ^{11}B of the metamorphic fluid at chlorite, garnet and biotite zone conditions is estimated to have been -4.1 ± 1.6 (1σ), -5.9 ± 2.0 and -10.3 ± 1.7‰, respectively, based on the average δ^{11}B values of the tourmalines from each grade. The differences in δ^{11}B between metamorphic fluid and whole-rock ($\Delta^{11}B_{f\text{–}t} = \delta^{11}B_{fluid} - \delta^{11}B_{whole\text{-}rock}$) are 7.1 ± 3.2, 5.3 ± 3.6 and 0.9 ± 3.3‰ for chlorite, garnet and biotite zone, respectively. Thus, an "apparent" B isotopic fractionation factor between metamorphic fluid and the whole-rock ($\alpha_{f\text{–}t} = (^{11}B/^{10}B)_{fluid}/(^{11}B/^{10}B)_{whole\text{-}rock}$) decreases from 1.007 ± 0.003 to 1.001 ± 0.003 with increasing metamorphic grade from the chlorite to the biotite zone.

These observations indicate that B isotopic fractionation is significant under the lower metamorphic conditions up to garnet zone metamorphism if the B liberated into the fluid is largely removed without formation of tourmaline. On the other hand, the higher-grade metamorphism beyond the biotite zone does not cause B isotope fractionation in the metasedimentary rocks even if B is removed together with fluid from the reacting rock system by the devolatilization of mica and chlorite minerals. The diagenetic and low-grade metamorphic tourmalines, developed by overgrowing the detrital tourmaline, are common in the sedimentary and metasedimentary rocks from many metamorphic terrains (e.g. Henry and Dutrow, 1996) as well as the tourmalines from the Shimanto Belt and the Sambagawa Metamorphic Belt of this study. This process may result in minimal isotopic fractionation of B between fluids and bulk rocks containing B in white-mica and tourmaline, making it possible to carry the B isotopic signatures of the initial sediments into the deep mantle, perhaps to form localized B isotope anomalies adjacent to convergent margins. It is thus essential to carefully document the presence or absence of tourmaline in rock samples used to investigate material recycling using B contents and isotopic compositions of subduction zone rocks.

5.4. Implications of B isotope systematics for crust–mantle recycling in subduction zones

Boron isotope systematics are generally believed to be useful for characterizing the subduction component and metamorphic fluid in subduction zones due to unique chemical characteristics of boron which is extremely soluble in aqueous phases and strongly partitions into melts as an incompatible element. However, geochemical applications of B isotopes to the studies of slab dehydration and mantle metasomatism in subduction zones are still relatively few (e.g. Palmer, 1991; Ishikawa and Nakamura, 1992, 1993; Chaussidon and Jambon, 1994; Chaussidon and Marty, 1995).

Based on the across-arc variations of concentrations and isotopic compositions of B, Pb and Li, it was inferred that a homogeneous slab fluid contributes to all Izu and Kurile volcanoes with the amount of fluid decreasing continuously with increasing depth to the slab (Ishikawa and Nakamura, 1994; Ishikawa and Tera, 1997; Moriguti and Nakamura, 1998). These authors assumed that unmetamorphosed sediments were representative of the subducted sedimentary components. However, the results obtained in this study indicate that B contents and isotopic compositions of metasedimentary rocks are somewhat lower than those of unmetamorphosed sediments. Such a modification of the end-component results in a slight decrease of the sedimentary B inventory compared to the altered MORB component ($\delta^{11}B = +2$ to $+9‰$; Spivack and Edmond, 1987; Ishikawa and Nakamura, 1992) which is one of major components derived from subducted oceanic crust. Thus, this leads to improving the arguments on the evolution of the source mantle for arc magmas in subduction zones.

Tourmaline is stable up to at least 6 GPa and 1000°C for the doravite component, and thus may remain stable in deeply subducted slabs (Werding and Schreyer, 1996). Tourmalines were found in ultra high-pressure metamorphic rocks from the Dora Maira Massif that have experienced *P–T* conditions of 3.7 GPa and 800°C (Schertle et al., 1991). If a significant proportion of the B liberated into aqueous fluids from hydrous minerals in the subducting sediments is retained in tourmalines, B can be effectively carried into the deep mantle, perhaps leading to a lower $\delta^{11}B$ in the mantle compared with that of MORB source mantle. Chaussidon and Marty (1995) reported a significantly lower $\delta^{11}B$ value ($-10 \pm 2‰$) for OIB source than that of a homogeneous MORB source ($-4‰$). Such a signature of B may suggest that the OIB source mantle was affected by subducted sediment component similar to the Sambagawa metasedimentary rocks.

Recently Peacock and Hervig (1999) reported B isotope compositions of minerals in subduction zone metamorphic rocks including metasediments, metabasalts and serpentinite, yielding $\delta^{11}B = -11$ to $-3‰$, suggesting that slab dehydration reactions significantly lower the $\delta^{11}B$ of subducted oceanic crust and sediments. These values are consistent with our whole-rock data for metasedimentary rocks from the Sambagawa in this study (-14.2 to $-8.8‰$), the low-grade Catalina Schist metasedimentary rocks (-16.3 to $-10.1‰$; Nakano and Nakamura, 1995), and the metabasites associated with metasedimentary rocks in the Sambagawa (-14.1 to $-0.9‰$; Nakano et al., 1996). Nakano et al. (1996; Nakano, 1998) inferred that the B isotopic compositions of the basic rocks in the Sambagawa were affected isotopically by the infiltration of metamorphic fluid from metasediments, resulting in a significantly lowering of $\delta^{11}B$ in the metabasaltic rocks. The shift in $\delta^{11}B$ to lower values demonstrated by Peacock and Hervig (1999) may thus also be a consequence of metasomatism caused by the infiltration of metamorphic fluid emitted from surrounding metasediments. According to Peacock

(1987), the serpentinite with the lowest $\delta^{11}B$ value in Peacock and Hervig (1999) was formed at temperatures ranging from 425 to 570°C by infiltration of fluid derived from subducting slab. This temperature range is equivalent to that of the garnet to biotite zone metamorphism in the Sambagawa. In this study, we determined the B isotopic fractionation between fluid and rocks to be very small (~1) at the given conditions. Furthermore, the original ultramafic source rock for the serpentinite should have an extremely low concentration of B, probably less than 0.01 ppm (Yoshikawa and Nakamura, 2000), and thus even a small addition of B with fluid to form serpentinite must largely inherit the significantly lower B isotope signature of the metasedimentary rocks. Therefore, such a low B isotopic composition of serpentinite is not a consequence of the isotopic fractionation governed by the dehydration of subducted slab.

Acknowledgements

We are grateful to T. Itaya for useful suggestions on the Sambagawa Metamorphic rocks and also for providing the rock samples used in this study. We also thank T. Ishikawa for much technical advice concerning B isotope analysis by TIMS. We are deeply indebted to T. Moriguti and A. Makishima for valuable discussion on this subject and encouragement in the course of this study. We also acknowledge H. Takei, C. Sakaguchi, and K. Kobayashi for analytical support in XRF, EDX–SEM and SIMS analysis, respectively, and all the other member of PML for useful discussions. We would like to express our appreciation to G. Bebout, M. Chaussidon, M. Palmer and D. Rubie for their efforts to improve the quality of this paper. We thank H. Asada for making thin sections. This work was supported by the Ministry of Education, Science, Sports and Culture (Monbusho) of Japanese Government and the Japan Society of the Promotion of Science (JSPS) to E. Nakamura.

References

Banno, S., 1964. Petrologic studies on Sanbagawa crystalline schists in the Bessi-Ino district, Central Shikoku, Japan. J. Fac. Sci., Univ. Tokyo 15, 203–319.

Banno, S., Sakai, C., 1989. Geology and metamorphic evolution of the Sanbagawa metamorphic belt, Japan. In: Daly, J.S., Cliff, R.A., Yardley, B.W.D. (Eds.), Evolution of metamorphic belts, Special Publication of the Geological Society of London, no. 43, pp. 519–531, Blackwell Scientific Publication, Oxford.

Banno, S., Sakai, C., Higashino, T., 1986. Pressure–temperature trajectory of the Sanbagawa metamorphism deduced from garnet zoning. Lithos 19, 51–63.

Bebout, G.E., 1995. The impact of subduction zone metamorphism on mantle–ocean chemical cycling. Chem. Geol. 126, 191–218.

Bebout, G.E., Ryan, J.G., Leeman, W.P., 1993. B–Be systematics in subduction-related metamorphic rocks: characterization of the subducted component. Geochim. Cosmochim. Acta 57, 2227–2237.

Bebout, G.E., Ryan, J.G., Leeman, W.P., Bebout, A.E., 1999. Fractionation of trace elements by subduction zone metamorphism — effect of convergent-margin thermal evolution. Earth Planet. Sci. Lett. 171, 63–81.

Chaussidon, M., Appel, P.W.U., 1997. Boron isotopic composition of tourmalines from the 3.8-Ga-old Isua supracrustals, West Greenland: implications on the $\delta^{11}B$ value of early Archean seawater. Chem. Geol. 136, 171–180.

Chaussidon, M., Jambon, A., 1994. Boron content and isotopic composition of oceanic basalts: geochemical and cosmochemical implications. Earth Planet. Sci. Lett. 121, 277–291.

Chaussidon, M., Marty, B., 1995. Primitive boron isotope composition of the mantle. Science 269, 383–386.

Enami, M., Wallis, S.R., Banno, Y., 1994. Paragenesis of sodic pyroxene-bearing quartz schists: implications for the *P–T* history of the Sanbagawa Belt. Contrib. Mineral. Petrol. 116, 182–198.

Ernst, W.G., Seki, Y., Onuki, H., Gilbert, M.C., 1970. Comparative study of low-grade metamorphism in California Coast Range and the outer metamorphic belt of Japan. Geol. Soc. Am. Mem. 124, 276.

Harder, H., 1970. Boron content of sediments as tool in facies analysis. Sediment. Geol. 4, 153–175.

Hart, S.R., Reid, M.R., 1991. Rb/Cs fractionation: a link between granulite metamorphism and the S-process. Geochim. Cosmochim. Acta 55, 2379–2383.

Henry, D.J., Dutrow, B.L., 1996. Metamorphic tourmaline and its petrologic applications. In: Grew, E.S., Anovitz, L.M. (Eds.), Boron Mineralogy, Petrology and Geochemistry — Reviews in Mineralogy, Vol. 33, pp. 503–557.

Henry, D.J., Guidotti, C.V., 1985. Tourmaline as a petrogenetic indicator mineral: an example from the staurolite-grade metapelites of NW Maine. Am. Mineral. 70, 1–15.

Higashino, T., 1975. Biotite zone of the Sanbagawa metamorphic terrain in the Shiragayama area, central Shikoku, Japan. J. Geol. Soc. Jpn. 81, 653–670 (in Japanese with English Abstract).

Higashino, T., 1990. Metamorphic zones of the Sambagawa Metamorphic Belt in central Shikoku, Japan. J. Geol. Soc. Jpn. 96, 703–718 (in Japanese with English Abstract).

Higashino, T., Sakai, C., Otsuki, M., Itaya, T., Banno, S., 1982. Electron microprobe analyses of rock-forming minerals from the Sanbagawa metamorphic rocks, Shikoku. Part I. Asemi River area. Sci. Rep. Kanazawa Univ. 26 (2), 73–122.

Hofmann, A.W., White, W.M., 1982. Mantle plumes from ancient oceanic crust. Earth Planet. Sci. Lett. 57, 421–436.

Ishikawa, T., Nakamura, E., 1992. Boron isotope geochemistry of the oceanic crust from DSDP/ODP Hole 504B. Geochim. Cosmochim. Acta 56, 1633–1639.

Ishikawa, T., Nakamura, E., 1993. Boron isotope systematics of marine sediments. Earth Planet. Sci. Lett. 117, 567–580.

Ishikawa, T., Nakamura, E., 1994. Origin of the slab component in arc lavas from across-arc variation of B and Pb isotopes. Nature 370, 205–208.

Ishikawa, T., Tera, F., 1997. Source, composition and distribution of the fluid in the Kurile mantle wedge: constraints from across-arc variations of B/Nb and B isotopes. Earth Planet. Sci. Lett. 152, 123–138.

Isozaki, Y., Itaya, T., 1990. Chronology of Sanbagawa metamorphism. J. Metamorphic Geol. 8, 401–411.

Itaya, T., 1981. Carbonaceous material in pelitic Schists of the Sanbagawa Metamorphic Belt in central Shikoku, Japan. Lithos 14, 215–224.

Itaya, T., Banno, S., 1980. Paragenesis of titanium-bearing accessories in pelitic schists of the Sanbagawa Metamorphic Belt, Central Shikoku, Japan. Contrib. Mineral. Petrol. 73, 267–276.

Kurata, H., Banno, S., 1974. Low-grade progressive metamorphism of pelitic schists of the Sanbagawa metamorphic terrain in central Shikoku, Japan. J. Petrol. 15, 361–382.

Malinko, S.V., Berman, I.B., Lisitsyn, A.Y., Stolyarova, A.N., 1979. Boron in rock-forming minerals based on data from a local radiographic analysis. Int. Geol. Rev. 21, 1274–1284.

Miyashiro, A., Banno, S., 1958. Nature of glaucophanic metamorphism. Am. J. Sci. 266, 968–979.

Moran, A.E., Sisson, V.B., Leeman, W.P., 1992. Boron depletion in subducted oceanic crust and sediments: effects of metamorphism and implications for arc magma compositions. Earth Planet. Sci. Lett. 111, 331–349.

Morgan, G.B., London, D., 1989. Experimental reactions of amphibolite with boron-bearing aqueous fluids at 200 MPa: implications for tourmaline stability and partial melting in mafic rocks. Contrib. Mineral. Petrol. 102, 281–297.

Moriguti, T., Nakamura, E., 1998. Across-arc variation of Li isotopes in lavas and implications for crust/mantle recycling at subduction zones. Earth Planet. Sci. Lett. 163, 167–174.

Morris, J.D., Leeman, W.P., Tera, F., 1990. The subducted component in island arc lavas: constraints from Be isotopes and B–Be systematics. Nature 344, 31–36.

Mottl, M.J., Holland, H.D., 1978. Chemical exchange during hydrothermal alteration of basalt by seawater. I. Experimental results for major and minor components of seawater. Geochim. Cosmochim. Acta 42, 1103–1115.

Nakajima, T., Banno, S., Suzuki, T., 1977. Reaction leading to the disappearance of pumpellyite in low-grade metamorphic rocks of Sanbagawa Metamorphic Belt in central Shikoku, Japan. J. Petrol. 18, 263–284.

Nakamura, E., Kushiro, I., 1998. Trace element diffusion in jadite and diopside melts at high pressures and its geochemical implication. Geochim. Cosmochim. Acta 62, 3151–3160.

Nakamura, E., Campbell, I.H., Sun, S.S., 1985. The influence of subduction processes on the geochemistry of Japanese alkaline basalt. Nature 316, 55–58.

Nakamura, E., Ishikawa, T., Birck, J.-L., Allégre, C.J., 1992. Precise boron isotopic analysis of natural rock samples using a boron–mannitol complex. Chem. Geol. (Isot. Geosci. Sect.) 94, 193–204.

Nakamura, E., Suzuki, T., Kobayashi, K., Makishima, A., Yoshikawa, M., 1998. Trace element partitioning between majorite and silicate melt under high pressures. In: Proceedings of the 1998 Annual Meeting of the Japan Earth and Planetary Science Joint Meeting. A-61, Tokyo.

Nakano, T., 1998. Boron isotope and trace element geochemistry of subduction-related metamorphic rocks: implications for the transportation of elements during metamorphism. Ph.D. Thesis, Okayama University, Japan.

Nakano, T., Nakamura, E., 1995. Boron isotope systematics of subduction-related metamorphic rocks from the Sambagawa Metamorphic Belt, Japan. In: Proceedings of the 1998 Annual meeting of the Japan Earth and Planetary Science Joint Meeting E22–23, Tokyo.

Nakano, T., Makishima, A., Nakamura, E., 1996. Fluid migration during subduction-related high-pressure metamorphism: evidence from boron isotopic compositional change in metamorphic rocks from the Sambagawa Metamorphic Belt, central Shikoku, Japan. In: Proceedings of the Joint Meeting Misasa Seminar on Evolutionary processes of Earth and Planetary Materials. ISEI, Okayama University at Misasa and Geophysical Laboratory, Carnegie Institution of Washington, pp. 31–32.

Palmer, M.R., 1991. Boron-isotope systematics of Halmahera arc (Indonesia) lavas: evidence for involvement of the subducted slab. Geology 19, 215–217.

Palmer, M.R., London, D., Morgan, G.B., Babb, H.A., 1992. Experimental determination of fractionation of $^{11}B/^{10}B$ between tourmaline and aqueous vapor: a temperature- and pressure-dependent isotopic system. Chem. Geol. 101, 123–129.

Peacock, S., 1987. Serpentinization and infiltration metasomatism in the Trinity peridotite, Klamath province, northern California: implications for subduction zone. Contrib. Mineral. Petrol. 95, 55–70.

Peacock, S.M., Hervig, R.L., 1999. Boron isotopic composition of subduction-zone metamorphic rock. Chem. Geol. 160, 281–290.

Perfit, M.R., Gust, D.A., Bence, A.E., Arculus, R.J., Taylor, S.R., 1980. Chemical characteristics of island arc basalts: implications for mantle sources. Chem. Geol. 30, 227–256.

Radvanec, M., Banno, S., Okamoto, K., 1994. Multiple stages of phengite formation in Sanbagawa schists. Mineral. Petrol. 51, 37–48.

Ryan, J.G., Langmuir, C.H., 1993. The systematics of boron abundances in young volcanic rocks. Geochim. Cosmochim. Acta 57, 1489–1498.

Sakuyama, M., Nesbitt, R.W., 1986. Geochemistry of the quaternary volcanic rocks of the northeast Japan arc. J. Volcanol. Geotherm. Res. 29, 413–450.

Saundaers, A.D., Tarney, J., Weaver, S.D., 1980. Transverse geochemical variations across the Antarctic Peninsula: implications for the genesis of calc-alkaline magmas. Earth Planet. Sci. Lett. 46, 344–360.

Schertle, H.P., Schreyer, W., Chopin, C., 1991. The pyrope–coesite rocks and their country rocks at parigi, Dora Maira Massif, western Alps: detailed petrography, mineral chemistry and P–T path. Contrib. Mineral. Petrol. 108, 1–21.

Seits, M.G., 1972. Boron mapping in synthetic and natural systems: crystal-melt assemblages, garnet, lherzolite, chondrites. Carnegie Inst. Wash. Yearb. 589.

Seyfried Jr., W.E., Janecky, D.R., Mottl, M.J., 1984. Alteration of the oceanic crust: implications for geochemical cycles of lithium and boron. Geochim. Cosmochim. Acta 48, 557–569.

Shibata, T., Nakamura, E., 1997. Across-arc variations of isotope and trace element compositions from quaternary basaltic volcanic rocks in northeastern Japan: implications for interaction between subducted oceanic slab and mantle wedge. J. Geophys. Res. 102, 8051–8064.

Spivack, A.J., Edmond, J.M., 1987. Boron isotopic exchange between seawater and the oceanic crust. Geochim. Cosmochim. Acta 51, 1033–1043.

Swihart, G.H., Moore, P.B., 1989. A reconnaissance of the boron isotopic composition of tourmaline. Geochim. Cosmochim. Acta 53, 911–916.

Taira, A., Katto, J., Tashiro, M., Okamura, M., Kodama, K., 1988. The Shimanto Belt in Shikoku, Japan — Evolution of cretaceous to miocene accretionary prism. Mod. Geol. 12, 5–46.

Werding, G., Schreyer, W., 1996. Experimental studies on borosilicates and selected borates. In: Grew, E.S., Anovitz, L.M. (Eds.), Boron Mineralogy, Petrology and Geochemistry — Reviews in Mineralogy, Vol. 33, pp. 117–163, Mineralogical Society of America, Washington, DC.

Xiao, Y.K., Beary, E.S., Fassett, J.D., 1988. An improved method for the high-precision isotopic measurement of boron by thermal ionization mass spectrometry. Int. J. Mass Spectrom. Ion Proc. 85, 203–213.

Yoshikawa, M., Nakamura, E., 2000. Geochemical evolution of the Horoman peridotite complex: implications for melt extraction, metasomatism, and compositional layering in the mantle. J. Geophys. Res. 105, 2879–2901.

ELSEVIER

Physics of the Earth and Planetary Interiors 127 (2001) 253–275

PHYSICS OF THE EARTH AND PLANETARY INTERIORS

www.elsevier.com/locate/pepi

Subduction, ultrahigh-pressure metamorphism, and regurgitation of buoyant crustal slices — implications for arcs and continental growth

W.G. Ernst*

Department of Geological and Environmental Sciences, Stanford University, Stanford, CA 94305-2115, USA

Received 13 October 1999; accepted 15 September 2000

Abstract

Intracontinental collisional belts that retain mineralogic relics of Phanerozoic ultrahigh-pressure (UHP) metamorphism are increasingly being recognized. Adjacent regions generally lack evidence of coeval calcalkaline volcanism/plutonism. Following consumption of intervening oceanic lithosphere, each orogen marks the site of a deeply subducted microcontinental promontory, fragments of which later returned to midcrustal levels; mafic and ultramafic rocks are volumetrically minor. Pressures in UHP complexes approached or exceeded 2.8 GPa at 600–900°C. Subduction zones involve low-T prograde gradients, and constitute the only plate-tectonic realm where such conditions exist. Although internal portions of descending plates may have even lower geothermal gradients, crustal upper margins are typified by P–T trajectories of 5–10°C/km. Attending underflow, clinoamphibole-rich metabasaltic rocks dehydrate at pressures >2.2–2.5 GPa. This accounts for the kinetically favorable solid state production of mafic eclogites and partial fusion of MORBs to form hydrous andesitic melts. Devolatilization of partly serpentinized mantle lithosphere also introduces H_2O to the magmagenic zone. As large masses of continental crust enter deeper portions of subduction zones, mineralogic transformations slow or cease because micas, the major OH-bearing phases in such rocks, retain volatile contents to pressures >4 GPa. Persistence of hydroxyl-bearing layer silicates may explain the metastable preservation of low-density assemblages in continental crust subjected to UHP conditions, and the dearth of calcalkaline magmatism in most collision belts.

Exhumed UHP terranes are intensely deformed and disaggregated. All consist dominantly of quartzofeldspathic compositions; mafic eclogites ± garnet peridotites constitute only a few volume percent. Eventual decoupling of the relatively low-density subducted crust promotes buoyancy-propelled transport to shallow depths along the subduction channel, with the hanging-wall lithosphere acting as a stress guide. Extensional collapse and erosional unroofing of rising masses result in the continued ascent of deeply subducted but buoyant crustal material. Although equidimensional massifs must return surfaceward, large volume-to-surface areas render the contained UHP phase assemblages virtually unquenchable. Surviving UHP terranes, thus, consist of relatively thin slabs of continental crust. Allochthons that retain UHP relics apparently lost heat by conduction across both upper, normal fault and lower, reverse fault contacts. Based on present exposures, estimated thicknesses of UHP sheets are: Qinling–Dabie–Sulu belt = 5 km; Kokchetav complex = 1–3 km; Dora Maira Massif = 1–2 km; Western Gneiss Region => 1 km. Slices evidently rose to midcrustal levels at exhumation rates approaching 10 mm per year. Back reaction attending decompression in all cases was nearly complete. Where UHP mineral inclusions persisted, retrogression

*Tel.: +1-650-7230185; fax: +1-650-7250979.
E-mail addresses: ernst@pangea.stanford.edu, ernst@geo.stanford.edu (W.G. Ernst).

PII: S0031-9201(01)00231-X

apparently was limited by the relative impermeability of the rocks to catalytic aqueous fluids and by declining temperatures; the coarse grain size of host phases also disfavored complete retrogression.

Keywords: Subduction; UHP metamorphism; Continental collision; Buoyant exhumation; Calcalkaline melts

1. Introduction

Compressional mountain belts owe their formation to subduction and loss of lithosphere. Two main orogenic types, Pacific and Alpine, are distinguished, although both reflect the same fundamental convergent plate-tectonic mechanism. During circum-Pacific subduction, vast tracts of oceanic crust-capped lithosphere sink into the upper mantle, carrying down moderately large volumes of poorly consolidated, structurally incompetent hemipelagic, trench, and fore-arc strata. Some of this section is recrystallized under relatively high-pressure (HP) conditions prior to offloading into the accretionary prism. In contrast, attending Alpine-style continental collision, consumption of an intervening oceanic plate allows the approach of a continent, microcontinent, or island arc — onstituting an integral part of the lithosphere — to the convergent plate junction; in some cases, a crustal promontory along the leading edge first enters the subduction zone. Because old, cool continental salients are firmly attached to the largely oceanic crust-capped lithosphere, lower density island arc, microcontinent, and/or passive margin rocks are carried down to great depths where HP and even ultrahigh-pressure (UHP) metamorphism can take place, prior to decoupling from the descending plate.

An increasing number of Eurasian intracratonic collisional orogens have been recognized to contain rare, scattered traces of subsolidus UHP recrystallization (Liou et al., 1994, 1996; Coleman and Wang, 1995; Schreyer, 1995; Harley and Carswell, 1995; Maruyama et al., 1996; Carswell and Zhang, 1999). Intensively investigated UHP tracts include the Qinling–Dabie–Sulu belt of eastern China, the Kokchetav complex of northern Kazakhstan, the Dora Maira Massif and the Zermatt area of the western Alps, and the Western Gneiss Region of southwestern Norway. These terranes (and others such as the Makbal and Atbashy complexes, western China, Sulawesi, Indonesia, the western Himalayas, the Bohemian Massif, the Maksyutov complex of the south Urals, and Mali, central Africa, summarized by Liou (1999)) occupy tectonic sutures juxtaposing markedly dissimilar continental entities. Slices of these crustal complexes contain traces of phases that testify to recrystallization at mantle depths of, or exceeding 100 km (Schreyer, 1988; Wang et al., 1995).

Mineralogic evidence includes the very rare preservation of UHP minerals and assemblages typified by coesite and/or diamond (both as micro-inclusions enclosed in strong, fluid-impermeable, unreactive mineralogic host grains, especially zircon and garnet), K- and/or Si-rich clinopyroxene, pyropic garnet, Al-rich rutile ± titanite, and volatile-bearing assemblages such as magnesite + diopside, coesite + dolomite, talc + kyanite, talc + jadeite, or talc + diopside, as well as relatively HP phases such as high-Si phengite, ellenbergerite, lawsonite, and glaucophane. Phase-equilibrium data and thermobarometric computations document temperatures of 700–900°C and pressures approaching 2.8–4 GPa. The *P–T* stability relations of some relevant phases and a generalized petrogenetic grid are summarized in Fig. 1. The discovery of Fe–Ti oxide exsolution lamellae in olivine grains in peridotite from Alpe Arami, southern Switzerland, and the provocative interpretation of mantle crystallization depths of about 300 km (Dobrzhinetskaya et al., 1996) raises the possibility that some rocks now at the surface may have ascended from depths far greater than currently recognized. The report of majorite component in garnet from a Western Gneiss Region peridotite (van Roermund et al., 1998), and allegedly very HP clinoenstatite in garnet lherzolites from the Sulu terrane (Zhang et al., 1998) as well as Alpe Arami (Bozhilov et al., 1999) are additional examples of potentially very HP assemblages. However, all of these >6 GPa occurrences are in small bodies of mantle material, and may be exotic in their present geologic setting.

As with relatively HP blueschist terranes, UHP conditions appear to have been generated through subduction (Ernst, 1971; Peacock, 1995). Low-temperatures at great depth reflect the fact that geologic materials

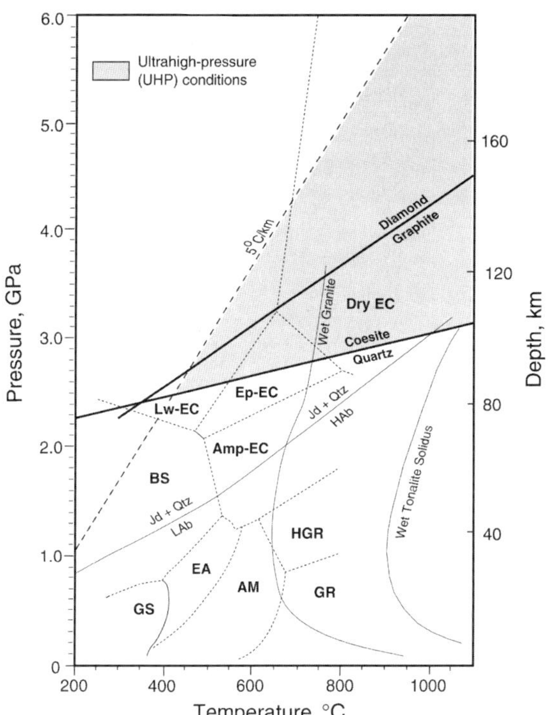

Fig. 1. Petrogenetic grid for metabasaltic bulk-rock compositions, after Liou et al. (1998) and Okamoto and Maruyama (1999). An extremely low subduction zone geothermal gradient of 5°C/km is also shown. For literature references to experimental data, both specific subsolidus phase transitions and the onset-of-melting curves, see Liou et al. (1998). Metamorphic facies abbreviations are as follows: AM, amphibolite; Amp–EC, amphibolite–eclogite; BS, blueschist; EA, epidote amphibolite; EC, eclogite; EP–EC, epidote–eclogite; LW–EC, Lawsonite–eclogite; GR, sillimanite-bearing granulite; GS, greenschist; HGR, kyanite-bearing granulite; P, prehnite; PA, pumpellyite-actinolite; PP, prehnite-pumpellyite; and ZE: zeolite.

are poor thermal conductors. No other plate-tectonic environment provides the appropriate conditions suitable for this type of metamorphism. Slab interiors may be attended by more extreme downwarping of the isotherms (Oxburgh and Turcotte, 1971), but the crustal upper margin of subducting lithosphere, juxtaposed against the mantle wedge of more normal thermal structure, is subjected to geothermal gradients ranging down to values of approximately 5–10°C/km (Schreyer, 1988). To explain the generation of UHP complexes, the deep underflow of coherent tracts of ancient, cool lithosphere are required (Ernst and Peacock, 1996). Thus, UHP orogenic belts attest to lithospheric descent prior to final suturing of the colliding continental crust-capped plates. The later decoupling and ascent of such complexes reflects the buoyancy of the continental crust (Cloos, 1993; Ernst and Liou, 1995). Prior to the ultradeep transformation of K-feldspar to hollandite, quartzofeldspathic rocks fully converted to the stable UHP phase assemblage would still remain of lower density than the mantle they displace, as illustrated in Table 1. Moreover, it is questionable whether low-density crust ever could be subducted to such great depths before achieving the brittle → ductile transition and decoupling from the downgoing slab. Additional strong support for the buoyancy propulsion mechanism is provided by the fact that, worldwide, dense rocks such as metabasaltic eclogites and garnet lherzolites represent but 1–10 percent by volume of UHP assemblages; only relatively low-density materials have returned surfaceward from such profound depths, generally in extensional settings.

In both Alpine- and Pacific-type suture zones, gradual heating/recrystallization (Rubie, 1983, 1990a) accompanying plate descent (reverse faulting) causes an increase in ductility of the downgoing slab, most substantially in the quartzofeldspathic segment overlying the mafic + ultramafic lithosphere. Due to buoyancy, some of this low-density continental material may decouple and ascend back up the subduction channel (normal faulting), with the overlying mantle wedge serving as a stress guide. Intrinsically weak trench sediments tend to shear off from the sinking lithosphere at relatively shallow depths, whereas more coherent, old continental crust may continue on to greater depths before decoupling from the descending plate. Accordingly, subduction zones marking circum-Pacific margins are characterized by exhumed HP blueschist belts, whereas the considerably higher pressure UHP eclogitic terranes are confined to continental collision sutures, as in the Alpine–Tethyan–Himalayan–Indonesian orogen. In either case, extensional collapse and/or erosional removal of the ascending crustal overburdon (Platt, 1986, 1993; Yin and Nie, 1996; Lin and Roecker, 1998; Beaumont et al., 1999) promotes continued exhumation of deeply subducted but buoyant continental material.

Post-subduction, near-adiabatic decompression results in the passage of rising UHP complexes

Table 1
Computed densities of major rock types transformed to UHP metamorphic mineral assemblages (Deer et al., 1966; Klein and Hurlbut, 1985)

Rock type	Mineral	Mineral density (g/cm^3)	Mode (vol.%)	Rock density (g/cm^3)
Garnet lherzolite	Enstatite	3.21	30	3.24
	Diopside	3.25	15	
	Olivine	3.22	50	
	Pyropic garnet	3.67	5	
Basaltic eclogite	Omphacite	3.34	60	3.68
	Garnet	4.17	35	
	Rutile	4.25	5	
Granitic gneiss	Jadeite	3.28	40	3.03
	Coesite	3.01	35	
	K-feldspar	2.56	15	
	Muscovite	2.85	10	

through markedly contrasting relatively high-*T*, low-*P* regimes. Complete annealing or even partial fusion at 700–900°C, 0.2–1 GPa is likely, and the rare occurrence of UHP relics suggests that phase assemblages have almost ubiquitously back reacted on return surfaceward. Surprisingly, because of the apparent short duration of individual UHP events during continental collision and prograde subduction zone metamorphism, metastable low-*P* mineral assemblages in some cases failed to recrystallize to the stable UHP configuration at the pressure maximum (Erambert and Austrheim, 1993; Harley and Carswell, 1995; Zhang and Liou, 1997; Katayama et al., 2000). Similarly, on later decompression, retrogression may not run to completion either (Rubie, 1986, 1990b; Hacker, 1990; Mosenfelder and Bohlen, 1997). Persistence of metastable phases seems to be a reflection of the kinetic effect of an aqueous fluid. Mineral transformations run rapidly when H_2O is present as a free phase, but rates are orders of magnitude slower when it is absent (e.g. Ernst et al., 1998).

2. Brief sketches of some HP/UHP complexes

2.1. The Qinling–Dabie–Sulu belt

The 2000 km long, 100–200 km wide Qinling–Dabie–Sulu orogen in eastern China defines the EW trending junction between the Sino-Korean craton to the north and the Yangtze craton to the south (Huang, 1978; Dong et al., 1986; Dong, 1993). The central segment of this tectonic zone consists of fault-bounded, penetratively deformed blocks. From south to north, these are: (1) Proterozoic, uppermost Proterozoic (Sinian), and lower Paleozoic metavolcanogenic, metaclastic, and metacarbonate strata, but including lesser amounts of bimodal metavolcanic + metasedimentary rocks of late Proterozoic age, all resting on latest Archean to early + middle Proterozoic, dominantly quartzofeldspathic gneisses of the Yangtze basement; (2) orthogneisses and migmatites, chiefly of early Cretaceous age of origin and/or metamorphism; and (3) feebly recrystallized metaclastic Foziling and spatially associated greenschists. The first two groups represent northerly portions of the Yangtze craton, whereas Foziling rocks accreted to the southern margin of the Sino-Korean craton (Ernst et al., 1991; Liou et al., 1995; Hacker et al., 1996). Except in the Dabie and Hong'n blocks, the dominant foliation dips steeply to the north, and folds verge to the south (Huang and Wu, 1992; Hacker et al., 1995). In these latter areas, late-stage deformation may be analogous to the back folding characteristic of the western Alps (e.g. Dal Piaz et al., 1972). Generalized geologic relations in the Dabie Shan + Hong'an region are depicted in Fig. 2.

Within the southern part of the Hong'an block, HP metamorphic rocks contain, from south to north, weakly reconstituted chlorite-actinolite schists, grading to transitional blueschist-greenschist facies assemblages; farther north, greenschist, epidote amphibolite, and eclogite facies assemblages progressively replace lower grade parageneses (Zhang and

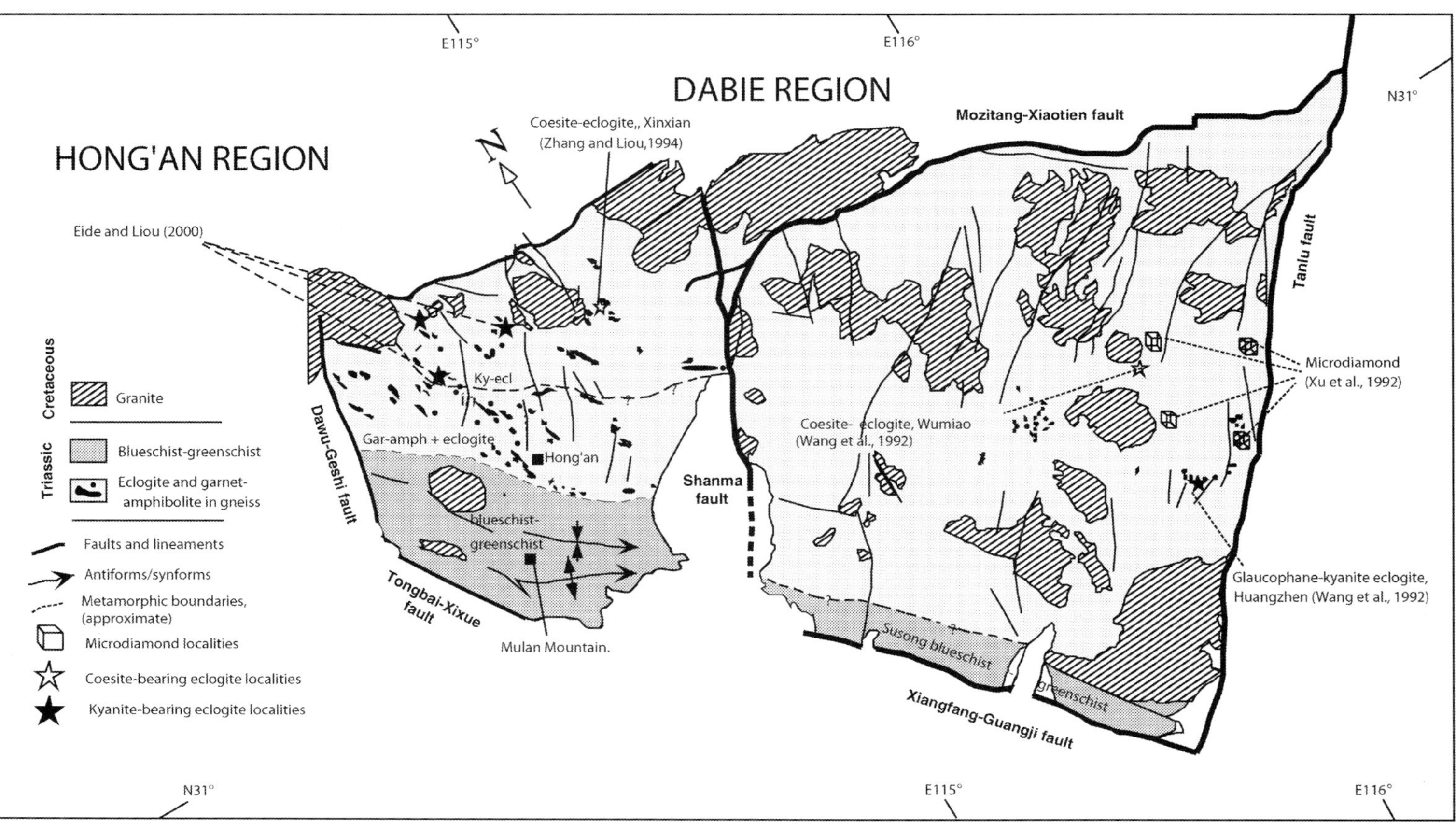

Fig. 2. Sketch of regional geologic relationships in the central sector of the Qinling–Dabie–Sulu belt, eastern China, modified after Eide and Liou (2000). Only igneous and metamorphic rocks of the Yangtze craton are shown.

Liou, 1994; Eide, 1995; Eide and Liou, 2000). Actinolite, crossite, and Mg-riebeckite associated with chlorite in the south give way to winchite and farther north, to barroisite or omphacite + garnet, possibly a synchronous paragenetic sequence. However, high-grade amphibolite-facies metamorphism in the northern part of the terrane has thoroughly overprinted precursor HP assemblages. Phase-equilibrium data, mineral chemistry, and thermobarometry indicate conditions for the earlier metamorphism (Liou et al., 1996): blueschist-greenschist = 325–425°C, 0.5–0.7 GPa; eclogites = 500–520°C, 1–1.2 GPa. Similar but lower grade parageneses are exposed to the WNW in the Qinling Mountains.

Coesite micro-inclusions in garnet, zoisite, zircon, kyanite, dolomite, and omphacite occur in the central part of the belt, confined to UHP eclogitic rocks of the Hong'an, Dabie Shan, and Sulu blocks (Okay et al., 1989; Wang et al., 1989; Zhang et al., 1994, 1995); associated paragneisses, schists, and calcsilicate rocks contain coesite inclusions in zircon (Sobolev et al., 1994; Tabata et al., 1998) and in dolomite (Schertl and Okay, 1994; Zhang and Liou, 1996), and coesite pseudomorphs in garnet (Wang and Liou, 1991). Very rare microdiamonds have been reported from garnetiferous UHP rocks (Xu et al., 1992; Okay, 1993). Phase-equilibrium data and thermobarometry indicate that Dabie coesite-bearing eclogites and enclosing country-rock gneisses ± marbles formed at pressures >2.8 GPa, diamond-bearing lithologies at pressures exceeding 3.5 GPa, both at temperatures of 750°C or more. Based on U–Pb, Sm–Nd, and $^{40}Ar/^{39}Ar$ geochronologic data, ages of the HP/UHP metamorphic mineral assemblages are Triassic, 240–245 Ma, with exhumation to crustal levels by about 230 Ma (Okay et al., 1993; Ames et al., 1996; Rowley et al., 1997; Hacker et al., 1998, 2000; Webb et al., 1999). This assemblage has been intensely overprinted by high-grade amphibolite-facies assemblages, and UHP phases are present as rare relics; midcrustal annealing apparently took place at the latest by about 200 ± 5 Ma. Variably deformed plutons, orthogneisses, and migmatites, principally of early Cretaceous age, lie directly north of, and intrude, the UHP belt (Hacker et al., 1996, 1998). Eastward, across the NNE-trending Tan-Lu fault, similar HP/UHP mafic and quartzofeldspathic assemblages crop out in the Sulu area (Hirajima et al., 1990; Wang et al., 1993; Ye et al., 1996; Zhang and Liou, 1998).

Blueschist + eclogite mineral parageneses, bulk-rock + phase compositions, and regional structural relations of the Qinling–Dabie–Sulu belt suggest formation attending northward underflow of the Yangtze continental margin beneath the Sino-Korean craton after the subduction of an intervening ocean basin. No fragments of this inferred ophiolitic crust are preserved, but remnant magnetic studies demonstrate the appropriate latitudinal closure (Yang et al., 1992; Nie and Rowley, 1994). Accretion of the Yangtze craton to the enlarging Asiatic continental assembly evidently took place in the late Triassic (Liou et al., 1996). Especially profound subduction that produced the UHP Qinling–Dabie–Sulu metamorphic complex apparently occurred as a consequence of plate descent involving a narrow, cool microcontinental salient (Yin and Nie, 1993) entrained with negatively buoyant ocean crust-capped lithosphere. Decoupling of the subducted continental crust at great depth ensued, followed by differential return toward the surface of deformed slices. Delayed separation from the sinking oceanic lithosphere could account for the high-temperatures of some of the recovered terranes; episodic ascent of individual low-density sheets after different storage intervals may explain the contrasting metamorphic facies and later overprinting in different segments of the orogen (Ernst and Liou, 1995). Moreover, Maruyama et al. (1994) and Maruyama (1997) postulated the subhorizontal extrusion of thin, fault-bounded allochthons of HP and UHP rocks to account for their present distribution in the Dabie Shan. Hacker et al. (1995) proposed a tectonic model involving the northward emplacement of a UHP nappe on the order of 5 km (10 km maximum) thickness, which arrived at midcrustal levels by early Jurassic time. In any case, all Dabie–Sulu tectonic models invoke relatively thin-slab geometries for the UHP and HP lithostratigraphic units, in order to account for the observed structural relationships and *P–T* histories (Hacker and Peacock, 1995; Ernst and Peacock, 1996; Hacker et al., 1998).

2.2. The Kokchetav complex

The Kokchetav complex is an isolated, 300 km × 150 km, fault-bounded massif within the central

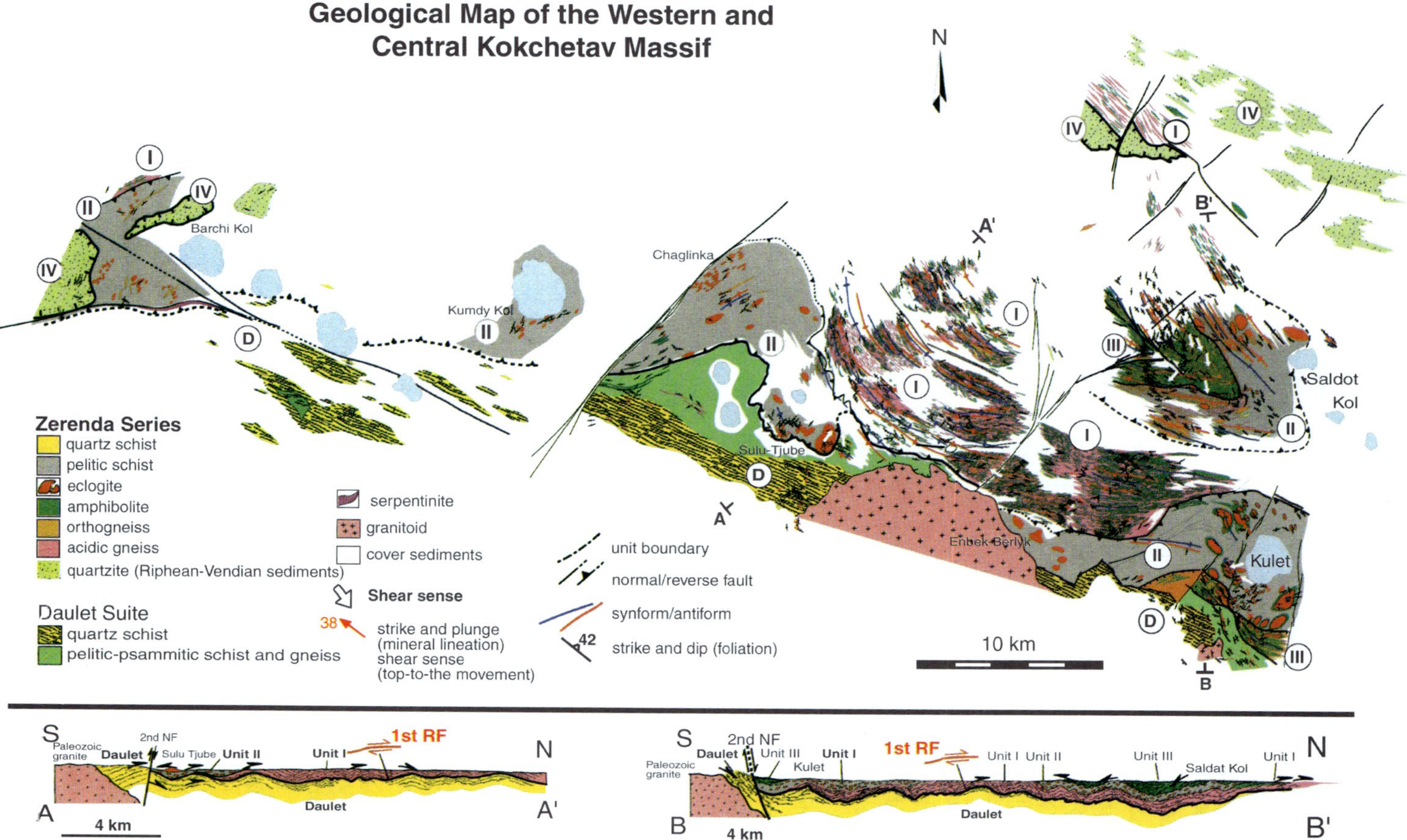

Fig. 3. (a) Provisional version of the regional geology and (b) north-south structural cross-sections AA′ and BB′ across the Kokchetav complex, northern Kazakhstan (Kaneko et al., 1998; Ishikawa et al., 1998; Maruyama and Parkinson, 2000), emphasizing imbricate nappe structures. RF, reverse fault; NF, normal fault. An alternative, contrasting structural scheme involving high-angle tectonic contacts has been presented by Dobretsov (1998) and Theunissen et al. (2000). In this map and cross-sections by Maruyama and Parkinson (2000), their Unit I is roughly equivalent to units I and II of Dobretsov (1998); the unrelated Daulet is illustrated here as D.

Asiatic accretionary collage. It lies between the Siberian platform to the NE, the Russian craton + Ural Mountains to the west, and the northern margin of the Kazakhstan–north Tienshan microcontinent on the south (Hamilton, 1970; Dobretsov and Sobolev, 1970, 1984; Zonenshain et al., 1990; Sengör et al., 1993). The complex comprises dislocated, reassembled fragments of pre-late Riphean continental crust, consisting of a penetratively deformed Precambrian basement core surrounded by a Phanerozoic fold belt. The amalgamated assembly of differentially exhumed blocks lies between two major WNW-trending transcurrent shear zones. Gneissic basement possesses 2.2–2.3 Ga Sm–Nd model ages (Dobretsov et al., 1995), and is overlain by upper Proterozoic–Vendian platform strata. REE patterns for the metasedimentary units suggest a passive margin setting (Dobrzhinetskaya, 1994). Middle and upper Paleozoic multicycle platform and arc/backarc strata surround the complex, reflecting the Devonian initiation of a postorogenic supracratonal basin. Syntectonic granitoids are chiefly of late Ordovician age, but include late Cambrian gabbroic precursors; younger anorogenic granitoids have invaded the complex, especially along its southern margin (Fig. 3).

Dobretsov et al. (1995) described a tectonic megamelange of pre-Ordovician units (I–VI) and including undifferentiated gneiss (VII), invaded by post-recrystallization granitoids and overlain by mid-Paleozoic and younger sedimentary strata. Diamond-bearing unit I crops out over an area of 15 km × 50 km. Detailed studies of the mineralogy and bulk-rock geochemistry of the HP/UHP rocks have been carried out by numerous investigators (Shatsky et al., 1989, 1991, 1995; Shatsky and Sobolev, 1993; Sobolev and Shatsky, 1990; Sobolev et al., 1986, 1991; Dobrzhinetskaya et al., 1994). Paraschists and paragneisses containing biogenic carbon were laid down at the Earth's surface (Sobolev and Sobolev, 1980; Chopin and Sobolev, 1995). Microdiamonds occur as inclusions in strong, refractory, neoblastic garnet, zircon, clinopyroxene, and kyanite in metasedimentary unit I. Rare coesite micro-inclusions and pseudomorphs have been discovered in zircon and garnet of the interlayered biotite gneiss and within eclogitic zircon (Sobolev et al., 1994; Dobretsov et al., 1995). Coesite has also been discovered in white schists of unit II (Zhang et al., 1997).

Metamorphism of paragneisses and associated units occurred in early Cambrian time, as indicated by Sm–Nd and U–Pb ages of 530–540 Ma for the UHP eclogitic and diamond-bearing metasedimentary rocks (Jagoutz et al., 1991; Claoué -Long et al., 1991; Shatsky et al., 1993). Magnesite + diopside pairs, coesite, pyropic garnet, potassic clinopyroxene, Si-rich phengite, barroisite and/or crossite, and Al-rich rutile ± titanite, and the assemblage talc + kyanite + pyropic garnet all testify to elevated pressures of crystallization in units I and II. *P–T* conditions involved pressures of about 4 GPa and temperatures exceeding 900°C (Sobolev et al., 1986, 1991; Sobolev and Shatsky, 1990; Shatsky et al., 1989, 1991, 1993, 1995, 1999). Adjacent units III, V, and, far to the south, unit VI, contain mineralogic evidence of lower *P–T* conditions. The metamorphosed lithotectonic units represent stages in what may initially have been a synchronous paragenetic series, but intense postmetamorphic tectonic disruption has resulted in a chaotic mixture. Only units I–III and probably parts of V and VI represent portions of this dismembered HP/UHP subduction zone paragenesis. The spatially associated sillimanite + cordierite-bearing Daulet is unrelated to the HP/UHP paragenesis.

Late Vendian–early Cambrian subduction of the Kazakhstan–north Tienshan microcontinental salient to depths exceeding 125 km, followed by decoupling from the descending oceanic crust-capped lithosphere is hypothesized to account for the UHP metamorphism of the Kokchetav complex (Dobretsov et al., 1995). Dimensions of the UHP unit are poorly constrained due to limited exposures, but judging from the outcrop pattern (Dobretsov, 1998), it represents a tectonic slice which cannot be thicker than ~5 km, and may be much thinner. Recent mapping has provided additional mineralogic, structural, and petrologic information (Kaneko et al., 1998; Ishikawa et al., 1998). Low-angle tectonic contacts have been documented, and are illustrated in Fig. 3. Provisional conclusions from the new field and laboratory studies are as follows (Maruyama and Parkinson, 2000): the HP/UHP terrane is structurally overlain by a weakly recrystallized unit, and is underlain by the low-pressure Daulet suite. These two major tectonic boundaries are subhorizontal, and are locally offset by high-angle normal faults. NE-SW and NW-SE trending, vertical strike-slip faults suggest post-orogenic transpiration. Fabric analyses confirm the presence of systematic normal faulting along the top and reverse faulting along the bottom of the HP/UHP slice. The structurally intermediate zone contains the highest pressure diamond-bearing assemblages. The HP/UHP complex apparently roots to the south. However, Dobretsov (1998) and Theunissen et al. (2000) have reported dominantly near-vertical tectonic contacts in the megamelange, so structural interpretations at this stage must be regarded as provisional.

Exhumation to midcrustal levels and cooling of the megamelange sheets evidently took place by about 515–517 Ma, according to $^{40}Ar/^{39}Ar$ ages on white mica and biotite from retrogressed garnetiferous paragneiss (Troesch and Jagoutz, 1993). Thus, the presently exposed complex decompressed relatively rapidly, and by late Cambrian time had been stabilized as part of the Kazakhstan–north Tienshan crustal collage.

2.3. The western Alps

The Alps constitute the world's most intensively studied contractional orogen. Large-scale structures and lithologies of the various units testify to the consumption of an intervening seaway, Mesozoic Tethys, and subduction of the European foreland beneath the stable, non-subducted African plate. Metamorphic assemblages reflect increasing depths of descent of the European continental margin and Tethyan oceanic crust progressing internally, toward and beneath the Po plain. The diachronous mineral-facies sequence ranges from anchimetamorphic to eclogitic, reflecting the southeastern polarity of subduction (Ernst, 1971; Frey et al., 1974). Throughout the Alps, the earlier HP parageneses have been largely overprinted by later greenschist-facies assemblages. Although UHP recrystallization has been recognized at isolated localities near Alpe Arami in the southernmost central Alps (Ernst, 1977; Evans et al., 1979; Becker, 1993), and the Zermatt area (Reinecke, 1991), the most thoroughly documented UHP tract crops out around the Dora Maira Massif (Chopin, 1984; Chopin et al., 1991; Schertl et al., 1991; Chopin and Schertl, 1999).

Similar to other Alpine internal complexes, the lowermost structural unit of the Dora Maira Massif consists of Hercynian (approximately 300 Ma) and older European granodioritic gneisses + metagranites (Monie and Chopin, 1991; Tilton et al., 1997), and an autochthonous cover series of Carboniferous graphitic and conglomeratic schists, Permo-Triassic metaclastics, and middle Triassic carbonate strata (Michard, 1965, 1967; Vialon, 1966; Borghi et al., 1985). The central massif is surrounded and overlain by allochthonous Pennine nappes, consisting of Mesozoic Tethyan oceanic crust and associated deep-water limestones + calcareous graywackes; slices of orthogneiss and polymetamorphic schist, constituting portions of the pre-Alpine basement, are also present. Tectonically higher Austroalpine nappes and the Sesia–Lanzo zone represent allochthonous nappes of the more southerly continent. The lowermost levels of both thrust sheets exhibit the effects of HP metamorphism; accordingly, at least basal slices of the African lithospheric plate must have decoupled from the parental slab and were dragged down with the more northerly subducted terrane (Dal Piaz et al., 1972; Ernst, 1973). Geologic and tectonic relationships are presented schematically in Fig. 4.

The Dora Maira Massif consists of several fault-bounded, subhorizontal lithotectonic packages (Henry, 1990; Chopin et al., 1991; Avigad, 1992). The basal Pinerolo nappe is made up of Carboniferous orthogneisses, Permo-Triassic and Triassic metasedimentary rocks + Permian diorites, incipiently recrystallized to blueschist-facies assemblages (Borghi et al., 1985). The Venasca complex, thrust over the basal nappe, consists of several juxtaposed slices of late Paleozoic, polymetamorphic schist and orthogneiss. The lower part of this unit includes boudinaged quartzite layers that contain coesite inclusions in pyropic garnets, but rare UHP relict assemblages also are scattered throughout fine-grained orthogneisses, and in pelitic schists + eclogitic pods (Michard et al., 1995). This 1–2 km thick, tabular allochthon has been juxtaposed against a structurally higher portion of the Venasca nappe; the latter contains HP eclogitic pods, but apparently is devoid of UHP phases. The overlying polymetamorphosed Dronero, similar to the Venasca in terms of protolith compositions, retains traces of pre-Alpine amphibolite-facies recrystallization (Compagnoni et al., 1995), and was later recrystallized under HP metamorphic conditions. Major P–T discontinuities, thus, characterize tectonic contacts of the Venasca complex with underlying and overlying nappes. Faulting has placed

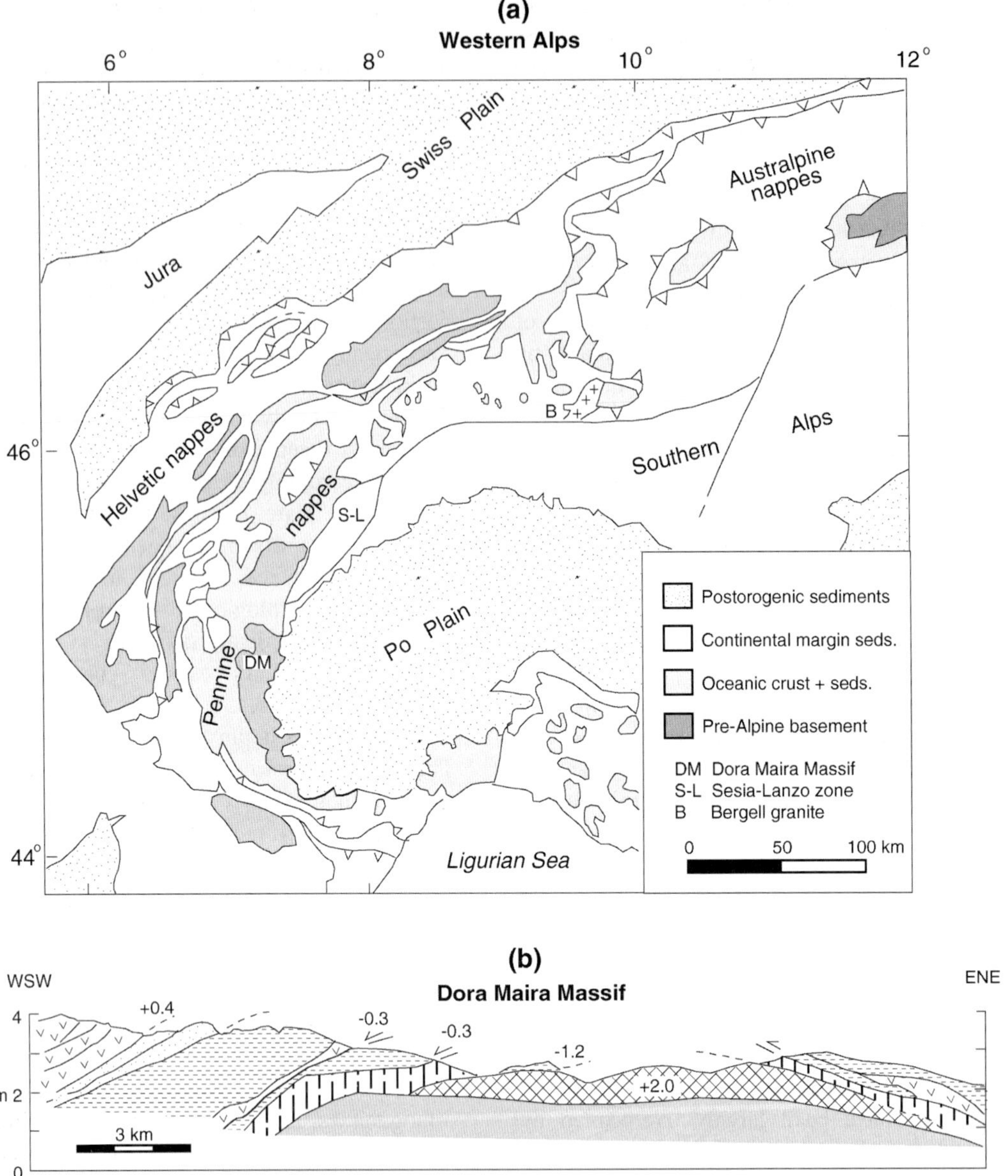

Fig. 4. (a) Regional and (b) local structural + metamorphic relationships in the Dora Maira Massif of the western Alps, simplified after Dietrich et al. (1974), Henry (1990), Compagnoni et al. (1995), Michard et al. (1995) and Coleman and Wang (1995). For the diagrammatic cross-section in the southern part of the Dora Maira Massif, numbers indicate the upward change in recorded pressure (GPa) relative to the adjacent underlying unit. Local nappe units are distinguished by different patterns in (b); see text for descriptions.

a thin slice of Pennine ophiolite of the Monte Viso complex + younger Mesozoic metasediments over the Venasca (Philippot, 1988; Henry, 1990; Blake et al., 1995). In turn, this oceanic nappe has been overthrust by a crustal sheet consisting of polymetamorphic schists and orthogneisses, associated with metaclastics of roughly mid-Mesozoic depositional age; both allochthons exhibit HP mineral parageneses.

Bulk-rock compositions of quartzite boudins that retain coesite and coesite pseudomorphs are highly magnesian. The UHP assemblage consists of pyrope + kyanite + talc + phengite + coesite (now mostly quartz). The quartzites occur within fine-grained gneisses consisting of the assemblage albite + microcline + phengite + biotite + quartz + clinozoisite, along with armored, back reacted grossular and rutile. Mafic eclogite assemblages within the Venasca nappe record the Alpine HP event, but lack coesite relics. Other phases compatible with the UHP assemblages in the quartzites, however, suggest coherence of the terrane during isofacial UHP Alpine recrystallization (Chopin, 1981, 1984; Kienast et al., 1991; Reinecke, 1991). Peak metamorphism of the Dora Maira Massif is estimated at 2.8–3.5 GPa and 700–750°C for the coesite-bearing quartzites (Schertl et al., 1991; Compagnoni et al., 1995). Thermobarometric measurements (Michard et al., 1995) suggest that eclogites of other Venasca units reflect conditions on the order of 1.7–2 GPa and 550–580°C. Eclogitic assemblages of the Monte Viso nappe recrystallized at ~1.2 GPa and 500–550°C. Fine-grained quartzofeldspathic gneisses enclosing the UHP layers of pyropic quartzite carry mineral assemblages indicating overprinting at about 0.3–0.6 GPa and 350–500°C. The earlier conditions reflect tectonic assembly of various lithostratigraphic units within a subduction zone, whereas the much lower pressures of retrograde metamorphism attest to cooling accompanying exhumation of the imbricated complex to midcrustal levels.

At least two distinct stages of relatively HP metamorphism are present in the Alps. Underflow and continental collision was initiated in the Cretaceous (90–125 Ma), and generated HP parageneses, especially in the western Alps (Paquette et al., 1989; Monie and Chopin, 1991). Oligocene–Miocene gneissic and granitoid domal uplifts record a widespread Tertiary thermal event. Relict UHP mineral assemblages also appear to have been produced at 35–40 Ma (Tilton et al., 1991; Gebauer et al., 1993, 1997). A brief Paleogene episode of intracontinental convergence evidently caused renewed recrystallization and the remarkably rapid uplift (5–10 million years) of deeply buried Alpine rocks (Gebauer, 1996).

2.4. The Western Gneiss Region

Along the SW coast of Norway, rocks of the Western Gneiss Region mark the zone of collision between the eclogite-bearing Greenland continental-crust-capped lithospheric plate and Fennoscandia during the Caledonian orogeny (Gilotti, 1993; Krogh and Carswell, 1995; Breuckner et al., 1998). The gneiss complex constitutes the basal tectonic member in a sequence of nappes that were thrust eastward over the Fennoscandian passive margin. These allochthonous sheets occupy an area 350 km in length and up to 150 km in width. Protoliths are middle Proterozoic or older, but polymetamorphic nappes as young as Vendian and/or early Paleozoic constitute portions of the imbricate stack (Kullerud et al., 1986). Two principal lithotectonic entities are recognized in the Western Gneiss Region (Krogh, 1977; Cuthbert et al., 1983; Griffin et al., 1985; Mörk and Krogh, 1987): (1) the HP/UHP metamorphosed Fjordane complex, a mixed assembly of supracrustals, anorthosites, and augen gneisses on the north and west; and (2), the Josuedl complex, an eastern, more coherent migmatitic orthogneiss unit that recrystallized under lower pressure conditions. Recent mapping in the north-central segment of the Western Gneiss Region by Robinson and Terry (1998) indicates the presence of a sequence of nappes, individual units being no more than 1–2 km thick.

Mafic phase assemblages are dominantly omphacite + garnet, with minor quartz, rutile, phengite, and/or kyanite; coesite has been documented as rare micro-inclusions within garnet and clinopyroxene in eclogite from westernmost exposures of the Fjordane complex (Smith, 1984; Smith and Lappin, 1989; Wain, 1997). Traces of microdiamond in kyanite + garnet + biotite gneiss exposed to the north on Fjörtoft Island (Dobrzhinetskaya et al., 1995; Smith, 1995) further supports the UHP nature of these rocks. Metamorphic conditions for the parautochthonous coesite ± diamond-bearing rocks of the Western Gneiss Region are estimated as 700–850°C at 3–3.5 GPa (Carswell et al., 1985, 1999). Associated garnet peridotites reflect comparable conditions of formation (Krogh and Carswell, 1995). Similar *P–T* values of 805 ± 40°C and 3.2 ± 0.2 GPa have been determined by van Roermund et al. (1998) for peridotites at Otroy; moreover, these authors described textural evidence suggesting exsolution of a precursor majoritic component in the garnet, and proposed an earlier stage of asthenospheric upwelling from depths exceeding 185 km prior to incorporation of mantle material in the UHP complex. Structural concordance between the HP and UHP lithologies in gneisses, metaperidotites, meta-anorthosites, and augen gneisses supports the concept of a regional metamorphic event within the Fjordane and Josuedl complexes. Estimated conditions attending the relatively HP event range from 500°C and 1–1.2 GPa in the east, to 775 ± 75°C at 3–3.5 GPa in the extreme west (Krogh, 1977; Griffin et al., 1985; Medaris and Wang, 1986; Smith and Lappin, 1989). Metamorphism accompanied collision of the Greenland and Fennoscandian cratons. Thermal softening promoted decoupling, and ascent of quartzofeldspathic crustal slices to midcrustal levels apparently was driven by buoyancy (Dewey et al., 1993). Overprinted mineral assemblages attest to a retrograde P–T trajectory involving nearly isothermal decompression from eclogite to granulite- or high-grade amphibolite-facies conditions; such relationships suggest a rather rapid, virtually adiabatic rise of the gneiss complex.

UHP recrystallization can be no older than the early Paleozoic polymetamorphic gneiss complexes, the youngest protoliths involved in the Caledonian continental collision. Ummetamorphosed upper Devonian strata fill fault-bounded basins along the Norwegian coast, providing a younger age limit to the metamorphism (Cuthbert et al., 1983). Based on geochronologic data employing various isotopic systems, Krogh and Carswell (1995) concluded that the metamorphic culmination (i.e. the thermal maximum) in the Western Gneiss Region took place at about 415 ± 10 Ma, with cooling and isotopic readjustment continuing until ∼395 Ma. These data suggest that approximately 20 million years elapsed between peak metamorphism and final unroofing of the eclogitic complex. Exhumation may have reflected imbricate slab rise, involving extensional ascent beneath the overlying, non-UHP Caledonian nappes, and low-angle reverse thrusting above the Fennoscandian basement (Dewey et al., 1993, 1998).

3. Physicochemical histories of UHP complexes

3.1. Constraints on subduction and exhumation provided by UHP terranes

Similarities among the four Eurasian UHP occurrences described above are listed in Table 2. These data support the following petrotectonic evolution. During Phanerozoic plate convergence, the underflow of old continental material — carried along as an integral part of chiefly oceanic lithosphere — resulted in

Table 2
Summary of UHP metamorphic complexes (modified after: Ernst and Peacock (1996); Robinson and Terry (1998); Hacker et al. (2000); Maruyama and Parkinson (2000))

	Qinling–Dabie–Sulu belt, coesite-eclogite	Kokchetav complex, UHP unit	Dora Maira Massif, lower Venasca	Western Gneiss Region, Fjordane complex
Protolith formation age (Ga)	1.3–2.9	2.2–2.3	∼300[a]	1.6–1.8
Temperature of UHP metamorphism (°C)	750 ± 75	900 ± 75	725 ± 50	775 ± 75
Depth of UHP metamorphism (km)	90–120	∼150	90–120	90–125
Time of UHP metamorphism (Ma)	240–245	530–540	35–40	410–420
Midcrustal annealing (Ma)	230 ± 5	515–517	30	∼395
Rise time to midcrust (million year)	15 ± 5	20 ± 5	5–10	20 ± 10
Exhumation rate[b] (mm per year)	5–10	7–8	9–12	5–7
Coesite inclusions	Relative abundant	Rare	Relative abundant	Rare
Diamond inclusions	Very rare	Relative abundant	Absent	Very rare
Blueschists	Present	Rare	Present	Absent
Areal extent (km)	400 × 75	∼100 × 15	225 × 60	350 × 70
Maximum thickness of UHP units (km)	5	1–3	1–2	>1

[a] This value is measured in Ma.

[b] Average exhumation rates were estimated by dividing depth of UHP metamorphism by time of ascent to about 10–15 km midcrustal depth (Coleman and Wang, 1995; Harley and Carswell, 1995).

the deep descent of low-density quartzofeldspathic ± metapelitic materials. Metamorphism of the crust took place at 90–125 km depth. The significance of individual phases and mineral parageneses described in previous sections can be evaluated using experimental phase-equilibrium data and thermobarometric constraints for studied bulk-rock compositions (Massone, 1995, 1999). Very low *P–T* gradients of 5–10°C/km are required for deeply buried rocks to enter the stability fields of some of the observed phases and phase associations. Such terrestrial HP and UHP environments are generated only in the vicinity of downgoing lithospheric plates, as shown by numerical computations (Oxburgh and Turcotte, 1971; Peacock, 1995). Even lower gradients typical of plate interiors are not characteristic of the upper boundary of the descending lithosphere because of frictional dissipation and thermal conduction from the overlying, non-subducted mantle wedge.

Times and rates of exhumation to midcrustal levels for the Qinling–Dabie–Sulu, Kokchetav complex, Dora Maira, and Western Gneiss terranes lie in the range 5–20 million years, 5–12 mm per year. Metamorphic overprinting at lower pressures probably reflects attainment of neutral buoyancy by the ascending subduction complexes, and prolonged storage at midcrustal levels. Exposed areas of the studied UHP terranes vary considerably. Although poorly constrained, evidence suggests that decompression of the most deeply subducted lithotectonic sections was relatively rapid, that exhumed tectonic units are comprised dominantly of old quartzofeldspathic crust, and that typically the recognizable UHP allochthons seem to be thin, tabular sheets. All supracrystal rocks subjected to UHP metamorphism appear to have been portions of pre-existing, relatively cool cratons, rather than segments of young, relatively warm oceanic crust or calcalkaline arcs.

Among the UHP terranes described above, diamond-bearing rocks of Kokchetav unit I reached the highest temperatures. Thus, it seems likely that the scarcity of coesite reflects elevated reaction rates during unloading at 900°C in the system SiO_2, compared with the more sluggish kinetics of the diamond-graphite transformation; in contrast, white schists of the lower grade Kokchetav unit II never reached temperatures exceeding 750°C, and accordingly retain relict coesite.

3.2. The role of aqueous fluid in subsolidus recrystallization

The systematic progression in subduction zone phase assemblages indicates that, in general, relict parageneses represent close approaches to chemical equilibrium at the time of formation. However, studies of several natural, coarse-grained, dry UHP assemblages have documented the astonishing metastable persistence of low-pressure mineral assemblages at great depths (Austrheim, 1990; Austerheim et al., 1997; Jamtveit et al., 1990; Hirajima et al., 1993; Wallis et al., 1997; Zhang and Liou, 1997; Katayama et al., 2000), and the metastable preservation of UHP phase assemblages and mineral inclusions during later decompression (Chopin, 1984; Smith, 1984; Sobolev and Shatsky, 1990; Liou and Zhang, 1996). Cases in which equilibrium was not achieved under extreme *P–T* conditions of recrystallization undoubtedly are due to the absence of a rate-enhancing aqueous fluid attending metamorphism (Rubie, 1986, 1990b; Brearley and Rubie, 1990; Mosenfelder and Bohlen, 1997). It seems likely that, for subducted rocks characterized by the presence of H_2O at depth, retrograde reactions would run to completion during exhumation, obliterating all evidence of precursor UHP assemblages. In contrast, dry lithologies might not recrystallize completely to the stable UHP assemblage during deep subduction, but where transformed, could well retain scattered UHP relics during ascent (Ernst et al., 1998; Parkinson, 2000). In addition, the great tensile strength of tough container minerals such as zircon ± garnet may allow the preservation of some UHP micro-inclusions on decompression of the host rock in part due to the "pressure-vessel" effect (Gillet et al., 1984; Van der Molen and van Roermund, 1986; Parkinson and Katayama, 1999).

3.3. The role of aqueous fluid in generation of arc magmas

Voluminous andesitic arcs and volcanic/plutonic continental margins are situated on the landward side of Pacific-type convergent plate junctions, but do not appear to have formed abundantly above most Alpine-type UHP intracontinental suture zones. Aqueous fluids driven off the downgoing lithospheric slab increase reaction rates, lower solids temperatures, and promote partial fusion of the warming subducted crust and/or the overlying, metasomatized mantle wedge, thereby generating calcalkaline magmas (Wyllie et al., 1996; Peacock and Wang, 1999). During shallow stages of plate descent, partial hydration of dry units occurs, and volatiles are expelled from wetter rocks, but the overall complement of volatiles decreases with increasing subduction depth (Frey, 1987). Phase equilibrium studies demonstrate that, under subduction zone geothermal gradients of 5–10°C/km, Ca- and/or Na-amphiboles are the principal volatile-bearing minerals in metabasaltic rocks to depths approaching 75–80 km (Liu et al., 1996). Other hydrous phases (e.g. lawsonite, chlorite, chloritoid, clinozoisite, biotite, and white micas) are absent or are of minor abundance in normal mafic lithologies (for a different interpretation, see Poli and Schmidt (1997)). Clinoamphiboles in rocks of MORB chemistry dehydrate at pressures in excess of 2.2–2.5 GPa under equilibrium conditions, but minor pressure overstepping is likely at the relatively low-temperatures characteristic of subducting plates. Therefore, mafic blueschists and amphibolites expel H_2O at greater depths, and commonly achieve the eclogitic assemblage of garnet + omphacite + rutile. Although the low density (about 2.5 g/cm^3) of fully serpentinized peridotite renders this rock essentially unsubductable, partly serpentinized mantle beneath the oceanic crust is carried down with the descending lithosphere; it devolatilizes at comparable to slightly higher pressures compared to amphiboles (Ulmer and Trommsdorff, 1995; Mysen et al., 1998), and may also contribute H_2O to the magmagenic zone.

In contrast, micas remain stable in granitic, metapelitic, and quartzofeldspathic gneisses to pressures far exceeding 4 GPa (Vielzeuf and Holloway, 1988; Vielzeuf and Montel, 1994; Luth, 1997). Thus, at depths substantially exceeding 100 km, micaceous lithologies that characterize the continental crust fail to evolve substantial amounts of H_2O during subduction and, because of sluggish reaction rates, may transform incompletely to the stable eclogite-facies assemblage of coesite + jadeite + K-feldspar + garnet + rutile + phengite ± biotite. Accordingly, although hydrous rock types expel volatiles during compaction/tectonic burial, very deep underflow involving hundreds — even thousands — of kilometers of hydrated oceanic lithosphere probably produces most of the UHP volatile flux along and above a subduction zone prior to continental collision. As continental masses enter a suture zone, devolatilization at deep levels severely diminishes, and due to buoyancy, underflow abruptly ceases or steps oceanward. Small amounts of peraluminous, muscovite-bearing, S-type anatectic melts may be produced, but I-type calcalkaline arcs owe their formation to the long-continued underflow of oceanic lithosphere, rather than being due to the briefer subduction of continental crust (Ernst, 1999). The contrasting roles played by aqueous fluids in metabasaltic ± serpentinized peridotite and quartzofeldspathic ± metapelitic protoliths seem to account for the general lack of calcalkaline arc rocks associated with UHP collisional orogens, as well as the kinetic effect involving more abundant production and preservation of UHP phases and mineral assemblages in mafic eclogites.

3.4. P–T time exhumation trajectories

In all four cases described earlier, a range of temperatures and pressures are preserved in the mineral assemblages, indicating stages of partial re-equilibration on ascent toward midcrustal levels. Incompleteness of back reaction testifies to absence of a rate-enhancing aqueous fluid, and/or the rapidity of exhumation. Rock and mineral oxygen isotope anomalies in eclogitic rocks from the Sulu block (Yui et al., 1995; Baker et al., 1997; Rumble, 1998) also indicate the total absence of an aqueous fluid during UHP metamorphism, and account for the failure of these rocks to totally re-equilibrate during uplift. Nevertheless, for sections undergoing nearly adiabatic decompression at 800 ± 100°C, residence times outside the stability fields of UHP minerals for tens of millions of years ought to have caused the total obliteration of such phases. Apparently, during initial stages of exhumation, some deeply subducted sections that retain traces of UHP precursors must have faithfully retraced the prograde *P–T* trajectory (Katayama et al., 2000). Such a retrograde *P–T* path could be realized by material moving back up the subduction channel-the zone between the overlying hanging wall and underlying footwall lithospheric plates-during continued plate descent (Ernst, 1988). For instance, in the Dora Maira Massif, talc + phengite remained stable during decompression from peak metamorphic conditions of 725 ± 50°C at depths of 90–120 km. This assemblage would dehydrate to phlogopite + kyanite + quartz + H_2O at approximately 600°C and 1 GPa, assuming $P_{\text{aqueous fluid}} = P_{\text{lithostatic}}$, and it should be relatively rapid running at these temperatures. The fact that the hydrous condensed assemblage remained stable constrains the maximum temperature of the Dora Maira terrane during decompression. Temperatures must have declined from a maximum of 725°C at 2.8–3.5 GPa to less than 500–600°C at 0.7–0.9 GPa accompanying greenschist-facies overprinting (Schertl et al., 1991; Chopin and Sobolev, 1995). The Dora Maira Massif evidently did not pass through conditions of the high-grade amphibolite or granulite facies during exhumation, however, other UHP terranes such as Kokchetav unit I (Dobretsov et al., 1995) and the Sulu block did (Wang et al., 1993).

3.5. Ultradeep subduction and partial preservation of buoyant regurgitated crustal slices

Researchers have demonstrated that profound subduction of continental crust accounts for the generation of UHP terranes. The petrotectonic challenge is to explain why these complexes were conducted so far down into the mantle, with portions containing preserved relics of the UHP phase assemblages then returned to shallow crustal levels. The "two-way street" nature of subduction zones was recognized long ago (Ernst, 1970; Suppe, 1972). Specifically to account for UHP terranes, salients, or peninsulas of old continental crust firmly attached to cool, oceanic-crust-capped plates, are conjectured to have descended rapidly due to net negative buoyancy of the lithospheric slab, generating the characteristic UHP mineralogy (Peacock, 1995; Ernst et al., 1997). Descent of extensive tracts of the complexes described earlier all represent tectonic regimes in which old continental crust constituted an integral part of the sinking oceanic crust-capped lithosphere.

The entrance of increasing volumes of strongly bonded, low-density quartzofeldspathic crust-capped lithosphere into the subduction zone would intensify the braking effect of buoyancy. This deceleration would promote loss of the high-density oceanic lithosphere pulling down the plate at intermediate upper mantle depths, a regime where the sinking lithosphere is in extension (Isacks et al., 1968). Slab breakoff (Sacks and Secor, 1990; von Blanckenburg and Davies, 1995) would enhance the buoyancy of the continental salient, or at least a slice thereof, allowing it to decouple from the descending lithosphere and slide back up the subduction channel. Exhumation would be enhanced by the reduction in shear forces acting along the base of the continental crust due to its more ductile behavior as it gradually warmed in the deep upper mantle. Because of continued subduction-induced refrigeration tectonically beneath the rising UHP complex, and extensional faulting against the overlying, cooler mantle wedge, relatively thin slices of UHP terranes should effectively lose heat by conduction along both upper and lower surfaces during ascent. Accordingly, some such complexes may nearly retrace the subduction zone P–T trajectory during decompression (Rubie, 1984; Ernst and Peacock, 1996). Generalized structural relationships are illustrated schematically in Fig. 5.

Even in favorable circumstances, only packets of rock sufficiently thin to be relatively efficiently cooled by thermal conduction along both upper and lower tectonic contacts would preserve evidence of an earlier UHP history. A mechanism involving ascent of thin tectonic slices is similar to the process advanced by Dobretsov (1991) and Dobretsov and Kirdyashkin (1994), as well as the tectonic-wedge-extrusion scenario proposed for the Dabie–Sulu UHP belt by Maruyama et al. (1994), and the contrasting Dabie Shan nappe models advanced by Okay et al. (1993) and by Hacker and Peacock (1995). Similar exhumation scenarios have been scale-modeled for Oman by Chemenda et al. (1996). Bulk-rock density contrasts in the various subducted units are more than adequate to promote the buoyancy-driven ascent of UHP metamorphosed quartzofeldspathic crust (Table 1). Substantially thicker masses of deeply subducted, low-density continental material would be even more readily propelled by body forces back to midcrustal levels, however, reflecting their large volume-to-surface-area ratios, such complexes would be characterized by thermal inertia. They undoubtedly would remain too hot during decompression to retain evidence of the pre-existing UHP conditions, except perhaps along their margins where conduction might lower temperatures sufficiently to allow retention of a border of UHP relics.

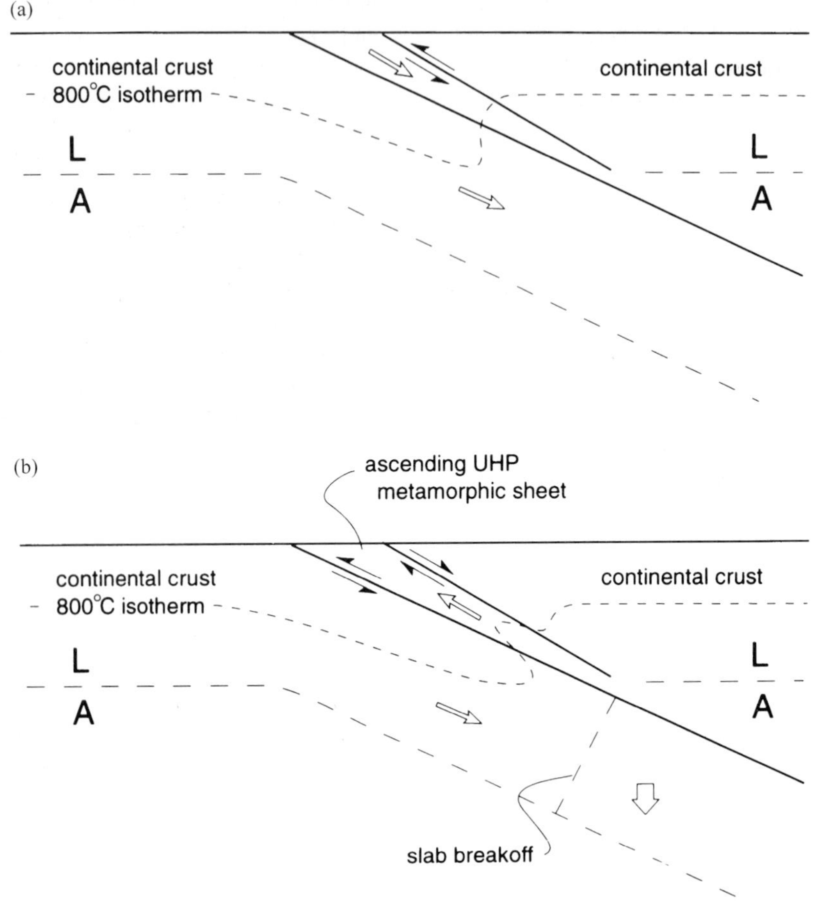

Fig. 5. Schematic representation of the thin-slab subduction–exhumation model proposed by Maruyama et al. (1996), Ernst and Peacock (1996), and Ernst et al. (1997). Diagrammatic thermal structure is illustrated: (a) accompanying lithospheric underflow (upper boundary of the UHP slab is a reverse fault); and (b), during continued plate subduction contemporaneous with ascent of a quartzofeldspathic crustal slice (upper boundary of the UHP slab is a normal fault; lower boundary is a reverse fault). L: lithosphere; A: asthenosphere. A similar but more detailed scale model has been presented by Chemenda et al. (1996).

4. Summary

In several well-documented Eurasian intracontinental collisional belts, promontories of continental crust, imbedded in oceanic lithosphere apparently have been carried down subduction zones at least 100 km or more attending plate consumption. Such crustal rocks were transformed, at least in part, to UHP phase assemblages at pressures exceeding 2.8 GPa and temperatures of ~750 ± 150°C. Rapid exhumation of thin sheets to midcrustal levels under dry conditions allowed the incomplete preservation of relict UHP minerals and textures. Geologic and structural relations, combined with mineral parageneses and experimental phase-equilibrium data, suggest several conclusions regarding the rates of prograde transformation, decompression recrystallization, and exhumation of portions of these UHP collisional complexes.

Subduction of oceanic and/or continental massifs at shallow depths results in the partial hydration of relatively dry units and in the evolution of volatiles from more hydrous lithologies. Attending this process, and the deeper, progressive evolution of volatiles, some H_2O diffuses into the overlying, non-subducted mantle wedge, whereas the rest of the aqueous fluid migrates back up the subduction channel, depending on the relative permeabilities of the lithologic units involved. Under anhydrous conditions, earlier assemblages may be preserved metastably. The presence or absence of a rate-enhancing aqueous fluid is, therefore, critical in determining the kinetics of prograde and retrograde reactions.

Within the subduction zone magmagenic environment, clinoamphiboles ± serpentine represent the principle sources of aqueous fluids expelled at depths greater than about 80 km. In contrast, layer silicates characteristic of granitic, quartzofeldspathic, and metapelitic materials remain stable to depths of at least 150–200 km. Thus, sustained underflow and devolatilization of oceanic-crust-capped lithosphere — not relatively brief consumption of continental crust — evidently is responsible for the generation of coeval andesitic arcs sited above the landward portions of convergent plate junctions. This H_2O-mediated process involves either partial melting of eclogitized oceanic crust or incipient fusion of the mantle hanging wall, or both. Pacific-type plate junctions, therefore, constitute the realm of true continental growth. Moreover, because the evolution of H_2O at UHP depths enhances transformation to the equilibrium phase configuration, MORB-type lithologies tend to more fully react to produce stable eclogitic assemblages whereas continental crust, being relatively dry, in some cases may metastably retain lower pressure minerals. These relationships help to account for the origin of circum-Pacific andesitic arcs, and the paucity of such belts in Alpine-type collisional orogens on the one hand, and the fact that UHP phases are especially rare in quartzofeldspathic lithologies on the other.

Salients of continental crust, integral portions of the downgoing, cool lithosphere, are sheared and disaggregated during thermal softening accompanying deep subduction, and portions may decouple from the oceanic plate in response to lithospheric slab break-off. The buoyancy of such slices drives them back toward midcrustal levels. Where these tectonically bounded entities are neither too thin to overcome frictional resistance accompanying differential ascent, nor too thick to cool efficiently by conduction along the faulted slab, UHP relics may survive the exhumation processes of decompression and annealing under more normal crustal P–T conditions. Suitable geometric configurations appear to consist of structural sheets 5±3 km thick. Their ascent velocities over approximately 5–20 million years to midcrustal levels involve average exhumation rates approaching10 mm per year or greater (see Perchuk et al., 1998; Webb et al., 1999). Similar rates appear to be characteristic of modern, active continental collisional culminations, as recently summarized by Blythe (1998). Because thick allochthons ascending from great depths would not cool efficiently by thermal conduction, the total obliteration of pre-existing UHP phase assemblages is to be expected. Thus, in collisional belts, only thin slabs retain fragmentary evidence of their antecedent deep subduction history.

Acknowledgements

This study was support by Stanford University, and by the US National Science Foundation through Grant no. EAR97-25347 to J.G. Liou. S. Maruyama and Chris Parkinson provided a prepublication copy of the structural geology of the Kokchetav complex. The first draft manuscript was reviewed by J.G. Liou, Ruth Zhang, and Mary Leech, the second draft paper by Paddy O'Brien and Jed Mosenfelder. I thank the above individuals and institutions for their help.

References

Ames, L., Zhou, G., Xiong, B., 1996. Geochronology and geochemistry of ultrahigh-pressure metamorphism with implications for collision of the Sino-Korean and Yangtze cratons, central China. Tectonics 15, 472–489.

Austrheim, H., 1990. The granulite-eclogite facies transition: a comparison of experimental work and a natural occurrence in the Bergen Arcs, western Norway. Lithos 25, 163–169.

Austerheim, H., Erambert, M., Engvik, A.K., 1997. Processing of crust in the root of the Caledonian continental collision zone: the role of eclogitization. Tectonophysics 273, 129–153.

Avigad, D., 1992. Exhumation of coesite-bearing rocks in the Dora Maira Massif (western Alps, Italy). Geology 20, 947–950.

Baker, J., Matthews, A., Mattey, D., Rowley, D.B., Xue, F., 1997. Fluid-rock interactions during ultrahigh-pressure metamorphism, Dabie Shan, China. Geochem. et Cosmochim. Acta 61, 1685–1696.

Beaumont, C., Ellis, S., Pfiffner, A., 1999. Dynamics of sediment subduction–accretion at convergent margins: short-term modes, long-term deformation, and tectonic implications. J. Geophys. Res. 104 (17), 573–17602.

Becker, H., 1993. Garnet peridotite and eclogite Sm–Nd mineral ages from the Lepontine dome (Swiss Alps): new evidence for Eocene high-pressure metamorphism in the central Alps. Geology 21, 599–602.

Blake Jr., M.C., Moore, D.E., Jayko, A.S., 1995. The role of serpentinite melanges in the unroofing of ultrahigh-pressure metamorphic rocks: an example from the western Alps. In: Coleman, R.G., Wang, X. (Eds.), Ultrahigh-Pressure Metamorphism. Cambridge University Press, New York, pp. 182–205.

Blythe, A.E., 1998. Active tectonics and ultrahigh-pressure rocks. In: Hacker, B.R., Liou, J.G. (Eds.), When Continents Collide: Geodynamics and Geochemistry of Ultrahigh-Pressure Rocks. Kluwer Academic Publishers, Dordrecht, pp. 275–295.

Borghi, A., Cadoppi, P., Porro, A., Sacchi, R., 1985. Metamorphism in the northern part of the Dora Maira Massif (Cottian Alps). Museo Regionale di Sci. Naturali di Torino, Bolletino 3, 369–380.

Bozhilov, K.N., Green, H.W., Dobrzhinetskaya, L., 1999. Clinoenstatite in Alpe Arami peridotite: additional evidence of very high-pressure. Science 284, 128–132.

Brearley, A.J., Rubie, D.C., 1990. Effects of H_2O on the disequilibrium breakdown of muscovite + quartz. J. Petrol. 31, 925–956.

Breuckner, H.K., Gilotti, J.A., Nutman, A., 1998. Caledonian eclogite-facies metamorphism of early Proterozoic protoliths from the north-east Greenland eclogite province. Contributions Mineral. Petrol. 130, 103–120.

Carswell, D.A., Zhang, R.Y., 1999. Petrographic characteristics and metamorphic evolution of ultrahigh-pressure eclogites in plate collision belts. Int. Geol. Rev. 41, 781–798.

Carswell, D.A., Krogh, E.J., Griffin, W.L., 1985. Norwegian orthopyroxene eclogites: calculated equilibration conditions and petrogenetic implications. The Caledonide Orogen. Wiley, New York, pp. 823–841.

Carswell, D.A., Cuthbert, S.J., Krogh Ravna, E.J., 1999. Ultrahigh-pressure metamorphism in the Western Gneiss Region of the Norwegian Caledonides. Int. Geol. Rev. 41, 955–966.

Chemenda, A.I., Mattauer, M., Bokun, A.N., 1996. Continental subduction and a mechanism for exhumation of high-pressure metamorphic rocks: new modelling, field data from Oman. Earth Planet. Sci. Lett. 143, 173–182.

Chopin, C., 1981. Talc-phengite: a widespread assemblage in high-grade pelitic blueschists of the western Alps. J. Petrol. 22, 628–650.

Chopin, C., 1984. Coesite and pure pyrope in high-grade blueschists of the western Alps: a first record and some consequences. Contributions Mineral. Petrol. 86, 107–118.

Chopin, C., Schertl, H.P., 1999. The UHP unit in the Dora Maira Massif, western Alps. Int. Geol. Rev. 41, 765–780.

Chopin, C., Sobolev, N.V., 1995. Principal mineralogic indicators of ultrahigh-pressures in crustal rocks. In: Coleman, R.G., Wang, X. (Eds.), Ultrahigh-Pressure Metamorphism. Cambridge University Press, New York, pp. 96–131.

Chopin, C., Henry, C., Michard, A., 1991. Geology and petrology of the coesite-bearing terrain, Dora Maira Massif, western Alps. Eur. J. Mineral. 3, 263–291.

Claoué -Long, J.C., Sobolev, N.V., Shatsky, V.S., Sobolev, A.V., 1991. Zircon response to diamond-pressure metamorphism in Kokchetav Massif. Geology 19, 710–713.

Cloos, M., 1993. Lithospheric buoyancy and collisional orogenesis: subduction of oceanic plateaus, continental margins, island arcs, spreading ridges, and seamounts. Geol. Soc. Am. Bull. 105, 715–737.

Coleman, R.G., Wang, X. (Eds.), 1995. Ultrahigh-Pressure Metamorphism. Cambridge University Press, New York, 528 pp.

Compagnoni, R., Hirajima, T., Chopin, C., 1995. UHPM rocks in the western Alps. In: Coleman, R.G., Wang, X. (Eds.), Ultrahigh-Pressure Metamorphism. Cambridge University Press, New York, pp. 206–243.

Cuthbert, S.J., Harvey, M.A., Carswell, D.A., 1983. A tectonic model for the metamorphic evolution of the Basal Gneiss complex, western south Norway. J. Metamorph. Geol. 1, 63–90.

Dal Piaz, G.V., Hunziker, J.C., Martinotti, G., 1972. La zona Sesia-Lanzo e l'evoluzione tettonico-metamorfica delle Alpi nordoccidentali interne. Memoir Soc. Geol. Italy 11, 433–460.

Deer, W.A., Howie, R.A., Zussman, J., 1966. An Introduction to the Rock-Forming Minerals. Wiley, New York.

Dewey, J.F., Ryan, P.D., Andersen, T.B., 1993. Orogenic uplift and collapse, crustal thickness, fabrics and metamorphic phase changes: the role of eclogites. In: Prichard, H.M., Alabaster, T., Harris, N.B.W., Neary, C.R. (Eds.), Magmatic Processes and Plate Tectonics. Geological Society of London, Special Publication No. 76, pp. 325–343

Dewey, J.F., Krabbendam, M., Wain, A., 1998. Transtensional exhumation of the UHP rocks of the Western Gneiss Region of southwest Norway. In: Proceedings of the International Workshop on Ultrahigh-Pressure Metamorphism and Exhumation. Stanford, CA, pp. A84–A85 (Abstract).

Dietrich, V., Vuagnat, M., Bertrand, J., 1974. Alpine metamorphism of mafic rocks. Schweizerisches Mineral. und Petrografisches Mittelungen 54, 291–333.

Dobretsov, N.L., 1991. Blueschists and eclogites: a possible plate tectonic mechanism for the emplacement from the upper mantle. Tectonophysics 186, 253–268.

Dobretsov, N.L. (Ed.), 1998. Geological Map of Kokchetav Region (northern Kazakhstan): Geophysics and Mineralogy. United Institute of Geology, Novosibirsk.

Dobretsov, N.L., Kirdyashkin, A.G., 1994. Blueschists of north Asia and models of subduction–accretion wedge. In: Coleman, R.G. (Ed.), Proceedings of the 29th International Geological Congress on Reconstruction of the Paleo-Asian Ocean, Part B. Japan, Kyoto, pp. 99–114.

Dobretsov, N.L., Sobolev, N.V., 1970. Eclogites from metamorphic complexes of the USSR. Phys. Earth Planet. Inter. 3, 401–424.

Dobretsov, N.L., Sobolev, N.V., 1984. Glaucophane schists and eclogites in the folded systems of northern Asia. Ofioliti 9, 401–424.

Dobretsov, N.L., Sobolev, N.V., Shatsky, V.S., Coleman, R.G., Ernst, W.G., 1995. Geotectonic evolution of diamondiferous paragneisses, Kokchetav complex, northern Kazakhstan: the geologic enigma of ultrahigh-pressure crustal rocks within a Paleozoic fold belt. The Island Arc 4, 267–279.

Dobrzhinetskaya, L.F., 1994. REE-patterns of metamorphic rocks from the Kokchetav microdiamond province of Kazakhstan and their implication for tectonics. Trans. Am. Geophys. Union 75, 701.

Dobrzhinetskaya, L.F., Braun, T.V., Sheshkel, G.G., Podkuiko, Y.A., 1994. Geology and structure of diamond-bearing rocks of the Kokchetav massif, Kazakhstan. Tectonophysics 233, 293–313.

Dobrzhinetskaya, L.F., Eide, E.A., Larsen, R.B., Sturt, B.A., Tronnes, R.G., Smith, D.C., Taylor, W.R., Posukhova, T.V., 1995. Microdiamond in high-grade metamorphic rocks of the Western Gneiss Region, Norway. Geology 23, 597–600.

Dobrzhinetskaya, L., Green, H.W., Wang, S., 1996. Alpe Arami: a peridotite massif from depths of more than 300 km. Science 271, 1841–1845.

Dong, S.B., 1993. Metamorphic and tectonic domains of China. J. Metamorph. Geol. 11, 465–481.

Dong, S.B., Shen, Q.H., Sun, D.Z., Lu, L.Z., 1986. Metamorphic Map of China (1:4,000,000) with Explanatory Text. Geological Publishing House, Beijing.

Eide, E.A., 1995. A model for the tectonic history of high- and ultrahigh-pressure metamorphic regions in east central China. In: Coleman, R.G., Wang, X. (Eds.), Ultrahigh-Pressure Metamorphism. Cambridge University Press, New York, pp. 391–426.

Eide, E.A., Liou, J.G., 2000. High-pressure blueschists and eclogites in Hong'an: a framework for addressing the evolution of ultrahigh-pressure rocks in central China. Lithos 52, 1–22.

Erambert, M., Austrheim, H., 1993. The effect of fluid and deformation on zoning and inclusion patterns in polymetamorphic garnets. Contributions Mineral. Petrol. 115, 204–214.

Ernst, W.G., 1970. Tectonic contact between the Franciscan melange and the Great Valley sequence, crustal expression of a late Mesozoic Benioff zone. J. Geophys. Res. 75, 886–901.

Ernst, W.G., 1971. Metamorphic zonations on presumably subducted lithospheric plates from Japan, California, and the Alps. Contributions Mineral. Petrol. 34, 43–59.

Ernst, W.G., 1973. Interpretive synthesis of metamorphism in the Alps. Geol. Soc. Am. Bull. 84, 2053–2078.

Ernst, W.G., 1977. Mineralogic study of eclogitic rocks from Alpe Arami, Lepontine Alps, southern Switzerland. J. Petrol. 18, 371–398.

Ernst, W.G., 1988. Tectonic history of subduction zones inferred from retrograde blueschist *P–T* paths. Geology 16, 1081–1084.

Ernst, W.G., 1999. Hornblende, the continent maker — evolution of H_2O during circum-Pacific subduction versus continental collision. Geology 27, 675–678.

Ernst, W.G., Liou, J.G., 1995. Contrasting plate-tectonic styles of the Qinling–Dabie–Sulu and Franciscan metamorphic belts. Geology 23, 353–356.

Ernst, W.G., Peacock, S.M., 1996. A thermotectonic model for preservation of ultrahigh-pressure phases in metamorphosed continental crust. Geophysical Monograph, Vol. 96. American Geophysical Union, pp. 171–178.

Ernst, W.G., Maruyama, S., Wallis, S., Buoyancy-driven, rapid exhumation of ultrahigh-pressure metamorphosed continental crust. In: Proceedings of the National Academy of Sciences, Vol. 94. 1997, pp. 9532–9537.

Ernst, W.G., Mosenfelder, J.L., Leech, M.L., Liu, J., 1998. H_2O recycling during continental collision: phase-equilibrium and kinetic considerations. In: Hacker, B.R., Liou, J.G. (Eds.), When Continents Collide: Geodynamics and Geochemistry of Ultrahigh-Pressure Rocks. Kluwer Academic Publishers, Dordrecht, pp. 275–295.

Ernst, W.G., Zhou, J.G., Liou, J.G., Eide, E., Wang, X., 1991. High-pressure and superhigh-pressure metamorphic terranes in the Qinling–Dabie mountain belt, central China; early- to mid-Phanerozoic accretion of the western Paleo-Pacific Rim. Pacific Sci. Assoc. Information Bull. 43, 6–15.

Evans, B.W., Trommsdorff, V., Richter, W., 1979. Petrology of an eclogite-metarodingite suite at Cima de Gagnone, Ticino, Switzerland. Am. Mineral. 64, 15–31.

Frey, M. (Ed.), 1987. Low-Temperature Metamorphism. Blackie Publishing, Glasgow, 351.

Frey, M., Hunziker, J.C., Frank, W., Bocquet, J., Dal Piaz, G.V., Jäger, E., Niggli, E., 1974. Alpine metamorphism of the Alps. Schweizerische Mineral. und Petrografisches Mittelungen 54, 248–290.

Gebauer, D., 1996. A *P–T* time path for an ultra high-pressure ultramafic/mafic rock-association and its felsic country-rocks based on SHRIMP-dating of magmatic and metamorphic zircon domains: example, Alpe Arami (central Swiss Alps). Geophysical Monograph, Vol. 95. American Geophysical Union, pp. 307–329.

Gebauer, D., Tilton, G.R., Schertl, H.P., Schreyer, W., 1993, Eocene/Oligocene ultrahigh-pressure (UHP) — metamorphism in the Dora Maira Massif (western Alps) and its geodynamic implications. In: Proceedings of the Fourth International Eclogite Conference on Terra Abstracts, Vol. 4. Calabria, Italy, pp. 10–11.

Gebauer, D., Schertl, H.P., Briz, M., Schreyer, W., 1997. 35 Ma old ultrahigh-pressure metamorphism and evidence for very rapid exhumation in the Dora Maira Massif, western Alps. Lithos 41, 5–24.

Gillet, P., Ingrin, J., Chopin, C., 1984. Coesite in subducted continental crust: *P–T* history deduced from an elastic model. Earth Planet. Sci. Lett. 70, 426–436.

Gilotti, J.A., 1993. Discovery of a medium-pressure eclogite province in the Caledonides of north-east Greenland. Geology 21, 523–526.

Griffin, W.L., Austerheim, H., Brastad, K., Bryhni, I., Krill, A.G., Krogh, E.A., Mörk, M.B.E., Qvale, H., Torudbakken, B., 1985. High-pressure metamorphism in the Scandinavian Caledonides. In: Gee, D.G., Sturt, B.A. (Eds.), The Caledonide Orogen. Wiley, New York, pp. 783–801.

Hacker, B.R., 1990. Amphibolite-facies to granulite-facies reactions in experimentally deformed unpowdered amphibolite. Am. Mineral. 75, 1349–1361.

Hacker, B.R., Peacock, S.M., 1995. Creation, preservation, and exhumation of UHPM rocks. In: Coleman, R.G., Wang, X. (Eds.), 1995. Ultrahigh-Pressure Metamorphism. Cambridge University Press, New York, pp. 159–181.

Hacker, B.R., Ratschbacher, L., Webb, L., Dong, S., 1995. What brought them up? Exhuming the Dabie Shan ultrahigh-pressure rocks. Geology 23, 743–746.

Hacker, B.R., Wang, X., Eide, E.A., Ratschbacher, L., 1996. The Qinling–Dabie ultrahigh-pressure collisional orogen. In: Yin, A., Harrison, T.M. (Eds.), The Tectonic Evolution of Asia. Cambridge University Press, New York, pp. 345–370.

Hacker, B.R., Ratschbacher, L., Webb, L., Ireland, T., Walker, D., Dong, S., 1998. U/Pb zircon ages constrain the architecture of the ultrahigh-pressure Quinling–Dabie orogen, China. Earth Planet. Sci. Lett. 161, 215–230.

Hacker, B.r., Ratschbacher, L., Webb, L., McWilliams, M.O., Ireland, T., Calvert, A., Dong, S., Wenk, H.-R., Chateigner, D., 2000. Exhumation of ultrahigh-pressure continental crust in east central China: late Triassic–early Jurassic tectonic unroofing. J. Geophys. Res. 105, 13339–13364.

Hamilton, W., 1970. The uralides and the motion of the Russian and Siberian platforms. Geol. Soc. Am. Bull. 81, 2553–2576.

Harley, S.L., Carswell, D.A., 1995. Ultradeep crustal metamorphism: a prospective review. J. Geophys. Res. 100, 8367–8380.

Henry, C., 1990. L'unite a coesite du Massif Dora Maira dans son cadre petrologique et structural (Alpes occidentales, Italie). Doctoral thesis, Université de Paris VI, Paris.

Hirajima, T., Ishiwatari, A., Cong, B., Zhang, R.Y., Banno, S., Nozaka, T., 1990. Coesite from Mengzhong eclogite at Dhonghai county, northeastern Jiangsu province, China. Mineral. Mag. 54, 579–583.

Hirajima, T., Wallis, S., Zhai, R., Ye, K., 1993. Eclogitized metagranitoid from the Su–Lu ultrahigh-pressure (UHP) province, eastern China. In: Proceedings of the Japan Academy, Section B, Vol. 69. pp. 249–254.

Huang, T.C., 1978. An outline of the tectonic characteristics of China. Eclogae Geologia Helvetiae 71, 611–635.

Huang, W., Wu, Z.W., 1992. Evolution of the Qinling orogenic belt. Tectonics 11, 371–380.

Isacks, B., Oliver, J., Sykes, L.R., 1968. Seismology and the new global tectonics. J. Geophys. Res. 73, 5855–5899.

Ishikawa, M., et al., 1998. Structures of ultrahigh-pressure terrain: Sulu-Tube, Kokchetav, Kazakhstan. Trans. Am. Geophys. Union 79, F983.

Jagoutz, E., Shatsky, V.S., Sobolev, N.V., 1991. The origin and history of ultrahigh *P–T* rocks from Kokchetav Massif. Terra Abstr. 3, 83.

Jamtveit, B., Bucher-Nurminen, K., Austrheim, H., 1990. Fluid controlled eclogitization of granulites in deep crustal shear zones, Bergen arcs, western Norway. Contributions Mineral. Petrol. 104, 184–193.

Kaneko, Y., et al., 1998. Geology of the Kokchetav UHP/HP Unit, northern Kazakhstan. Trans. Am. Geophys. Union 79, F983.

Katayama, I., Zayachkovsky, A.A., Maruyama, S., 2000, Progressive *P–T* records from inclusions in zircons from UHP/HP rocks of the Kokchetav Massif, northern Kazakhstan. The Island Arc, 9, 417–427.

Kienast, J.R., Lombardo, B., Biino, G., Pinardon, J.L., 1991. Petrology of very high-pressure eclogitic rocks from the Brossasco–Isasca complex, Dora Maira Massif, Italian western Alps. J. Metamorph. Geol. 9, 19–34.

Klein, C., Hurlbut, C.S., Jr., 1985. Manual of Mineralogy. Wiley, New York.

Krogh, E.J., 1977. Evidence for a Precambrian continent-continent collision in western Norway. Nature 267, 17–19.

Krogh, E.J., Carswell, D.A., 1995. HP and UHP eclogites and garnet peridotites in the Scandinavian Caledonides. In: Coleman, R.G., Wang, X. (Eds.), Ultrahigh-Pressure Metamorphism. Cambridge University Press, New York, pp. 244–298.

Kullerud, L., Tirudbakken, B.O., Ilebekk, S., 1986. A compilation of radiometric age determinations from the Western Gneiss Region, south Norway. Norges Geol. Undersökelse Bull. 406, 17–42.

Lin, C.H., Roecker, S.W., 1998. Active crustal subduction and exhumation in Taiwan. In: Hacker, B.R., Liou, J.G. (Eds.), When Continents Collide: Geodynamics and Geochemistry of Ultrahigh-Pressure Rocks. Kluwer Academic Publishers, Dordrecht, pp. 1–25.

Liou, J.G., 1999. Petrotectonic summary of less-intensively studied UHP regions. Int. Geol. Rev. 41, 571–586.

Liou, J.G., Zhang, R.Y., 1996. Occurrences of intergranular coesite in ultrahigh-*P* rocks from the Sulu region, eastern China: implications for lack of fluid during exhumation. Am. Mineral. 81, 1217–1221.

Liou, J.G., Zhang, R.Y., Ernst, W.G., 1994. An introduction to ultrahigh-pressure metamorphism. The Island Arc 3, 1–24.

Liou, J.G., Zhang, R.Y., Ernst, W.G., 1995. Occurrence of hydrous and carbonate phases in ultrahigh-pressure rocks from east-central China — implications for the roles of volatiles deep in cold subduction zones. The Island Arc 4, 362–375.

Liou, J.G., Zhang, R.Y., Wang, X., Eide, E.A., Ernst, W.G., Maruyama, S., 1996. Metamorphism and tectonics of high-pressure and ultrahigh-pressure belts in the Dabie–Sulu region, China. In: Yin, A., Harrison, T.M. (Eds.), The Tectonic Evolution of Asia. Cambridge University Press, New York, pp. 300–344.

Liou, J.G., Zhang, R.Y., Ernst, W.G., Rumble III, D., Maruyama, S., 1998. High-pressure minerals from deeply subducted metamorphic rocks. Rev. Mineral. 37, 33–96.

Liu, J., Bohlen, S.R., Ernst, W.G., 1996. Stability of hydrous phases in subducting oceanic crust. Earth Planet. Sci. Lett. 143, 161–171.

Luth, R.W., 1997. Experimental study of the system phlogopite-diopside from 3.5 to 17 GPa. Am. Mineral. 82, 1198–1209.

Maruyama, S., 1997. Pacific-type orogeny revisited: Miyashiro-type orogeny proposed. The Island Arc 6, 91–120.

Maruyama, S., Parkinson, C.D., 2000. Overview of the geology, petrology and tectonic framework of the HP/UHP metamorphic belt of the Kokchetav Massif, Kazakhstan. The Island Arc, 9, 439–455.

Maruyama, S., Liou, J.G., Zhang, R., 1994. Tectonic evolution of the ultrahigh-pressure (UHP) and high-pressure (HP) metamorphic belts from central China. The Island Arc 3, 112–121.

Maruyama, S., Liou, J.G., Terabayashi, M., 1996. Blueschists and eclogites of the world, and their exhumation. Int. Geol. Rev. 38, 485–594.

Massone, H.J., 1995. Experimental and petrogenetic study of UHPM. In: Coleman, R.G., Wang, X. (Eds.), Ultrahigh-Pressure Metamorphism. Cambridge University Press, New York, pp. 33–95.

Massone, H.-J., 1999. Experimental aspects of UHP metamorphism in pelitic systems. Int. Geol. Rev. 41, 623–638.

Medaris, L.G.J., Wang, H.F., 1986. A thermal-tectonic model for high-pressure rocks in the Basal Gneiss complex of western Norway. In: Proceedings of the 2nd International Eclogite Conference on Lithos. pp. 323–332.

Michard, A., 1965. Une nappe de socle dans les Alpes cottiennes internes? Implications paleogeographiques et role eventuel des movements cretaces. Comptes Rendu Acad. des Sci. de Paris D-260, 4012–4015.

Michard, A., 1967. Etudes Geologiques dans les Zones Internes des Alpes cottiennes. CNRS, Paris, 447 pp.

Michard, A., Henry, C., Chopin, C., 1995. Structures in UHPM rocks: a case study from the Alps. In: Coleman, R.G., Wang, X. (Eds.), 1995. Ultrahigh-Pressure Metamorphism. Cambridge University Press, New York, pp. 132–158.

Monie, P., Chopin, C., 1991. $^{40}Ar/^{39}Ar$ dating in coesite-bearing and associated units of the Dora Maira Massif, western Alps. Eur. J. Mineral. 3, 239–262.

Mörk, M.B.E., Krogh, E.J., 1987. Excursion Guide for the Eclogite Field Symposium in western Norway. Special publication University of Tromso, Norway, 115 pp.

Mosenfelder, J.L., Bohlen, S.R., 1997. Kinetics of the coesite to quartz transformation. Earth Planet. Sci. Lett. 153, 133–147.

Mysen, B.O., Ulmer, P., Konzett, J., Schmidt, M.W., 1998. The upper mantle near convergent plate boundaries. Rev. Mineral. 37, 97–138.

Nie, S., Rowley, D.B., 1994. Comment on paleomagnetic constraints on the geodynamic history of the major blocks of China from the Permian to the present. In: Enkin, R.J., et al. (Eds.), J. Geophys. Res. 99, 18,035–18,042.

Okamoto, K., Maruyama, S., 1999. The high-pressure synthesis of lawsonite in the MORB + H_2O system. Am. Mineral. 84, 362–373.

Okay, A.I., 1993. Petrology of a diamond and coesite-bearing metamorphic terrane, Dabie Shan, China. Eur. J. Mineral. 5, 659–673.

Okay, A.I., Shutong, X., Sengör, A.M.C., 1989. Coesite from the Dabie Shan eclogites, central China. Eur. J. Mineral. 1, 595–598.

Okay, A.I., Sengör, A.M.C., Satir, M., 1993. Tectonics of an ultrahigh-pressure metamorphic terrane: the Dabie Shan/Tongbai Shan orogen, China. Tectonics 12, 1320–1334.

Oxburgh, E.R., Turcotte, D.L., 1971. Origin of paired metamorphic belts and crustal dilation in island arc regions. J. Geophys. Res. 76, 1315–1327.

Paquette, J.R., Chopin, C., Peucat, J.J., 1989. U–Pb zircon, Rb–Sr and Sm–Nd geochronology of high- to very high-pressure meta-acidic rocks from the western Alps. Contributions Mineral. Petrol. 101, 280–289.

Parkinson, C.D., 2000. Coesite inclusions and prograde compositional zonation of garnet in whiteschist of the HP/UHP Kokchetav massif, Kazakhstan: a record of progressive UHP metamorphism. Lithos 52, 215–233.

Parkinson, C.D., Katayama, I., 1999. Present-day ultrahigh-pressure conditions of coesite inclusions in zircon and garnet: evidence from laser Raman microspectroscopy. Geology 27, 979–982.

Peacock, S.M., 1995. Ultrahigh-pressure metamorphic rocks and the thermal evolution of continent collision belts. The Island Arc 4, 376–383.

Peacock, S.M., Wang, K., 1999. Seismic consequences of warm versus cool subduction metamorphism: examples from southwest and northeast Japan. Science 286, 937–939.

Philippot, P., 1988. Deformation et eclogitization progressives d'une croute oceanique subductee: le Monviso, Alpes occidentales. Contraintes cinematiques durant la collision alpine, Vol. 19. Travaux Centre Geologique et Geophysiques, Montpelier, 269 pp.

Platt, J.P., 1986. Dynamics of orogenic wedges and the uplift of high-pressure metamorphic rocks. Geol. Soc. Am. Bull. 97, 1037–1053.

Platt, J.P., 1993. Exhumation of high-pressure rocks: a review of concepts and processes. Terra Nova 5, 119–133.

Poli, S., Schmidt, M.W., 1997. The high-pressure stability of hydrous phases in orogenic belts: an experimental approach on eclogite-forming processes. Tectonophysics 273, 169–184.

Perchuk, A.L., Yaspaskurt, V.O., Podlesskii, S.K., 1998. Genesis and exhumation dynamics of eclogites in the Kokchetav Massif near Mount Sulu–Tyube, Kazakhstan. Geochem. Int. 36, 877–885.

Reinecke, T., 1991. Very high-pressure metamorphism and uplift of coesite-bearing metasediments from the Zermatt–Saas zone, western Alps. Eur. J. Mineral. 3, 7–17.

Robinson, P., Terry, M.P., 1998. Evolution of amphibolite-facies structural features during transition from convergence to extensional collapse, Western Gneiss Region, Norway. In: Proceedings of the International Workshop on UHP Metamorphism and Exhumation. Stanford University, Stanford, CA, pp. A79–A83.

Rowley, D.B., Xue, F., Tucker, R.D., Peng, Z., Baker, J., Davis, A., 1997. Ages of ultrahigh-pressure metamorphism and protolith orthogneisses form the eastern Dabie Shan: U/Pb zircon geochronology. Earth Planet. Sci. Lett. 151, 191–204.

Rubie, D.C., 1983. Reaction-induced ductility: the role of solid–solid univariant reactions in deformation of the crust and mantle. Tectonophysics 96, 331–352.

Rubie, D.C., 1984. A thermal-tectonic model for high-pressure metamorphism in the Sesia zone, western Alps. J. Geol. 92, 21–36.

Rubie, D.C., 1986. The catalysis of mineral reactions by water and restrictions on the presence of aqueous fluid during metamorphism. Mineral. Mag. 50, 399–415.

Rubie, D.C., 1990a. Mechanisms of reaction-enhanced deformability in rocks and minerals. In: Barber, D.J., Meredith, P.G. (Eds.), Deformation Processes in Minerals, Ceramics and Rocks. Unwin Hyman, London, pp. 262–295.

Rubie, D.C., 1990b. Role of kinetics in the formation and preservation of eclogites. In: Carswell, D.A. (Ed.), Eclogite Facies Rocks. Chapman and Hall, New York, pp. 111–140.

Rumble, D., 1998. Stable isotope geochemistry of ultrahigh-pressure rocks. In: Hacker, B.R., Liou, J.G. (Eds.),

When Continents Collide: Geodynamics and Geochemistry of Ultrahigh-Pressure Rocks. Kluwer Academic Publishers, Dordrecht, pp. 241–259.

Sacks, P.E., Secor Jr, D.T., 1990. Delaminations in collisional orogens. Geology 18, 999–1002.

Schertl, H.P., Schreyer, W., Chopin, C., 1991. The pyrope-coesite rocks and their country rocks at Parigi, Dora Maira Massif, western Alps: detailed petrography, mineral chemistry and P–T path. Contributions Mineral. Petrol. 108, 1–21.

Schertl, H.P., Okay, A.I., 1994. Coesite inclusion in dolomite of Dabieshan, China: petrological and rheological significance. Eur. J. Mineral. 6, 995–1000.

Schreyer, W., 1988. Experimental studies on metamorphism of crustal rocks under mantle pressures. Mineral. Mag. 53, 1–26.

Schreyer, W., 1995. Ultradeep metamorphic rocks: the retrospective viewpoint. J. Geophys. Res. 100, 8353–8366.

Sengör, A.M.C., Natal'in, B.A., Burtman, V.S., 1993. Evolution of the Altaid tectonic collage and Palaeozoic crustal growth in Eurasia. Nature 364, 299–307.

Shatsky, V.S., Sobolev, N.V., 1993. Some specific features of the origin diamonds in metamorphic rocks. Doklady Akadamiya Nauk SSSR 331, 217–219 (in Russian).

Shatsky, V.S., Sobolev, N.V., Gilbert, A.E., 1989. Eclogites of Kokchetav Massif. In: Sobolev, N.V. (Ed.), Eclogites and Glaucophane Schists in Folded Belts. Nauka Press, Novosibirsk, pp. 54–82 (in Russian).

Shatsky, V.S., Sobolev, N.V., Vavilov, M.A., 1995. Diamond-bearing metamorphic rocks of the Kokchetav massif (northern Kazakhstan). In: Coleman, R.G., Wang, X. (Eds.), Ultrahigh-Pressure Metamorphism. Cambridge Press, New York, pp. 427–455.

Shatsky, V.S., Sobolev, N.V., Zayachkovsky, A.A., Zorin, Yu. M., Vavilov, M.A., 1991. A new find of the microdiamonds occurrence as an evidence of the regional ultrahigh-pressure metamorphism at the Kokchetav Massif. Doklady Akadamiya Nauk SSSR 321, 189–193 (in Russian).

Shatsky, V.S., Jagoutz, E., Kozmenko, O.A., Blinchik, T.M., Sobolev, N.V., 1993. The age and origin of eclogites from Kokchetav massif, north Kazakhstan. Geol. i Geofizika 34 (12), 47–58 (in Russian).

Shatsky, V.S., Jagoutz, E., Sobolev, N.V., Kozmenko, O.A., Parkhomenko, V.S., Troesch, M., 1999. Geochemistry and age of ultrahigh-pressure metamorphic rocks from the Kokchetav massif (northern Kazakhstan). Contributions Mineral. Petrol. 137, 185–205.

Smith, D.C., 1984. Coesite in clinopyroxene in the Caledonides and its implications for geodynamics. Nature 310, 641–644.

Smith, D.C., 1995. Microcoesites and microdiamonds in Norway: an overview. In: Coleman, R.G., Wang, X. (Eds.), 1995. Ultrahigh-Pressure Metamorphism. Cambridge University Press, New York, pp. 299–355.

Smith, D.C., Lappin, M.A., 1989. Coesite in the Straumen kyanite-eclogite pod Norway. Terra Res. 1, 47–56.

Sobolev, N.V., Shatsky, V.S., 1990. Diamond inclusions in garnets from metamorphic rocks: a new environment for diamond formation. Nature 343, 742–746.

Sobolev, V.S., Sobolev, N.V., 1980. New proof on very deep subsidence of eclogitized crustal rocks. Doklady Akadamiya Nauk SSSR 250, 683–685 (in Russian).

Sobolev, N.V., Dobretsov, N.L., Bakirov, A.B., Shatsky, V.S., 1986. Eclogites from various types of metamorphic complexes in the USSR and problems of their origin. In: Evans, B.W., Brown, E.H. (Eds.), Blueschists and Eclogites, Vol. 164. Geological Society of America Memoir, pp. 349–363.

Sobolev, N.V., Shatsky, V.S., Vavilov, M.A., 1991. Inclusions of microdiamonds and coexisting minerals in garnets and zircons from metamorphic rocks of Kokchetav Massif USSR. Terra Abstr. 3, 83.

Sobolev, N.V., Shatsky, V.S., Vavilov, M.A., Gorgainov, S.V., 1994. Zircon from ultrahigh-pressure metamorphic rocks of folded regions as an unique container of inclusions of diamond, coesite, and coexisting minerals. Doklady Akadamiya Nauk 334, 488–492 (in Russian).

Suppe, J. Interrelationships of high-pressure metamorphism, deformation and sedimentation in Franciscan tectonics, USA. In: Proceedings of the 24th International Geological Congress on Montreal Reports Section, Vol. 3. 1972, pp. 552–559.

Tabata, H., Yamauchi, K., Maruyama, S., Liou, J.G., 1998. Tracing the extent of a UHP metamorphic terrane: mineral-inclusion study of zircons in gneisses from the Dabie Shan. In: Hacker, B.R., Liou, J.G. (Eds.), When Continents Collide: Geodynamics and Geochemistry of Ultrahigh-Pressure Rocks. Kluwer Academic Publishers, Dordrecht, pp. 261–273.

Theunissen, K., Dobretsov, N.L., Travin, A., Shatsky, V.S., Korsakov, A., Smirnova, L., Boven, A., 2000. Two contrasting petrotectonic domains in the Kokchetav megamelange (north Kazakhstan): difference in exhumation mechanisms of ultrahigh-pressure crustal rocks, or a result of subsequent deformation. The Island Arc, 9, 284–303.

Tilton, G.R., Schreyer, W., Schertl, H.P., 1991. Pb–Sr–Nd isotopic behavior of deeply subducted crustal rocks from the Dora Maira Massif, western Alps-II: what is the age of the ultrahigh-pressure metamorphism? Contributions Mineral. Petrol. 108, 22–33.

Tilton, G.R., Ames, L., Schertl, H.P., Schreyer, W., 1997. Reconnaissance isotopic investigations on rocks of an undeformed granite contact within the coesite-bearing unit of the Dora Maira Massif. Lithos 41, 25–36.

Troesch, M., Jagoutz, E., 1993. Mica cooling ages of a diamond-bearing gneiss from the Kokchetav Massif, Kazakhstan. Terra Abstr. 7, 396.

Ulmer, P., Trommsdorff, V., 1995. Serpentine stability to mantle depths and subduction-related magmatism. Science 268, 858–861.

Van der Molen, I., van Roermund, H.L.M., 1986. The pressure path of solid inclusions in minerals: the retention of coesite inclusions during uplift. Lithos 19, 317–324.

van Roermund, H.L.M., Drury, M.R., Barnhoorn, A., de Ronde, A., 1998. Ultrahigh-pressure ($P > 6$ GPa) garnet peridotites in western Norway: implications for exhumation processes of mantle rocks. In: Proceedings of the International Workshop on Ultrahigh-Pressure Metamorphism and Exhumation. Stanford, CA, pp. A9–A12 (Abstract).

Vialon, P., 1966. Etude geologique du massif cristallin Dora Maira, Alpes cottiennes internes, Italie. Thesis d'Etat, Université de Grenoble, Grenoble, 293 pp.

Vielzeuf, D., Holloway, J.R., 1988. Experimental determination of the fluid-absent melting relations in the pelitic system: consequences for crustal differentiation. Contributions Mineral. Petrol. 98, 257–276.

Vielzeuf, D., Montel, J.M., 1994. Partial melting of metagraywackes. 1. Fluid-absent experiments and phase relationships. Contributions Mineral. Petrol. 117, 375–393.

von Blanckenburg, F., Davies, J.H., 1995. Slab break-off: a model for syncollision magmatism and tectonics in the Alps. Tectonics 14, 120–131.

Wain, A., 1997. New evidence for coesite in eclogite and gneisses: defining an ultrahigh-pressure province in the Western Gneiss Region of Norway. Geology 25, 927–930.

Wallis, S.R., Ishiwatari, A., Hirajima, T., Ye, K., Guo, J., Nakamura, D., Kato, T., Zhai, M., Enami, M., Cong, B., Banno, S., 1997. Occurrence and field relationships of UHP meta-granitoid and coesite eclogite in the Su–Lu terrane, eastern China. J. Geol. Soc. London 154, 45–55.

Wang, Q., Ishiwatari, A., Zhao, Z., Hirajima, T., Enami, M., Zhai, M., Li, J., Cong, B.L., 1993. Coesite-bearing granulite retrograded from eclogite in Wehai, eastern China: a preliminary study. Eur. J. Mineral. 5, 141–152.

Wang, X., Liou, J.G., 1991. Regional ultrahigh-pressure metamorphic terrane in central China: evidence from coesite-bearing eclogite, marble, and metapelite. Geology 19, 933–936.

Wang, X., Liou, J.G., Mao, H.K., 1989. Coesite-bearing eclogites from the Dabie Mountains in central China. Geology 17, 1085–1088.

Wang, X., Zhang, R., Liou, J.G., 1995. UHPM terrane in east central China. In: Coleman, R.G., Wang, X. (Eds.), Ultrahigh-Pressure Metamorphism. Cambridge University Press, New York, pp. 356–390.

Webb, L.E., Hacker, B.R., Ratschbacher, L., McWilliams, M.O., Dong, S., 1999. $^{40}Ar/^{39}Ar$ thermochronologic constraints on deformation and cooling history of high and ultrahigh-pressure rocks in the Qinling–Dabie orogen. Tectonics 18, 621–638.

Wyllie, P.J., Wolf, M.B., van der Laan, S.R., 1996. Conditions for formation of tonalites and trondhjemites: magmatic sources and products. In: de Wit, M.J., Ashwahl, L.D. (Eds.), Tectonic Evolution of Greenstone Belts. Oxford Press, London, pp. 256–266.

Xu, S., Okay, A.I., Ji, S., Sengör, A.M.C., Su, W., Liu, Y., Jiang, L., 1992. Diamond from Dabie Shan eclogite and its implication for tectonic setting. Science 256, 80–82.

Yang, Z., Courtillot, V., Besse, J., 1992. Jurassic paleomagnetic constraints in the collision of the north and south China blocks. Geophys. Res. Lett. 19, 577–580.

Ye, K., Hirajima, T., Ishiwatari, A., Guo, J., Zhai, M., 1996. Finding of coesite in eclogite in Qingdao and its implications. Chin. Sci. Bull. 11, 1407–1408 (in Chinese).

Yin, A., Nie, S., 1993. An indentation model for the north and south China collision and the development of the Tan-Lu and Honam fault systems, eastern Asia. Tectonics 12, 801–813.

Yin, A., Nie, S., 1996. A Phanerozoic palinspastic reconstruction of China and its neighboring regions. In: Harrison, T.M., Yin, A. (Eds.), The Tectonic Development of Asia. Cambridge University Press, New York, pp. 442–485.

Yui, T.F., Rumble, D., Lo, C.H., 1995. Unusually low ^{18}O ultrahigh-pressure metamorphic rocks from the Sulu terrain, eastern China. Geochem. et Cosmochim. Acta 59, 2859–2864.

Zhang, R.Y., Liou, J.G., 1994. Coesite-bearing eclogite in Henan province, central China: detailed petrography, glaucophane stability and P–T path. Eur. J. Mineral. 6, 217–233.

Zhang, R.Y., Liou, J.G., 1996. Coesite inclusions in dolomite from eclogite in the southern Dabie Mountains, China: the significance of carbonate minerals in UHPM rocks. Am. Mineral. 81, 181–186.

Zhang, R.Y., Liou, J.G., 1997. Partial transformation of gabbro to coesite-bearing eclogite from Yangkou, the Sulu terrane, eastern China. J. Metamorph. Geol. 15, 183–202.

Zhang, R.Y., Liou, J.G., 1998. Petrochemical constraints for dual origin of garnet peridotites of the Dabie–Sulu UHP terrane, China. In: Proceedings of the International Workshop on Ultrahigh-Pressure Metamorphism and Exhumation. Stanford, CA, pp. A134–A137 (Abstract).

Zhang, R.Y., Liou, J.G., Cong, B., 1994. Petrogenesis of garnet-bearing ultramafic rocks and associated eclogites in the Su–Lu ultrahigh-pressure metamorphic terrane, China. J. Metamorph. Geol. 12, 169–186.

Zhang, R.Y., Liou, J.G., Cong, B., 1995. Talc-, magnesite-, and Ti-clinohumite-bearing ultrahigh-pressure meta-mafic and ultramafic complex in the Dabie Mountains, China. J. Petrol. 36, 1011–1037.

Zhang, R.Y., Liou, J.G., Ernst, W.G., Coleman, R.G., Sobolev, N.V., Shatsky, V.S., 1997. Metamorphic evolution of diamond-bearing and associated rocks from the Kokchetav Massif, northern Kazakhstan. J. Metamorph. Geol. 15, 479–496.

Zhang, R.Y., Shau, Y.H., Liou, J.G., Lo, C.H., 1998. Discovery of clinoenstatite in the Sulu garnet pyroxenite and its implication for continental subduction. In: Proceedings of the International Workshop on Ultrahigh-Pressure Metamorphism and Exhumation. Stanford, CA, pp. A16–A19 (Abstract).

Zonenshain, L.P., Kuzmin, M.I., Natapov, L.M., 1990. Geology of the USSR: a plate-tectonic synthesis. In: Page, B.M. (Ed.), Am. Geophys. Union Geodyn. Ser. 21, 242.

ELSEVIER

Physics of the Earth and Planetary Interiors 127 (2001) 277–291

PHYSICS OF THE EARTH AND PLANETARY INTERIORS

www.elsevier.com/locate/pepi

Subduction followed by collision: Alpine and Himalayan examples

Patrick J. O'Brien*

Bayerisches Geoinstitut, Universität Bayreuth, D-95440 Bayreuth, Germany

Received 3 August 2000; accepted 20 October 2000

Abstract

The high mountains of the Himalaya and Alps are both the result of ongoing continental collision following closure, by subduction, of Neo-Tethys and related ocean basins during the Mesozoic and Tertiary. The high topography is associated, in both areas, with a thickened crustal root of up to 70 km thickness as deduced from various geophysical investigations. In the Himalayan region, where several thousand kilometres of ocean crust was consumed below the Lhasa block (southern margin of the Asian continent), both continental (Trans-Himalayan batholith) and oceanic (Kohistan–Ladakh arc, accreted to Asia in the Jurassic) magmatic arcs are recognisable. In the Alps, where much smaller oceanic rifts and basins were consumed, such arc sequences are absent in the area of the high mountains. Regardless of this major difference in volume of subducted oceanic lithosphere, seismic tomography in both regions has identified several anomalous regions in the mantle interpreted as the remnants of detached slabs of oceanic crust. Metamorphosed rocks of continental crust sequences in both the Himalaya and Alps record evidence of subduction to depths exceeding 90 km (coesite stability field) as well as rapid exhumation to shallower depths at rates of 4 mm/a or more. This subduction-related metamorphism of the eclogite and blueschist-facies is clearly distinguishable from subsequent, younger, lower pressure, greenschist–amphibolite-facies metamorphic overprints related to imbricate stacking of continental crust slices, thermal relaxation, and final exhumation. Dragging of buoyant upper crust down to upper mantle depths requires an attached root of oceanic lithosphere but the rapid exhumation stage may well be a consequence of slab break-off. This process appears to have been equally effective in the Alps and the NW Himalaya, despite the significant differences in duration of subduction and the volumes of oceanic lithosphere involved.

Keywords: Subduction; Alpine; Himalaya

1. Introduction

The subduction of oceanic lithosphere at present-day plate margins and concomitant arc magmatism are important indicators of the dynamic evolution of the earth's surface due to plate tectonics. Over time, subduction leads to complete closure of ocean basins and eventually to collision of continental plates. The records of such subduction and collision processes are found in belts of metamorphic rocks of various ages throughout the world. In ancient collision belts, such as the Caledonian and Variscan belts in Europe, continental crust has returned to a normal thickness, high mountains are absent, and the lithosphere is tectonically relatively inactive. In contrast, regions where collision is still going on, such as in the Alps or Himalaya, are typified by high mountains, overthickened crust, and a high degree of tectonic activity reflected in the frequency and magnitude of earthquakes. Thus, modern collision belts provide an active laboratory

* Corresponding author. Present address: Institut fuer Geowissenschaften, Universitaet Potsdam, Postfach 601553, D 14415 Potsdam, Germany.

PII: S0031-9201(01)00232-1

for the testing of dynamic models for the evolution of the continental lithosphere over time. In addition, such active areas must also act as a lid to the many kilometres of oceanic lithosphere that have been consumed, thus, the detection of subducted oceanic crust in these regions can provide clues as to the rate of assimilation of such material back into the mantle, or its possible role in post-collisional magmatic activity.

In general, the high mountains of the European Alps and the Himalaya result from the approach and collision of major lithospheric plates, as well as intervening minor plate fragments and arc units, from late Mesozoic times to the present-day. Permo-Carboniferous break-up of the supercontinent Pangea and the formation of the Tethys ocean between the Caribbean and Himalayan areas (Suess, 1885–1888) resulted in the formation of numerous small continental blocks, wide areas of shallow, epicontinental seas, and subsided continental margins that became covered by thick carbonate sequences during the Triassic and Jurassic. As the direction of plate movement changed in the Cretaceous, major ocean basins were closed by subduction of oceanic lithosphere, remnants of ocean crust (ophiolites) were obducted, both large and small continental fragments collided leading to imbricate stacking of the thinned continental margins, regional metamorphism took place and high mountains were formed in some locations.

A detailed reconstruction of the events leading to the present-day configuration of mountain belts, active volcanoes, subduction zones and active earthquake zones requires identification of the individual blocks, the boundaries of the blocks and the rates and directions of horizontal and/or vertical movement of the blocks over time. This information comes partly from field studies and partly from geophysical investigations. The greater accessibility of the Alps, from both political and practical standpoints, means that the mapping of geological and structural information is much more advanced than in the Himalaya where even today the majority of the area has only been reconnaissance-mapped. However, both regions have been the target of recent intensive geophysical investigations (EGT, Blundell et al., 1992; NRP-20, Pfiffner et al., 1997; Zhao and Nelson, 1993; Nelson et al., 1996) and so the knowledge of deep structure is of comparable quality. In the following sections, the evolution of the Alpine and Himalayan orogenic belts will be outlined in turn. I will concentrate on the plate tectonic setting, the number and location of plate boundaries and subduction zones, and the characteristic features of the metamorphic and structural evolution.

2. The European Alps

The European Alps comprise a series of mountain chains forming a 1200-km long arc stretching from the French–Italian Mediterranean coast via Switzerland and southern Germany to the Vienna basin of Austria (Fig. 1a and b). Following the Carboniferous Variscan orogeny in Europe, the break-up of the Pangea supercontinent, by rifting and wrench faulting, led to the formation of a number of continental fragments, of different size, separated by shallow, epicontinental seas (Dewey et al., 1973; Arthaud and Matte, 1977). Rift development is recorded by changes in facies of sedimentary rocks deposited on the crystalline basement as well as subsidence rates of these basins (e.g. Stampfli et al., 1998, and references therein). Closure of these rift basins in the late Mesozoic and Tertiary led to collision of the microplates and subduction and, in some cases, obduction of the intervening oceanic crust. The reconstruction of the pre-collision history of the Alps has been undertaken by locating the former plate margins by (1) identifying ophiolite-bearing suture zones (e.g. Dietrich, 1979) and (2) by documenting facies changes in deformed and metamorphosed Mesozoic sediments forming the cover to basement slices (e.g. Stampfli et al., 1991). In fact, it is variations in these Mesozoic series that have been utilised in the basic subdivision of the Alpine crust.

Geologically, the Alps are split into western, central and eastern parts following the general strike of the structures from NS in the western Alps to EW in the eastern Alps as well as considering the major tectonic units in the areas. A further subdivision, based on the nature of the Mesozoic sediments, distinguishes the Helvetic/Sub-Alpine, Penninic, Austro-Alpine and South-Alpine units (Fig. 1a and b). The significance of this subdivision is realised by reference to a simplified plate–tectonic evolution for the Mediterranean area as shown in Fig. 2 (derived from Stampfli et al., 1998). The Helvetic or Sub-Alpine units represent the former submerged margin of the European continent that underwent only weak metamorphism (if any) and mild tectonism to form a fold-and-thrust belt. Detachment of post-Triassic sedimentary cover occurred along Triassic evaporite layers but crystalline basement (in the so-called external basement massifs) was also affected (e.g. Aiguilles Rouge, Mt. Blanc, Aar, Gotthard Massifs). The Penninic units represent the oceanic crust of the Alpine Tethys (Piemont–Ligurian and Valais oceans), the southernmost, thinned, distal margin of the European plate, and units representing the promontory of the Iberian plate (Briançonnais). The oceanic, ophiolitic units clearly represent the suture zones between the former separated continental plates. Austro-Alpine units, dominant as the name suggests in the Austrian (eastern) part of the Alps, comprise the terranes on the southern margin of the Alpine Tethys that were dissected and accreted to Europe as the Meliata (already consumed below the southern margin of the Austro-Alpine terrane in Fig. 2a) and Vardar oceans closed, and the Apulian microplate collided with Europe. Only small klippen of the structurally uppermost Austro-Alpine nappes remain in the central Alps, whereas in the eastern Alps Austro-Alpine units dominate and the underlying Penninic units are exposed only in tectonic windows (e.g. Engadine, Tauern). The Penninic and Austro-Alpine units are separated from the South-Alpine units by a major dextral-transpressive fault (Schmid et al., 1989) known as the Insubric Line (more properly Peri-Adriatic Lineament: also marked by the presence of ophiolitic units, see Fig. 1a). The rocks of the southern

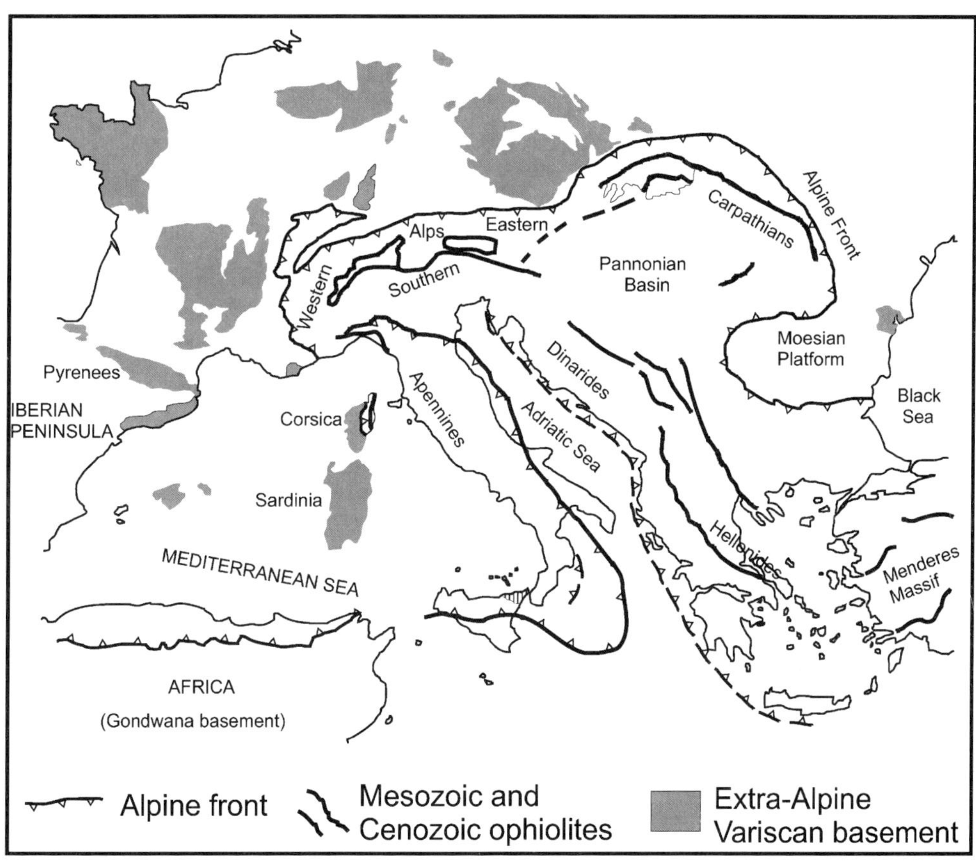

(a)

Fig. 1. (a) Map of the Mediterranean region showing the location of ophiolites and the limits of the Alpine deformation. (b) Simplified geological map of the European Alps showing the main subdivisions. The high mountains are located in an arc between the Tertiary molasse basins (redrawn after Marchant and Stampfli, 1997).

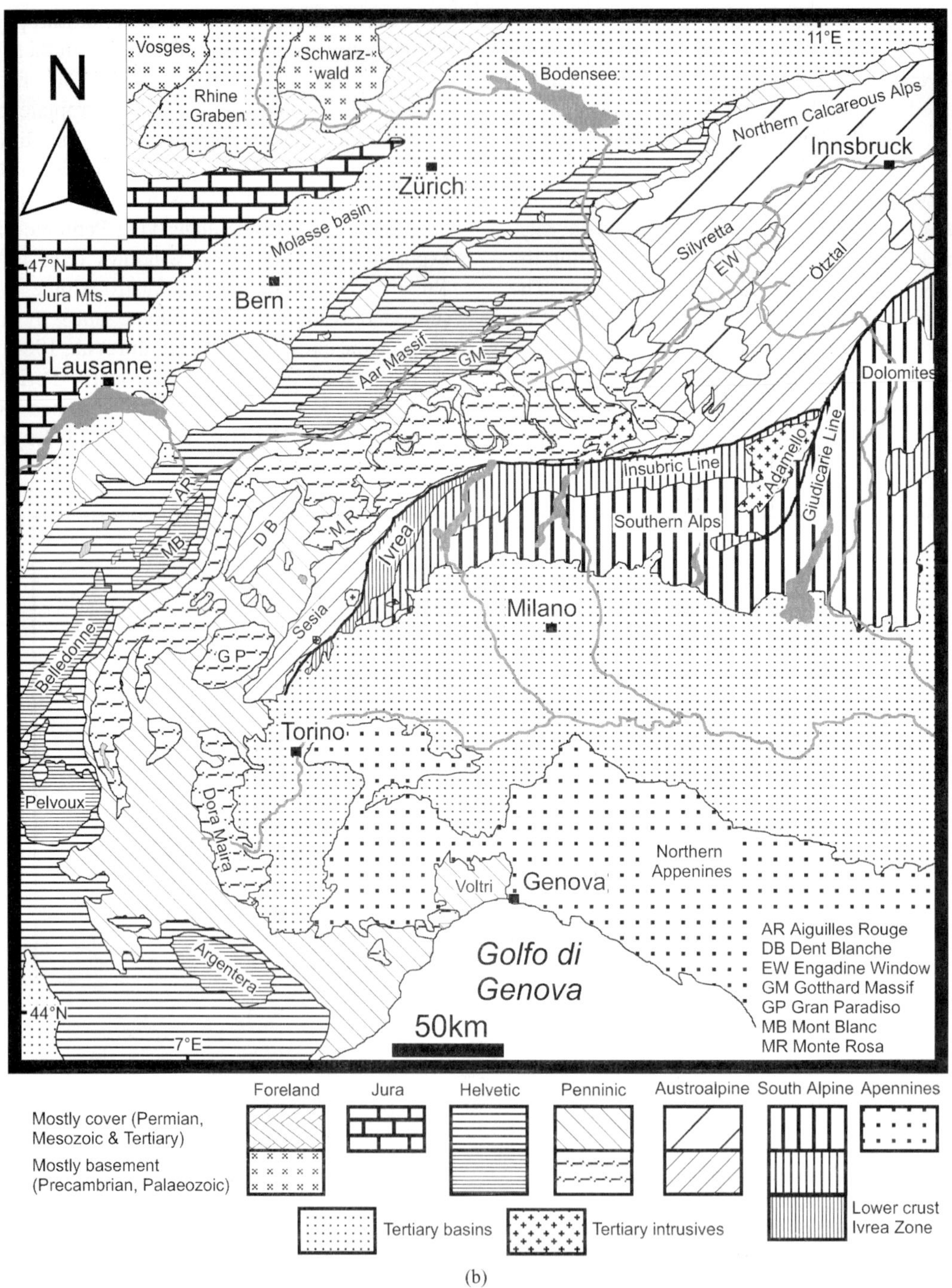

(b)

Fig. 1. (*Continued*).

Alps, part of the southern, Adriatic plate margin, form a south-vergent fold and thrust belt. Also belonging to the South-Alpine unit is a tilted cross-section through the mafic–ultramafic lower crust of the Adriatic plate known as the Ivrea (or Ivrea-Verbano) Zone. This area contains granulite-to-amphibolite-facies metamorphic rocks that are essentially unaffected by the Alpine metamorphism.

The plate tectonic reconstruction of Fig. 2 shows clearly that there was no simple, single-stage collision

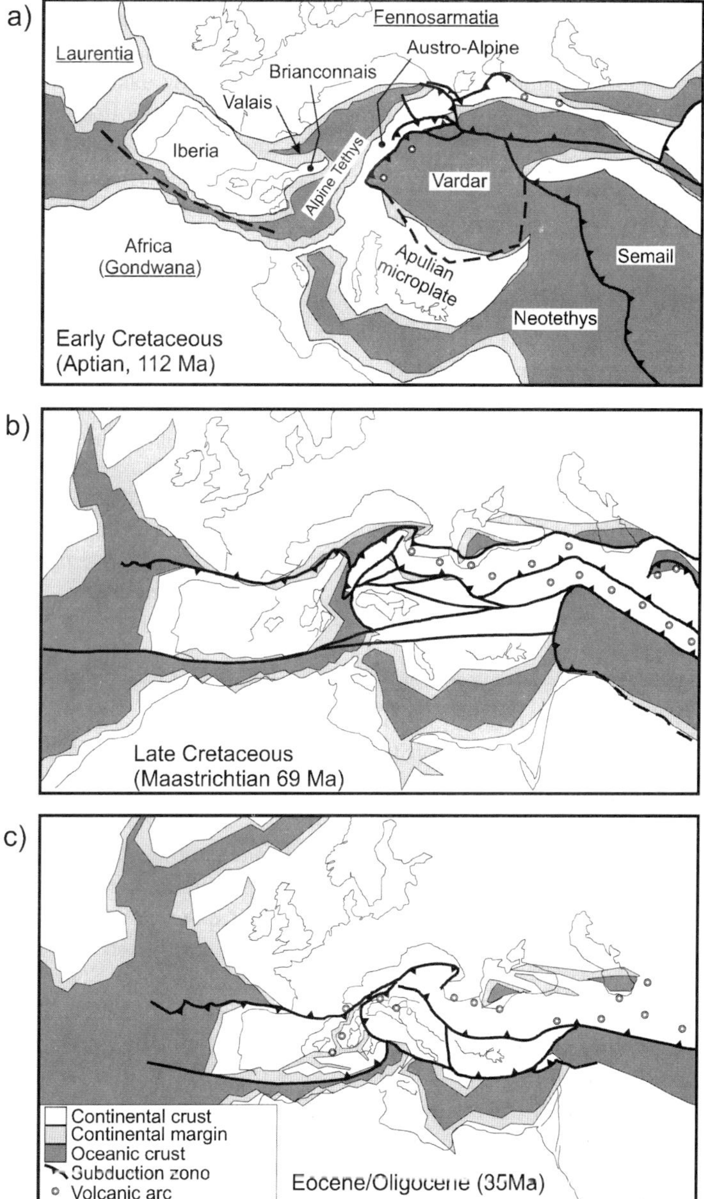

Fig. 2. Plate–tectonic reconstructions at 112, 69 and 35 Ma for the Alpine realm showing interpreted locations of microplates, ocean basins, subduction zones and volcanic arcs (redrawn after Stampfli et al., 1998; Marchant and Stampfli, 1997).

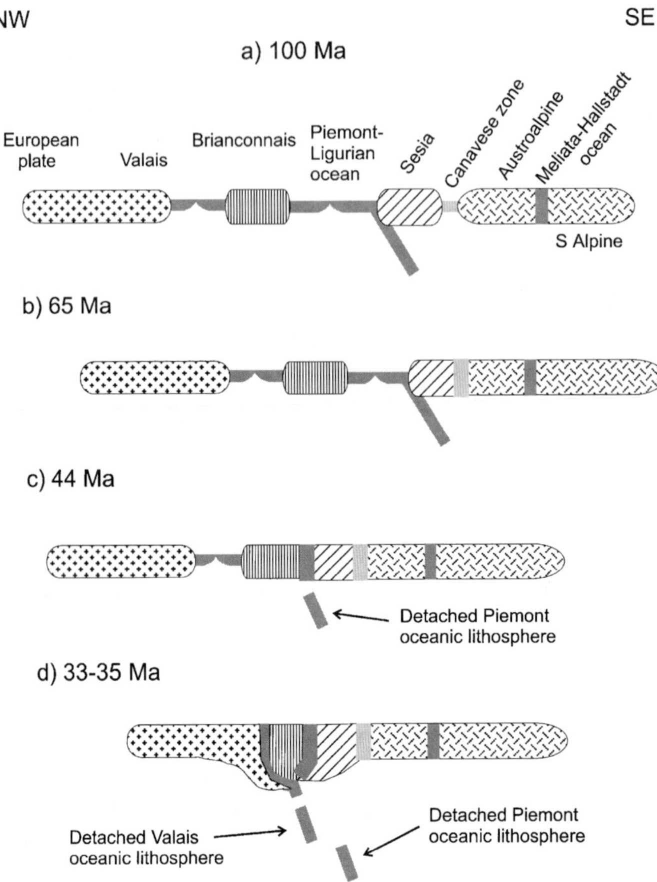

Fig. 3. Cartoon showing the main crustal blocks, the intervening oceanic domains, and the relative timing of subduction, basin closure and continental collision during the Alpine orogenic evolution (redrawn after Gebauer, 1999).

in the Mediterranean region, but rather a series of opening and closing episodes involving different oceanic basins. As each collision led to deformation and related metamorphism in the respective continental blocks and microplates, it is no surprise that there is a wide range of metamorphic ages for the different units of the Alps. This evolution is summarised in the cartoon of Fig. 3. The metamorphic evolution of the Alps has been recently reviewed for the publication of a new version of the metamorphic map of the Alps (Frey et al., 1999 and refs therein). In summary, the Penninic units of the eastern Alps underwent a Cretaceous subduction-zone metamorphism reaching eclogite-facies conditions (70–85 km depth) followed by a later greenschist-to-amphibolite-facies overprint in Oligocene to Miocene times. Austro-Alpine units show a similar two-stage evolution, albeit with a lower peak pressure stage. In the western and central Alps, subduction-related metamorphism (eclogite–blueschist-facies) is recorded in the Austro-Alpine Sesia block at ca. 65 Ma and in the Penninic units, where continental crust reached depths of 100 km, at 35–40 Ma (Gebauer, 1999). The deeply subducted units were rapidly returned to shallow levels because the high-pressure assemblages are widely overprinted by lower pressure greenschist parageneses and then cut by post-orogenic plutons by 30–32 Ma (Gebauer, 1999). This post-nappe-stacking overprint is best developed in the Lepontine area of the central Alps where the amphibolite-facies is reached in the centre of a dome-like feature (Frey and Ferreiro Mählmann, 1999).

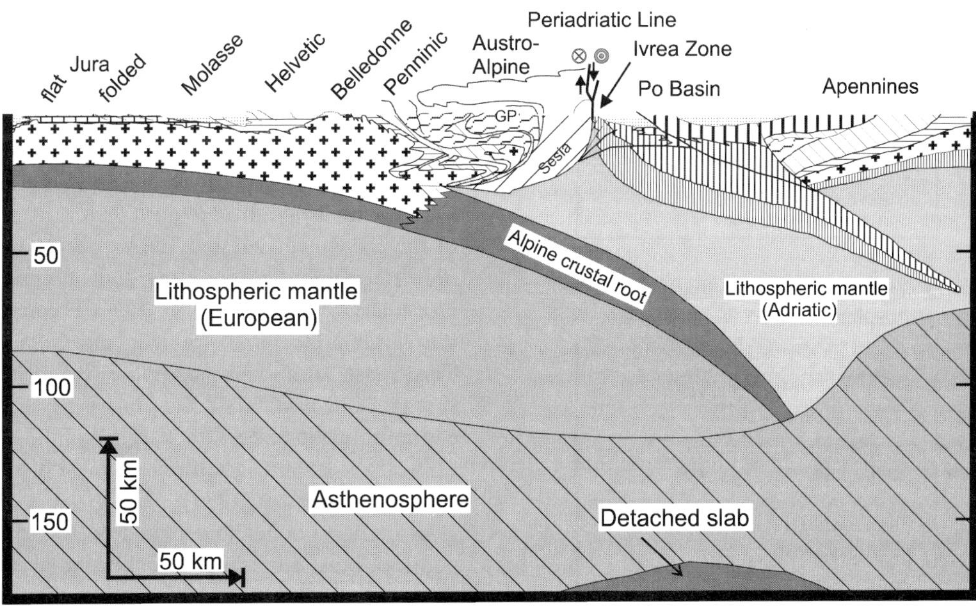

Fig. 4. Deep structure of the lithosphere beneath the Alps interpreted from deep seismic surveys. The section runs from NW (left side, upper-left of Fig. 1b) to SE (right side, lower-right side of Fig. 1b) (redrawn after Marchant and Stampfli, 1997).

The deep structure of the Alps, revealed by seismic and other geophysical investigations (Blundell et al., 1992; Pfiffner et al., 1997), shows that a deep crustal root exists beneath the Alps today (Fig. 4). The fact that high-pressure, metamorphic rocks of deep subduction-zone character (coesite-bearing eclogite-facies rocks) are already exposed at the surface means that some of the deeply subducted crust must have detached from the down-going slab and, on the basis of geochronological data, must have rapidly returned to shallow levels. Modelling of the type of processes necessary for this kind of evolution has been increasingly undertaken in recent years (e.g. Chemenda et al., 1995, 2000) and it appears that the buoyancy of subducted crust, despite transformation to high-pressure mineralogy, is the driving force for the detachment of the deeply subducted crust from the plunging slab.

Ongoing convergence of the Eurasian and African plates causes abundant seismic activity, and the distribution as well as focal mechanisms of earthquakes have been an important source of data for seismic tomography studies of the deep mantle below the Mediterranean region (Spakman, 1991; Spakman et al., 1993). These studies of the velocity structure below the Alpine region have shown that anomalies, interpreted as detached slabs of former oceanic lithosphere, occur in the location of ancient and present-day subduction zones (Wortel and Spakman, 1992). In Fig. 5, anomalies in the mantle deduced from seismic tomography are shown for bodies attributed to be the subducted remnants of the Valais and Piemont oceans. The relative position of these fragments fit with the schematic collision scenario depicted in Fig. 3. Certainly, a detachment of Valais oceanic lithosphere would have allowed the rapid ascent of the part of the deeply subducted continental crust now found at the surface. This is an important aspect to consider when trying to reconstruct the lithospheric processes that allow deep subduction of continental crust as well as its exceedingly rapid return to shallow depths.

3. The Himalayan region

The spectacular mountains of the Himalayan mountain chain and the high relief of the adjacent Tibetan Plateau result from collision of the Indian plate with the Asian margin and intervening microplates that started about 50 Ma and is still going on today (Le

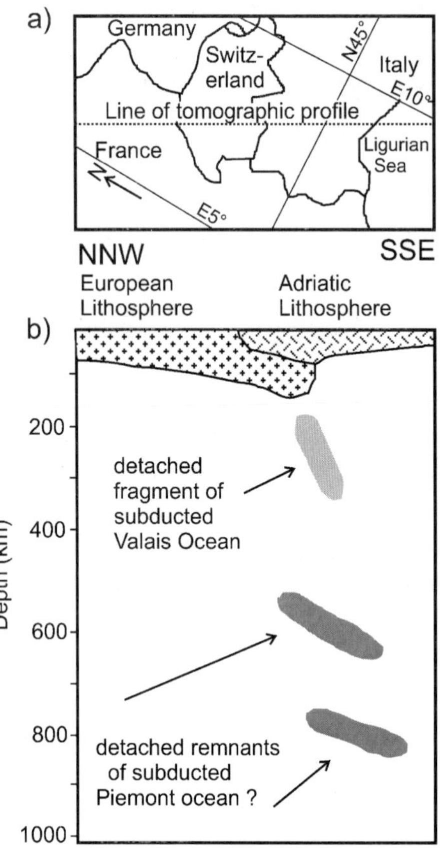

Fig. 5. Location of mantle regions below the Alps with anomalous seismic velocities (b), deduced from seismic tomography studies, interpreted as detached oceanic lithosphere. The line of the profile is shown in (a).

Fort, 1975). The 2500-km long Himalayan chain of mountains (Fig. 6) forms an arc between the syntaxes of Namche Barwa (7756 m) in SE Tibet and Nanga Parbat (8125 m) in NW Pakistan (Wadia, 1931). Both the Tsangpo/Brahmaputra (in the east) and the Indus (in the west), very important rivers for transporting material eroded from the high mountains and the plateau, have courses paralleling the chain until cutting southwards after rounding these syntaxial zones. The Himalaya, with the first 26 in the list of highest mountain peaks of the world, is often seen as the archetypal continent–continent collision zone and, as such, presents an excellent natural laboratory for the study of orogeny and the testing of tectonometamorphic models.

Following the Permo-Carboniferous break-up of Pangea, the landmass later to become India found itself located deep in the southern hemisphere with several thousand kilometres of the Neo-Tethys ocean separating it from the nearest land to the north (Dercourt et al., 1993). North of Neo-Tethys, the Cimmerian continental array drifted northwards by subduction of the Palaeo (or Meso-) Tethys ocean, finally closing in Early Cretaceous time: the ophiolite-bounded Lhasa and Qiangtang blocks (Fig. 6) represent part of this Cimmerian block (Şengör and Natal'in, 1996). The most prominent ophiolite-rich suture zones (Fig. 6), however, mark the boundaries between this earlier-consolidated Asian margin and the Indian plate as well as, in the northwest, the Kohistan–Ladakh island arc (Gansser, 1979).

In order to better understand the evolution of the collision belt, it is useful to delineate the main structural, magmatic and metamorphic features of the Himalayan region as determined from many years of detailed field studies. This fundamental tectonic framework can then be used as a basis to explain the more recent lithosphere-scale investigations such as those of the International Deep Profiling of Tibet and the Himalaya (more commonly INDEPTH) program (Zhao and Nelson, 1993; Nelson et al., 1996). The main geological components of the Himalaya are depicted in Fig. 6 and abundant literature on this subject can be found in the recent papers of Mattauer et al. (1999) and Hodges (2000).

As already mentioned, the northern boundary of the Himalaya comprises a series of continental blocks that accreted to the Asian margin during the Palaeozoic and Mesozoic, each shows ophiolitic sutures along its boundaries. The most important with respect to the Himalayan evolution is the Lhasa block. This collided with the northern Qiangtang block in Jurassic times along the Bangong suture. The southern margin of the Lhasa block then became an Andean-type continental margin, with subduction of a considerable volume of oceanic crust, reflected in the development of the 2500-km long calc-alkaline intrusive continental arc complex, the Trans-Himalayan batholith, with its associated andesites, rhyolites and ignimbrites. At the end of the Cretaceous, an oceanic island arc, the Kohistan–Ladakh arc found in the northwest Himalaya, collided with Asia, and was then itself intruded by granites of the Andean-type arc.

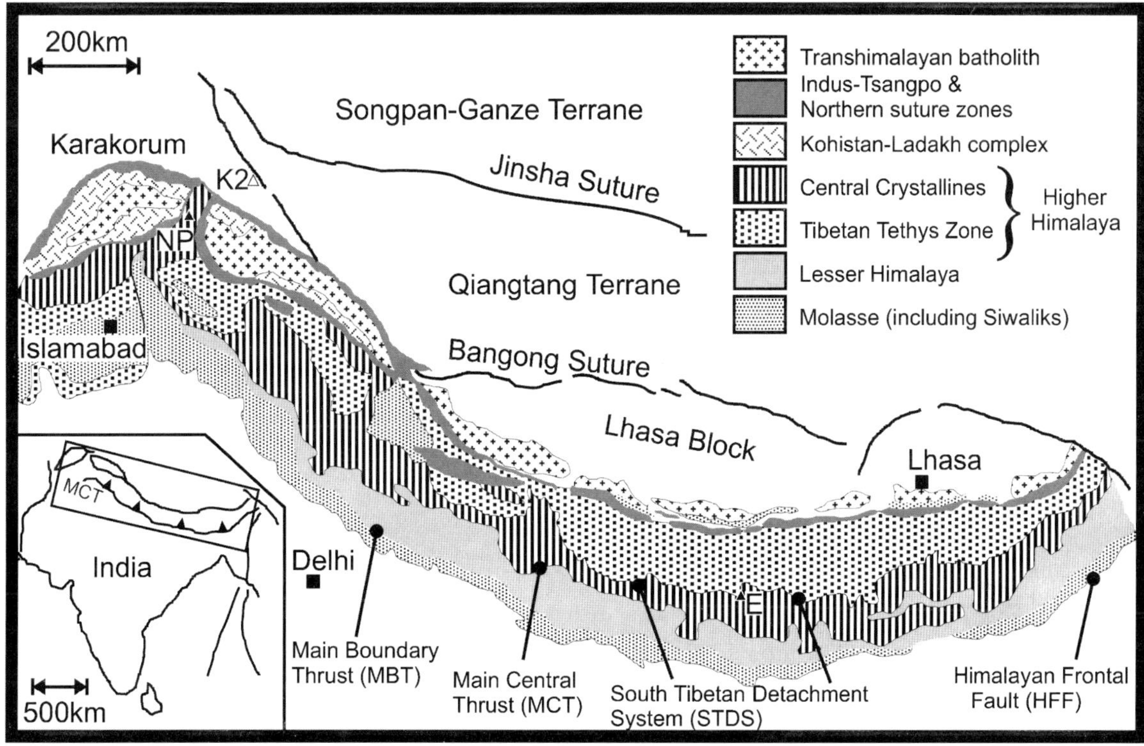

Fig. 6. Simplified geological subdivision of the Himalaya showing the main tectonic units and the thrusts or ophiolitic sutures separating them. E: Everest; NP: Nanga Parbat.

Collision of India with the Lhasa block occurred in the Eocene (around 50 Ma) as documented (Searle et al., 1987, 1997) by (1) the end of marine sedimentation in the suture zone (Rowley, 1996), (2) the end of major I-type magmatism of the Trans-Himalayan batholith (Honegger et al., 1982; Petterson and Windley, 1985), (3) the presence of the first S-type anatectic granites and migmatites in the Lhasa block, (4) a major deformation event causing south-directed thrusting in the sediments of the Lhasa block and suture zone and (5), a change in Indian plate movement. The northward drift of India, recorded in the palaeomagnetic record of the Indian Ocean floor, shows a dramatic decrease from 15–20 to ca. 5 cm/a after about 50 Ma (Patriat and Achache, 1984; Klootwijk et al., 1992). Another change in velocity took place at 65 Ma that is also reflected in a change in deep oceanic sedimentation (Klootwijk et al., 1992) and appears also to correspond to a phase of ophiolite obduction in western Pakistan argued by some authors as the start of collision (Beck et al., 1995, 1996).

The main continental collision caused: crustal imbrication and stacking of the Indian plate margin (e.g. Coward et al., 1988); underthrusting of the Indian plate beneath the southern Asian margin (Tibet) and the already accreted Kohistan–Ladakh arc (Coward et al., 1986; Zhao and Nelson, 1993) and strike-slip movement along faults well away from the collision zone in N Tibet and SE Asia, allowing eastward extrusion of Tibet (Tapponier et al., 1986). At least 1000 km of shortening has occurred since collision at 50 Ma (Dewey et al., 1989). The main thrusts imbricating the margin of the Indian plate (Figs. 6 and 7) delineate important boundaries between units of different metamorphic character. Immediately south of the Indus Tsangpo suture are the Palaeozoic and Mesozoic sediments deposited on the northern margin of India — the Tethyan Himalaya — that show folding and south-directed thrusting, but generally only

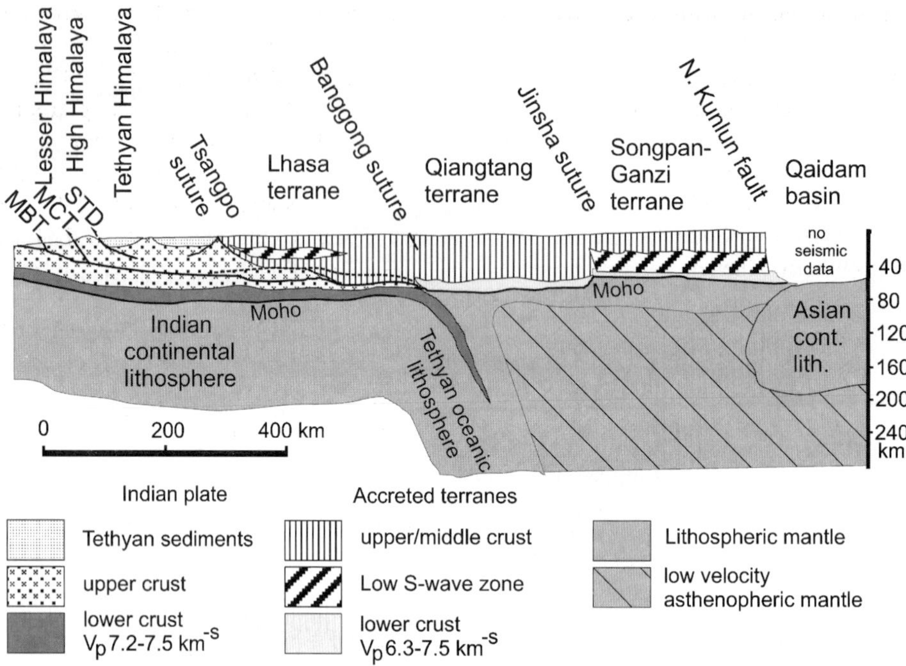

Fig. 7. Cross-section through the main units of the Himalaya from the Indian plate in the south (left) to the Asia margin in the north (right) showing the proposed deep underthrusting of Indian lithosphere below Asia at least as far as the northern margin of the Lhasa block (redrawn after Owens and Zandt, 1997).

very weak metamorphism. This unit is bounded by an important normal fault, the South Tibetan Detachment (STD) that marks a major break in metamorphic grade to the high amphibolite-facies gneisses of the Higher Himalayan Crystalline Nappes (HHC) in the footwall (Burchfiel et al., 1992). Further south, a generally lower grade greenschist-facies unit, the Lesser Himalaya (LH), has an upper boundary with the HHC along the main central thrust (MCT) and itself overrides the underlying molasse sediments along the main boundary thrust (MBT). On the basis of isotopic studies, the rocks of the HHC represent upper proterozoic to lower palaeozoic cover sequences stripped from an older basement that is exposed in the LH (and Nanga Parbat) (Parrish and Hodges, 1996; Whittington et al., 1999). It is interesting to note that no rock typical of lower continental crust has been exhumed south of the Indus suture despite metamorphic assemblages indicating upper mantle depths.

Rare eclogites in the northwest HHC, in some cases coesite-bearing, prove a deep subduction (80–100 km) of the leading edge of the Indian plate which, from geochronological studies, appears to have occurred between 55 and 45 Ma (Tonarini et al., 1993; Spencer and Gebauer, 1996; Sigoyer de et al., 1998; O'Brien et al., 1999). These rocks are significantly younger than the rare blueschists (80–100 Ma) found along the Indus suture zone (e.g. Honegger et al., 1989) or the granulite-facies rocks exhumed from the deepest part of the Kohistan arc. The main metamorphism in the HHC, however, comprises an early Barrovian kyanite-bearing stage (6–10 kbar), mostly older than 30 Ma, that is overprinted by a younger, lower pressure but higher temperature metamorphism (4–8 kbar), characterised by sillimanite growth and migmatisation, occurring around 15–22 Ma (see review by Guillot et al., 1999). The second stage of metamorphism is closely linked to the production of leucogranites, emplaced close to the STD, that formed by decompression melting of the HHC metasediments driven by major movements along the MCT and STD thrust-and-fault system (e.g. Harris and Massey, 1994). At the same time as this high-temperature event, rocks of the LH initially underwent a greenschist or lower grade metamorphism, with grade increasing structurally upwards (reaching amphibolite-facies) towards the MCT (e.g. Pêcher, 1989).

Continued continental convergence in the Himalaya is reflected in intracontinental earthquakes and thrust-type motions at depth linked to the MBT (Ni and Barazangi, 1984). Seismic investigations (Zhao and Nelson, 1993) have identified a major reflector beneath southern Tibet at a depth of 28 km (south) to 40 km (north) interpreted as the thrust plane along which India is underthrusting Asia today — the so-called 'main Himalayan thrust'. The MCT, MBT and MFT all link to this thrust surface at depth (Fig. 7). Recent seismic studies (Zhao and Nelson, 1993; Nelson et al., 1996) have confirmed the ca. 70 km depth of the Moho below southern Tibet deduced by Hirn et al. (1984) — i.e. a thickened continental crust resulting from the collision. Importantly, this present-day crustal configuration is probably significantly different from that at the time of coesite–eclogite formation that required crust to have been taken to a depth of 100 km or more. A possible analogue situation occurs in the Hindu Kush (northwest of the Himalaya) where seismic studies indicate the presence of a very young (<7 Ma) slice of material with continental crust characteristics located at a depth of over 170 km (Chatelain et al., 1980; Baranowski et al., 1984).

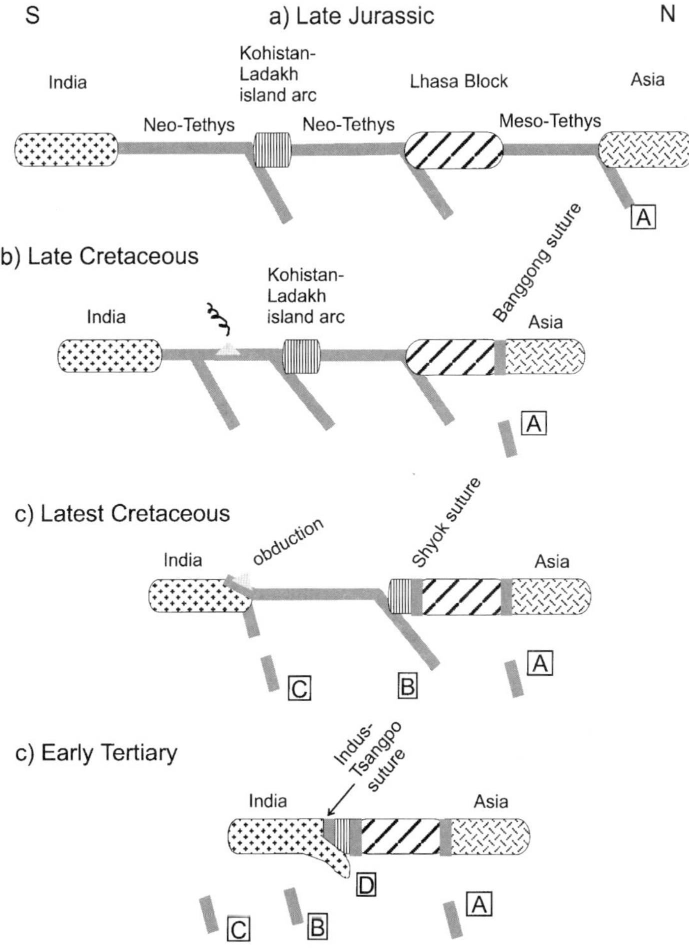

Fig. 8. Cartoon showing the main crustal blocks, the intervening oceanic domains, and the relative timing of subduction, basin closure and continental collision during the Himalayan orogenic evolution (redrawn after Van der Voo et al., 1999; Mahéo et al., 2000).

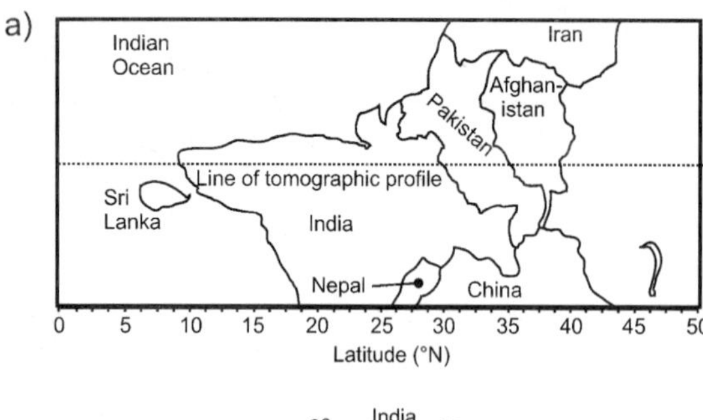

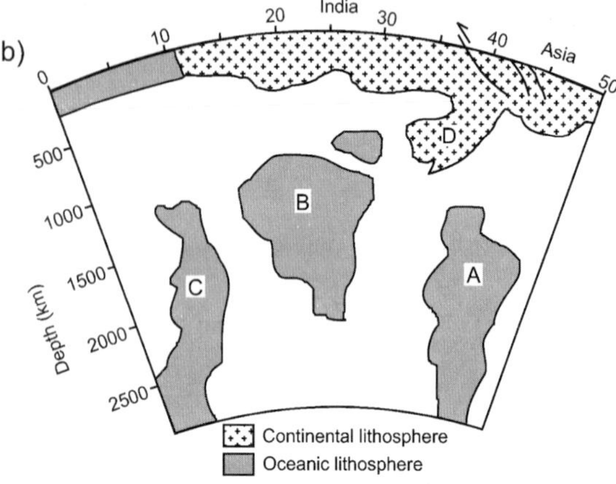

Fig. 9. Location of mantle regions below India and the Himalaya with anomalous seismic velocities (b), deduced from seismic tomography studies, interpreted as detached oceanic lithosphere (redrawn after Van der Voo et al., 1999). The line of the profile is shown in (a).

A simplified cartoon illustrating the India–Asia collision is shown in Fig. 8. The important features to recognise are the number of continental blocks and micro-plates involved and the corresponding intervening sutures and overridden oceanic domains. Recent seismic tomography studies of Asia (Van der Voo et al., 1999) have identified a number of anomalous regions in the deep mantle as well as a major roll-over structure at the base of the lithosphere under northern Pakistan (Fig. 9). The deep mantle anomalies have been attributed to the detachment of oceanic lithosphere from the continental crust as a result of collision and basin closure. Comparison of Figs. 8 and 9 shows that the most northerly anomaly is linked to the Bangong suture whereas both southern anomalies are attributed to subduction south of the Kohistan–Ladakh arc (Van der Voo et al., 1999; Mahéo et al., 2000).

4. Discussion

There are many similarities between the Alps and the Himalaya, but also several important differences. Both areas were formed as a result of closure of Neo-Tethys and related oceanic domains in Mesozoic and Tertiary times, so the ages and timescales of geological processes are similar. Both areas contain high mountains but a considerable difference exists between the 4810 m elevation of the highest peak of the Alps and the *average* 5000 m elevation of the Tibetan plateau. Regardless, geophysical evidence points to a similar 70 km maximum depth to the Moho, and thus a similar maximum thickness of continental crust in both areas. The difference is that 70-km thick crust extends over a much wider region in the Himalaya-Tibetan Plateau than in the Alps where it is a narrow zone below the Periadriatic Lineament. The distribution of metamorphic rocks of amphibolite grade or higher is restricted to a relatively narrow zone in both cases. However, it is apparent from the geophysical results (e.g. Owens and Zandt, 1997) that currently high temperatures prevail at shallow depths of the south Tibetan crust, indicating that high-grade metamorphism is currently in progress. It is important to realise that the thickened crust of both the Alps and Himalaya is not a stable feature and will eventually reequilibrate, perhaps through exhumation of rocks now at deeper levels and complete erosion of the rocks now exposed at the surface. Regardless of the possible future metamorphic pattern, it is clear that differences exist in the proportion of high-pressure, low-temperature rocks between the Alps (abundant) and Himalaya (scarce), although both show coesite-bearing rocks representing the extremes of subduction-zone metamorphism. Both areas show a lower pressure overprint on an earlier, generally higher pressure metamorphism; the difference is that this is much more widespread in the Himalaya, reflecting a longer period for thermal reequilibration following crustal imbrication.

A major difference is seen in the volume of oceanic crust consumed. The Himalayan region shows well-developed continental (Trans-Himalayan batholith) and oceanic (Kohistan–Ladakh arc, accreted to Asia in the Jurassic) magmatic arcs, reflecting the effect of several thousand kilometres of ocean crust that was consumed below the Asian margin. In stark contrast, much smaller oceanic rifts and basins were consumed in the Alps, and thus such arc sequences are absent in the area of the high mountains. However, in both cases, seismic tomography has identified anomalous regions in the mantle interpreted as the remnants of detached slabs of oceanic crust relating to the subduction of Tethys. Despite the differences in duration of oceanic subduction, it was still possible for the oceanic lithosphere to drag down the continental crust to ca. 100 km. The rapid exhumation of these rocks to much shallower levels, combined with the likely detachment of the oceanic lithosphere, strongly supports the type of lithospheric collision mechanism proposed by Chemenda et al. (1995, 2000). These models require an attached root of oceanic lithosphere to drag the buoyant upper crust down to upper mantle depths, but can only accommodate rapid exhumation of this crust if slab break-off occurs. Slab detachment and rapid return of buoyant continental slabs has also been emphasised in models for the evolution of coesite–eclogites from the Dabie–Sulu area, central China (Ernst et al., 1997; Ernst, 1999). This is only a part of the whole evolution, however. The degree of subsequent, shallower, continental imbrication and the duration of burial before exhumation will then determine what the major metamorphic pattern will be. It is only through good fortune that some of the ultrahigh-pressure relics survive and tell us of the initial subduction style. Continental collision can therefore lead to a variety of processes that differ in duration and extent as well as degree of preservation. Thus, although mountain collision zones have numerous similarities, the fine detail makes every example unique.

Acknowledgements

I wish to thank Gary Ernst, Dave Rubie and an anonymous reviewer for their comments. I take full responsibility for remaining prejudices and misconceptions.

References

Arthaud, F., Matte, P., 1977. Late Paleozoic strike-slip faulting in southern Europe and northern Africa. Geol. Soc. Am. Bull. 88, 1305–1320.

Baranowski, J., Armbuster, J., Seeber, L., Molnar, P., 1984. Focal depths and fault plane solutions of earthquakes and active tectonics of the Himalaya. J. Geophys. Res. 89, 6918–6928.

Beck, R.A., Burbank, D.W., Sercombe, W.J., Riley, G.W., Barndt, J.K., Berry, J.R., Afzal, J., Khan, A.M., Jurgen, H., Metje, H., Cheema, A., Shafique, N.A., Lawrence, R.D., Khan, M.A., 1995. Stratigraphic evidence for an early collision between northwest India and Asia. Nature 373, 55–58.

Beck, R.A., Burbank, D.W., Sercombe, W.J., Khan, A.M., Lawrence, R.D., 1996. Late Cretaceous ophiolite obduction and Palaeocene India–Asia collision in the westernmost Himalaya. Geodyn. Acta 9, 114–144.

Blundell, D., Freeman, R., Mueller, S., 1992. A Continent Revealed: The European Geotraverse. Cambridge University Press, Cambridge.

Burchfiel, B.C., Chen, Z., Hodges, K.V., Liu, Y., Royden, L., Deng, C., Xu, J., 1992. The South Tibetan Detachment, Himalayan Orogen: Extension Contemporaneous with and Parallel to Shortening in a Collisional Mountain Belt. Geological Society of America Special Paper, Vol. 269, 41 pp.

Chatelain, J.L., Roecker, S.W., Hatzfeld, D., Molnar, P., 1980. Microearthquake seismicity and fault plane solutions in the Hindu Kush region and their tectonic implications. J. Geophys. Res. 85, 1365–1387.

Chemenda, A.I., Mattauer, M., Malavieille, J., Bokum, A.N., 1995. A mechanism for syncollisional rock exhumation and associated normal faulting: results from physical modelling. Earth Planet. Sci. Lett. 132, 225–232.

Chemenda, A.I., Burg, J.P., Mattauer, M., 2000. Evolutionary model of the Himalaya–Tibet system: geopoem based on new modelling, geological and geophysical data. Earth Planet. Sci. Lett. 174, 397–409.

Coward, M.P., Windley, B.F., Broughton, R.D., Luff, I.W., Petterson, M.G., Pudsey, C.J., Rex, D.C., Khan, M.A., 1986. Collision tectonics in the NW Himalayas. In: Coward, M.P., Ries, A. (Eds.), Collision Tectonics. Geological Society, London, Special Publication, Vol. 19, pp. 205–221.

Coward, M.P., Butler, R.W.H., Chambers, A.F., Graham, R.H., Izatt, C.N., Khan, M.A., Knipe, R.J., Prior, D.J., Treloar, P.J., Williams, M.P., 1988. Folding and imbrication of the Indian crust during Himalayan collision. Phil. Trans. Roy. Soc., Ser. A 326, 89–116.

Dewey, J.F., Pitman III, W.C., Ryan, W.B.F., Bonnin, J., 1973. Plate tectonics and the evolution of the Alpine system. Geol. Soc. Am. Bull. 84, 3137–3180.

Dewey, J.F., Cande, S., Pitman, W.C., 1989. Tectonic evolution of the India/Eurasia collision zone. Eclogae Geol. Helv. 82, 717–734.

Dercourt, J., Ricou, L.E., Vrielynck, B., 1993. Atlas of Tethys Palaeoenvironmental Maps, Gauthier–Villars, Paris, 307 pp.

Dietrich, V.J., 1979. Ophiolitic belts of the central Mediterranean. In: International Atlas of Ophiolites. Geological Society of America Map and Chart Series, MC-33.

Ernst, W.G., 1999. Metamorphism, partial preservation, and exhumation of ultrahigh-pressure belts. The Island Arc 8, 125–153.

Ernst, W.G., Maruyama, S., Wallis, S., 1997. Buoyancy-driven, rapid exhumation of ultrahigh-pressure metamorphosed continental crust. Proc. Natl. Acad. Sci., U.S.A. 94, 9532–9537.

Frey, M., Ferreiro Mählmann, R., 1999. Alpine metamorphism of the central Alps. Schweizerische Mineralogische und Petrographische Mitteilungen 79, 135–154.

Frey, M., Desmons, J., Neubauer, F., 1999. The new Metamorphic Map of the Alps. Schweizerische Mineralogische und Petrographische Mitteilungen, Vol. 79, Heft 1.

Gansser, A., 1979. Ophiolitic belts of the Himalayan and Tibetan region. In: International Atlas of Ophiolites. Geological Society of America Map and Chart Series, MC-33.

Gebauer, D., 1999. Alpine geochronology of the central and western Alps: new constraints for a complex geodynamic evolution. Schweizerische Mineralogische und Petrographische Mitteilungen 79, 191–208.

Guillot, S., Cosca, M., Allemand, P., Le Fort, P., Contrasting Metamorphic and Geochronologic Evolution along the Himalaya Belt. Geological Society of America Special Paper, Vol. 328, pp. 117–128.

Harris, N., Massey, J., 1994. Decompression and anatexis of Himalayan metapelites. Tectonics 13, 1537–1546.

Hirn, A., Sapin, M., Jobert, G., Xin, X.Z., Yuan, G.E., Wen, T.J., 1984. Lhasa block and bordering sutures a continuation of the 500-km Moho traverse through Tibet.. Nature 307, 25–27.

Hodges, K.V., 2000. Tectonics of the Himalaya and southern Tibet from two perspectives. GSA Bull. 112, 324–350.

Honegger, K., Dietrich, V., Frank, W., Gansser, A., Thoni, M., Trommsdorf, V., 1982. Magmatism and metamorphism in the Ladakh–Himalaya (the Indus–Tsangpo suture zone). Earth Planet. Sci. Lett. 60, 253–292.

Honegger, K., Le Fort, P., Mascle, G., Zimmermann, J.-L., 1989. The blueschists along the Indus Suture Zone in Ladakh NW Himalaya. J. Metamorph. Geol. 7, 57–72.

Klootwijk, C.T., Gee, J.S., Pierce, J.W., Smith, G.M., McFadden, P.L., 1992. An early India–Asia contact: palaeomagnetic constraints from Ninetyeast Ridge ODP Leg 121. Geology 20, 395–398.

Le Fort, P., 1975. Himalayas: the collided range, present knowledge of the continental arc. Am. J. Sci. 275A, 1–44.

Mahéo, G., Bertrand, H., Guillot, S., Mascle, G., Pecher, A., Picard, C., de Sigoyer, J., 2000. Temoins d'un arc immature tethysien dans les ophiolites du Sud Ladakh (NW Himalaya, India). Compte Rendus de la Academie des Sciences, Paris, Sciences de la Terre et des planetes 330, 289–295.

Marchant, R.H., Stampfli, G.M., 1997. Crustal and lithospheric structures of the western Alps: geodynamic significance. In: Pfiffner, O.A., Lehner, P., Heitzman, P.Z., Mueller, S., Steck, A. (Eds.), Deep Structure of the Swiss Alps — Results from NRP20. Birkhaüser AG, Basel, pp. 326–337.

Mattauer, M., Matte, P., Olivet, J.L., 1999. A 3D model of the India–Asia collision at plate scale. Compte Rendus de la Academie des Sciences, Paris, Sciences de la Terre 328, 499–508.

Nelson, K.D., Zhao, W., Brown, L.D., Kuo, J., Che, J., Liu, X., Klemperer, S.L., Makovsky, Y., Meissner, R., Mechie, J., Kind, R., Wenzel, F., Ni, J., Nabelek, J., Chen, L., Tan, H., Wei, W., Jones, A.G., Booker, J., Unsworth, M., Kidd, W.S.F., Hauck, M., Alsdorf, D., Ross, A., Cogan, M., Wu, C., Sandvol, E., Edwards, M., 1996. Partially molten middle crust beneath southern Tibet: synthesis of Project INDEPTH results. Science 274, 1684–1688.

Ni, J., Barazangi, M., 1984. Seismotectonics of the Himalayan collision zone: geometry of the underthrusting Indian plate beneath the Himalaya. J. Geophys. Res. 89, 1147–1163.

O'Brien, P.J., Zotov, N., Law, R., Khan, M.A., Jan, M.Q., 1999. Coesite in eclogite from the Upper Kaghan Valley, Pakistan: a first record and implications. Terra Nostra 99–102, 109–111.

Owens, T.J., Zandt, G., 1997. Implications of crustal property variations for models of Tibetan plateau evolution. Nature 387, 37–43.

Parrish, R.R., Hodges, K.V., 1996. Isotopic constraints on the age and provenance of the Lesser and Greater Himalaya sequences, Nepalese Himalaya. Geol. Soc. Am. Bull. 108, 904–911.

Pêcher, A., 1989. The metamorphism in central Himalaya. J. Metamorph. Geol. 7, 31–41.

Patriat, P., Achache, J., 1984. India–Asia collision chronology and implications for crustal shortening and driving mechanism of plates. Nature 311, 615–621.

Petterson, M.G., Windley, B.F., 1985. Rb–Sr dating of the Kohistan batholith in the Trans-Himalaya of N Pakistan and tectonic implications. Earth Planet. Sci. Lett. 74, 45–57.

Pfiffner, O.A., Lehner, P., Heitzman, P.Z., Mueller, S., Steck, A., 1997. Deep Structure of the Swiss Alps — Results from NRP20. Birkhaüser AG, Basel, 380 pp.

Rowley, D.B., 1996. Age of initiation of collision between India and Asia: a review of stratigraphic data. Earth Planet. Sci. Lett. 145, 1–13.

Schmid, S.M., Aebli, H.R., Heller, F., Zingg, A., 1989. The role of the Periadriatic Line in the tectonic evolution of the Alps. In: Coward, M.P., Dietrich, D., Park, R.G. (Eds.), Alpine Tectonics. Geological Society, London, Special Publication, Vol. 45, pp. 153–171.

Searle, M.P., Windley, B.F., Coward, M.P., Cooper, D.J.W., Rex, A.J., Rex, D., Li, T.D., Xiao, X.C., Jan, M.Q., Thakur, V.C., Kumar, S., 1987. The closing of Tethys and the tectonics of the Himalaya. Geol. Soc. Am. Bull. 98, 678–701.

Searle, M.P., Corfield, R.I., Stephenson, B., McCarron, J., 1997. Structure of the North Indian continental margin in the Ladakh–Zanskar Himalayas: implications for the timing of obduction of the Spontang ophiolite, India–Asia collision and deformation events in the Himalaya. Geol. Mag. 134, 297–316.

Şengör, A.M.C., Natal'in, B.A., 1996. Paleotectonics of Asia: fragments of a synthesis. In: Yin, A., Harrison, M. (Eds.), The Tectonic Evolution of Asia. Cambridge University Press, Cambridge, pp. 486–640.

Sigoyer de, J., Villa, I., Chavagnac, V., Guillot, S., Mascle, G., Multichronometry of Tso Morari eclogites: Ordovician plutonism, Tertiary eclogitisation and inheritance. In: Proceedings of the 13th HKT Workshop. Peshawar, Pakistan, abstract volume. Geological Bulletin, University of Peshawar (Special Issue), Vol. 31, pp. 185–187.

Spakman, W., 1991. Delay time tomography of the upper mantle below Europe, the Mediterranean, and Asia Minor. Geophys. J. Int. 107, 309–332.

Spakman, W., van der Lee, S., van der Hilst, R.D., 1993. Travel-time tomography of the European–Mediterranean mantle down to 1400 km. Phys. Earth Planet. Interiors 79, 3–74.

Spencer, D.A., Gebauer, D., 1996. SHRIMP evidence for a Permian protolith age and a 44 Ma age for the Himalayan eclogites, Upper Kaghan, Pakistan: implications for the subduction of Tethys and the subdivision terminology of the NW Himalaya. In: Proceedings of the 11th HKT Workshop. Flagstaff, AZ, USA, abstract volume, pp. 147–150.

Stampfli, G.M., Marcoux, J., Baud, A., 1991 Tethyan margins in space and time. In: Channel, J.E.T., Winterer, E.L., Jansa, L.F. (Eds.), Paleogeography and Paleoceanography of the Tethys. Palaeogeography, Palaeoclimatology and Palaeoecology, Vol. 87, pp. 373–410.

Stampfli, G.M., Mosar, J., Marquer, D., Marchant, R., Baudin, T., Borel, G., 1998. Subduction and obduction processes in the Swiss Alps. Tectonophysics 296, 159–204.

Suess, E. 1885–1888. Das Antlitz der Erde, Temsky, Vienna.

Tapponier, P., Peltzer, G., Armijo, R., 1986. On the mechanics of the collision between India and Asia. In: Coward, M.P., Ries, A. (Eds.), Collision Tectonics. Geological Society, London, Special Publication, Vol. 19, pp. 115–157.

Tonarini, S., Villa, I., Oberli, M., Meier, F., Spencer, D.A., Pognante, U., Ramsey, J.G., 1993. Eocene age of eclogite metamorphism in Pakistan Himalaya: implications for India–Eurasia collision. Terra Nova 5, 13–20.

Van der Voo, R., Spakman, W., Bijwaard, H., 1999. Tethyan subducted slabs under India. Earth Planet. Sci. Lett. 171, 7–20.

Wadia, D.N., 1931. The syntaxis of the northwest Himalayas: its rocks, tectonics and orogeny. Records Geol. Surv. India 65, 189–220.

Whittington, A., Foster, G., Harris, N., Vance, D., Ayres, M., 1999. Lithostratigraphic correlations in the western Himalaya — an isotopic approach. Geology 27, 585–588.

Wortel, M.J.R., Spakman, W., 1992. Structure and dynamics of subducted lithosphere in the Mediterranean region. Proc. Koninklijke Nederlandse Akademie van Wetenschappen 95, 325–347.

Zhao, W., Nelson, K.D., Project INDEPTH Team, 1993. Deep seismic reflection evidence for continental underthrusting beneath southern Tibet. Nature 366, 557–559.